Immune-Mediated Diseases

Immune-Mediated Diseases

From Theory to Therapy

Edited by

Michael R. Shurin
Division of Clinical Immunopathology
Department of Pathology
University of Pittsburgh Medical Center
Pittsburgh, Pennsylvania

Yuri S. Smolkin
Department of Immunology
Institute for Continuing Medical Education
and
Clinical and Research Center for Allergy and Immunology
Moscow, Russia

Editors:
Michael R. Shurin
Associate Professor of Pathology
and Immunology
Associate Director, Division of Clinical
Immunopathology
5725 Children's Hospital of Pittsburgh
Main Tower
200 Lothrop Street
Pittsburgh, PA 15213
shurinmr@upmc.edu

Yuri S. Smolkin
Professor of Immunology
Institute for Continuing Medical
Education
Vise-President, Russian Association of
Pediatric Allergology and Immunology
Director, Clinical and Research Center
of Allergy and Immunology
Ostrovitianova St., Bldg. 6
Moscow 117513, Russia
smolking@dataforce.net

Library of Congress Control Number: 2007928315

ISBN 978-0-387-72004-3 e-ISBN 978-0-387-72005-0

Printed on acid-free paper.

9 8 7 6 5 4 3 2 1

springer.com

Contents

News in Immunodiagnostics and Immunomonitoring

Immunotherapy and Vaccines

PREFACE AND ACKNOWLEDGMENTS

The idea for the International Immune-Mediated Diseases: From Theory to Therapy (IMD) Congress was born in 2003 as a result of our discussions about the uneven development of immunological ideas and rate of progress in immunology research in different countries during the last century. Consequently, the First International meeting was held in Moscow, Russia, in October 2005 with the goal to bring the best world immunologists and clinicians to Moscow to participate in plenary sessions, symposia, and educational workshops designed to expose basic and clinical immunologists as well as practicing clinicians to the newest developments in the field. First IMD Congress consisted of 8 Plenary Sessions, 40 Symposia, and 25 Workshops. The meeting attracted almost 2000 participants and speakers from 15 countries. After this great success of the First IMD Congress, the Organizing Committee decided to convene the meeting every other year, and the Second International IMD Congress was planned for September 2007. Special attention will be given to innovative therapeutic modalities, allergy and asthma, immunodeficiency, tumor and transplant immunology, immunotherapy and vaccines, HIV and infectious diseases, autoimmune diseases, inflammation, immunomonitoring and immunodiagnostics.

The slogan of the IMD Congress is "BETTER LIVING THROUGH IMMUNO-LOGY," which can be achieved by: (i) promoting basic and clinical research in immunology, (ii) integrating the immunological research and educational activities, (iii) disseminating information and encouraging international collaborations, (iv) identifying ways to improve the quality of immunological research and clinical care, (v) encouraging and providing training and continuous medical and biomedical education, and (vi) promoting novel immune-based preventive, diagnostic, and therapeutic modalities and good patient care.

This volume includes contributions from the speakers of the Second IMD Congress (September 10–15, 2007; Moscow, Russia), who are eager to share some of the academic and clinical enthusiasm that defines the IMD meetings. Unfortunately, due to the publication deadline, only a limited number of contributions are presented in this volume. We apologize for not being able to accommodate all the outstanding talks and lectures that will make the IMD Congress 2007 such an attractive, interesting, and unique scientific event.

We express our appreciation to Dr. Anatoli Malyguine (NIH, NCI, Bethesda, MD, USA), Dr. Svetlana Bykovskaia (Institute for Rheumatology, Moscow, Russia), and Lori Perez (University of Pittsburgh, Pittsburgh, PA, USA) for their exceptional and generous help in organizing the IMD Congress 2007 and for the preparation of this book. It is a great gift to have friends as colleagues and colleagues as friends!

Olivera Finn and Rakhim M. Khaitov
IMD Congress Co-Chairs
Michael R. Shurin and Yuri S. Smolkin
IMD Congress Co-Organizers, Editors
Spring 2007

List of Contributing Speakers

Maryam Aalamian-Matheis Department of Medicine, Johns Hopkins University, Baltimore, MD, USA

Michael Anderson Tumor Immunity and Tolerance Section, Laboratory of Molecular Immunoregulation, Cancer and Inflammation Program, CCR, NIH, Frederick, MD, USA

Ron N. Apte Department of Microbiology and Immunology and Cancer Research Center, Ben-Gurion University of the Negev, Beer-Sheva, Israel

Natalia Aptsiauri Department of Clinical Analyes, University Hospital Virgen de las Nieves, Granada, Spain

Irina Baldueva Department of Biotherapy and Bone Marrow, N.N. Petrov Research Institute of Oncology, Transplantation, Saint-Petersburg, Russia

Anatoly Yu. Baryshnikov Department of Experimental Diagnosis and Biotherapy of Tumors, Russian Cancer Research Center, Moscow, Russia

Nadzeya Barysik Department of Pediatric Pathology, University of Mainz, Mainz, Germany

Michael Baseler Applied and Developmental Research Support Program, SAIC-Frederick, Inc., NCI-Frederick, Frederick, MD, USA

Dhurba J. Bharali Department of Immunology, Roswell Park Cancer Institute, Buffalo, NY, USA

Andrey Bologov Russian Clinical Children's Hospital and Institute for Clinical Genetics, Moscow, Russia

Anna A. Borunova N.N. Blokhin Russian Cancer Research Center, Russian Academy of Medical Sciences, Moscow, Russia

Damien Bresson Department of Developmental Immunology, La Jolla Institute for Allergy and Immunology, La Jolla, CA, USA

Anna Budikhina National Research Center Institute of Immunology, FMBA, Moscow, Russia

Svetlana N. Bykovskaia Laboratory of Cell Monitoring, Institution of Rheumatology, Russian Academy of Medical Sciences, Moscow, Russia

Teresa Cabrera Department of Clinical Analyes, University Hospital Virgen de las Nieves, Granada, Spain

Yaron Carmi Department of Microbiology and Immunology and Cancer Research Center, Ben-Gurion University of the Negev, Beer-Sheva, Israel

Gurkamal S. Chatta Department of Medicine, University of Pittsburgh Cancer Institute, Pittsburgh, PA, USA

Denis Chuvirov Federal Educational Establishment Advanced Study Institute, FMBA, Moscow, Russia

Alexey Danilov Department of Biotherapy and Bone Marrow Transplantation, N.N. Petrov Research Institute of Oncology, Saint-Petersburg, Russia

Anna Danilova Department of Biotherapy and Bone Marrow Transplantation, N.N. Petrov Research Institute of Oncology, Saint-Petersburg, Russia

Sophie Dessureault Department of Interdisciplinary Oncology, University of South Florida, H. Lee Moffitt Cancer Center & Research Institute, Tampa, FL, USA

Natalya I. Egorova N.I. Lobachevsky State University of Nizhny Novgorod, Nizhny Novgorod, Russia

Amos Etzioni Meyer Children Hospital and the Rappaport School of Medicine, Technion, Haifa, Israel

Irina V. Evsegneeva I.M. Sechenov Medical Academy of Moscow, Moscow, Russia

Thomas A. Fleisher Department of Laboratory Medicine, NIH Clinical Center, National Institutes of Health, DHHS, Bethesda, MD, USA

Georgia Fousteri Department of Developmental Immunology, La Jolla Institute for Allergy and Immunology, La Jolla, CA, USA

Dmitry I. Gabrilovich H. Lee Moffitt Cancer Center, University of South Florida, Tampa, FL, USA

Elizaveta Galkina Russian Clinical Children's Hospital, Moscow, Russia

Ludmila V. Gankovskaya Department of Immunology, Russian State Medical University, Moscow, Russia

Oksana A. Gankovskaya Department of Viral Infection Diagnostics, Mechnikov Research Institute of Vaccines and Sera, Moscow, Russia

Angel Garcia-Lora Department of Clinical Analyses, University Hospital Virgen de las Nieves, Granada, Spain

Federico Garrido Department of Clinical Analyes, University Hospital Virgen de las Nieves, Granada, Spain

Valentin Gerein Department of Pediatric Pathology, Institute of Pathology, University of Mainz, Mainz, Germany

Ronald E. Gress Experimental Transplantation and Immunology Branch, National Cancer Institute, National Institutes of Health, Bethesda, MD, USA

Matthias von Herrath Department of Developmental Immunology, La Jolla Institute for Allergy and Immunology, La Jolla, CA, USA

Edith Huland Department of Urology, University Clinics Hamburg-Eppendorf, Hamburg, Germany

Hartwig Huland Department of Urology, University Clinics Hamburg-Eppendorf, Hamburg, Germany

Arthur A. Hurwitz Tumor Immunity and Tolerance Section, Laboratory of Molecular Immunoregulation, Cancer and Inflammation Program, CCR, NIH, Frederick, MD, USA

Anna Ilinskaya National Research Center Institute of Immunology, Russian Federal Medical Biological Agency, Moscow, Russia

Betina P. Iliopoulou Departments of Pharmacology and Experimental Therapeutics; Internal Medicine and Biochemistry; Immunology Program, Tufts University school of Medicine and Tufts-New Engineering Center, Boston, MA, USA

Evgeny Imyanitov Department of Molecular Genetics, N.N. Petrov Research Institute of Oncology, Saint-Petersburg, Russia

Zaira G. Kadagidze N.N. Blokhin Russian Cancer Research Center, Russian Academy of Medical Sciences, Moscow, Russia

Alexandr V. Karaulov Department of Immunology, I.M. Sechenov Medical Academy of Moscow, Moscow, Russia

Dmitry B. Kazansky Institute of Carcinogenesis, Blokhin Cancer Research Center, Moscow, Russia

Duraisamy Kempuraj Departments of Pharmacology and Experimental Therapeutics; Internal Medicine and Biochemistry, Immunology Program, Tufts University school of Medicine and Tufts-New Engineering Medical Center, Boston, MA, USA

Rakhim M. Khaitov National Research Center Institute of Immunology, Russian Federal Medical Biological Agency, Moscow, Russia

Thomas W. Klein Department of Molecular Medicine, University of South Florida, Tampa, FL, USA

Irina Kondratenko Russian Clinical Children's Hospital, Moscow, Russia

Leonid V. Kovalchuk Department of Immunology, Russian State Medical University, Moscow, Russia

Georgii Yu. Kurnikov N.I. Lobachevsky State University of Nizhny Novgorod, Nizhny Novgorod, Russia

Vyacheslav F. Lavrov Department of Viral Infection, Mechnikov Research Institute of Vaccines and Sera, Diagnostics, Moscow, Russia

Joel de León Department of Vaccines, Center of Molecular Immunology, Havana, Cuba

Valerii Livov National Research Center Institute of Immunology, Russian Federal Medical Biological Agency, Moscow, Russia

Elena Yu. Lyssuk Laboratory of Cell Monitoring, Institution of Rheumatology, Russian Academy of Medical Sciences, Moscow, Russia

Anatoli Malyguine Applied and Developmental Research Support Program, SAIC-Frederick, Inc., NCI-Frederick, Frederick, MD, USA

Alagar Manickam Department of Biotechnology, Government College of Technology, Tamil Nadu, India

Tatiana Markova Federal Educational Establishment Advanced Study Institute, FMBA, Moscow, Russia

Rosa Mendez Department of Clinical Analyes, University Hospital Virgen de las Nieves, Granada, Spain

Chandra Mohan Department of Internal Medicine and the Center for Immunology, University of Texas, Southwestern Medical School, Dallas, TX, USA

Shaker A. Mousa Pharmaceutical Research Institute at Albany, Albany College of Pharmacy, Albany, NY, USA

Vladimir Moiseyenko Department of Biotherapy and Bone Marrow Transplantation, N.N. Petrov Research Institute of Oncology, Saint-Petersburg, Russia

Fanny Monneaux Institute for Molecular and Cellular Biology, Strasbourg, France

Sylviane Muller Institute for Molecular and Cellular Biology, Strasbourg, France

Srinivas Nagaraj H. Lee Moffitt Cancer Center, University of South Florida, Tampa, FL, USA

Evgeny L. Nassonov Laboratory of Cell Monitoring, Institution of Rheumatology, Russian Academy of Medical Sciences, Moscow, Russia

Catherine A. Newton Department of Molecular Medicine, University of South Florida, Tampa, FL, USA

Natalija Novak Department of Dermatology, University of Bonn, Bonn, Germany

Victor V. Novikov N.I. Lobachevsky State University of Nizhny Novgorod, Nizhny Novgorod, Russia

Hans D. Ochs University of Washington and Children's Hospital and Regional Medical Center Division of Immunology, Seattle, WA, USA

Natalia Oliferuk National Research Center Institute of Immunology, Russian Federal Medical Biological Agency, Moscow, Russia

Joost J. Oppenheim Laboratory of Molecular Immunoregulation, Center for Cancer Research, NCI-Frederic, MD, USA

Jordan S. Orange Department of Pediatrics, University of Pennsylvania School of Medicine, The Children's Hospital of Philadelphia, Philadelphia, PA, USA

Vladimir Pak National Research Center Institute of Immunology, FMBA, Moscow, Russia

Olga Paschenko Russian Clinical Children's Hospital, Moscow, Russia

Mikhail Pashenkov National Research Center Institute of Immunology, FMBA, Moscow, Russia

Wenming Peng Department of Dermatology, University of Bonn, Bonn, Germany

Lori Perez Department of Pathology, University of Pittsburgh Medical Center, Pittsburgh, PA, USA

Boris Pinegin National Research Center Institute of Immunology, FMBA, Moscow, Russia

Alexandr Polyakov Russian Clinical Children's Hospital, Moscow, Russia

Miroslava Potapenko Department of Pathology, University of Pittsburgh, Pittsburgh, PA, USA

Gonzalo de la Rosa Laboratory of Molecular Immunoregulation, Center for Cancer Research, NCI-Frederic, MD, USA

Francisco Ruiz-Cabello Department of Clinical Analyes, University Hospital Virgen de las Nieves, Granada, Spain

Thomas Sayers Cancer and Inflammation Program, SAIC-Frederick Inc., NCI-Frederick, Frederick, MD, USA

Boris Semenov Mechnikov Research Institute of Vaccines and Sera, Moscow, Russia

Alexandra L. Sevko R.E. Kavetsky Institute of Experimental Pathology, Oncology and Radiobiology Ukrainian Academy of Sciences, Kyiv, Ukraine

Kimberly Shafer-Weaver Applied and Developmental Research Support Program, SAIC-Frederick, Inc., NCI-Frederick, Frederick, MD, USA

Anil Shanker Cancer and Inflammation Program, SAIC-Frederick Inc., NCI-Frederick, Frederick, MD, USA

Galina V. Shurin Department of Pathology, University of Pittsburgh, Pittsburgh, PA, USA

Michael R. Shurin Departments of Pathology and Immunology, University of Pittsburgh, Pittsburgh, PA, USA

Alexander B. Sigalov Department of Pathology, University of Massachusetts Medical School, Worcester, MA, USA

Muthukumaran Sivanandham Department of Biotechnology, Sri Venkateswara College of Engineering, Tamil Nadu, India

Yuri S. Smolkin Clinical and Research Center for Allergy and Immunology, Moscow, Russia

Sergey K. Soloviev Laboratory of Cell Monitoring, Institution of Rheumatology, Russian Academy of Medical Sciences, Moscow, Russia

Claude Sportès Experimental Transplantation and Immunology Branch, National Cancer Institute, National Institutes of Health, Bethesda, MD, USA

Richard E. Stiehm Mattel Children's Hospital and David Geffen School on Medicine, UCLA, Los Angeles, CA, USA

Susan Strobl Applied and Developmental Research Support Program, SAIC-Frederick, Inc., NCI-Frederick, Frederick, MD, USA

Kathleen E. Sullivan Department of Pediatrics, Division of Allergy & Immunology, Children's Hospital of Philadelphia, Philadelphia, PA, USA

Poonam Tewary Laboratory of Molecular Immunoregulation, Center for Cancer Research, NCI-Frederic, MD, USA

Yasmin Thanavala Department of Immunology, Roswell Park Cancer Institute, Buffalo, NY, USA

Theoharis C. Theoharides Departments of Pharmacology and Experimental Therapeutics; Internal Medicine and Biochemistry; Immunology Program, Tufts University School of Medicine and Tufts-New England Medical Center, Boston, MA, USA

Patricia L. Thompson Department of Interdisciplinary Oncology, H. Lee Moffitt Cancer Center & Research Institute, University of South Florida, Tampa, FL, USA

Anna V. Torgashina Laboratory of Cell Monitoring, Institution of Rheumatology, Russian Academy of Medical Sciences, Moscow, Russia

Troy R. Torgerson Division of Immunology, University of Washington and Children's Hospital and Regional Medical Center, Seattle, WA, USA

Irina L. Tourkova Department of Pathology, University of Pittsburgh Medical Center, Pittsburgh, PA, USA

Joseph A. Trapani Peter MacCallum Cancer Centre, St. Andrew's Place, East Melbourne, Australia

Anna Valujskikh Department of Immunology, The Cleveland Clinic Foundation, Cleveland, OH, USA

Amalia Vartanian Department of Experimental Diagnosis and Biotherapy of Tumors, Russian Cancer Research Center, Moscow, Russia

Elena Voronov Department of Microbiology and Immunology and Cancer Research Center, Ben-Gurion University of the Negev, Beer-Sheva, Israel

Ilia Voskoboinik Peter MacCallum Cancer Centre, St. Andrew's Place, East Melbourne, Australia

De Yang Basic Research Program, SAIC-Frederick, Inc., Frederic, MD, USA

Xi Yang Laboratory for Infection and Immunity, Departments of Medical Microbiology and Immunology, University of Manitoba, Winnipeg, Manitoba, Canada

Chunfeng Yu Department of Dermatology, University of Bonn, Bonn, Germany

Tatiana N. Zabotina N.N. Blokhin Russian Cancer Research Center, Russian Academy of Medical Sciences, Moscow, Russia

Liubov Zaritskaya Applied and Developmental Research Support Program, SAIC-Frederick, Inc., NCI-Frederick, Frederick, MD, USA

Jiankun Zhu Department of Internal Medicine, University of Texas Southwestern Medical School, Dallas, TX, USA

Vitaly Zverev Mechnikov Research Institute of Vaccines and Sera, Moscow, Russia

Immune-Mediated Diseases

1

Immune-Mediated Diseases: Where Do We Stand?

Michael R. Shurin[1] and Yuri S. Smolkin[2]

[1] Departments of Pathology and Immunology, University of Pittsburgh Medical Center, Pittsburgh, Pennsylvania, USA, shurinmr@upmc.edu
[2] Research and Clinical Center for Allergy and Clinical Immunology, Moscow, Russia

Abstract. The progress in basic immunology during the past 50–60 years has been associated with the emergence of clinical immunology as a new discipline in the 1970s. It was defined as the application of basic immunology principles to the diagnosis and treatment of patients with diseases in which immune-mediated mechanisms play an etiological role. Immune-mediated diseases such as autoimmune diseases, allergic diseases, and asthma are important health challenges in the United States and worldwide. For instance, autoimmune diseases afflict 5–8% of the US population; asthma and allergic diseases combined represent the sixth leading cause of chronic illness and disability in the United States and the leading cause among children. As shocking as these numbers and other data in this chapter are, they cannot adequately reflect the physical and emotional devastation to individuals, families, and communities coping with hundreds of immune-mediated disorders nor do they capture the enormous deleterious impact of these diseases on the economies of countries, nations, and indeed entire planet.

1. Immune-Mediated Diseases: Diversity

The marvelous progress in basic immunology during the past 50–60 years has been associated with a synchronous application of new knowledge and techniques to the diagnosis and treatment of human diseases. As a result, clinical immunology was born as a new discipline in the 1970s. It was defined as the application of the principles of basic immunology to the clinical detection, diagnosis, and treatment of patients with immunologically mediated diseases. It became clear that the immune system plays a central role not only in fighting infections but also in many other diseases and medical conditions including cancer, AIDS, and organ transplantation. In addition, the immune system is a key player in the etiology and pathogenesis of various hypersensitive illnesses, such as asthma, certain dermatitis, and other allergies, as well as systemic and organ-specific autoimmune disorders, such as multiple sclerosis, lupus, rheumatoid arthritis, and diabetes.

Involvement of the immune system in certain diseases or groups of diseases was not seen as obvious and apparent as it is now. For instance, for 50 years, the theory on autoimmunity was paralyzed due to the Nobel Prize winner, Paul Ehrlich, who received the prize for creating antisyphilis therapy and the development of the concept of autoimmunity (Shoenfeld and Zandman-Goddard 2003). He believed it was not possible that autoantibodies could exist in the body and that such conditions should be associated with immediate death. At that time, no one dared further to investigate the feasibility of autoimmune diseases. However, as the years passed, understanding of autoimmune disorders was developed, and Nobel Prizes were awarded to those who discovered the mechanisms of autoimmune diseases (Shoenfeld and Zandman-Goddard 2003). Today, autoimmune diseases include about 80 different disorders and while some autoimmune diseases are rare—affecting fewer than 200,000 individuals in the United States—collectively these diseases afflict millions of people, an estimated 4% of the population or 12 million people in the United States and approximately 90,000,000 worldwide. Other reports suggest that these numbers should be doubled and estimate that autoimmune diseases afflict 5–8% of the US population. These chronic and disabling diseases include insulin-dependent diabetes mellitus (type 1 diabetes), multiple sclerosis, systemic lupus erythematosus (SLE), rheumatoid arthritis, inflammatory bowel disease (IBD), psoriasis, uveitis, autoimmune thyroid disease, and other disorders. Autoimmune diseases are among the leading causes of death among young and middle-aged women in the United States. Incidence rates vary amongst the autoimmune diseases, with estimates ranging from less than one newly diagnosed case of systemic sclerosis to more than 20 cases of adult-onset rheumatoid arthritis per 100,000 person/year. Prevalence rates range from less than 5 per 100,000 (e.g., chronic active hepatitis, and uveitis) to more than 500 per 100,000 (e.g., Graves' disease, rheumatoid arthritis, and thyroiditis). From the incidence data, it is estimated that about 250,000– 400,000 Americans develop an autoimmune disease annually. The social and financial burden of these chronic, debilitating diseases is immense and includes poor quality of life, high health care costs, and substantial loss of productivity. These disorders result in direct and indirect annual costs of more than $100 billion in the United States alone.

Rheumatoid arthritis accounts for greater than 2 million cases in the United States, including 30,000–50,000 children. On an annual basis, it results in 25,000 hospitalizations and 2.1 million lost workdays. Rheumatoid arthritis accounts for 22% of all deaths from arthritis and other rheumatic conditions. Rheumatic diseases are the leading cause of disability among adults aged 65 and older. Type 1 diabetes is described in about 0.5–0.8 million cases in the United States, including approximately 150,000 younger than 20 years of age. Recently compiled data show that approximately 15 million people have type I diabetes mellitus worldwide and that this number may well double by the year 2025. Prevalence of multiple sclerosis is 250,000–500,000 cases in the United States, resulting in 30,000 hospitalizations per year. In Europe, prevalence of multiple sclerosis is approximately 1 million cases and it may be greater than 200,000 in Russia. Multiple sclerosis is potentially the most common cause of neurological disability in young adults. SLE affects approximately 1.5 million Americans and may be more than 3 million in Europe. Its prevalence in Russia may be estimated as greater than 700,000 cases. More than 800,000 patients suffer from IBD in the United States: more than two-thirds due to ulcerative colitis and approximately one-third due to Crohn's disease. IBD is associated with

2.3 million outpatient visits per year and more than 31,000 restricted-activity days annually (Jacobson et al. 1997; Lawrence et al. 1998; Cooper and Stroehla 2003; Sacks et al. 2004; Hirtz et al. 2007)

2. Immune-Mediated Diseases: Statistics

According to the modern and common point of view, immune-mediated diseases are represented by a big group of the immune system diseases and even a large group of the diseases directly or indirectly associated with the immune system. The diseases of the immune system include (i) immunodeficiencies (primary or inherited and secondary or acquired) and (ii) immunoproliferative disorders, such as malignancies of the immune system (i.e., multiple myeloma, lymphomas, and leukemias), autoimmune diseases, and immune hypersensitivities (i.e., allergies). The disorders where the immune system is not the primary cause of a disease, although plays an obvious role in the pathogenesis, include, for instance, cancer and infectious diseases (Figure 1).

Primary immunodeficiency diseases were once thought to be rare, mostly because only the more severe forms were recognized. Today, we realize that primary immunodeficiencies are not uncommon, may be relatively mild, and can occur in teenagers and adults as often as in infants and children (Fleisher 2006). Although very serious inherited immunodeficiencies become apparent almost as soon as after a baby is born, there are some inherited immune deficiencies that never produce symptoms. The exact number of persons with primary immunodeficiencies is not known. It is estimated that each year about 400 children are born in the United States with a serious primary immunodeficiency. The number of Americans now living with a primary immunodeficiency is estimated to be between 25,000 and 50,000. Worldwide, the estimate is 1 million or more (not including selective IgA deficiency), most of whom will die prematurely, without ever having a diagnosis made. Approximately one out of 600 individuals has the most common primary immunodeficiency — selective IgA deficiency, or almost 500,000 people in the United States. Among those with this disease, people of European ancestry greatly outnumber those of other ethnic groups. Importantly, as new laboratory tests become more widely available, more cases of primary immunodeficiencies, are being recognized. At the same time, new types of these disorders are being discovered and described. Currently, the World Health Organization lists over 70 primary immunodeficiencies, and the numbers are increasing (Ochs and Notarangelo 2004; Shearer et al. 2004).

Allergic diseases or immune hypersensitivity-related conditions are one of the biggest groups of the immune system diseases. It is roughly estimated that 300 million people of all ages and all ethnic backgrounds suffer from asthma. More than 60 million people in America have asthma or allergies, costing the US economy over $20 billion each year in health care costs, such as hospitalizations, medical services, lost productivity at work or school, and more. A recent nationwide survey found that more than half (54.6%) of all US citizens test positive to one or more allergens, and allergies are the sixth leading cause of chronic disease. Asthma accounts for nearly 500,000 hospitalizations, 2 million emergency department visits, and 5000 deaths in

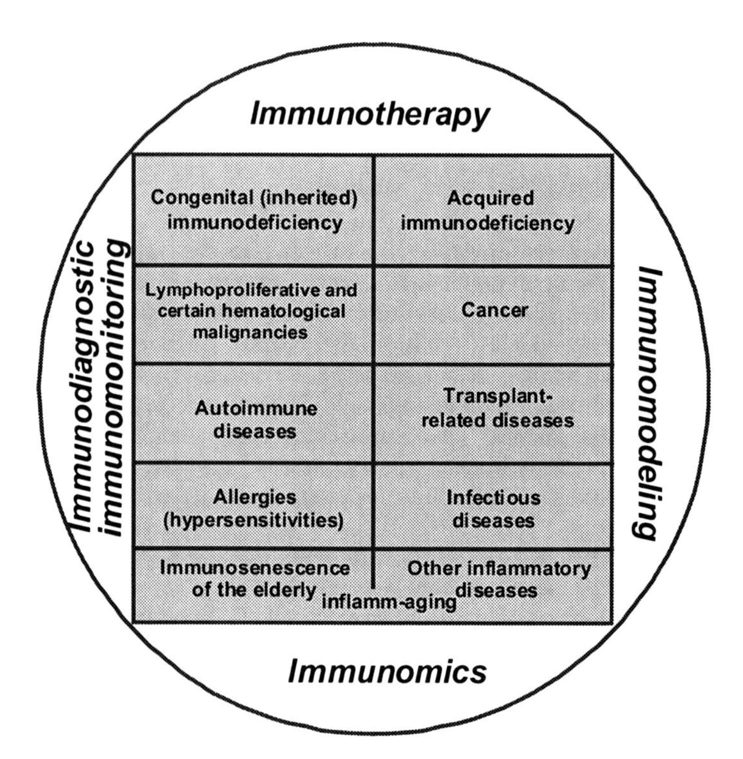

FIGURE 1. Schematic presentation and classification of immune-mediated diseases, their research, diagnosis, and therapy. Left column combines diseases of the immune system, such as primary immunodeficiencies, autoimmune diseases (rheumatoid arthritis, multiple sclerosis, etc.), allergies (asthma, food allergy, etc.), and immunological malignancies (myltiple myeloma, lymphomas, leukemias, etc.). Right column represents diseases where the immune system is involved in the pathogenesis of the disease, such as infectious diseases, cancer, and conditions related to organ or cell transplantation. Additional groups of related diseases include certain inflammatory disorders (brain trauma, Alzheimer's disease, etc.) and conditions associated with the immunosenescence of the elderly.

the United States each year. It affects an estimated 5 million children (under age 18) out of an estimated 70 million children. Atopic dermatitis is a common inflammatory skin condition that affects up to 20% of children and up to 3% of adults in Western countries, developing in almost 60% of patients during the first year of life and in an additional 25% between the ages of 1 and 5 years. At least 36 million people in the United States have seasonal allergic rhinitis accounting for approximately 17 million office visits and totaled $6 billion/year. Furthermore, sinusitis develops in approximately 35 million Americans each year and overall health care expenditures attributable to sinusitis in the United States were estimated to be over $6 billion (Hartert and Peebles 2000; Chandra 2002; Bloom et al. 2006; Enriquez et al. 2007).

Infectious and malignant diseases represent immune-mediated diseases with not only the highest level of incidence, but, probably, the highest levels of morbidity and mortality. Incidence rate of flu infection reaches up to 1 in 3 or 36.00% or greater than 100 million people in the United States annually. Flu-associated illnesses are responsible for an average of 20,000 to 40,000 deaths per year in the United States. In Africa, infectious and parasitic diseases account for about 5,625,000 deaths annually. Approximately 40 million people worldwide are living with HIV/AIDS. In 2004 alone, an estimated 5 million people worldwide were newly infected with HIV—about 14,000 each day, or approximately 600 per hour, or approximately 10 per minute. Tuberculosis is a leading cause of death among people who are HIV-positive. It accounts for about 13% of AIDS deaths worldwide. Someone in the world is newly infected with tuberculosis bacilli every second. Overall, one-third of the world's population is currently infected with the tuberculosis bacillus. Every year, 300–500 million people are affected by malaria, and between 1 and 3 million die. Tuberculosis, malaria and HIV/AIDS together account for nearly 18 % of the disease burden in the poorest countries. Cancer accounts for 7 million deaths per year worldwide. About 1,300,000 new cases of cancer are diagnosed yearly in the United States and every year about 560,000 Americans are expected to die of cancer; more than 1500 people a day (Dunavan 2005; Johnson and Decker 2006; Zielonka 2006).

3. Immune-Mediated Diseases: Correlations

Immune-mediated inflammatory diseases, or IMIDs, is another term, which is used to describe immune-mediated diseases associated with inflammatory pathogenesis mechanisms. IMIDs are characterized by immune dysregulation that results in acute or chronic inflammation, causing organ or tissue damage. One causal manifestation of immune deregulation is the inappropriate expression of proinflammatory cytokines, such as IL-1, IL-6 and TNF-α as well as Th1/Th2 cytokine disbalance leading to pathological consequences. This is in agreement with the "cytokine theory of disease," which states that an overproduction of cytokines can cause the clinical manifestations of disease (Nathan 2002; Frieri 2003; Elenkov et al. 2005; Tracey 2005). For instance, Th1-associated IMIDs include rheumatoid arthritis, Crohn's disease, multiple sclerosis, SLE, type I diabetes mellitus, psoriasis, sarcoidosis, ankylosing spondylitis, uveitis, and pathologic conditions associated with lung transplantation. Th2-associated disorders are asthma and allergies, pulmonary fibrosis, and ulcerative colitis. Diseases related to inappropriate expression of inflammatory cytokine are represented by osteoarthritis, chronic obstructive pulmonary disease (COPD), traumatic brain and spinal cord, and Alzheimer's disease. Furthermore, autoinflammatory diseases can be specified as inborn errors of the innate immune system. The main component of autoinflammatory diseases is the group of hereditary periodic fevers that are characterized by intermittent bouts of clinical inflammation with focal organ involvement mainly: abdomen, musculoskeletal system, and skin (Grateau 2006). Finally, new diseases are constantly added to the IMID group. For instance, recent data suggest that many risk factors leading to the development of osteoporosis, a major cause of morbidity in older people, could exert their effects through immunologically mediated modulation of bone remodeling. "Inflamm-ageing" plays a role in bone remodeling through proinflammatory cytokines, which are known to influence osteoclasts and osteoblasts (De Martinis et al. 2006). Thus, immune-mediated inflammation may play an important role in determining bone resorption and support the appearance of new discipline, osteoimmunology (Walsh et al. 2006).

Interestingly, an excessive Th1-type proinflammatory response is believed to be a common base for autoimmune disorders, whereas excessive Th2-type cytokine production (IL-4, IL-5, and IL-13) is associated with supporting of IgE and eosinophilic responses and is believed to be the cause of allergic disorders (Singh et al. 1999). An increasing number of scientific publications provide proof for the concept that an impairment of immune-tolerance mechanisms might be causally related to the development of unwanted Th2-driven, allergen-induced airway diseases. Impaired expansion of natural and/or adaptive Tregs is hypothesized to lead to the development of allergy and asthma (Umetsu and DeKruyff 2006). However, the results of the relationship between Th2-related atopic disorders and Th1-related autoimmune diseases are conflicting. Some epidemiologic studies concluded that Th2-weighted imbalance that favors allergic response, might be protective against Th1-related autoimmune disorders. Indeed, atopic eczema was shown to be associated with a lower risk for type 1 diabetes mellitus in children (Stene and Joner 2004). Similarly, another report demonstrated reduced frequency of allergic symptoms in children with type 1 diabetes (Caffarelli et al. 2004). In contrary, increased incidence of asthma in patients with type 1 diabetes, rheumatoid arthritis, and IBD was also reported (Kero et al. 2001; Simpson, Anderson et al. 2002). These and other findings suggest that the significance and the role of Th1/Th2 paradigm in supporting different immune-mediated diseases might be oversimplified and should be further investigated (Sheikh et al. 2003).

The link between other major groups of immune-mediated diseases, particularly cancer and allergies, has been discussed for years with a variety of supporting and disproving evidence. Theoretically, two contradictory hypotheses may be advanced: either tumor immunosurveillance may operate more efficiently in those individuals who have allergies or the alterations of the immunologic system can enhance inflammatory response and favor the tumor onset (Carrozzi and Viegi 2005). Many studies suggest that the association between allergy and cancer is complex and depends on the specific allergy and the specific organ site under consideration. For instance, following the cohort of patients for 6 years, Mills et al. reported that prostate and breast cancer risk were elevated in persons who reported any type of allergic history, as was risk of lymphatic or hematopoietic cancers and sarcoma. For each of these types of cancer, risk increased with increasing numbers of allergies. However, ovarian cancer risk was decreased in persons with any allergic history, and increasing numbers of allergies was associated with decreasing risk of this form of cancer (Mills et al. 1992). Similarly, evaluating the associations among certain allergic disorders, atopy upon skin-prick testing, and specific cancers in a prospective study, Talbot-Smith et al. reported that having a skin reaction to house dust mites nearly tripled the risk of prostate cancer: however, history of asthma and hay fever was associated with a trend toward a reduced risk of colorectal cancer and increased risk of leukemia, but no association was found between breast and lung cancers and allergic disorders or atopy (Talbot-Smith et al. 2003).

Analyzing available published reports, Wang and Diepgen concluded that despite the mixed results, the emerging picture from most of the currently available epidemiological data indicates that atopic disease is associated with a reduced risk for cancer (Wang and Diepgen 2005). On the other hand, there seems to be a clear association of lung cancer with asthma, and thus, asthmatic subjects have to be considered "at risk" for lung cancer (Carrozzi and Viegi 2005). But, the result of another examination reveals that there is significant lower frequency of allergic diseases in patients with lung cancer in comparison with frequency of allergic diseases in control adult population (Tolak et al. 2006). Another study demonstrated a reduced risk of primary malignant brain tumors in people reporting asthma, hay fever, and other allergic conditions (Schwartzbaum et al. 2005). In contrast to most previous studies, Soderberg et al. (2004) study suggests that allergic conditions might increase the risk of some hematological malignancies. The study showed that people with hives and asthma were about twice as likely to develop leukemia and those who had eczema are under an increased risk of non-Hodgkin's lymphoma. At the same time, a possible connection between allergy and cancer has been suspected, but allergy-related conditions or atopy have been inconsistently associated with reduced risks of non-Hodgkin lymphoma, suggesting that allergy may not be causally associated with the risk of non-Hodgkin lymphoma (Melbye et al. 2007). Thus, the problem of whether allergy is a "risk" or a "protective" factor for cancer is far to be solved.

Another recent report suggests that autoimmunity and malignancy frequently coexist and they may share etiological and pathological mechanisms (Toubi and Shoenfeld 2007). For instance, natural protective autoantibodies against tumor-antigens were isolated from patients and healthy donors reflecting the development of naturally occurring B cell responses during the process of cancer evolvement. They fulfill the definition of autoantibodies since they are self-reactive, and they also bind altered self-antigens such as tumor cells. In this regard, various autoantibodies such as anti-dsDNA and anti-Fas autoantibodies were found to be significantly higher in patients with various carcinomas, thus playing a role for their improved survival. Finally, clinical use of a number of biologic response modifiers including cytokines, antibodies, and other immunomodulators might be associated with an increased risk of malignancy during the treatment of allergic or autoimmune diseases and vice versa (Kong et al. 2006).

In spite of the controversy of positive and negative relationships between incidences of certain immune-mediated diseases, it is clear that there are both etiopathogenetic and clinical consequence links between many of these disorders. For instance, certain immune-mediated diseases, such as type 1 diabetes or arthritis, are not just very common but result in more than 19,000 organ-transplant recipients each year in the United States. Even though the first reference to the concept of organ transplantation and replacement for therapeutic purposes may be dated to approximately 200 AD, when Hua-To in China replaced diseased organs with healthy ones, only in 1954, the kidney was the first human organ to be transplanted successfully. Liver, heart, and pancreas transplants were successfully performed by the late 1960s, while lung and intestinal organ-transplant procedures were begun in the 1980s. Increasing numbers of transplants in patients with immune-mediated diseases, such as autoimmune and malignant diseases, was responsible for a new group of immune-mediated diseases, that is, diseases associated with organ, tissue, or cell transplantation, for example, graft-versus-host disease or graft rejection. On the other hand, transplantation is the leading form of treatment for many forms of end-stage organ failure and has saved and enhanced the lives of more than 300,000 people in the United States. With this success, however, has come increasing demand for donated organs. Today, approximately 85,000 people are awaiting transplants nationwide with almost 50,000 people awaiting kidney transplant. Sadly, approximately 17 patients die every day while waiting for organ for transplantation, that is, one person every 85 min. More than 10,000 people are on the Eurotransplant waiting list in a similar situation.

4. Concluding Remarks

Thus, immune-mediated diseases represent an enormous medical, social, and economical problem and require serious and instant attention of clinicians, scientists, pharmacists, biotech professionals, and politicians as well. The study of the immune system has led to significant findings in many fields of medicine and biology and resulted in the discovery of novel and powerful tools and reagents that are now used for diagnosis and therapy of a variety of diseases. The development of prophylactic and therapeutic vaccines against many infectious agents, as well as innovative immunotherapeutic and immune gene engineering approaches for patients with cancer, allergy, and immunodeficiency, has substantially decreased mortality and morbidity and improved the life style, life expectancy, and well being of the millions of patients with immune-mediated and immune-associated diseases. With a strong research background and the availability of powerful modern research tools, we can anticipate that our basic and applied research and clinical programs will further strengthen our fight against allergic, malignant, infectious, and autoimmune diseases and improve our defenses against potential bioterroristic attacks. Thus, the field of immune-mediated diseases is one of continuously escalating significance, as well as scientifically, economically, and socially important.

References

Bloom, B., Dey, A.N. and Freeman, G. (2006) Summary health statistics for U.S. children: National Health Interview Survey, 2005. Vital Health Stat. 10, 1–84.

Caffarelli, C., Cavagni, G., Pierdomenico, R., Chiari, G., Spattini, A. and Vanelli, M. (2004) Coexistence of IgE-mediated allergy and type 1 diabetes in childhood. Int. Arch. Allergy Immunol. 134, 288–294.

Carrozzi, L. and Viegi, G. (2005) Allergy and cancer: a biological and epidemiological rebus. Allergy 60, 1095–1097.

Chandra, R.K. (2002) Food hypersensitivity and allergic diseases. Eur. J. Clin. Nutr. 56 (Suppl. 3), S54–S56.

Cooper, G.S. and Stroehla, B.C. (2003) The epidemiology of autoimmune diseases. Autoimmun. Rev. 2, 119–125.

De Martinis, M., Di Benedetto, M.C., Mengoli, L.P. and Ginaldi, L. (2006) Senile osteoporosis: is it an immune-mediated disease? Inflamm. Res. 55, 399–404.
Dunavan, C.P. (2005) Tackling malaria. Sci. Am. 293, 76–83.
Elenkov, I.J., Iezzoni, D.G., Daly, A., Harris, A.G. and Chrousos, G.P. (2005) Cytokine dysregulation, inflammation and well-being. Neuroimmunomodulation 12, 255–269.
Enriquez, R., Hartert, T. and Persky, V. (2007) Trends in asthma prevalence and recommended number of childhood immunizations are not parallel. Pediatrics 119, 222–223.
Fleisher, T.A. (2006) Back to basics: primary immune deficiencies: windows into the immune system. Pediatr. Rev. 27, 363–372.
Frieri, M. (2003) Neuroimmunology and inflammation: implications for therapy of allergic and autoimmune diseases. Ann. Allergy Asthma Immunol. 90, 34–40.
Grateau, G. (2006) Autoinflammatory diseases. Acta Clin. Belg. 61, 264–269.
Hartert, T.V. and Peebles, R.S., Jr. (2000) Epidemiology of asthma: the year in review. Curr. Opin. Pulm. Med. 6, 4–9.
Hirtz, D., Thurman, D.J., Gwinn-Hardy, K., Mohamed, M., Chaudhuri, A.R. and Zalutsky, R. (2007) How common are the "common" neurologic disorders? Neurology 68, 326–337.
Jacobson, D.L., Gange, S.J., Rose, N.R. and Graham, N.M. (1997) Epidemiology and estimated population burden of selected autoimmune diseases in the United States. Clin. Immunol. Immunopathol. 84, 223–243.
Johnson, M.D. and Decker, C.F. (2006) Tuberculosis and HIV infection. Dis. Mon. 52, 420–427.
Kero, J., Gissler, M., Hemminki, E. and Isolauri, E. (2001) Could TH1 and TH2 diseases coexist? Evaluation of asthma incidence in children with coeliac disease, type 1 diabetes, or rheumatoid arthritis: a register study. J. Allergy Clin. Immunol. 108, 781–783.
Kong, J.S., Teuber, S.S. and Gershwin, M.E. (2006) Potential adverse events with biologic response modifiers. Autoimmun. Rev. 5, 471–485.
Lawrence, R.C., Helmick, C.G., Arnett, F.C., Deyo, R.A., Felson, D.T., Giannini, E.H., Heyse, S.P., Hirsch, R., Hochberg, M.C., Hunder, G.G., Liang, M.H., Pillemer, S.R., Steen, V.D. and Wolfe, F. (1998) Estimates of the prevalence of arthritis and selected musculoskeletal disorders in the United States. Arthritis Rheum. 41, 778–799.
Melbye, M., Smedby, K.E., Lehtinen, T., Rostgaard, K., Glimelius, B., Munksgaard, L., Schollkopf, C., Sundstrom, C., Chang, E.T., Koskela, P., Adami, H.O. and Hjalgrim, H. (2007) Atopy and risk of non-Hodgkin lymphoma. J. Natl. Cancer. Inst. 99, 158–166.
Mills, P.K., Beeson, W.L., Fraser, G.E. and Phillips, R.L. (1992) Allergy and cancer: organ site-specific results from the Adventist Health Study. Am. J. Epidemiol. 136, 287–295.
Nathan, C. (2002) Points of control in inflammation. Nature 420, 846–852.
Ochs, H.D. and Notarangelo, L.D. (2004) X-linked immunodeficiencies. Curr. Allergy Asthma Rep. 4, 339–348.
Sacks, J.J., Helmick, C.G. and Langmaid, G. (2004) Deaths from arthritis and other rheumatic conditions, United States, 1979–1998. J. Rheumatol. 31, 1823–1828.
Schwartzbaum, J., Ahlbom, A., Malmer, B., Lonn, S., Brookes, A.J., Doss, H., Debinski, W., Henriksson, R. and Feychting, M. (2005) Polymorphisms associated with asthma are inversely related to glioblastoma multiforme. Cancer Res. 65, 6459–6465.
Shearer, W.T., Cunningham-Rundles, C. and Ochs, H.D. (2004) Primary immunodeficiency: looking backwards, looking forwards. J. Allergy Clin. Immunol. 113, 607–609.
Sheikh, A., Smeeth, L. and Hubbard, R. (2003) There is no evidence of an inverse relationship between TH2-mediated atopy and TH1-mediated autoimmune disorders: lack of support for the hygiene hypothesis. J. Allergy Clin. Immunol. 111, 131–135.
Shoenfeld, Y. and Zandman-Goddard, G. (2003). The History of Autoimmune Disease. In *Autoimmune Diseases "The Enemy from Within"*, Bio-Rad Pub., Hercules, CA, USA: 65.
Simpson, C.R., Anderson, W.J., Helms, P.J., Taylor, M.W., Watson, L., Prescott, G.J., Godden, D.J. and Barker, R.N. (2002) Coincidence of immune-mediated diseases driven

by Th1 and Th2 subsets suggests a common aetiology. A population-based study using computerized general practice data. Clin. Exp. Allergy 32, 37–42.
Singh, V.K., Mehrotra, S. and Agarwal, S.S. (1999) The paradigm of Th1 and Th2 cytokines: its relevance to autoimmunity and allergy. Immunol. Res. 20, 147–161.
Soderberg, K.C., Hagmar, L., Schwartzbaum, J. and Feychting, M. (2004) Allergic conditions and risk of hematological malignancies in adults: a cohort study. BMC Public Health 4, 51.
Stene, L.C. and Joner, G. (2004) Atopic disorders and risk of childhood-onset type 1 diabetes in individuals. Clin. Exp. Allergy 34, 201–206.
Talbot-Smith, A., Fritschi, L., Divitini, M.L., Mallon, D.F. and Knuiman, M.W. (2003) Allergy, atopy, and cancer: a prospective study of the 1981 Busselton cohort. Am. J. Epidemiol. 157, 606–612.
Tolak, K., Omernik, A., Zdan, O., Ragus, D., Domagala-Kulawik, J., Rudzinski, P. and Chazan, R. (2006) The frequency of allergy in lung cancer patients. Pneumonol. Alergol. Pol. 74, 144–148.
Toubi, E. and Shoenfeld, Y. (2007) Protective autoimmunity in cancer (Review). Oncol. Rep. 17, 245–251.
Tracey, K.J. (2005). *Fatal Sequence: The Killer Within.* Washington, DC, Dana Press.
Umetsu, D.T. and DeKruyff, R.H. (2006) The regulation of allergy and asthma. Immunol. Rev. 212, 238–255.
Walsh, M.C., Kim, N., Kadono, Y., Rho, J., Lee, S.Y., Lorenzo, J. and Choi, Y. (2006) Osteo-immunology: interplay between the immune system and bone metabolism. Annu. Rev. Immunol. 24, 33–63.
Wang, H. and Diepgen, T.L. (2005) Is atopy a protective or a risk factor for cancer? A review of epidemiological studies. Allergy 60, 1098–1111.
Zielonka, T.M. (2006) Tuberculosis in Poland, Europe, and world. Part I--prevalence. Pol. Merkur. Lekarski. 21, 243–252.

Immunodeficiencies: News and Updates

2

The Four Most Common Pediatric Immunodeficiencies

Richard E. Stiehm

Mattel Children's Hospital and David Geffen School on Medicine, UCLA, Los Angeles, CA, USA, estiehm@mednet.ucla.edu

Abstract. Other than the physiologic hypogammaglobulinemia of infancy, 80% of the confirmed immunodeficiencies consist of four syndromes: transient hypogammaglobulinemia of infancy (THI), IgG subclass deficiency, partial antibody deficiency with impaired polysaccharide responsiveness (IPR), and selective IgA deficiency IgAD. None are life threatening, all can be readily managed, and many recover spontaneously. An exact incidence of these disorders is not known. A summary of immunodeficiency registries in four countries listed IgAD in 27.5% of the patients, IgG subclass deficiency in 4.8%, and THI in 2.3%. The 1999 US survey of primary immunodeficiencies conducted by the Immune Deficiency Foundation found that 17.5% of these patients had IgAD and 24% had IgG subclass deficiency, while THI and IPR were not listed. The Jeffrey Modell Foundation (2005) survey of their global centers in 2004 reported IgAD in 15.5%, subclass deficiencies in 8%, and THI in 2% of their patients.

1. Transient Hypogammaglobulinemia of Infancy

1.1. Definition and History

This disorder was first described in 1956 (Gitlin and Janeway 1956). It is classically considered a prolongation of physiologic hypogammaglobulinemia that occurs from age 3 to 6 months as a result of disappearance of maternal transplacental IgG and slow increase of the infants' own IgG levels (Roifman 2004; Dalal and Roifman 2006). Despite their low IgG levels, most infants are able to respond to vaccine antigens during the first 6 months of life.

Although some use low levels of either IgG, IgM, or IgA below 2 SD from the mean as diagnostic criteria, transient hypogammaglobulinemia of infancy (THI) is best defined as low levels of IgG with or without depression of IgA and/or IgM in an infant beyond 6 months of age in which other primary immunodeficiencies have been excluded. The condition can persist up to the age of 5 years (Roifman 2004; Dalal and Roifman 2006). This definition will include many infants who have no increased susceptibility to infection and have normal antibody responses to vaccine antigens. Such infants rarely come to the attention of the immunologist. Thus, a clinically significant THI occurs in that subgroup of THI infants with frequent infections and/or poor antibody responses to one or more vaccine antigens.

1.2. Etiology

Various causes of this disorder have been proposed. An early study suggested that IgG genetic allotypes (Gm types) of the fetal IgG induced anti-Gm antibodies in the mother that crossed the placenta and suppressed fetal immunoglobulin production (Fudenberg and Fudenberg 1964). This was not confirmed in another study (Nathenson 1971). A genetic cause was suggested that THI represented heterozygosity for genetic hypogammaglobulinemia based on family studies (Nathenson 1971). Another suggested that a T-helper deficiency caused THI (Siegel et al. 1981). A final study suggested cytokine imbalance (Kowalczyk et al. 1997). Another cause is seen among infants, usually premature, who have had prolonged stays in the neonatal ICU for a variety of illnesses. Their IgG is often low because of stress, loss of plasma into the GI or respiratory tract, steroid use, and frequent blood draws. Finally, many of these children may simply be at the low end of the normal range for IgG and/or at the low end of the normal progression of immune maturation as reflected by IgG levels, thus have no immunodeficiency, and are like the asymptomatic patients with low CD4 cells who are designated as idiopathic CD4 lymphopenia.

1.3. Clinical Features

THI is more common in males, who usually are identified at an earlier age than females (Whelan et al. 2006). About 25% of the symptomatic patients are identified before the age of 6 months, another 50 % from ages 6 to 12 months, and the rest after age 12 months.

Two groups of THI patients are recognized. The first group of infants are asymptomatic who have immunoglobulins done routinely or because a family member has an immunodeficiency. Most of these infants remain asymptomatic, have normal responses to vaccine antigen, and grow out of their hypogammaglobulinemia after several years. The second group is infants identified because of recurrent or severe infection often starting in the first weeks of life. The majority of their infections are respiratory-otitis, sinusitis, and bronchitis. Other children have recurrent diarrhea, or prolonged oral candidacies.

We recently identified seven infants with severe eczema with low levels of IgG (Lin, Roberts, and Stiehm, unpublished observations), possibly associated with protein loss through the skin. Positive skin or Radioallergosorbent tests (RAST) and elevated IgE levels are often seen in these patients. Hematological abnormalities are often present, such as mild neutropenia and less commonly thrombocytopenia. Tonsils and lymph nodes are present but may be small.

1.4. Laboratory Features

In addition to decreased levels of IgG, low levels of IgM and IgA are noted in over half of the patients (Whelan et al. 2006). There is no molecular test for THI. Antibody titers to tetanus, diphtheria, hepatitis A and B, *Hemophilus influenza,* and pneumococcal vaccine antigens are variable. About 15% of the patients have non-protective antibody titers to one or more of these antigens, most commonly 1 or more of the serotypes present in the conjugated pneumococcal vaccine. Under these circumstances, a booster immunization is recommended followed by repeat titers after one month. Lymphocyte subsets (CD3, CD4, and CD8 T cells), CD19 (B cells), and CD16/CD56 (NK cells) are usually normal. Very low numbers of B cells suggest X-linked agammaglobulinemia. As noted, allergy tests including IgE levels may be abnormal. Repeated respiratory infections may be associated with chronic sinusitis as identified by sinus film or limited CT scan.

1.5. Management and Prognosis

Infants in group 1 require no treatment. In symptomatic patients, a conservative approach is initially warranted, such as removing the infant from day care, prompt treatment of respiratory infections, and occasionally prophylactic antibiotic therapy. In the rare THI patient who has severe infections or very poor antibody responses to vaccine antigens, Intravenous immunoglobulin (IVIG) substitution at a dose of 400–500 mg/kg q3 or 4 weeks is used, usually for a 6–12 month period. IVIG is then stopped, and immunoglobulin levels and antibody responses are retested 3 months later.

By definition, all of these patients eventually recover. Most patients recover by age 2, but in some patients low IgG levels may persist until age 5. Patients with a combined IgG and IgA deficiency (IgGD and IgAD, respectively) may develop selective IgAD. Other patients will develop an IgG subclass deficiency, while others with poor antibody responses may normalize their IgG levels but have persistent impaired polysaccharide responsiveness (IPR). There was a striking difference in immune maturation, with females having persistently slower recovery (Whelan et al. 2006).

2. IgG Subclass Immunodeficiency

2.1. Definition and History

Schur et al. in 1970 first described IgG subclass deficiencies in three adult patients (Schur et al. 1970). Since then, numerous publications have identified subclass deficiencies in many patients, particularly children. Indeed, it is perhaps the most common immunodeficiency described, and certainly the one for which IVIG is most often used and misused.

The definition of an IgG subclass deficiency is the finding that one or more IgG subclasses are <2 SD below the mean for age with normal or near normal total IgG levels (Stiehm et al. 2004; Lemmon and Knutsen 2006). Up to 20% of the population will thus have an IgG subclass deficiency of one or more IgG subclasses. Since most of these subjects are asymptomatic, the previously mentioned definition only defines a clinical laboratory finding, not a disease. A clinically significant IgG subclass immunodeficiency is associated with recurrent infection and a significant defect in antibody responsiveness. Many such patients present with recurrent respiratory infections. Others, often with more serious infections, may have a subclass deficiency in association with another primary immunodeficiency (e.g., selective IgAD and DiGeorge syndrome), or with a secondary immunodeficiency (e.g., HIV infection and cirrhosis), or with an autoimmune disease, for example, immune thrombocytopenia and lupus (Stiehm et al. 2004).

2.2. Etiology

The four IgG subclasses are defined by unique structures of the constant region of their heavy chains. They make up about 70%, 20%, 7%, and 3% of the total IgG levels. Each has unique structural, antigenic, and biologic characteristics (Stiehm et al. 2004). The most significant biologic difference is that IgG2 contains the preponderance of antibodies to polysaccharide antigens. Other differences include activation of the classical complement pathway by IgG1 and IgG3, a shorter half-life for IgG3, and less placental passage for IgG2. Other antibodies are not evenly distributed, and mostly in the IgG4 subclass.

Since each subclass is encoded by a different heavy chain segment, gene deletions may be responsible for some subclass deficiencies, particularly those associated with a complete absence of a subclass (Lefranc et al.1983; Migone et al. 1984). Other abnormalities may include transcriptional defects or genetic association with certain genetic IgG allotypes (Gm types). The most common subclass deficiencies among patients presenting with recurrent infections are IgG4 deficiency (40%), IgG2 deficiency (28%), IgG3 deficiency (17%), and IgG1 deficiency (14%). Isolated IgG1 deficiency is rare. Combinations of one subclass deficiency with another IgG subclass or an IgAD are common, notably IgG2 and IgG4, IgG4 and IgA, IgG3 and IgA, IgG2 and IgG4 and IgA, and IgG2 and IgA combinations in that order of frequency (Stiehm et al. 2004).

2.3. Clinical Features

Subclass deficiencies are heterogeneous and rarely familial. IgG4 subclass deficiency may occur in as many as 20% of both adults and children, depending on the sensitivity of the assay and thus is of rare clinical significance. In adults, IgG3 deficiency is the second most common deficiency, and females are more likely to be affected. Adults with IgG subclass deficiency generally have more severe infections, and some of them may be developing common variable immunodeficiency (CVID).

In children, males make up 75% of the cases, and IgG2 is the second most common deficiency. Children under the age of 6 years may be recovering from transient hypogamma-globulinemia, so it is difficult to diagnose an IgGD deficiency in children before age 4.

Selective IgG1 may be associated with more severe infections and is more common in adults. Selective IgG2 deficiency is the most common subclass disorder associated with recurrent infection and may be accompanied by IgA and/or IgG4 deficiencies. Many of these patients have IPR as discussed in Impaired Polysaccharide Responsiveness. IgG2 deficiency may resolve with time. Most symptomatic IgG3- deficient subjects have an associated deficiency of another class. Familial IgG3 deficiency has been recorded. As noted, isolated IgG4 is common and not usually of clinical significance. However, recurrent pneumonia has been described, suggesting that it is a marker for these illnesses rather than the cause of them.

2.4. Laboratory Features

IgG subclass determinations are indicated for patients with documented antibody defects, in patients with IgA deficiencies, and in patients in which early CVID is suspected. Levels must be compared with age-matched controls, especially in the first 2 years of age. For children aged 4–10 years, an IgG1 level less than 250 mg/dl, an IgG2 level less than 50 mg/dl, an IgG3 level less than 15 mg/dl, and an IgG4 level less than 1 mg/dl are abnormal. For subjects older than age 10, an IgG1 level less than 300 mg/dl, an IgG2 level less than 75 mg/dl, an IgG3 level less than 25, and an IgG4 level less than 1 mg/dl are abnormal.

A clinically significant IgG subclass deficiency must be established by measuring the antibody response to a vaccine antigen, particularly pneumococcal polysaccharide vaccine. A deficient response is defined as non-protective titers to a majority of the 12 serotypes tested or failure to exhibit a twofold rise in titer to serotypes for which there were non-protective titers. Tests for cellular immunity, complement activity, and phagocytic function should be done as necessary.

2.5. Management and Prognosis

Many patients do well with prompt medical management of each infectious episode, and the use of antibiotics early in the course of respiratory exacerbation is of value. Some patients with recurrent infections or chronic infections do well on prophylactic antibiotics. Vaccines should be kept current unless there is complete absence of antibody responses. A failure of prolonged antibiotics, severe symptoms, and persistent radiographic abnormalities may occasionally require IVIG therapy (Buckley 2002). The presence of a subclass deficiency alone is not an indication for IVIG. Nevertheless, it is a common practice to give IVIG under such circumstances, which is a costly misuse of a scarce and potentially harmful form of therapy and labels the patients as chronically ill and uninsurable. Its use in young children has the potential of inhibiting normal immune maturation.

Most patients do well on conservative therapy outlined above, but treatment may be prolonged and in some lifelong. Children under 10 may recover from a subclass deficiency spontaneously, particularly if there is not a complete absence of a subclass. By contrast, symptomatic adults may progress to CVID.

3. Impaired Polysaccharide Responsiveness

3.1. Definition and History

IPR is characterized by recurrent bacterial respiratory infection, an absent or subnormal response to a majority of polysaccharide antigens, normal or elevated immunoglobulins and IgG subclasses, and intact antibody responses to protein antigens in subjects over age 2 (Stiehm et al. 2004; Sorensen and Paris 2006).

As early as 1968, patients were described who had normal immunoglobulin levels, but had profound deficiencies of antibodies to both protein and polysaccharide antigens (Blecher et al. 1968; Saxon et al. 1980). This entity, *antibody deficiency with normal immunoglobulins*, should not be confused with IPR since such patients are considerably more susceptible to infections and more akin to CVID in their prognosis.

IPR was identified in the late 1980s following the introduction of unconjugated *H. influenza* type B polysaccharide vaccines (Granoff et al. 1986b). This large collaborative study identified children greater than 2 years of age that had a poor response to this vaccine despite normal responses to other vaccines. Subsequent studies indicated that sometimes this deficiency was familial and occurred more frequently among certain ethnic groups such as Apache Native Americans and Alaskan Eskimos (Stiehm et al. 2004). Adults with IPR were first described in 1987 (Ambrosino et al. 1987). Since unconjugated *H. influenza* vaccine has been replaced by a protein-conjugated vaccine that is immunogenic in infants, IPR is usually identified by a deficient response to the pneumococcal polysaccharide vaccine (Pneumovax) in children greater than 2 years of age and adults. Sorenson and Paris have identified this disorder as the most common immunodeficiency among children presenting with increased susceptibility to infection (Sorensen and Paris 2006).

3.2. Etiology

IPR is a heterogeneous illness with several postulated causes. In younger children aged 2–6, it may be an exaggeration of the physiologic non-responsiveness to polysaccharide vaccines of infants less than 2 years of age; these children recover spontaneously with time. Some of these youngsters have had THI with poor antibody responses. Most polysaccharide antibodies are in the IgG2 subclass, so that selective IgG2 subclass deficiency with or without selective IgAD must be sought. IPR may be part of another primary immunodeficiency such as Wiskott–Aldrich syndrome, DiGeorge syndrome, and mucocutaneous candidiasis, secondary deficiencies associated with aging, HIV disease, immunosuppressive drugs, genetic syndromes, chronic lung disease, and absence or deficiency of the spleen (Stiehm et al. 2004). IPR may be genetic in some families and linked to certain Gm and Km IgG allotypes (Granoff 1986a). A defect in the B cell repertoire, similar to certain mice strains, has been postulated (Ambrosino et al. 1987) and Ambrosino et al. have suggested a splenic defect in the marginal zone where dendritic cells interact with B cells. One adult IPR male with few circulating B cells had a BTK mutation associated with X-linked agammaglobulinemia (Wood et al. 2001).

3.3. Clinical Features

IPR is more common in males and may be more common in certain ethnic groups. Most of the patients are children aged 2–7, and most of these patients recover completely from their illness. Clinically, IPR patients resemble patients with IgG subclass and selective IgA deficiencies. They usually have recurrent bacterial respiratory infections such as sinusitis, otitis, and bronchitis: less common are systemic infections including pneumonia, sepsis, or meningitis. Many patients have asthma or wheezing with infections, often because of chronic sinusitis. Many have had ear tubes placed, tonsillectomy, a history of decreased hearing, and multiple courses of antibiotics. The physical examination often suggests chronic respiratory allergy, with circles under the eyes, pallor, gaping mouth, post-nasal drip, purulent nasal discharge, and moderate cervical adenopathy.

3.4. Laboratory Features

IPR patients by definition have normal IgG, IgM, and IgA levels and IgG subclass levels, as well as normal responses to protein antigens such as tetanus, diphtheria, and *H. influenza* type B-conjugated vaccines. Thus, the diagnosis rests on their antibody response to pneumococcal polysaccharide vaccine. Pre-immunization titers are recommended followed by retesting 1 month following vaccination. A normal response is development of protective titers (>1.3 μg/ml) to a majority of the subtypes tested. Sorenson and Paris (2006) suggested that a normal response for children under 6 consists of a majority of responses to be protective and for older individuals at least 70% of the serotypes should be protective. Three types of responses are noted. In the severe form, there is essentially no response to any or just one or two of the serotypes, and then the titers are low but protective. In the mild/moderate form, there is some but less than expected responses. Some of these patients, as well as some patients with an adequate response initially proceed to the third form associated with poor immunologic memory. These patients, upon retesting in 6–12 months, have lost some or most of their previously protective titers and continue to have recurrent infections.

All of these patients should be tested for T cell, phagocyte, and complement deficiencies. In young children, repeat antibody testing is recommended at yearly intervals, because spontaneous recovery often ensues. Symptomatic adults should be followed periodically to look for progression to CVID.

3.5. Treatment and Prognosis

Many of these patients do well with prompt and vigorous treatment of each respiratory infection. Ancillary treatment with inhaled steroids, bronchodilators, and decongestants are often of value. Vaccines, particularly influenza vaccine, should be updated. We often give these patients two doses of the pediatric-conjugated pneumococcal vaccine to boost their immunity to the serotypes in this vaccine. If these measures are ineffective, a course of prophylactic antibiotics is tried, usually for at least 6 months. If there is persistent infection, a trial of IVIG should be considered in full therapeutic doses. Only an occasional patient requires such therapy, but a good response has been documented in some series (Herrod 1993).

Children with partial responses usually recover with time. Patients with the severe form of disease usually have lifelong problems. Some of these patients may develop IgG subclass deficiency or CVID.

4. Selective IgAD Deficiency

4.1. Definition and History

Selective IgAD is defined as the absence or very low levels (<7 mg/dl) of serum IgA with normal levels of total IgG and IgM and no other major immune defects. This definition excludes young infants since they have low levels of IgA physiologically and do not obtain adult levels until 3 years of age. IgAD is the most common primary immunodeficiency, and in many individuals, it is unassociated with any illness including an increased susceptibility to infection. IgAD was first described in ataxia-telangiectasia, then in patients with frequent infections (West et al. 1962) and in normal subjects (Rockey et al. 1964). Low or absent IgA is a variable component of many other primary immunodeficiencies, notably all forms of agammaglobulinemia, ataxia-telangiectasia, mucocutaneous candidiasis (Kalfa et al. 2003) and IgG2 subclass deficiency (Oxelius et al. 1981). The frequency of IgAD varies in different ethnic groups, and is as high as 1:142 in Arabs (al-Attas and Rahi 1998) and as low as 1:15,000 in Japanese (Kanoh et al. 1986). Among Europeans and Americans, the frequency is about 1:500 (Hostoffer 2006).

4.2. Etiology

IgA is the second most abundant immunoglobulin synthesized; it appears not only in the serum but also in the secretions, including colostrums and breast milk. Much of the serum and all of the secretory IgA are synthesized in plasma cells of glandular tissues. Locally produced monomer IgA combines with an epithelial cell-synthesized secretory component and a joining chain and is secreted as a dimer (secretory IgA) into the lumen of the gland. Serum IgA does not get into the secretions; most of the serum IgA is synthesized locally and enters the blood stream in the monomeric form.

Selective IgAD deficiency is occasionally familial (Koistinen 1976), and in some families, there is a shared propensity of relatives of IgAD patients to have CVID. Genetic defects of a tumor necrosis factor receptor family member termed Transmembrane activator and calcium-modulator and cyclophilin-ligand interactor (TACI) have been identified in a few patients with IgAD and CVID, possibly causing defects in isotype switching (Castigli et al. 2005). There is an association of IgAD deficiency with certain HLA types, suggesting a linkage to the major histocompatibility complex (Lakhanpal et al. 1988). Genetic defects involving deletions of chromosome 14 and abnormalities of chromosome 18 and chromosome 4 are also associated with IgAD (Hostoffer 2006).

Certain drugs, notably penicillamine, gold, fenclofenac, and valporate, may cause depression of serum IgA levels, which sometimes are permanent. Congenital rubella and Epstein–Barr virus infections have been implicated in a few cases of acquired IgAD.

4.3. Clinical Features

Up to 90% of IgAD patients are asymptomatic. Indeed a few patients, particularly children under age 5, with low but measurable IgA levels outgrow the illness; this is most unusual in adults and in patients without detectable IgA (Blum et al. 1982).

Most symptomatic IgAD patients have frequent respiratory infections similar to patients with IgG subclass deficiencies or IPR. Many of these patients have a concomitant IgG2 deficiency and/or IPR. The most common infections are otitis, sinusitis, or bronchitis. A few develop recurrent pneumonia, obstructive lung disease, or bronchiectasis or other chronic lung diseases.

Atopic patients have a high incidence of IgAD. This includes patients with asthma, rhinitis, hives, and eczema. A search for chronic sinusitis should be done in those with IgAD and asthma. Food intolerance, particularly milk allergy, is common in infants with IgAD; some of these patients have high titers of milk antibodies. Other causes of gastrointestinal symptoms in patient with IgAD include celiac disease, inflammatory bowel disease, nodular lymphoid hyperplasia, hepatitis, and others. Autoimmune disease is common among IgAD patients, notably rheumatoid arthritis, lupus erythematosus, and thyroiditis. At least 16 other autoimmune disorders have been identified in patients with IgAD, involving all organ systems such as the skin, central nervous system (mental retardation), and the hematopoietic system, (immune thrombocytopenic purpura and autoimmune hemolytic anemia). One theory is that absence of IgA in the serum permits cross-reactive antigens to enter the circulation and initiate autoimmune reactions. A very rare IgAD patient may have an anaphylactic reaction to a blood product containing IgA. Such patients have developed anti-IgA antibodies to a previous infusion and recognize IgA as a neo-antigen. Thus, blood products should be avoided in IgAD patients, and they should wear a Medic Alert badge to this effect. If IVIG is needed, a product low in IgA should be used, with caution and pre-medication (Cunningham-Rundles 2004).

4.4. Laboratory Features

Immunoglobulin and IgG subclass levels are the primary diagnostic tests. If IgAD is present, its absence should be confirmed by repeat sampling. Next, antibody titers to vaccine antigens should be done to determine whether there is a concomitant functional antibody deficiency. Assays for cellular immunity, phagocyte function, and complement ment are usually normal. The presence of autoimmune antibodies such as antinuclear antibodies (ANA) and antithyroid antibodies are common. Allergy tests are often positive. Milk antibodies and celiac antibodies should be done if there is evidence of food intolerance or malabsorption. Anti-IgA antibodies can be assessed but do not correlate well with intolerance to IVIG.

4.5. Treatment and Prognosis

Treatment of infections consists of prolonged or even prophylactic antibiotics. Vaccination status should be kept updated. A Medic-Alert badge is recommended for some patients to warn about administration of IVIG or blood products. A rare patient with refractory infection may require IVIG.

Prognosis is largely dependent on the presence of antibody deficiency, allergy, or autoimmune disease. IgAD is usually lifelong but not associated with life-threatening infections. Rare instances of spontaneous recovery have been recorded, particularly in young patients with measurable IgA. If the patient has been on a medication known to cause IgAD, its discontinuance may lead to recovery.

5. Conclusions

The four most common immunodeficiencies in pediatric patients are THI, IgG subclass deficiency, IPR, and selective IgAD. All four illnesses are characterized by recurrent bacterial respiratory infections such as purulent rhinitis, sinusitis, otitis, and bronchitis. Except for some IgA-deficient patients, the molecular basis for these illnesses is not known, and indeed each syndrome is heterogeneous, with multiple causes. The chronic respiratory infections present are rarely life threatening but often require prolonged antibiotics. Only a few of these cases require the use of IVIG, and the outlook for long life is excellent.

References

al-Attas, R.A. and Rahi, A.H. (1998) Primary antibody deficiency in Arabs: first report from eastern Saudi Arabia. J. Clin. Immunol. 18, 368–371.

Ambrosino, D.M., Siber, G.R., Chilmonczyk, B.A., Jernberg, J.B. and Finberg, R.W. (1987) An immunodeficiency characterized by impaired antibody responses to polysaccharides. N. Engl. J. Med. 316, 790–793.

Blecher, T.E., Soothill, J.F., Voyce, M.A. and Walker, W.H. (1968) Antibody deficiency syndrome: a case with normal immunoglobulin levels. Clin. Exp. Immunol. 3, 47–56.

Blum, P.M., Hong, R. and Stiehm, E.R. (1982) Spontaneous recovery of selective IgA deficiency. Additional case reports and a review. Clin. Pediatr. 21, 77–80.

Buckley, R.H. (2002) Immunoglobulin G subclass deficiency: fact or fancy? Curr. Allergy Asthma Rep. 2, 356–360.

Castigli, E., Wilson, S.A., Garibyan, L., Rachid, R., Bonilla, F., Schneider, L. and Geha, R.S. (2005) TACI is mutant in common variable immunodeficiency and IgA deficiency. Nat. Genet. 37, 829–834.

Cunningham-Rundles, S. (2004) The effect of aging on mucosal host defense. J. Nutr. Health Aging 8, 20–25.

Dalal, I. and Roifman, C.H. (2006). Transient hypogammaglobulinemia of infancy, UpToDate. Available at www.uptodate.com.

Fudenberg, H.H. and Fudenberg, B.R. (1964) Antibody to hereditary human gamma-globulin (Gm) factor resulting from maternal-fetal incompatibility. Science 145, 170–171.

Gitlin, D. and Janeway, C.A. (1956) Agammaglobulinemia, congenital, acquired and transient forms. Prog. Hematol. 1, 318–329.

Granoff, D.M., Shackelford, P.G., Pandey, J.P. and Boies, E.G. (1986a) Antibody responses to Haemophilus influenzae type b polysaccharide vaccine in relation to Km(1) and G2m(23) immunoglobulin allotypes. J. Infect. Dis. 154, 257–264.

Granoff, D.M., Shackelford, P.G., Suarez, B.K., Nahm, M.H., Cates, K.L., Murphy, T.V., Karasic, R., Osterholm, M.T., Pandey, J.P. and Daum, R.S. (1986b) Hemophilus influenzae type B disease in children vaccinated with type B polysaccharide vaccine. N. Engl. J. Med. 315, 1584–1590.

Herrod, H.G. (1993) Management of the patient with IgG subclass deficiency and/or selective antibody deficiency. Ann. Allergy 70, 3–8.

Hostoffer, R. (2006). IgA deficiency, UpToDate. Available at www.uptodate.com.

Kalfa, V.C., Roberts, R.L. and Stiehm, E.R. (2003) The syndrome of chronic mucocutaneous candidiasis with selective antibody deficiency. Ann. Allergy Asthma Immunol. 90, 259–264.

Kanoh, T., Mizumoto, T., Yasuda, N., Koya, M., Ohno, Y., Uchino, H., Yoshimura, K., Ohkubo, Y. and Yamaguchi, H. (1986) Selective IgA deficiency in Japanese blood donors: frequency and statistical analysis. Vox Sang. 50, 81–86.

Koistinen, J. (1976) Familial clustering of selective IgA deficiency. Vox. Sang 30, 181–190.

Kowalczyk, D., Mytar, B. and Zembala, M. (1997) Cytokine production in transient hypogammaglobulinemia and isolated IgA deficiency. J. Allergy Clin. Immunol. 100, 556–562.

Lakhanpal, S., O'Duffy, J.D., Homburger, H.A. and Moore, S.B. (1988) Evidence for linkage of IgA deficiency with the major histocompatibility complex. Mayo Clin. Proc. 63, 461–465.

Lefranc, G., Chaabani, H., Van Loghem, E., Lefranc, M.P., De Lange, G. and Helal, A.N. (1983) Simultaneous absence of the human IgG1, IgG2, IgG4 and IgA1 subclasses: immunological and immunogenetical considerations. Eur. J. Immunol. 13, 240–244.

Lemmon, J.K. and Knutsen, A.P. (2006). Clinical manifestations, diagnosis and treatment of IgG subclass deficiency, UpToDate. Available at www.uptodate.com.

Migone, N., Oliviero, S., de Lange, G., Delacroix, D.L., Boschis, D., Altruda, F., Silengo, L., DeMarchi, M. and Carbonara, A.O. (1984) Multiple gene deletions within the human immunoglobulin heavy-chain cluster. Proc. Natl. Acad. Sci. U.S.A. 81, 5811–5815.

Nathenson, G. (1971) Development of Gm antibodies following injection of anti-Rh gamma globulin. Transfusion 11, 302–306.

Oxelius, V.A., Laurell, A.B., Lindquist, B., Golebiowska, H., Axelsson, U., Bjorkander, J. and Hanson, L.A. (1981) IgG subclasses in selective IgA deficiency: importance of IgG2-IgA deficiency. N. Engl. J. Med. 304, 1476–1477.

Rockey, J.H., Hanson, L.A., Heremans, J.F. and Kunkel, H.G. (1964) Beta-2a Aglobulinemia in two healthy men. J. Lab. Clin. Med. 63, 205–212.

Roifman, C.M. (2004). Immunodeficiency disorders: general considerations. In: E.R. Stiehm, H.D. Ochs and J.A. Winkelstein (Eds), *Immunologic Disorders in Infants and Children*. Philadelphia, Elsevier: 391–393.

Saxon, A., Kobayashi, R.H., Stevens, R.H., Singer, A.D., Stiehm, E.R. and Siegel, S.C. (1980) In vitro analysis of humoral immunity in antibody deficiency with normal immunoglobulins. Clin. Immunol. Immunopathol. 17, 235–244.

Schur, P.H., Borel, H., Gelfand, E.W., Alper, C.A. and Rosen, F.S. (1970) Selective gamma-g globulin deficiencies in patients with recurrent pyogenic infections. N. Engl. J. Med. 283, 631–634.

Siegel, R.L., Issekutz, T., Schwaber, J., Rosen, F.S. and Geha, R.S. (1981) Deficiency of T helper cells in transient hypogammaglobulinemia of infancy. N. Engl. J. Med. 305, 1307–1313.

Sorensen, R.U. and Paris, K. (2006). Selective antibody deficiency with normal immunoglobulins (polysaccharide non-responses), UpToDate. Available at www.uptodate.com.

Stiehm, E.R., Ochs, H.D. and Winkelstein, J.A. (2004a). IgG subclass deficiencies. In: E.R. Stiehm, H.D. Ochs and J.A. Winkelstein (Eds), *Immunologic Disorders in Infants and Children*. Philadelphia, Elsevier: 393–398.

Stiehm, E.R., Ochs, H.D. and Winkelstein, J.A. (Eds)(2004b). Immunodeficiency disorders: general considerations. In *Immunologic Disorders in Infants and Children*. Philadelphia, Elsevier: 289–355.

Stiehm, E.R., Ochs, H.D. and Winkelstein, J.A. (Eds)(2004). Impaired polysacchride responsiveness (selective antibody deficiency). In *Immunologic Disorders in Infants and Children*. Philadelphia, Elsevier: 398–401.
West, C.D., Hong, R. and Holland, N.H. (1962) Immunoglobulin levels from the newborn period to adulthood and in immunoglobulin deficiency states. J. Clin. Invest. 41, 2054–2064.
Whelan, M.A., Hwan, W.H., Beausoleil, J., Hauck, W.W. and McGeady, S.J. (2006) Infants presenting with recurrent infections and low immunoglobulins: characteristics and analysis of normalization. J. Clin. Immunol. 26, 7–11.
Wood, P.M., Mayne, A., Joyce, H., Smith, C.I., Granoff, D.M. and Kumararatne, D.S. (2001) A mutation in Bruton's tyrosine kinase as a cause of selective anti-polysaccharide antibody deficiency. J. Pediatr. 139, 148–151.

3

Immune Dysregulation, Polyendocrinopathy, Enteropathy, X-Linked Inheritance: Model for Autoaggression

Hans D. Ochs and Troy R. Torgerson

University of Washington and Children's Hospital and Regional Medical Center Division of Immunology, Seattle, WA, USA, allgau@u.washington.edu

Abstract. Patients with the rare X-linked syndrome, immune dysregulation, polyendocrinopathy, enteropathy (IPEX) may present early in life with type I diabetes, hyperthyroidism, chronic enteropathy, villous atrophy, dermatitis, autoimmune hemolytic anemia, and antibody- induced neutropenia and thrombocytopenia. Of the reported families with IPEX, most affected boys died before the age of 3 years of malabsorbtion, failure to thrive, infections, or other complications. Characteristic findings at autopsy include lymphocytic infiltrates affecting the lungs, endocrine organs, such as pancreas and thyroid and skin, and increased lymphoid elements in lymph nodes and spleen. Although symptomatic therapy with immunosuppressive drugs provides some beneficial effects, the only curative treatment is hematopoietic stem cell transplantation.

1. Introduction

More than 20 years before, the initial description of immune dysregulation, polyendocrinopathy, enteropathy (IPEX), a spontaneously occurring mutant mouse strain called scurfy, was identified that affected newborn male mice. Scurfy has many phenotypic similarities to IPEX, including X-lined inheritance, polyendocrinopathy, enteropathy resulting in failure to thrive, and dermatitis. Death occurs invariably during the first 3–4 weeks of life. Using positional cloning, the gene responsible for the scurfy syndrome was identified and found to be a transcription factor, Foxp3, belonging to the forkhead-winged helix family. The human ortholog, FOXP3, was subsequently recognized as the causative gene for IPEX. The DNA- binding protein, FOXP3, plays an essential role in generating $CD4^{+}$ $CD25^{+}$ regulatory T (Treg) cells in the thymus. Naturally occurring mutations of FOXP3 result in the absence of Treg cells and the generation of autoaggressive lymphocyte clones that are directly responsible for IPEX in humans and scurfy in mice. Experiments in knock-in mice have clearly demonstrated the importance of Treg cells in preventing autoimmune disease throughout the life of a mouse. Thus, the study of IPEX and Scurfy has provided important insight into the mechanisms of immunosuppression, autoimmunity, and tolerance.

2. Clinical and Pathologic Manifestations of IPEX

The first description of IPEX was provided in 1982 when a large family with multiple affected males in a five-generation pedigree was reported with early onset of polyendocrinopathy, enteropathy, dermatitis/eczema, and premature death (Powell et al. 1982). Following the discovery of the genetic defect in IPEX (Bennett et al. 2001b; Wildin et al. 2001), more than 60 unrelated families with FOXP3 mutations have been identified. In addition to patients with the classic triad of severe diarrhea and failure to thrive, dermatitis/eczema, and early onset diabetes, patients with milder phenotypes were observed, often with delayed onset, which were not readily recognized as IPEX. The oldest patient in our series was 24 years of age when the diagnosis of IPEX was considered and subsequently confirmed by mutation analysis. The clinical findings and laboratory abnormalities observed in a cohort of 50 IPEX patients with identified FOXP3 mutations are summarized in Table 1.

2.1. Gastrointestinal Symptoms and Failure to Thrive

Severe diarrhea associated with villous atrophy and extensive lymphocytic infiltrates of the small bowel mucosa is the most prominent clinical finding in patients with the IPEX phenotype. With one exception, all patients in our series presented with gastrointestinal symptoms of watery or mucoid-bloody diarrhea with poor response to dietary

TABLE 1. Clinical and laboratory abnormalities in IPEX patients with FOXP3 mutations

Clinical features (n = 51)	No. affected	Percent
Enteropathy	49	98
Skin pathology	47	94
Endocrinopathy	37	74
Diabetes (type 1)	27	54
Thyroid disease	16	32
Cytopenia (RBC, neutrophils, and platelets)	24	48
Nephropathy	15	30
Hepatitis	11	22
Neurological diseases	11	22
Serious infections	31	62
Lymphadenopathy	4	8
Arthritis/vasculitis	4	8
Laboratory features		
Elevated IgE (n = 27)	25	93
Elevated IgA (n = 30)	20	67
Outcome		
Alive	28	56
BMTransplantation	12	24

manipulation. Total parenteral nutrition (TPN) is frequently a life-saving intervention. Treatment with immunosuppressive drugs may improve gastrointestinal symptoms in some patients (Bindl et al. 2005).

2.2. Autoimmune Endocrinopathy

The second most common complication of IPEX is autoimmune endocrinopathy. Early onset (sometimes present at birth) insulin-dependent type I diabetes is almost a pathognomonic finding originally reported by Powell et al. (1982). The diabetes is difficult to control, and affected males often have anti-islet cell autoantibodies. At autopsy, the pancreas shows chronic interstitial inflammation with lymphocytic infiltrates and absence of islet cells (Wildin et al. 2002). Thyroid disease, initially presenting as either hypo- or hyper-thyroidism, is a common complication (Powell et al. 1982; Nieves et al. 2004), and may be associated with elevated thyroid stimulation hormone levels or anti-thyroid microsomal antibodies (Wildin et al. 2002).

2.3. Autoimmune Hematologic Disorders

Coombs-positive hemolytic anemia, autoimmune thrombocytopenia, or neutropenia with early or late onset are frequently observed complications (Powell et al. 1982; Wildin et al. 2002; Nieves et al. 2004); anti-red blood cell, anti-platelet, and anti-neutrophil autoantibodies can be frequently demonstrated.

2.4. Dermatologic Abnormalities

Lesions of the skin are common clinical findings (Powell et al. 1982; Nieves et al. 2004). During infancy, the lesions may be erythematous involving the entire body. Older patients often develop chronic eczema or localized psoriasiform dermatitis. Histologically, the psoriasiform lesions show irregular hyperplasia of the epidermis with overlying parakeratosis and lymphocytic infiltrates (Nieves et al. 2004). Alopecia universalis has been observed in several patients. Treatment with steroid ointment often improves the skin lesions.

2.5. Infections

The increased susceptibility to infections may be a direct function of the genetic defect, as was postulated by Powell et al. (1982), or it may be secondary to the decreased barrier function of the skin and gut or iatrogenic due to prolonged immunosuppressive therapy. In our series of over 100 patients with the IPEX phenotype, approximately one half had serious infections including sepsis, meningitis, pneumonia, and osteomyelitis. These infections were observed in many cases prior to the initiation of immunosuppressive therapy. Sepsis as a direct result of line infections is a common complication. Neutropenia, if present, may contribute to susceptibility to bacterial infections. The most common pathogens identified in our series of IPEX patients were *Enterococcus* and *Staphylococcus* species, cytomegalovirus, and Candida (Gambineri et al. 2003).

2.6. Other Clinical Manifestations

Renal disease, often described as glomerulonephropathy or interstitial nephritis, has been reported (Powell et al. 1982). In our own series of IPEX patients, more than 50% with demonstrated FOXP3 mutations had renal abnormalities. In some instance, renal disease might be directly caused or worsened by treatment with cyclosprin A or FK506. Renal disease, however, has been described in IPEX patients not receiving immunosuppressive drugs. Splenomegaly and lymphadenopathy due to extensive lymphocytic infiltrates have been reported in patients at autopsy (Wildin et al. 2002). Unexpectedly, almost half of the IPEX patients in our own series had neurologic problems including seizures and mental retardation.

2.7. Laboratory Findings and Histopathology

The immunologic evaluation of IPEX patients is unremarkable except for elevated serum levels of IgE and IgA and marked eosinophilia. Functional analysis of the immune system is difficult, since most patients are on systemic immunosuppressive therapy at the time of testing. The presence of autoantibodies is a hallmark of the syndrome. Most patients with insulin-dependent diabetes have autoantibodies against pancreatic islet cells (Baud et al. 2001; Wildin et al. 2002). Circulating autoantibodies against human intestinal enterocytes and an autoantibody specific for a 75-kDa gut and kidney-specific antigen (AIE-75) have been observed (Kobayashi et al. 1998). Circulating lymphocytes expressing CD4 and CD25 markers are present in IPEX patients with FOXP3 mutations, but none of these cells express the FOXP3 protein.

The intestinal tract abnormalities are characterized by a loss of villi in the small bowel associated with mucosal erosions and lymphocytic infiltrates in the lamina propria and submucosa. Lymphocytic infiltrates may also be present in the colon. The pancreas of infants with insulin-dependant diabetes shows lymphocytic infiltrates and loss of Langerhans cells. The thyroid, if affected, demonstrates extensive lymphocytic infiltrates (Wildin et al. 2002). Microscopic evaluations of skin biopsies are consistent with eczema; some lesions resemble psoriasiform dermatitis, showing hyperplasia of the epidermis and parakeratosis. The dermis may contain large numbers of infiltrating lymphocytes that consist predominantly of $CD4^+$ and $CD8^+$ T cells (Nieves et al. 2004). Spleen, liver, and lymph nodes often show lymphocytic infiltrates (Wildin et al. 2002). A hypotrophic thymus, often observed at autopsy, may be the result of chronic illness or prolonged immunosuppressive therapy. Renal abnormalities include interstitial nephritis, focal tubular atrophy, membranous glomerulopathy, and lymphocytic infiltrates.

3. Molecular Basis of IPEX

Based on family studies, the gene responsible for IPEX was mapped to Xp11.23–Xq13.3. Following the discovery that mutation of a forkhead/winged-helix DNA-binding protein (Foxp3) resulted in a lymphoproliferative syndrome in mice similar to the IPEX syndrome in humans (Brunkow et al. 2001), it was recognized that symptomatic males from IPEX families had mutations in the *FOXP3* gene (Bennett et al. 2001; Wildin et al. 2001). The human *FOXP3* gene is located at the short arm of the X chromosome and consists of 11 translated exons that encode a protein of 431 amino acids. The gene is expressed predominantly in lymphoid tissue, particularly in $CD4^+$, $CD25^{bright}$ T cells. The forkhead protein FOXP3 is a member of the P subfamily of Fox transcription factors which are characterized by the presence of a highly conserved winged-helix/forkhead DNA-binding domain. Proteins bearing a forkhead DNA-binding motif comprise a large family of related molecules that play an important role in embryonic patterning, development, and metabolism (Carlsson and Mahlapuu 2002). Only a small subset of these transcription factor family members are crucial for the development and maintenance of normal immune responses and thymic development (Foxn1), for lineage commitment (Foxp3), and for the function of lymphocytes (Foxj1 and Foxo3). The FOXP3 protein consists of a proline-rich domain at the N-terminus, a C2H2 zinc finger and a leucine zipper in the central portion thought to be involved in protein/protein interaction, and a forkhead DNA-binding domain at the C-terminus.

4. Function of FOXP3

Transcription factors have two essential functions: nuclear import and direct binding to DNA. If the forkhead domain is deleted or mutated, nuclear import and DNA binding are completely abolished (Lopes et al. 2006). To be functionally active, FOXP3 has to be in the homodimer formation, which depends on the intact leucine zipper sequence. Mutations occurring in this region interfere with homodimerization. A novel functional domain within the N-terminal region is required for FOXP3-mediated repression of NFAT-controlled gene transcription (Lopes et al. 2006). Based on in vivo and in vitro experiments, FOXP3 is considered a transcriptional repressor of cytokine promoters by functioning as a specific repressor for the two pivotal transcription factors, NF-κB and NFAT which play a key role in the expression of multiple cytokine genes (Schubert et al. 2001; Bettelli et al. 2005; Lopes et al. 2006; Wu et al. 2006). Transgenic mice expressing multiple copies of the *Foxp3* gene have a dramatic suppression of immune responses characterized by markedly decreased numbers of $CD4^+$ T cells in the peripheral blood and decreased cellularity in lymph nodes and spleen. These mice were hyporesponsive to stimulation both in vitro and in vivo. In this model, Foxp3 functions as a rheostat of the immune system, with activation responses being inversely proportional to the amount of the Foxp3 protein being expressed by $CD4^+$ T cells. Taken together, FOXP3 has an important function in the regulation of peripheral $CD4^+$ T cells as well as in the generation of Treg cells in the thymus.

5. Foxp3 and Treg Cells

A small subset of $CD4^+$ T cells expressing the low-affinity IL-2 receptor α-chain (CD25) has been shown to be anergic. Upon activation, these cells suppress proliferation and IL-2 production of naïve and memory effector T cells through a contact-dependent cytokine-independent mechanism (Itoh et al. 1999; Bacchetta et al. 2005). Although the precise molecular events that lead to the production of Treg cells are unknown, it was recently shown that in mice, Foxp3 plays a crucial role in the generation of Tregs (Fontenot et al. 2003). It is presently unknown how this population of $CD4^+$ $CD25^+$ Treg cells exerts its potent suppressive effect on the immune response and how Treg cells themselves are generated and modulated. There is, however, no question that Tregs play a critical role in establishing and maintaining self-tolerance and immune homeostasis (Fontenot and Rudensky 2005). Treg cells have been recognized as leading factors in the control of chronic human disease. They play a major role in transplantation tolerance (Wood and Sakaguchi 2003) and seem to be low in numbers (and expression of FOXP3) in patients with chronic graft versus host disease following bone marrow transplantation (Zorn et al. 2005). The lack of Treg cells has been associated with autoimmune diseases in both humans and mice, and as expected, treatment strategies to increase Treg function have improved transplantation tolerance and autoimmune symptoms (Loser et al. 2005; Randolph and Fathman 2006). Experiments in mice have demonstrated that $CD4^+$ $CD25^+$ Treg cells are required for induction of oral tolerance (Dubois et al. 2003) and that administration of oral antigen dramatically induces the number and function of antigen-specific $CD4^+$ $CD25^+$ Tregs (Zhang et al. 2001). Increased numbers and activities of Tregs have been associated with tumor progression (Ormandy, Hillemann, Wedemeyer, Manns, Greten, and Korangy 2005), and new anti-tumor therapies are being explored to reduce Treg activity.

6. Fox Mutation Analysis in Patients with the IPEX Phenotype

In an effort to define the clinical and immunologic phenotype of IPEX and to explore a possible phenotype–genotype correlation, we have evaluated more than 100 patients from over 60 families who presented with the clinical phenotype of IPEX for mutations in FOXP3. To date, we have identified over 30 novel mutations, including missense mutations in the proline-rich domain, the leucine zipper, and the forkhead domain in addition to deletions and splice-site mutations. In one family with multiple affected members, we found a large deletion upstream of exon 1 resulting in failure to initiate normal splicing. Two families were found to have a point mutation affecting the first canonical polyadenylation region (Bennett et al. 2001). In our experience, only 60% of patients presenting with the IPEX phenotype have identifiable mutation of FOXP3.

7. Diagnoses and Treatment of IPEX

The diagnoses of IPEX should be considered in any young male patient presenting with intractable diarrhea, villous atrophy, and failure to thrive. The presence of an erythematous rash, eczema, or psoriasiform dermatitis strongly supports this diagnosis. Early onset of type 1 diabetes in a male patient with gastrointestinal symptoms and eczema is highly suspicious of IPEX. Autoimmune hemolytic anemia, thrombocytopenia, or neutropenia are not always present or may occur at a later age. The diagnoses of IPEX is highly suspect by demonstrating the absence of $CD4^+$ $CD25^+$ $FOXP3^+$ Treg cells and is confirmed by mutation analysis of the *FOXP3* gene. If the mutation in a given family is known, carrier females can be identified and prenatal diagnosis performed in a male fetus by sequence analysis of *FOXP3* using DNA extracted from chorionic villous biopsies or cultured amniocytes.

It is important to initiate early aggressive therapy including TPN. Red blood cell and platelet transfusions may be necessary. Long-term immunosuppression has proven effective in some patients, but usually only partially and for a limited period. Cyclosporin A or tacrolimus often in combination with steroids has been used with some success. Some patients respond to sirolimus (rapamycin), which seems to be less nephrotoxic. Other immunosuppressive medications, including infliximab and retuximab, have been tried alone or in combination but with limited success. Chronic immunosuppressive therapy may facilitate opportunistic infections and cause secondary renal damage. Hematopoietic stem cell transplantation is currently the only effective cure for IPEX. Some patients have achieved complete remission of symptoms following bone marrow transplantation (Baud et al. 2001; Mazzolari et al. 2005; Rao et al. 2007). Both full and reduced intensity conditioning have been reported to be successful. Generally, the prognosis for patients with IPEX is poor, and if untreated, most die at an early age.

8. Animal Models

The original scurfy (sf) mutation, a fatal X-linked condition, has occurred spontaneously in a partially inbred strain of mice (Russell et al. 1959). Shortly after birth, affected male mice present with a scaly skin rash and severe runting secondary to chronic diarrhea and malabsorption. Characteristically, the mice exhibited lymphadenopathy, splenomegaly, massive lymphocytic infiltrates of the skin, liver, and lungs and developed hemolytic anemia associated with a positive Coombs test, suggesting that the sf mutation causes a generalized autoimmune-like syndrome. The gene responsible for the sf mutation was identified and designated as Foxp3 (Brunkow et al. 2001) consisting of a two-base-pair insertion in exon 8, resulting in a frame shift that leads to a truncated protein product lacking the carboxy-terminal fork-head domain. This sf mouse model was instrumental in the discovery of Treg cells (Fontenot et al. 2003). Using Foxp3-negative gene-targeted mice and a GFP-Foxp3 fusion-protein-reporter knock-in allele, it could be shown that expression of Foxp3 was highly restricted to αβ $CD4^+$ T cells and irrespective of CD25 expression correlated with suppressor activity (Fontenot et al. 2005). Acute in vivo ablation of Treg cells demonstrated that Treg cells play vital function in neonatal and adult mice suggesting that self-reactive T cells are continuously suppressed by Treg cells. When this suppression is relieved, self-reactive T cells become activated and facilitate accelerated maturation of dendritic cells and events that lead to catastrophic autoimmunity for the lifespan of the mice (Kim et al. 2007).

9. Conclusions

Foxp3 is the key mediator of Treg development in the thymus. Naturally occurring mutations of FOXP3 interfere with this process, resulting in the generation of autoaggressive T lymphocyte clones that are directly responsible for the IPEX syndrome in humans and scurfy in mice; both are lethal diseases. Hematopoietic stem cell transplantation is the only cure for patients with IPEX. It is anticipated that exploitation of the scurfy mouse model and further evaluation of patients with IPEX will lead to a better understanding of the function of FOXP3 as a transcription factor, and will facilitate the identification of genes that are regulated by FOXP3. Investigations along these lines will provide important insight into the mechanisms of immunosuppression, autoimmunity, and tolerance and may lead to novel strategies to treat patients with autoimmune diseases, graft versus host disease, and cancer.

References

Bacchetta, R., Gregori, S. and Roncarolo, M.G. (2005) CD4+ regulatory T cells: mechanisms of induction and effector function. Autoimmun. Rev. 4, 491–496.

Baud, O., Goulet, O., Canioni, D., Le Deist, F., Radford, I., Rieu, D., Dupuis-Girod, S., Cerf-Bensussan, N., Cavazzana-Calvo, M., Brousse, N., Fischer, A. and Casanova, J.L. (2001) Treatment of the immune dysregulation, polyendocrinopathy, enteropathy, X-linked syndrome (IPEX) by allogeneic bone marrow transplantation. N. Engl. J. Med. 344, 1758–1762.

Bennett, C.L., Brunkow, M.E., Ramsdell, F., O'Briant, K.C., Zhu, Q., Fuleihan, R.L., Shigeoka, A.O., Ochs, H.D. and Chance, P.F. (2001a) A rare polyadenylation signal mutation of the FOXP3 gene (AAUAAA-->AAUGAA) leads to the IPEX syndrome. Immunogenetics 53, 435–439.

Bennett, C.L., Christie, J., Ramsdell, F., Brunkow, M.E., Ferguson, P.J., Whitesell, L., Kelly, T.E., Saulsbury, F.T., Chance, P.F. and Ochs, H.D. (2001b) The immune dysregulation, polyendocrinopathy, enteropathy, X-linked syndrome (IPEX) is caused by mutations of FOXP3. Nat. Genet. 27, 20–21.

Bettelli, E., Dastrange, M. and Oukka, M. (2005) Foxp3 interacts with nuclear factor of activated T cells and NF-kappa B to repress cytokine gene expression and effector functions of T helper cells. Proc. Natl. Acad. Sci. U.S.A. 102, 5138–5143.

Bindl, L., Torgerson, T., Perroni, L., Youssef, N., Ochs, H.D., Goulet, O. and Ruemmele, F.M. (2005) Successful use of the new immune-suppressor sirolimus in IPEX (immune dysregulation, polyendocrinopathy, enteropathy, X-linked syndrome). J. Pediatr. 147, 256–259.

Brunkow, M.E., Jeffery, E.W., Hjerrild, K.A., Paeper, B., Clark, L.B., Yasayko, S.A., Wilkinson, J.E., Galas, D., Ziegler, S.F. and Ramsdell, F. (2001) Disruption of a new forkhead/winged-helix protein, scurfin, results in the fatal lymphoproliferative disorder of the scurfy mouse. Nat. Genet. 27, 68–73.

Carlsson, P. and Mahlapuu, M. (2002) Forkhead transcription factors: key players in development and metabolism. Dev. Biol. 250, 1–23.
Dubois, B., Chapat, L., Goubier, A. and Kaiserlian, D. (2003) CD4+CD25+ T cells as key regulators of immune responses. Eur. J. Dermatol. 13, 111–116.
Fontenot, J.D., Gavin, M.A. and Rudensky, A.Y. (2003) Foxp3 programs the development and function of CD4+CD25+ regulatory T cells. Nat. Immunol. 4, 330–336.
Fontenot, J.D., Rasmussen, J.P., Williams, L.M., Dooley, J.L., Farr, A.G. and Rudensky, A.Y. (2005) Regulatory T cell lineage specification by the forkhead transcription factor foxp3. Immunity 22, 329–341.
Fontenot, J.D. and Rudensky, A.Y. (2005) A well adapted regulatory contrivance: regulatory T cell development and the forkhead family transcription factor Foxp3. Nat. Immunol. 6, 331–337.
Gambineri, E., Torgerson, T.R. and Ochs, H.D. (2003) Immune dysregulation, polyendocrinopathy, enteropathy, and X-linked inheritance (IPEX), a syndrome of systemic autoimmunity caused by mutations of FOXP3, a critical regulator of T cell homeostasis. Curr. Opin. Rheumatol. 15, 430–435.
Itoh, M., Takahashi, T., Sakaguchi, N., Kuniyasu, Y., Shimizu, J., Otsuka, F. and Sakaguchi, S. (1999) Thymus and autoimmunity: production of CD25+CD4+ naturally anergic and suppressive T cells as a key function of the thymus in maintaining immunologic self-tolerance. J. Immunol. 162, 5317–5326.
Kim, J.M., Rasmussen, J.P. and Rudensky, A.Y. (2007) Regulatory T cells prevent catastrophic autoimmunity throughout the lifespan of mice. Nat. Immunol. 8, 191–197.
Kobayashi, I., Imamura, K., Yamada, M., Okano, M., Yara, A., Ikema, S. and Ishikawa, N. (1998) A 75-kD autoantigen recognized by sera from patients with X-linked autoimmune enteropathy associated with nephropathy. Clin. Exp. Immunol. 111, 527–531.
Lopes, J.E., Torgerson, T.R., Schubert, L.A., Anover, S.D., Ocheltree, E.L., Ochs, H.D. and Ziegler, S.F. (2006) Analysis of FOXP3 reveals multiple domains required for its function as a transcriptional repressor. J. Immunol. 177, 3133–3142.
Loser, K., Hansen, W., Apelt, J., Balkow, S., Buer, J. and Beissert, S. (2005) In vitro-generated regulatory T cells induced by Foxp3-retrovirus infection control murine contact allergy and systemic autoimmunity. Gene Ther. 12, 1294–1304.
Mazzolari, E., Forino, C., Fontana, M., D'Ippolito, C., Lanfranchi, A., Gambineri, E., Ochs, H., Badolato, R. and Notarangelo, L.D. (2005) A new case of IPEX receiving bone marrow transplantation. Bone Marrow Transplant. 35, 1033–1034.
Nieves, D.S., Phipps, R.P., Pollock, S.J., Ochs, H.D., Zhu, Q., Scott, G.A., Ryan, C.K., Kobayashi, I., Rossi, T.M. and Goldsmith, L.A. (2004) Dermatologic and immunologic findings in the immune dysregulation, polyendocrinopathy, enteropathy, X-linked syndrome. Arch. Dermatol. 140, 466–472.
Ormandy, L.A., Hillemann, T., Wedemeyer, H., Manns, M.P., Greten, T.F. and Korangy, F. (2005) Increased populations of regulatory T cells in peripheral blood of patients with hepatocellular carcinoma. Cancer Res. 65, 2457–2464.
Powell, B.R., Buist, N.R. and Stenzel, P. (1982) An X-linked syndrome of diarrhea, polyendocrinopathy, and fatal infection in infancy. J. Pediatr. 100, 731–737.
Randolph, D.A. and Fathman, C.G. (2006) Cd4+Cd25+ regulatory T cells and their therapeutic potential. Annu. Rev. Med. 57, 381–402.
Rao, A., Kamani, N., Filipovich, A., Lee, S.M., Davies, S.M., Dalal, J. and Shenoy, S. (2007) Successful bone marrow transplantation for IPEX syndrome after reduced-intensity conditioning. Blood 109, 383–385.
Russell, W.L., Russell, L.B. and Gower, J.S. (1959) Exceptional inheritance of a sex-linked gene in the mouse explained on the basis that the X/O sex-chromosome constitution is Female. Proc. Natl. Acad. Sci. U.S.A. 45, 554–560.

Schubert, L.A., Jeffery, E., Zhang, Y., Ramsdell, F. and Ziegler, S.F. (2001) Scurfin (FOXP3) acts as a repressor of transcription and regulates T cell activation. J. Biol. Chem. 276, 37672–37679.

Wildin, R.S., Ramsdell, F., Peake, J., Faravelli, F., Casanova, J.L., Buist, N., Levy-Lahad, E., Mazzella, M., Goulet, O., Perroni, L., Bricarelli, F.D., Byrne, G., McEuen, M., Proll, S., Appleby, M. and Brunkow, M.E. (2001) X-linked neonatal diabetes mellitus, enteropathy and endocrinopathy syndrome is the human equivalent of mouse scurfy. Nat. Genet. 27, 18–20.

Wildin, R.S., Smyk-Pearson, S. and Filipovich, A.H. (2002) Clinical and molecular features of the immunodysregulation, polyendocrinopathy, enteropathy, X linked (IPEX) syndrome. J. Med. Genet. 39, 537–545.

Wood, K.J. and Sakaguchi, S. (2003) Regulatory T cells in transplantation tolerance. Nat. Rev. Immunol. 3, 199–210.

Wu, Y., Borde, M., Heissmeyer, V., Feuerer, M., Lapan, A.D., Stroud, J.C., Bates, D.L., Guo, L., Han, A., Ziegler, S.F., Mathis, D., Benoist, C., Chen, L. and Rao, A. (2006) FOXP3 controls regulatory T cell function through cooperation with NFAT. Cell 126, 375–387.

Zhang, X., Izikson, L., Liu, L. and Weiner, H.L. (2001) Activation of CD25(+)CD4(+) regulatory T cells by oral antigen administration. J. Immunol. 167, 4245–4253.

Zorn, E., Kim, H.T., Lee, S.J., Floyd, B.H., Litsa, D., Arumugarajah, S., Bellucci, R., Alyea, E.P., Antin, J.H., Soiffer, R.J. and Ritz, J. (2005) Reduced frequency of FOXP3+ CD4+CD25+ regulatory T cells in patients with chronic graft-versus-host disease. Blood 106, 2903–2911.

4

DiGeorge Syndrome/Velocardiofacial Syndrome: The Chromosome 22q11.2 Deletion Syndrome

Kathleen E. Sullivan

Department of Pediatrics, Division of Allergy & Immunology Children's Hospital of Philadelphia, Philadelphia, PA, USA, sullivak@mail.med.upenn.edu

Abstract. Chromosome 22q11.2 deletion (CH22qD) syndrome is also known as DiGeorge syndrome or velocardiofacial syndrome. This deletion syndrome is extremely common with nearly one in 4000 children being affected. Recent advances and a holistic approach to patients have improved the care and well-being of these patients. This review will summarize advances in understanding the health needs and immune system of patients with CH22qD syndrome. Patients will most often need interventions directed at maximizing function for many organ systems but can ultimately have a high level of functioning.

1. Nomenclature

Chromosome 22q11.2 deletion (CH22qD) syndrome is an umbrella term referring to a number of syndromes, which have in common a hemizygous deletion of chromosome 22. Approximately 90% of patients carrying the clinical diagnosis of DiGeorge syndrome and 80% of patients carrying the clinical diagnosis of velocardiofacial syndrome carry the hemizygous deletion. Additional patients with CHARGE and conotruncal anomaly face syndrome have been demonstrated to carry the deletion. Importantly, not all patients with DiGeorge syndrome have the deletion, and it is important to recognize those patients as well as the more common patients who have the deletion but have features which are not consistent with DiGeorge syndrome. The term CH22qD will be utilized when referencing data related to patients known to have the deletion and specific syndromic nomenclature utilized when the resource data relied on clinical features in this review.

Most patients with the deletion have a conotruncal cardiac anomaly and a mild to moderate immune deficiency. Developmental delay, palatal dysfunction, and feeding problems are also seen in a majority of infants. Beyond these few relatively consistent features, the remaining clinical features are seen in fewer than 50% of patients (Table 1). The diversity of clinical features requires that the term CH22qD be used when the deletion is present and the syndromic terms be used only when the genetic basis is not known.

TABLE 1. Clinical findings in patients with chromosome 22q11.2 deletion syndrome

Finding	%
Cardiac anomalies	49–83%
Tetralogy of fallot	17–22%
Interrupted aortic arch	14–15%
Ventriculoseptal defect	13–14%
Truncus arteriosus	7–9%
Hypocalcemia	17–60%
Growth hormone deficiency	4%
Palatal anomalies	69–100%
Cleft palate	9–11%
Submucous cleft palate	5–16%
Velopharyngeal insufficiency	27–92%
Bifid uvula	5%
Renal anomalies	36–37%
Absent/dysplastic	17%
Obstruction	10%
Reflux	4%
Ophthalmologic abnormalities	7–70%
Tortuous retinal vessels	58%
Posterior embryotoxon (anterior segment dysgenesis)	69%
Neurologic	8%
Cerebral atrophy	1%
Cerebellar hypoplasia	0.4%
Dental: delayed eruption and enamel hypoplasia	2.5%
Skeletal abnormalities	17–19%
Cervical spine anomalies	40–50%
Vertebral anomalies	19%
Lower extremity anomalies	15%
Speech delay	79–84%
Developmental delay in infancy	75%
Developmental delay in childhood	45%
Behavior/psychiatric problems	9–50%
Attention deficit hyperactivity disorder	25%
Schizophrenia	6–30%

Taken from Gerdes et al. 1999; McDonald-McGinn et al. 1999; Moss et al. 1999; Motzkin et al. 1993; Ryan et al. 1997; Shprintzen et al. 1992; Swillen et al. 1997; Vantrappen et al. 1999; Wang et al. 2000; Yan et al. 1998.

2. Genetics

CH22qD is felt to occur in approximately 1:4000 births (Botto et al. 2003; Wilson et al. 1993). The frequency may be increasing as patients who had life-saving cardiac surgery in the 1980s are now reaching reproductive age, but the spontaneous deletion rate is quite high. The deletion event is mediated by low copy number repeats (LCRs) (Dunham et al. 1999). Four blocks of LCRs are found in this region (Figure 1), and each block consists of a number of modules of repeats, which are found in various lengths and orientations within a block (Shaikh et al. 2000). The LCRs on chromosome

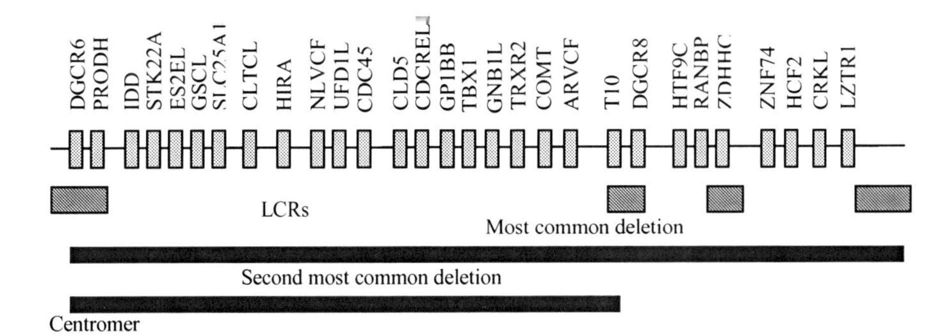

FIGURE 1. The genes within the commonly deleted region of chromosome 22. The light gray boxes represent the genes. The darker gray boxes below the schematic chromosome represent the low copy number repeats (LCRs). The most common deletion, which is 3Mb, occurs between the two most distant LCRs. Eight percent of the patients have a slightly smaller 1.5 Mb deletion, and the remainder of patients have a variety of deletions with one breakpoint in an LCR.

22q11.2 are larger, more complex, and have higher homology than any of the other LCRs in the genome associated with human chromosomal deletion syndromes, perhaps explaining why this syndrome is so common.

Within the commonly deleted region of chromosome 22q11.2, there are over 35 genes. LoxP Cre deletions in mice were constructed, which mimicked the deletions in humans and defined *TBX1* as the most likely gene contributing to the cardiac phenotype (Lindsay et al. 1999). A *TBX1* knockout mouse confirmed the importance of this gene in cardiac development. Initially, the deletion mice were felt to have an isolated cardiac phenotype. When the deletion was bred onto other strains, the parathyroid and thymic phenotypes were revealed. Studies of humans also suggest that genetic background may subtly alter the phenotype (Jiang et al. 2005; Munoz et al. 2001).

In mice, *TBX1* is expressed in the pharyngeal mesenchyme and endodermal pouches. These segmentation structures ultimately develop into the thorax and facial structures. Haplosufficiency for *TBX1* leads to smaller third and fourth pouches/arches due to decreased proliferation of endoderm cells in the branchial arches (Xu et al. 2005; Zhang et al. 2005). These small arches subsequently lead to compromised development of the facial structures, parathyroid, and thymus. *TBX1* is also expressed in the secondary heart field. Cells of the secondary heart field are derived from the pharyngeal mesoderm and give rise to the right ventricle and the outflow tract. These are the regions primarily affected in CH22qD.

Patients with Ch22qD syndrome have a variety of malformations which cannot be attributed to the effects on the branchial arches. Behavioral, cognitive, and psychiatric disturbances are extremely common, and distal skeletal anomalies, vertebral anomalies, and renal anomalies are seen in a minority of patients. *TBX1* is expressed in the early brain mesoderm under the influence of the protein sonic hedgehog and in the sclerotome which gives rise to structures in the spinal column (Mahadevan et al. 2004). It is felt that haplosufficiency for *TBX1* impairs appropriate development in these regions as well.

There is substantial interest in identifying the specific functions of TBX1 in the hope that an intervention could be developed to ameliorate the effects of haplosufficiency for *TBX1*. One potential step in that direction arises from investigation of a different syndrome. Fetal isotretinoin exposure causes a syndrome with remarkable similarity to CH22qD (Cipollone et al. 2006). Retinoic acid is a repressor of *TBX1* expression and therefore exerts its effects via the same pathway (Roberts et al. 2005). Manipulation of this pathway could normalize levels of TBX1 if detected early enough in affected infants. Identification of modifier genes could also provide a framework for developing meaningful interventions (Lawrence et al. 2003; Stalmans et al. 2003).

3. Management

The diagnosis of CH22qD is simple and widely available. The fluorescent in situ hybridization (FISH) method is used currently. Efforts to develop a rapid PCR-based method are underway and may yield a rapid commercial test soon (Chen et al. 2006; Fernandez et al. 2005; Vorstman et al. 2006). The issue of diagnosis is more difficult when a patient with classic features has no evidence of a deletion by FISH. There are several potential explanations: a point mutation in *TBX1*, which has been described in a very small number of patients (Yagi et al. 2003), a deletion which is smaller than normal and not detected by standard FISH, or a non-chromosome 22 explanation. Deletions of chromosome 10, mutations in *CHD7*, and patients with prenatal exposure to isotretinoin or elevated glucose can be associated with a phenotype similar to that seen in CH22qD (Coberly et al. 1996; Digilio et al. 1995; Novak and Robinson 1994; Theodoropoulos 2003; Van Esch et al. 1999). There are also patients with velocardiofacial syndrome (VCFS) or DiGeorge syndrome and no known etiology. This is a significant clinical issue because the recurrence risk in these kindred is not known.

There are limited data and fewer recommendations regarding appropriate use of the FISH test. Not all cardiac anomalies are equally associated with CH22qD, and other clinical features have varying predictive values (Table 2). In infants with a congenital heart defect and no syndromic features, the incidence of CH22qD is reported to be very low (0–1%) (Frohn-Mulder et al. 1999). Thus, it is the combination of an appropriate cardiac anomaly and any other consistent feature, which is most strongly predictive of the deletion. The population most difficult to identify consists of patients with only developmental delay or speech delay. One study has shown that physicians trained to recognize the subtle facial features are more likely to identify patients correctly; however, most primary care clinicians have only one or two patients under their care (Becker et al. 2004).

The management of patients with CH22qD is completely dependent on age and phenotype (Figure 2). Patients with CH22qD may present at any age although the majority of patients are identified shortly after birth due to the presence of a cardiac anomaly. In newborns, a physical examination, laboratory studies, and radiographic studies should target problems that are likely to

TABLE 2. The frequency of the chromosome 22q11.2 deletion in various patient populations

Phenotypic feature	Frequency of deletion (%)
Any cardiac lesion	1.1
Conotruncal cardiac anomaly	7–50
Interrupted aortic arch	50–60
Pulmonary atresia	33–45
Aberrant subclavian	25
Tetralogy of Fallot	11–17
Velopharyngeal insufficiency	64
Velopharyngeal insufficiency post-adenoidectomy	37
Neonatal hypocalcemia	74
Schizophrenia	0.3–6.4

require immediate intervention such as cardiac anomalies, hypocalcemia, severe immunodeficiency, or intestinal malrotation. Feeding problems are most often seen early in life and can be extremely distressing to parents (Rommel et al. 1999). Ages 2–4 years require attention to development and speech while the school age years require additional attention to cognitive development and growth. Behavioral and psychiatric disorders are seen in teenagers and adults. Each of these organ systems will in turn be reviewed.

Cardiac anomalies are seen in approximately 75% of all patients with CH22qD and are the major cause of death. The range of cardiovascular anomalies is enormous although conotruncal defects predominate. Most, but not all, patients with a cardiac defect will require surgery, and the surgical risk is probably comparable to that of other children. The two issues, which require attention prior to surgery, are monitoring of serum calcium levels and identification of a serious immunodeficiency. Low T cell numbers are seen in 75–80% of infants with Ch22qD although few patients have a severe immune deficiency (Ryan et al. 1997). These patients are rare but do require protection from infection, live viral vaccines, and blood products.

It is not uncommon for patients to require cardiac surgery prior to an immunologic evaluation. Several strategies may limit risk to the patient. Many large centers in the United States simply irradiate all blood products provided to infants less than 1 year of age. Another strategy is to stratify risk according to the absolute lymphocyte count from a complete blood count although this is not very accurate.

Individuals with Ch22qD frequently have a small hypoplastic thymus. The size of the thymus does not predict circulating T cell counts partly due to microscopic rests of thymic epithelial cells (Bale and Sotelo-Avila 1993). Early laboratory evaluations of T cell counts and T cell function are typically recommended to stratify the risk of infection. Patients with severely impaired immunologic function are susceptible to the same types of infections and complications as patients with severe

	Low		High →
Birth	Calcium	Feeding	Cardiac
1 year	Cardiac	Feeding	Development
2 years	Feeding	Infection	Development
3 years	Infection	Behavior	Development
Early school	Infection	Behavior	Cognitive
Teens	Infection	Behavior	Cognitive
Adults	Infection	Psychiatric	Employment

FIGURE 2. The dynamic nature of health concerns in patients with chromosome 22q11.2 deletion (CH22qD). The level of concern is indicated schematically across the top, and age is indicated on the y-axis. The relative levels of concern are roughly indicated to emphasize how the needs of patients change with age.

combined immunodeficiency. For example, graft-versus-host disease, severe adenoviral disease, disseminated cytomegalovirus (CMV), disseminated parainfluenza, extraintestinal or prolonged rotaviral infections, and B cell lymphomas have been described. These patients require thymic tissue transplantation or fully matched sibling hematopoietic stem cell transplantation for survival (Markert et al. 1999; Markert et al. 2003; Markert et al. 1995). Even with appropriate treatment, mortality rates are high when the T cell defect is severe.

Most patients have a mild to moderate decrease in the mean number of CD3+ T cells and CD4+ T-helper cells compared to age-matched controls (Chinen et al. 2003; Jawad et al. 2001; Junker and Driscoll 1995; Kanaya et al. 2006; Kornfeld et al. 2000; Sediva et al. 2005; Sullivan et al. 1999). T cell proliferation is normal in the majority of patients and when decreased is simply reflecting the low frequency of responder cells (Bastian et al. 1989; Chinen et al. 2003; Junker et al. 1995; Kornfeld et al. 2000; Sullivan et al. 1999). The majority of patients with the deletion are modestly immunocompromised and do not develop opportunistic infections. Viral infections can be prolonged and abnormal palatal anatomy may lead to compromised drainage and an increased susceptibility to upper airway bacterial infections.

To gain further insight into the T cell defect in patients with CH22qD, studies have been performed to evaluate subtle functional defects, which might contribute to the recurrent infection pattern seen in many patients. Cytokine production appears to be normal (Kanaya et al. 2006); however, proliferation deteriorates with age, and the decreased proliferative ability may relate to shortened telomeres (Piliero et al. 2004). Telomere critical length is seen in patients with CH22qD in middle age and may prevent full proliferative potential. This compromise in proliferation is associated with mild defects in repertoire diversity.

Serum IgG and IgM levels, as well as specific IgG to diphtheria and tetanus, are usually normal, although there is a high rate of IgA deficiency (Chinen et al. 2003; Finocchi et al. 2006; Gennery et. al. 2002; Junker et al. 1995; Smith et al. 1998). Given the modest laboratory defects found in the majority of patients with CH22qD, it is germane to ask whether there are any health consequences. With the exception of very compromised children, live viral vaccines may be safely given. No special precautions need to be taken to prevent graft versus host disease or opportunistic infections except in those patients with very severe immunocompromise. Having said that, the more typical patients are not immunologically normal. Adults and children over the age of 9 years have significant infections. Only 40% of patients with the deletion over the age of nine were thought to be as healthy as others their age. Approximately one-quarter to one-third of the patients had either recurrent sinusitis or otitis media and 4–7% had recurrent lower airway infections (Jawad et al. 2001). These patients also have a high rate of autoimmune disease as a consequence of their dysregulated immune system. The autoimmune diseases do not correlate with severe T cell dysfunction and include a range of pediatric autoimmune diseases. Autoimmune thrombocytopenia and juvenile rheumatoid arthritis appear to be the most common and may occur 20–100 times more frequently than in the general population (DePiero et al. 1997; Gennery et al. 2002; Levy et al. 1997; Pinchas-Hamiel et al. 1994; Rasmussen et al. 1996). The juvenile rheumatoid arthritis is often polyarticular and may be difficult to manage (Davies et al. 2001). Autoimmune thyroid disease and other autoimmune processes have also been described, and it is likely that the T cell defect acts synergistically with other predisposing factors to cause autoimmune disease. Diminished CD4/CD25 regulatory T cells have been described and may also contribute to autoimmunity (Sullivan, McDonald-McGinn, and Zackai 2002).

Speech, hearing, and vision issues are typically addressed in early childhood. Hearing is important for language development, and approximately 10% of patients with the deletion will have a sensorineural hearing loss and 45% of patients have a conductive loss (Digilio et al. 1999). Speech is one of the most troubling aspects of CH22qD for most parents. Phonation can be aberrant due to anatomical issues including laryngeal webs, velopharyngeal insufficiency, or vocal cord paralysis. Hoarseness and hypernasality respond to surgical intervention, but phonation remains somewhat abnormal in many cases (Losken et al. 2003; Losken et al. 2006). Expressive language and speech skills are typically delayed, while receptive skills are near normal. Management of speech delay in this syndrome is very controversial. Proponents of sign language feel that the ability to communicate is of paramount importance, and signing allows the child to communicate (Solot et al. 2001, 2000). Proponents of natural speech feel that signing delays acquisition of speech (Golding-Kushner et al. 1985). There have been no direct comparisons. Ultimately, the vast majority of patients learn to speak and communicate effectively.

The degree of developmental delay also varies widely. The mean full scale IQ is approximately 70, with a range from normal to moderately disabled (Gerdes et al. 1999; Moss et al. 1999; Swillen et al. 1997, 1999). Most patients have reasonable skills with comprehension and social rules. Motor skills are intermediate, and visuo-perceptual abilities and planning tend to be the weakest areas (Gerdes et al. 2001; Moss et al. 1999). School-based interventions have been successfully developed for children with non-verbal learning disabilities, which are thought to be appropriate for children with CH22qD although no studies have attempted to define the optimal learning strategy.

Nearly half of patients have microcephaly with the parietal lobe most often involved (Barnea-Goraly et al. 2005; Campbell et al. 2006). A small vermis is detected in patients with CH22qD similar to that seen in autistic spectrum disorder (Bish et al. 2006). The posterior vermis appears to control social drive, and this may explain some of the social awkwardness seen in certain patients with CH22qD. Functional MRI studies have shown that the patterns of brain use during mathematical tasks are different in patients with CH22qD compared to controls (Barnea-Goraly et al. 2005; Simon et al. 2005a,b). Other functional studies may shed light on the pathophysiology of some cognitive features seen in patients with CH22qD.

The behavioral aspects of CH22qD include attention-deficit hyperactivity disorder, poor social interaction skills, and impulsivity (Antshel et al. 2006; Lajiness-O'Neill et al. 2006; Niklasson et al. 2005; Parissis and Milonas 2005). Bipolar disorder, autistic spectrum disorder, and schizophrenia/schizo-affective disorder are found in 10–30% of older patients, which is clearly above the frequency seen in other syndromes associated with developmental delay. Thus far, patients with behavioral problems and frank psychiatric disturbances have been treated with conventional modalities.

4. Conclusions

Patients with Ch22qD require a multidisciplinary approach to management. Most patients will require care from a variety of subspecialists. The immunologist is often the coordinator of care and plays an important role in guiding the family and other subspecialists. The immune system often requires no specific intervention; however, immunomonitoring and vigilance for autoimmune disease are critical.

References

Antshel, K.M., Fremont, W., Roizen, N.J., Shprintzen, R., Higgins, A.M., Dhamoon, A. and Kates, W.R. (2006) ADHD, major depressive disorder, and simple phobias are prevalent psychiatric conditions in youth with velocardiofacial syndrome. J. Am. Acad. Child Adolesc. Psychiatry 45, 596–603.

Bale, P.M. and Sotelo-Avila, C. (1993) Maldescent of the thymus: 34 necropsy and 10 surgical cases, including 7 thymuses medial to the mandible. Pediatr. Pathol. 13, 181–190.

Barnea-Goraly, N., Eliez, S., Menon, V., Bammer, R. and Reiss, A.L. (2005) Arithmetic ability and parietal alterations: a diffusion tensor imaging study in velocardiofacial syndrome. Brain Res. Cogn. Brain Res. 25, 735–740.

Bastian, J., Law, S., Vogler, L., Lawton, A., Herrod, H., Anderson, S., Horowitz, S. and Hong, R. (1989) Prediction of persistent immunodeficiency in the DiGeorge anomaly. J. Pediatr. 115, 391–396.

Becker, D.B., Pilgram, T., Marty-Grames, L., Govier, D.P., Marsh, J.L. and Kane, A.A. (2004) Accuracy in identification of patients with 22q11.2 deletion by likely care providers using facial photographs. Plast. Reconstr. Surg.114, 1367–1372.

Bish, J.P., Pendyal, A., Ding, L., Ferrante, H., Nguyen, V., McDonald-McGinn, D., Zackai, E. and Simon, T.J. (2006) Specific cerebellar reductions in children with chromosome 22q11.2 deletion syndrome. Neurosci. Lett. 399, 245–248.

Botto, L.D., May, K., Fernhoff, P.M., Correa, A., Coleman, K., Rasmussen, S.A., Merritt, R.K., O'Leary, L.A., Wong, L.Y., Elixson, E.M., Mahle, W.T. and Campbell, R.M. (2003) A population-based study of the 22q11.2 deletion: phenotype, incidence, and contribution to major birth defects in the population. Pediatrics 112, 101–107.

Campbell, L.E., Daly, E., Toal, F., Stevens, A., Azuma, R., Catani, M., Ng, V., van Amelsvoort, T., Chitnis, X., Cutter, W., Murphy, D.G. and Murphy, K.C. (2006) Brain and behaviour in children with 22q11.2 deletion syndrome: a volumetric and voxel-based morphometry MRI study. Brain 129, 1218–1228.

Chen, Y.F., Kou, P.L., Tsai, S.J., Chen, K.F., Chan, H.H., Chen, C.M. and Sun, H.S. (2006) Computational analysis and refinement of sequence structure on chromosome 22q11.2 region: application to the development of quantitative real-time PCR assay for clinical diagnosis. Genomics 87, 290–297.

Chinen, J., Rosenblatt, H.M., Smith, E.O., Shearer, W.T. and Noroski, L.M. (2003) Long-term assessment of T cell populations in DiGeorge syndrome. J. Allergy Clin. Immunol. 111, 573–579.

Cipollone, D., Amati, F., Carsetti, R., Placidi, S., Biancolella, M., D'Amati, G., Novelli, G., Siracusa, G. and Marino, B. (2006) A multiple retinoic acid antagonist induces conotruncal anomalies, including transposition of the great arteries, in mice. Cardiovasc. Pathol. 15, 194–202.

Coberly, S., Lammer, E. and Alashari, M. (1996) Retinoic acid embryopathy: case report and review of literature. Pediatr. Pathol. Lab. Med. 16, 823–836.

Davies, K., Stiehm, E.R., Woo, P. and Murray, K.J. (2001) Juvenile idiopathic polyarticular arthritis and IgA deficiency in the 22q11 deletion syndrome. J. Rheumatol. 28, 2326–2334.

DePiero, A.D., Lourie, E.M., Berman, B.W., Robin, N.H., Zinn, A.B. and Hostoffer, R.W. (1997) Recurrent immune cytopenias in two patients with DiGeorge/velocardiofacial syndrome. J. Pediatr. 131, 484–486.

Digilio, M.C., Marino, B., Formigari, R. and Giannotti, A. (1995) Maternal diabetes causing DiGeorge anomaly and renal agenesis Am. J. Med. Genet. 55, 513–514.

Digilio, M.C., Pacifico, C., Tieri, L., Marino, B., Giannotti, A. and Dallapiccola, B. (1999) Audiological findings in patients with microdeletion 22q11 (di George/velocardiofacial syndrome). Br. J. Audiol. 33, 329–333.

Dunham, I., Shimizu, N., Roe, B.A., Chissoe, S., Hunt, A.R., Collins, J.E., Bruskiewich, R., Beare, D.M., Clamp, M., Smink, L.J., Ainscough, R., Almeida, J.P., Babbage, A., Bagguley, C., Bailey, J., Barlow, K., Bates, K.N., Beasley, O., Bird, C.P., Blakey, S., Bridgeman, A.M., Buck, D., Burgess, J., Burrill, W.D., O'Brien, K.P., et al. (1999) The DNA sequence of human chromosome 22. Nature 402, 489–495.

Fernandez, L., Lapunzina, P., Arjona, D., Lopez Pajares, I., Garcia-Guereta, L., Elorza, D., Burgueros, M., De Torres, M.L., Mori, M.A., Palomares, M., Garcia-Alix, A. and Delicado, A. (2005) Comparative study of three diagnostic approaches (FISH, STRs and MLPA) in 30 patients with 22q11.2 deletion syndrome. Clin. Genet. 68, 373–378.

Finocchi, A., Di Cesare, S., Romiti, M.L., Capponi, C., Rossi, P., Carsetti, R. and Cancrini, C. (2006) Humoral immune responses and CD27+ B cells in children with DiGeorge syndrome (22q11.2 deletion syndrome). Pediatr. Allergy Immunol. 17, 382–388.

Frohn-Mulder, I.M., Wesby Swaay, E., Bouwhuis, C., Van Hemel, J.O., Gerritsma, E., Niermeyer, M.F. and Hess, J. (1999) Chromosome 22q11 deletions in patients with selected outflow tract malformations. Genet. Couns. 10, 35–41.
Gennery, A.R., Barge, D., O'Sullivan, J.J., Flood, T.J., Abinun, M. and Cant, A.J. (2002) Antibody deficiency and autoimmunity in 22q11.2 deletion syndrome. Arch. Dis.Chil. 86, 422–425.
Gerdes, M., Solot, C., Wang, P.P., McDonald-McGinn, D.M. and Zackai, E.H. (2001) Taking advantage of early diagnosis: preschool children with the 22q11.2 deletion. Genet. Med. 3, 40–44.
Gerdes, M., Solot, C., Wang, P.P., Moss, E., LaRossa, D., Randall, P., Goldmuntz, E., Clark, B.J., 3rd, Driscoll, D.A., Jawad, A., Emanuel, B.S., McDonald-McGinn, D.M., Batshaw, M.L. and Zackai, E.H. (1999) Cognitive and behavior profile of preschool children with chromosome 22q11.2 deletion. Am. J. Med. Genet. 85, 127–133.
Golding-Kushner, K.J., Weller, G. and Shprintzen, R.J. (1985) Velo-cardio-facial syndrome: language and psychological profiles. J. Craniofac. Genet. Dev. Biol. 5, 259–266.
Jawad, A.F., McDonald-McGinn, D.M., Zackai, E. and Sullivan, K.E. (2001) Immunologic features of chromosome 22q11.2 deletion syndrome (DiGeorge syndrome/velocardiofacial syndrome). J. Pediatr. 139, 715–723.
Jiang, L., Duan, C., Chen, B., Hou, Z., Chen, Z., Li, Y., Huan, Y. and Wu, K.K. (2005) Association of 22q11 deletion with isolated congenital heart disease in three Chinese ethnic groups. Int. J. Cardiol. 105, 216–223.
Junker, A.K. and Driscoll, D.A. (1995) Humoral immunity in DiGeorge syndrome. J. Pediatr. 127, 231–237.
Kanaya, Y., Ohga, S., Ikeda, K., Furuno, K., Ohno, T., Takada, H., Kinukawa, N. and Hara, T. (2006) Maturational alterations of peripheral T cell subsets and cytokine gene expression in 22q11.2 deletion syndrome. Clin. Exp. Immunol. 144, 85–93.
Kornfeld, S.J., Zeffren, B., Christodoulou, C.S., Day, N.K., Cawkwell, G. and Good, R.A. (2000) DiGeorge anomaly: a comparative study of the clinical and immunologic characteristics of patients positive and negative by fluorescence in situ hybridization. J. Allergy Clin. Immunol. 105, 983–987.
Lajiness-O'Neill, R., Beaulieu, I., Asamoah, A., Titus, J.B., Bawle, E., Ahmad, S., Kirk, J.W. and Pollack, R. (2006) The neuropsychological phenotype of velocardiofacial syndrome (VCFS): relationship to psychopathology. Arch. Clin. Neuropsychol. 21, 175–184.
Lawrence, S., McDonald-McGinn, D.M., Zackai, E. and Sullivan, K.E. (2003). Thrombocytopenia in patients with chromosome 22q11.2 deletion syndrome. J. Pediatr 143, 277–278.
Levy, A., Michel, G., Lemerrer, M. and Philip, N. (1997) Idiopathic thrombocytopenic purpura in two mothers of children with DiGeorge sequence: a new component manifestation of deletion 22q11? Am. J. Med. Genet. 69, 356–359.
Lindsay, E.A., Botta, A., Jurecic, V., Carattini-Rivera, S., Cheah, Y.C., Rosenblatt, H.M., Bradley, A. and Baldini, A. (1999) Congenital heart disease in mice deficient for the DiGeorge syndrome region. Nature 401, 379–383.
Losken, A., Williams, J.K., Burstein, F.D., Malick, D. and Riski, J.E. (2003) An outcome evaluation of sphincter pharyngoplasty for the management of velopharyngeal insufficiency. Plast. Reconstr. Surg. 112, 1755–1761.
Losken, A., Williams, J.K., Burstein, F.D., Malick, D.N. and Riski, J.E. (2006) Surgical correction of velopharyngeal insufficiency in children with velocardiofacial syndrome. Plast. Reconstr. Surg. 117, 1493–1498.
Mahadevan, N.R., Horton, A.C. and Gibson-Brown, J.J. (2004) Developmental expression of the amphioxus Tbx1/ 10 gene illuminates the evolution of vertebrate branchial arches and sclerotome. Dev. Genes Evol. 214, 559–566.

Markert, M.L., Boeck, A., Hale, L.P., Kloster, A.L., McLaughlin, T.M., Batchvarova, M.N., Douek, D.C., Koup, R.A., Kostyu, D.D., Ward, F.E., Rice, H.E., Mahaffey, S.M., Schiff, S.E., Buckley, R.H. and Haynes, B.F. (1999) Transplantation of thymus tissue in complete DiGeorge syndrome. N. Engl. J. Med. 341, 1180–1189.

Markert, M.L., Sarzotti, M., Ozaki, D.A., Sempowski, G.D., Rhein, M.E., Hale, L.P., Le Deist, F., Alexieff, M.J., Li, J., Hauser, E.R., Haynes, B.F., Rice, H.E., Skinner, M.A., Mahaffey, S.M., Jaggers, J., Stein, L.D. and Mill, M.R. (2003) Thymus transplantation in complete DiGeorge syndrome: immunologic and safety evaluations in 12 patients. Blood 102, 1121–1130.

Markert, M.L., Watson, T., McLaughlin, T., McGuire, C.A., Ward, F.E., Kostyu, D., Hale, L.P., Buckley, R.H., Schiff, S.E., Bleesing, J.J.H., Broome, C.B., Ungarleider, R.M., Mahaffey, S., Oldman, K.T. and Haynes, B. (1995) Development of T cell function after post-natal thymic transplanation for DiGeorge syndrome. Am. J. Hum. Genet. 57, A14.

McDonald-McGinn, D.M., Kirschner, R., Goldmuntz, E., Sullivan, K., Eicher, P., Gerdes, M., Moss, E., Solot, C., Wang, P., Jacobs, I., Handler, S., Knightly, C., Heher, K., Wilson, M., Ming, J.E., Grace, K., Driscoll, D., Pasquariello, P., Randall, P., Larossa, D., Emanuel, B.S. and Zackai, E.H. (1999) The Philadelphia story: the 22q11.2 deletion: report on 250 patients. Genet. Couns. 10, 11–24.

Moss, E.M., Batshaw, M.L., Solot, C.B., Gerdes, M., McDonald-McGinn, D.M., Driscoll, D.A., Emanuel, B.S., Zackai, E.H. and Wang, P.P. (1999) Psychoeducational profile of the 22q11.2 microdeletion: a complex pattern. J. Pediatr. 134, 193–198.

Motzkin, B., Marion, R., Goldberg, R., Shprintzen, R. and Saenger, P. (1993) Variable phenotypes in velocardiofacial syndrome with chromosomal deletion. J. Pediatr. 123, 406–410.

Munoz, S., Garay, F., Flores, I., Heusser, F., Talesnik, E., Aracena, M., Mellado, C., Mendez, C., Arnaiz, P. and Repetto, G. (2001) Heterogeneidad de la presentacion clinica del sindrome de microdelecion del cromosoma 22, region q11. Revista Medica de Chile 129, 515–521.

Niklasson, L., Rasmussen, P., Oskarsdottir, S. and Gillberg, C. (2005) Attention deficits in children with 22q.11 deletion syndrome. Dev. Med. Child. Neurol. 47, 803–807.

Novak, R.W. and Robinson, H.B. (1994) Coincident DiGeorge anomaly and renal agenesis and its relation to maternal diabetes Am. J. Med. Genet. 50, 311–312.

Parissis, D. and Milonas, I. (2005) Chromosome 22q11.2 deletion syndrome: an underestimated cause of neuropsychiatric impairment in adolescence. J. Neurol. 252, 989–990.

Piliero, L.M., Sanford, A.N., McDonald-McGinn, D.M., Zackai, E.H. and Sullivan, K.E. (2004) T cell homeostasis in humans with thymic hypoplasia due to chromosome 22q11.2 deletion syndrome. Blood 103, 1020–1025.

Pinchas-Hamiel, O., Engelberg, S., Mandel, M. and Passwell, J.H. (1994) Immune hemolytic anemia, thrombocytopenia and liver disease in a patient with DiGeorge syndrome. Isr. J. Med. Sci. 30, 530–532.

Rasmussen, S.A., Williams, C.A., Ayoub, E.M., Sleasman, J.W., Gray, B.A., Bent-Williams, A., Stalker, H.J. and Zori, R.T. (1996) Juvenile rheumatoid arthritis in velo-cardio-facial syndrome: coincidence or unusual complication. Am. J. Med. Genet. 64, 546–550.

Roberts, C., Ivins, S.M., James, C.T. and Scambler, P.J. (2005) Retinoic acid down-regulates Tbx1 expression in vivo and in vitro. Dev. Dyn. 232, 928–938.

Rommel, N., Vantrappen, G., Swillen, A., Devriendt, K., Feenstra, L. and Fryns, J.P. (1999) Retrospective analysis of feeding and speech disorders in 50 patients with velo-cardio-facial syndrome. Genet. Couns. 10, 71–78.

Ryan, A.K., Goodship, J.A., Wilson, D.I., Philip, N., Levy, A., Seidel, H., Schuffenhauer, S., Oechsler, H., Belohradsky, B., Prieur, M., Aurias, A., Raymond, F.L., Clayton-Smith, J., Hatchwell, E., McKeown, C., Beemer, F.A., Dallapiccola, B., Novelli, G., Hurst, J.A., Ignatius, J., Green, A.J., Winter, R.M., Brueton, L., Brondum-Nielson, K., Stewart, F.,

Van Essen, T., Patton, M., Paterson, J. and Scambler, P.J. (1997) Spectrum of clinical features associated with interstitial chromosome 22q11 deletions: a European collaborative study. J. Med. Genet. 34, 798–804.
Sediva, A., Bartunkova, J., Zachova, R., Polouckova, A., Hrusak, O., Janda, A., Kocarek, E., Novotna, D., Novotna, K. and Klein, T. (2005) Early development of immunity in di-George syndrome. Med. Sci. Monit. 11, CR182–CR187.
Shaikh, T.H., Kurahashi, H., Saitta, S.C., O'Hare, A.M., Hu, P., Roe, B.A., Driscoll, D.A., McDonald-McGinn, D.M., Zackai, E.H., Budarf, M.L. and Emanuel, B.S. (2000) Chromosome 22-specific low copy repeats and the 22q11.2 deletion syndrome: genomic organization and deletion endpoint analysis. Hum. Mol. Genet. 9, 489–501.
Shprintzen, R.J., Goldberg, R., Golding-Kushner, K.J. and Marion, R.W. (1992) Late-onset psychosis in the velo-cardio-facial syndrome. Am. J. Med. Genet. 42, 141–142.
Simon, T., Bearden, C., Mc-Ginn, D. and Zackai, E. (2005a) Visuospatial and numerical cognitive deficits in children with chromosome 22q11.2 deletion syndrome. Cortex 41, 145–155.
Simon, T.J., Ding, L., Bish, J.P., McDonald-McGinn, D.M., Zackai, E.H. and Gee, J. (2005b) Volumetric, connective, and morphologic changes in the brains of children with chromosome 22q11.2 deletion syndrome: an integrative study. Neuroimage 25, 169–180.
Smith, C.A., Driscoll, D.A., Emanuel, B.S., McDonald-McGinn, D.M., Zackai, E.H. and Sullivan, K.E. (1998) Increased prevalence of immunoglobulin A deficiency in patients with the chromosome 22q11.2 deletion syndrome (DiGeorge syndrome/velocardiofacial syndrome). Clin. Diagn. Lab. Immunol. 5, 415–417.
Solot, C.B., Gerdes, M., Kirschner, R.E., McDonald-McGinn, D.M., Moss, E., Woodin, M., Aleman, D., Zackai, E.H. and Wang, P.P. (2001) Communication issues in 22q11.2 deletion syndrome: children at risk.. Genet. Med. 3, 67–71.
Solot, C.B., Knightly, C., Handler, S.D., Gerdes, M., McDonald-McGinn, D.M., Moss, E., Wang, P., Cohen, M., Randall, P., Larossa, D. and Driscoll, D.A. (2000) Communication disorders in the 22Q11.2 microdeletion syndrome. J. Commun. Disord. 33, 187–203.
Stalmans, I., Lambrechts, D., De Smet, F., Jansen, S., Wang, J., Maity, S., Kneer, P., von der Ohe, M., Swillen, A., Maes, C., Gewillig, M., Molin, D.G., Hellings, P., Boetel, T., Haardt, M., Compernolle, V., Dewerchin, M., Plaisance, S., Vlietinck, R., Emanuel, B., Gittenberger-de Groot, A.C., Scambler, P., Morrow, B., Driscol, D.A., Moons, L., Esguerra, C.V., Carmeliet, G., Behn-Krappa, A., Devriendt, K., Collen, D., Conway, S.J. and Carmeliet, P. (2003) VEGF: a modifier of the del22q11 (DiGeorge) syndrome? Nat. Med. 9, 173–182.
Sullivan, K.E., McDonald-McGinn, D., Driscoll, D., Emanuel, B.S., Zackai, E.H. and Jawad, A.F. (1999) Longitudinal analysis of lymphocyte function and numbers in the first year of life in chromosome 22q11.2 deletion syndrome (DiGeorge syndrome/velocardiofacial syndrome). Clin. Lab. Diagn. Immunol. 6, 906–911.
Sullivan, K.E., McDonald-McGinn, D. and Zackai, E.H. (2002) CD4(+) CD25(+) T cell production in healthy humans and in patients with thymic hypoplasia. Clin. Lab. Diagn. Immunol. 9, 1129–1131.
Swillen, A., Devriendt, K., Legius, E., Eyskens, B., Dumoulin, M., Gewillig, M. and Fryns, J.P. (1997) Intelligence and psychosocial adjustment in velocardiofacial syndrome: a study of 37 children and adolescents with VCFS. J. Med. Genet. 34, 453–458.
Swillen, A., Devriendt, K., Legius, E., Prinzie, P., Vogels, A., Ghesquiere, P. and Fryns, J.P. (1999) The behavioural phenotype in velo-cardio-facial syndrome (VCFS): from infancy to adolescence. Genetic. Couns. 10, 79–88.
Theodoropoulos, D.S. (2003) Immune deficiency in CHARGE association. Clin. Med. Res. 1, 43–48.

Van Esch, H., Groenen, P., Fryns, J.P., Van de Ven, W. and Devriendt, K. (1999) The phenotypic spectrum of the 10p deletion syndrome versus the classical DiGeorge syndrome. Genet. Couns. 10, 59–65.

Vantrappen, G., Devriendt, K., Swillen, A., Rommel, N., Vogels, A., Eyskens, B., Gewillig, M., Feenstra, L. and Fryns, J.P. (1999) Presenting symptoms and clinical features in 130 patients with the velo-cardio-facial syndrome. The Leuven experience. Genet. Couns. 10, 3–9.

Vorstman, J.A., Jalali, G.R., Rappaport, E.F., Hacker, A.M., Scott, C. and Emanuel, B.S. (2006) MLPA: a rapid, reliable, and sensitive method for detection and analysis of abnormalities of 22q. Hum. Mutat. 27, 814–821.

Wang, P.P., Woodin, M.F., Kreps-Falk, R. and Moss, E.M. (2000) Research on behavioral phenotypes: velocardiofacial syndrome (deletion 22q11.2). Dev. Med. Child. Neurol. 42, 422–427.

Wilson, D.I., Burn, J., Scambler, P. and Goodship, J. (1993) DiGeorge syndrome: part of CATCH 22. J. Med. Genet. 30, 852–856.

Xu, H., Cerrato, F. and Baldini, A. (2005) Timed mutation and cell-fate mapping reveal reiterated roles of Tbx1 during embryogenesis, and a crucial function during segmentation of the pharyngeal system via regulation of endoderm expansion. Development 132, 4387–4395.

Yagi, H., Furutani, Y., Hamada, H., Sasaki, T., Asakawa, S., Minoshima, S., Ichida, F., Joo, K., Kimura, M., Imamura, S., Kamatani, N., Momma, K., Takao, A., Nakazawa, M., Shimizu, N. and Matsuoka, R. (2003) Role of TBX1 in human del22q11.2 syndrome. Lancet 362, 1366–1373.

Yan, W., Jacobsen, L.K., Krasnewich, D.M., Guan, X.Y., Lenane, M.C., Paul, S.P., Dalwadi, H.N., Zhang, H., Long, R.T., Kumra, S., Martin, B.M., Scambler, P.J., Trent, J.M., Sidransky, E., Ginns, E.I. and Rapoport, J.L. (1998) Chromosome 22q11.2 interstitial deletions among childhood-onset schizophrenics and "multidimensionally impaired". Am. J. Med. Genet. 81, 41–43.

Zhang, Z., Cerrato, F., Xu, H., Vitelli, F., Morishima, M., Vincentz, J., Furuta, Y., Ma, L., Martin, J.F., Baldini, A. and Lindsay, E. (2005) Tbx1 expression in pharyngeal epithelia is necessary for pharyngeal arch artery development. Development 132, 5307–5315.

5

Leukocyte Adhesion Deficiencies: Molecular Basis, Clinical Findings, and Therapeutic Options

Amos Etzioni

Meyer Children Hospital and the Rappaport School of Medicine, Technion, Haifa, Israel, etzioni@rambam.health.gov.il

Abstract. Leukocyte trafficking from bloodstream to tissue is important for the continuous surveillance for foreign antigens, as well as for rapid leukocyte accumulation at sites of inflammatory response or tissue injury. Leukocyte interaction with vascular endothelial cells is a pivotal event in the inflammatory response and is mediated by several families of adhesion molecules. The crucial role of the β2-integrin subfamily in leukocyte emigration was established after leukocyte adhesion deficiency (LAD) I was discovered. Patients with this disorder suffer from life-threatening bacterial infections, and in its severe form, death usually occurs in early childhood unless bone marrow transplantation is performed. The LAD II disorder clarifies the role of the selectin receptors and their fucosylated ligands. Clinically, patients with LAD II suffer from a less severe form of disease, resembling the moderate phenotype of LAD I. LAD III emphasizes the importance of the integrin activation phase in the adhesion cascade. Although the primary defect is still unknown, it is clear that all hematopoietic integrin activation processes are defective, which lead to severe infection as observed in LAD I and to marked increase tendency for bleeding problems.

1. Introduction

Migration of leukocytes from the bloodstream to the tissue occurs in several distinct steps (Hogg et al. 1999; Luster et al. 2005). Leukocytes in the circulation must resist shear forces in order to arrest along the vascular endothelium. Under normal conditions, leukocytes move rapidly in the laminar flow stream of blood and do not adhere to the endothelium. However, with activation of the endothelium due to local trauma or inflammation, leukocytes immediately begin to roll along the venular wall. The endothelial selectins will bind to the leukocytes through their carbohydrate ligands and thus, cause the leukocytes to roll on the endothelium. This first transient and reversible step is a prerequisite for the next stage, the activation of leukocytes. Binding to the selectins tethers the leukocytes, exposing them to stimuli in the local microenvironment mainly chemokines, which will activate the integrins. This process will increase their affinity (through conformational changes) and avidity (through

TABLE 1. Leukocyte adhesion deficiency syndromes

	LAD I	LAD II	LAD III
Clinical manifestation			
Recurrent severe infections	+++	+	+++
Periodontitis	++	++	?
Skin infection	++	+	++
Delayed separation of the umbilical cord	+++	No	+++
Developmental abnormalities	No	+++	No
Bleeding tendency	No	No	+++
Laboratory findings			
Neutrophilia	+++	+++	+++
CD18 expression	↓ or no	N	N
SLeX expression	N	No	N
Neutrophil rolling	N	↓	↓
Neutrophil adherence	↓	↓	↓
Platelet aggregation	N	N	↓
Primary genetic defect	+ITGB2	+FUCT1	?

LAD, leukocyte adhesion deficiency; N, normal; No, absent; ?, unknown.

clastering) (Kinashi 2005). Integrin activation is followed by firm adhesion, that is, sticking which is an extremely rapid phase, after which transmigration occurs (Shamri et al. 2005). Each of these steps involves different adhesion molecules and can be differentially regulated. Several years after reporting the structure of the leukocyte integrin molecule, a genetic defect in the subunit of the molecule (ITGB2) was discovered. This syndrome, now called leukocyte adhesion deficiency (LAD) I (OMIM 116920), has been described in more than 300 children and is characterized by delayed separation of the umbilical cord, recurrent soft tissue infections, chronic periodontitis, marked leukocytosis, and a high mortality rate at early age.

Ten years later, in 1992, a second defect, LAD II (OMIM 266265), was discovered and was found to be due to the defect in the synthesis of selectin ligands. The clinical course with respect to infectious complications is a milder one than LAD I. However, LAD II patients present other abnormal features, such as growth and mental retardation. The primary defect is mutation in the specific fucose transporter to the Golgi apparatus (FUCT1).

Recently, a third rare LAD syndrome has been described. Patients with LAD III suffer from severe recurrent infections, similar to LAD I, and a severe bleeding tendency. Although integrin structure is intact, a defect in integrins activation is the primary abnormality in LAD III (Table 1).

2. Leukocyte Adhesion Deficiency I

LAD I is an autosomal recessive disorder caused by mutations in the common chain (CD18) of the β_2 integrin family. Up to now, several hundreds of patients have been reported worldwide. The prominent clinical feature of these patients is recurrent bacterial infections, primarily localized to skin and mucosal surfaces. Sites of infection often progressively enlarge, and they may lead to systemic spread of the bacteria. Infections are usually apparent from birth onward, and a common presenting infection is omphalitis with delayed separation of the umbilical cord. The most frequently encountered bacteria are *Staphylococcus aureus* and gram-negative enteric organisms, but fungal infections are also common. The absence of pus formation at the sites of infection is one of the hallmarks of LAD I. Severe gingivitis and periodontitis are major features among all patients who survive infancy. Impaired healing of traumatic or surgical wounds is also characteristic of this syndrome (Anderson and Smith 2001).

The recurrent infections observed in affected patients result from a profound impairment of leukocyte mobilization into extravascular site of inflammation. Skin windows yield few, if any, leukocytes, and biopsies of infected tissues demonstrate inflammation totally devoid of neutrophils. These findings are particularly striking considering that marked peripheral blood leukocytosis (5 to 20 times normal values) is consistently observed during infections. In contrast to their difficulties in defense against bacterial and fungal microorganisms, LAD I patients do not exhibit a marked increase in susceptibility to viral infections (Etzioni 1996).

The severity of clinical infectious complications among patients with LAD I appear to be directly related to the degree of CD18 deficiency. Two phenotypes, designated severe deficiency and moderate deficiency, have been defined (Fischer et al. 1988). Patients with less than 1% of the normal surface expression exhibited a severe form of disease with earlier, frequent, and serious episodes of infection, often leading to death in infancy, whereas patients with some surface expression of CD18 (2.5–10%) manifested a moderate to mild phenotype with fewer serious infectious episodes and survival into adulthood.

The defective migration of neutrophils from patients with LAD I was observed in studies in vivo as well as in vitro. Neutrophils failed to mobilize to skin sites in the in vivo Rebuck skin-window test. In vitro studies demonstrated a marked defect in random migration as well as chemotaxis to various chemoattractant substances. Adhesion and transmigration through endothelial cells were found to be severely impaired. With the use of an intravital microscopy assay, it was found that fluorescein-labeled neutrophils from a LAD I boy rolled normally on inflamed rabbit venules, suggesting that they were capable of initiating adhesive interactions with inflamed endothelial cells (von Andrian et al. 1993). However, these cells failed to perform activation-dependent, β2-integrin-mediated adhesion steps and did not stick or emigrate when challenged with a chemotactic stimulus.

Patients with LAD I exhibit neutrophilia in the absence of overt infection with marked granulocytosis with neutrophils in peripheral blood reaching levels of up to 100,000/μl during acute infections. Early studies showed that patients with this disorder were uniformly deficient in the expression of all three leukocyte integrins (Mac-1, LFA-1, p150, and 95), suggesting that the primary defect was in the common β2-subunit, which is encoded by a gene located at the tip of the long arm of chromosome 21q22.3. Subsequently, several LAD I variants were reported in which there was a defect in β2-integrin adhesive functions despite normal surface expression of CD18. A child with classical LAD I features with normal surface expression of CD18 was reported (Hogg et al. 1999), in whom a mutation in CD18 was found to lead to a non-functional molecule.

The molecular basis for CD18 deficiency varies (Roos et al. 2002). In some cases, it is due to the lack or diminished expression of CD18 mRNA. In other cases, there is expression of mRNA or protein precursors of aberrant size with both larger and smaller CD18 subunits. Analysis at the gene level has revealed a degree of heterogeneity, which reflects this diversity. A number of point mutations have been reported, some of which lead to the biosynthesis of defective proteins with single amino acid substitutions, while others lead to splicing defects, resulting in the production of truncated and unstable proteins.

Notably, a high percentage of CD18 mutations identified in LAD I is contained in the extracellular domain of the CD18 (on exon 9), which is a highly conserved region. Domains within this segment are presumably required for association and biosynthesis of precursors and may represent critical contact sites between the α-subunit and β-subunit precursors. Thus, LAD I can be caused by a number of distinct mutational events, all resulting in the failure to produce a functional leukocyte β2-subunit. Recently, a case of somatic mosaicism due to in vivo reversion of the mutant CD18 was reported (Tone et al. 2007).

In any infant male or female with recurrent soft tissue infection and a very high leukocyte count, the diagnosis of LAD I should be considered. The diagnosis is even more suggestive if a history of delayed separation of the umbilical cord is present. To confirm the diagnosis, absence of the α- and β-subunits of the β2-integrin complex must be demonstrated. This can be accomplished with the use of the appropriate CD11 and CD18 monoclonal antibodies by flow cytometry. Sequence genetic analysis to define the exact molecular defect in the β2-subunit is a further option.

As leukocytes express CD18 on their surface at 20 weeks of gestation, cordocentesis performed at this age can establish a prenatal diagnosis. In families in whom the exact molecular defect has been previously identified, an earlier prenatal diagnosis is possible by chorionic biopsy and mutation analysis. Furthermore, recently, pre implantation diagnosis was also performed (Lorusso et al. 2006). Patients with the moderate LAD I phenotype usually respond to conservative therapy and the prompt use of antibiotics during acute infectious episodes. Prophylactic antibiotics may reduce the risk of infections. Although granulocyte transfusions may be life saving, their use is limited because of difficulties in supply of daily donors and immune reactions to the allogeneic leukocytes.

At present, the only corrective treatment that should be offered to all cases with the severe phenotype is bone marrow transplantation (Thomas et al. 1995). The absence of host LFA-1 may be advantageous in these transplants because graft rejection appears to be in part dependent upon the CD18 complex. The introduction of a normal β2-subunit gene (ITGB2) into hematopoietic stem cells has the potential to cure children with LAD I (Bauer and Hickstein 2000). Recently, successful gene therapy has been reported in canine model of LAD I (Bauer et al. 2006), applying this procedure may be beneficial also in humans.

3. Leukocyte Adhesion Deficiency II

LAD II syndrome results from a general defect in fucose metabolism, causing the absence of SLeX and other fucosylated ligands for the selectins. LAD II was first described in two unrelated Arab consanguineous parents (Etzioni et al. 1992). This is an extremely rare condition with only six patients reported (Yakubenia and Wild 2006).

Affected children were born after uneventful pregnancies with normal height and weight. No delay in the separation of the umbilical cord was observed. They have severe mental retardation, short stature, a distinctive facial appearance, and the rare Bombay (hh) blood phenotype. From early life, they have suffered from recurrent episodes of bacterial infections, mainly pneumonia, periodontitis, otitis media, and localized cellulitis. During times of infections, the neutrophil count increases up to 150,000/μl. Several mild to moderate skin infections, without obvious pus have also been observed (Wild et al. 2002). The infections have not been life-threatening events and are usually treated in the outpatient clinic. Interestingly, after the age of 3 years, the frequency of infections has decreased and the children no longer need prophylactic antibiotics. At older age, their main infectious problem is severe periodontitis as is also observed in patients with LAD I (Etzioni et al. 1998).

Overall, the infections in LAD II appear to be comparable to the moderate rather than the severe phenotype of LAD I. It is possible that the ability of LAD II neutrophils to adhere and transmigrate via β2-integrin under conditions of reduced shear forces (von Andrian et al. 1993) may permit some neutrophils to emigrate at sites of severe inflammation, where flow may be impaired, thereby allowing some level of neutrophil defense against bacterial infections. Rolling, the first step in neutrophil recruitment to site of inflammation, is mediated primarily by the binding of the selectins to their fucosylated glycoconjugate ligands. Using intravital microscopy, it was observed that the rolling fraction of normal donor neutrophils in this assay was around 30%, and LAD I neutrophils behaved similarly. In contrast, LAD II neutrophils rolled poorly (only 5%) and failed to emigrate (von Andrian et al. 1993).

Since the first two LAD II patients identified were the offspring of first-degree relatives and since the parents were clinically unaffected, autosomal recessive inheritance was assumed. In addition to the Bombay phenotype (absence of the H antigen), the cells of LAD II patients were also found to be Lewis A- and B-negative, and the patients were non-secretors. The three blood phenotypes (Bombay, Lewis A, and B) have in common a lack of fucosylation of glycoconjugates. These facts suggested that the primary defect in LAD II must instead be a general defect in fucose production. After the observation that the defect in the Arab patients may be localized in the de novo GDP-1-fucose biosynthesis pathway (Karsan et al. 1998), the two enzymes involved with this pathway, GMD and FX protein, were measured and were found to be normal with no mutation in cDNA isolated from LAD II patients. Another child, from a Turkish origin, was also described with LAD II in whom decreased GDP-1-fucose transport into the Golgi vesicles was detected (Lubke et al. 1999).

Using the complementation cloning technique, the human gene encoding the fucose transporter was found to be located on chromosome 11. The Turkish child was found to be homozygous for a mutation at amino acid 147 in which arginine is changed to cysteine, while the two Arab patients examined were found to have a mutation in amino acid 308 in which threonine is changed to arginine (Lubke et al. 2001). Both mutations are located in highly conserved transmembrane domains through evolution. LAD II is thus one of the group of congenital disorders of glycosylation (CDG) and is classified as CDG-IIc. In some cases, the mutated (non-functional) transporter is correctly located in the Golgi apparatus, while in two patients, the truncated transporter was unable to localize to the Golgi complex (Helmus et al. 2006). Although only four mutations were described, some genotype-phenotype correlation can be observed.

From the biochemical aspect, once the primary defect was found, several studies were carried out to clarify the defect. As growth and mental retardation are prominent features of LAD II, and Notch protein which are important in normal development contain fucose, Sturla et al. (2003) looked at the fucosylation process in LAD II. Fractionation and analysis of the different classes of glycans indicated that the decrease in fucose incorporation is not generalized and is mainly confined to terminal fucosylation of N-linked oligosaccharides. In contrast, the total levels of protein O-fucosylation, including that observed in Notch protein, were unaffected. Indeed, it was recently observed that the O-fucosylation process take place in the endoplasmatic compartment and not in the Golgi apparatus (Luo and Haltiwanger 2005). Thus, it is still unclear what leads to the severe developmental delay observed in LAD II.

LAD II is a very rare syndrome described so far only in six children. As the clinical phenotype is very striking, the diagnosis can be made based on the presence of recurrent, albeit mild infections, marked leukocytosis, and the Bombay blood group, in association with mental and growth retardation. An analysis of peripheral blood leukocytes by flow cytometry using a CD15s monoclonal antibody should be performed to determine SLeX expression. To confirm the diagnosis, sequence analysis of the gene encoding the GDP-fucose transporter should be performed.

Each of the patients described so far with LAD II suffered from several episodes of infections, which responded well to antibiotics. No serious consequences were observed, and prophylactic antibiotic is not needed. The patients' main chronic problem has been periodontitis, a condition that is especially difficult to treat in children with severe mental retardation (Etzioni et al. 1998). The oldest LAD II patient is now 19 years and has a severe psychomotor retardation with mild infectious problems.

Because of the proposed defect in fucose production, supplemental administration of fucose to the patients has been suggested. Indeed, fucose supplementation caused a dramatic improvement in the condition of the Turkish child (Marquardt et al. 1999). A marked decrease in leukocyte count with improved neutrophil adhesion was noted. Unfortunately, while using exactly the same protocol, no improvement in laboratory data or clinical features was seen in the two Arab children (Etzioni and Tonetti 2000). This difference may be due to the fact that the genetic defect in the Turkish child leads to a decreased affinity of the transporter for fucose, and thus an increase in the cytosolic concentration of fucose would be expected to overcome, at least in part, the defect in fucose transport.

4. Leukocyte Adhesion Deficiency III

Recently, a rare autosomal recessive LAD syndrome that is distinct from LAD I has been reported (Alon et al. 2003). Although leukocyte integrin expression and intrinsic adhesive activities to endothelial integrin ligands were normal, in situ activation of all major leukocyte integrins, including LFA-1, Mac -1 and VLA-4, by endothelial-displayed chemokines or chemoattractants was severely impaired in patient-derived lymphocytes and neutrophils. Although LAD leukocytes rolling on endothelial surfaces was normal, they failed to arrest on endothelial integrin ligands in response to endothelial-displayed chemokines. G-protein-coupled receptor signaling on these cells appeared to be normal, and the ability of leukocyte to migrate toward a chemotactic gradient was not impaired. The key defect in this syndrome was attributed to a genetic loss of integrin activation by rapid chemoattractant-stimulated G-protein-coupled receptor (GPCR) signals (Alon and Etzioni 2003). This novel LAD shows significant similarities to three previous cases commonly referred to as LAD I variants (Kuijpers et al. 1997; Harris et al. 2001; McDowall et al. 2003). All four cases had similar clinical symptoms, characterized by severe recurrent infections, bleeding tendency, and marked leukocytosis. In all cases, integrin expression and structure were intact with a defect in integrin activation by physiological inside-out stimuli. As these events cannot take place in LAD I leukocyte, it was proposed to designate this group of integrin activation disorders as LAD III (Alon and Etzioni 2003). The term LAD I variant, which has been ascribed to these syndromes, is inaccurate because these syndromes do not evolve from structural defects in leukocyte or platelet integrins.

In all of the LAD III cases, defects in receptor-mediated integrin activation are also accompanied by variable defect in non-receptor-mediated inside-out activation of leukocyte integrins. A growing body of evidence implicates the Ras-related GTPase, Rap-1, as a key regulator of integrin activation by these and other inside-out stimuli. Rap-1 was also implicated in the activation of platelet and megakaryocyte GpIIbIIIa, which is defective in LAD III. It is thus highly attractive that one or more of the new LAD III cases involve either direct or indirect defect in Rap-1 activation of leukocyte and platelet integrins. Although lymphocytes from two cases expressed normal level of Rap-1, in one case studied, an aberrant activation pattern was observed (Kinashi et al. 2004). However, the ubiquitous expression of Rap-1 in most tissues and its highly diverse functions in non-hematopoietic cells (Bos et al. 2001) make it unlikely that Rap-1 is structurally mutated in any of the new LAD III cases. It is possible that specific hematopoietic effector of Rap-1 activity is defective in LAD III cases. Indeed, a mouse model in which this activator was knockout resembled the bleeding tendency seen in LAD III (Crittenden et al. 2004).

These patients need prophylactic antibiotics as well as repeated blood transfusion. The only curative treatment is bone marrow transplantation, which should be performed as early as possible.

5. Conclusions

Leukocyte trafficking from bloodstream to tissue is important for the continuous surveillance for foreign antigens, as well as for rapid leukocyte accumulation at sites of inflammatory response or tissue injury. Leukocyte emigration to sites of inflammation is a dynamic process, involving multiple steps in an adhesion cascade. These steps must be precisely orchestrated to ensure a rapid response with only minimal damage to healthy tissue (Springer 1994; von Andrian and Mackay 2000). Leukocyte interaction with vascular endothelial cells is a pivotal event in the inflammatory response and is mediated by several families of adhesion molecules.

The crucial role of the β2-integrin subfamily in leukocyte emigration was established after LAD I was discovered. Patients with this disorder suffer from life-threatening bacterial infections (Anderson and Smith 2001). In its severe form, death usually occurs in early childhood unless bone marrow transplantation is performed.

The LAD II disorder clarifies the role of the selectin receptors and their fucosylated ligands such as SLeX. In vitro as well as in vivo studies establish that this family of adhesion molecules is essential for neutrophil rolling, the first step in neutrophil emigration through the blood vessel (Etzioni et al. 1992). Clinically, patients with LAD II suffer from a less severe form of disease, resembling the moderate phenotype of LAD I. This may be due in part to the ability of LAD II neutrophils to emigrate when blood flow rate is reduced and to the observed normal T and B lymphocyte function in LAD II as opposed to LAD I.

LAD III emphasizes the importance of the integrin activation phase in the adhesion cascade. Although the primary defect is still unknown, it is clear that all hematopoietic integrin molecules activation processes are defective leading to severe infection as observed in LAD I and to marked increase tendency for bleeding problems (Alon and Etzioni 2003).

References

Alon, R., Aker, M., Feigelson, S., Sokolovsky-Eisenberg, M., Staunton, D.E., Cinamon, G., Grabovsky, V., Shamri, R. and Etzioni, A. (2003) A novel genetic leukocyte adhesion deficiency in subsecond triggering of integrin avidity by endothelial chemokines results in impaired leukocyte arrest on vascular endothelium under shear flow. Blood 101, 4437–4445.

Alon, R. and Etzioni, A. (2003) LAD-III, a novel group of leukocyte integrin activation deficiencies. Trends Immunol. 24, 561–566.

Anderson, D.C. and Smith, C.W. (2001). *Leukocyte adhesion deficiency and other disorders of leukocyte adherence and motility*. New York, McGraw-Hill.

Bauer, T.R., Jr., Hai, M., Tuschong, L.M., Burkholder, T.H., Gu, Y.C., Sokolic, R.A., Ferguson, C., Dunbar, C.E. and Hickstein, D.D. (2006) Correction of the disease phenotype in canine leukocyte adhesion deficiency using ex vivo hematopoietic stem cell gene therapy. Blood 108, 3313–3320.

Bauer, T.R., Jr. and Hickstein, D.D. (2000) Gene therapy for leukocyte adhesion deficiency. Curr. Opin. Mol. Ther. 2, 383–388.

Bos, J.L., de Rooij, J. and Reedquist, K.A. (2001) Rap1 signalling: adhering to new models. Nat. Rev. Mol. Cell. Biol. 2, 369–377.

Crittenden, J.R., Bergmeier, W., Zhang, Y., Piffath, C.L., Liang, Y., Wagner, D.D., Housman, D.E. and Graybiel, A.M. (2004) CalDAG-GEFI integrates signaling for platelet aggregation and thrombus formation. Nat. Med. 10, 982–986.

Etzioni, A. (1996) Adhesion molecules- role in health and disease. Pediat. Res. 39, 191–198.

Etzioni, A., Frydman, M., Pollack, S., Avidor, I., Phillips, M.L., Paulson, J.C. and Gershoni-Baruch, R. (1992) Brief report: recurrent severe infections caused by a novel leukocyte adhesion deficiency. N. Engl. J. Med. 327, 1789–1792.

Etzioni, A., Gershoni-Baruch, R., Pollack, S. and Shehadeh, N. (1998) Leukocyte adhesion deficiency type II: long-term follow-up. J. Allergy Clin. Immunol. 102, 323–324.

Etzioni, A. and Tonetti, M. (2000) Fucose supplementation in leukocyte adhesion deficiency type II. Blood 95, 3641–3643.

Fischer, A., Lisowska-Grospierre, B., Anderson, D. and Springer, T. (1988) Leukocyte adhesion deficiency: molecular basis and functional consequences. Immunol. Rev. 1, 39–54.

Harris, E.S., Shigeoka, A.O., Li, W., Adams, R.H., Prescott, S.M., McIntyre, T.M., Zimmerman, G.A. and Lorant, D.E. (2001) A novel syndrome of variant leukocyte adhesion deficiency involving defects in adhesion mediated by β1 and β2 integrins. Blood 97, 767–776.

Helmus, Y., Denecke, J., Yakubenia, S., Robinson, P., Luhn, K., Watson, D.L., McGrogan, P.J., Vestweber, D., Marquardt, T. and Wild, M.K. (2006) Leukocyte adhesion deficiency II patients with a dual defect of the GDP-fucose transporter. Blood 107, 3959–3966.

Hogg, N., Stewart, M.P., Scarth, S.L., Newton, R., Shaw, J.M., Law, S.K. and Klein, N. (1999) A novel leukocyte adhesion deficiency caused by expressed but nonfunctional beta2 integrins Mac-1 and LFA-1. J. Clin. Invest. 103, 97–106.

Karsan, A., Cornejo, C.J., Winn, R.K., Schwartz, B.R., Way, W., Lannir, N., Gershoni-Baruch, R., Etzioni, A., Ochs, H.D. and Harlan, J.M. (1998) Leukocyte adhesion deficiency type II is a generalized defect of de novo GDP-fucose biosynthesis. Endothelial cell fucosylation is not required for neutrophil rolling on human nonlymphoid endothelium. J. Clin. Invest. 101, 2438–2445.

Kinashi, T. (2005) Intracellular signalling controlling integrin activation in lymphocytes. Nat. Rev. Immunol. 5, 546–559.

Kinashi, T., Aker, M., Sokolovsky-Eisenberg, M., Grabovsky, V., Tanaka, C., Shamri, R., Feigelson, S., Etzioni, A. and Alon, R. (2004) LAD-III, a leukocyte adhesion deficiency syndrome associated with defective Rap1 activation and impaired stabilization of integrin bonds. Blood 103, 1033–1036.

Kuijpers, T.W., Van Lier, R.A., Hamann, D., de Boer, M., Thung, L.Y., Weening, R.S., Verhoeven, A.J. and Roos, D. (1997) Leukocyte adhesion deficiency type 1 (LAD-1)/variant. A novel immunodeficiency syndrome characterized by dysfunctional beta2 integrins. J. Clin. Invest. 100, 1725–1733.

Lorusso, F., Kong, D., Jalil, A.K., Sylvestre, C., Tan, S.L. and Ao, A. (2006) Preimplantation genetic diagnosis of leukocyte adhesion deficiency type I. Fertil. Steril. 85, 494 e415–e498.

Lubke, T., Marquardt, T., Etzioni, A., Hartmann, E., von Figura, K. and Korner, C. (2001) Complementation cloning identifies CDG-IIc, a new type of congenital disorders of glycosylation, as a GDP-fucose transporter deficiency. Nat. Genet. 28, 73–76.

Lubke, T., Marquardt, T., von Figura, K. and Korner, C. (1999) A new type of carbohydrate-deficient glycoprotein syndrome due to a decreased import of GDP-fucose into the golgi. J. Biol. Chem. 274, 25986–25989.

Luo, Y. and Haltiwanger, R.S. (2005) O-fucosylation of notch occurs in the endoplasmic reticulum. J. Biol. Chem. 280, 11289–11294.

Luster, A.D., Alon, R. and von Andrian, U.H. (2005) Immune cell migration in inflammation: present and future therapeutic targets. Nat. Immunol. 6, 1182–1190.

Marquardt, T., Luhn, K., Srikrishna, G., Freeze, H., Harms, E. and Vestweber, D. (1999) Correction of LAD type II with oral fucose. Blood 94, 3976–3985.

McDowall, A., Inwald, D., Leitinger, B., Jones, A., Liesner, R., Klein, N. and Hogg, N. (2003) A novel form of integrin dysfunction involving beta1, beta2, and beta3 integrins. J Clin. Invest. 111, 51–60.
Roos, D., Meischl, C., de Boer, M., Simsek, S., Weening, R.S., Sanal, O., Tezcan, I., Gungor, T. and Law, S.K. (2002) Genetic analysis of patients with leukocyte adhesion deficiency: genomic sequencing reveals otherwise undetectable mutations. Exp. Hemat. 30, 252–261.
Shamri, R., Grabovsky, V., Gauguet, J.M., Feigelson, S., Manevich, E., Kolanus, W., Robinson, M.K., Staunton, D.E., von Andrian, U.H. and Alon, R. (2005) Lymphocyte arrest requires instantaneous induction of an extended LFA-1 conformation mediated by endothelium-bound chemokines. Nat. Immunol. 6, 497–506.
Springer, T.A. (1994) Traffic signals for lymphocyte recirculation and leukocyte emigration: the multistep paradigm. Cell 76, 301–314.
Sturla, L., Rampal, R., Haltiwanger, R.S., Fruscione, F., Etzioni, A. and Tonetti, M. (2003) Differential terminal fucosylation of N-linked glycans versus protein O-fucosylation in leukocyte adhesion deficiency type II (CDG IIc). J. Biol. Chem. 278, 26727–26733.
Thomas, C., Le Deist, F., Cavazzana-Calvo, M., Benkerrou, M., Haddad, E., Blanche, S., Hartmann, W., Friedrich, W. and Fischer, A. (1995) Results of allogeneic bone marrow transplantation in patients with leukocyte adhesion deficiency. Blood 86, 1629–1635.
Tone, Y., Wada, T., Shibata, F., Toma, T., Hashida, Y., Kasahara, Y., Koizumi, S. and Yachie, A. (2007) Somatic revertant mosaicism in a patient with leukocyte adhesion deficiency type 1. Blood 109, 1182–1184.
von Andrian, U.H., Berger, E.M., Ramezani, L., Chambers, J.D., Ochs, H.D., Harlan, J.M., Paulson, J.C., Etzioni, A. and Arfors, K.E. (1993) In vivo behavior of neutrophils from two patients with distinct inherited leukocyte adhesion deficiency syndromes. J. Clin. Invest. 91, 2893–2897.
von Andrian, U.H. and Mackay, C.R. (2000) T cell function and migration. Two sides of the same coin. N. Engl. J. Med. 343, 1020–1034.
Wild, M.K., Luhn, K., Marquardt, T. and Vestweber, D. (2002) Leukocyte adhesion deficiency II: therapy and genetic defect. Cells Tissues Organs 172, 161–173.
Yakubenia, S. and Wild, M.K. (2006) Leukocyte adhesion deficiency II. Advances and open questions. FEBS J. 273, 4390–4398.

6

Nijmegen Breakage Syndrome

Irina Kondratenko, Olga Paschenko, Alexandr Polyakov, and Andrey Bologov

Russian Clinical Children's Hospital and Institute for Clinical Genetics, Moscow, Russia, ikondratenko@rambler.ru

Abstract. Nijmegen breakage syndrome (NBS) is a rare autosomal recessive disease, characterized by microcephaly, growth retardation, immunodeficiency, chromosome instability, radiation sensitivity, and a strong predisposition to lymphoid malignancy. The gene responsible for the development of this syndrome (*NBS1*) was mapped on chromosome 8q21. The product of this gene—nibrin—is a protein with 95 kDa molecular weight (p95). The same mutation in the *NBS1* gene (deletion 657del5) was detected in most of the evaluated patients. In this chapter, we describe the analysis of the literature and our results on clinical and immunological features and genetic evaluation of 21 NBS patients.

1. Introduction

Nijmegen breakage syndrome (NBS, OMIM #251260) is a rare hereditary disease, characterized by immune deficiency, microcephaly, and an extremely high incidence of lymphoid tissue malignancies. Other clinical signs are low birth weight, growth retardation, and face skeletal abnormalities (Weemaes et al. 1993; The International Nijmegen Breakage Syndrome Study Group 2000). NBS, as a separate disease form, was first described by Weemaes in 1981 (Weemaes et al. 1981).

The gene mutated in NBS, *NBS1*, was mapped to 8q21 chromosome in 1997 (Matsuura et al 1997). The product of this gene is the protein with molecular weight of 95 kDa termed nibrin. It was found that nibrin in a complex with hRAD50 and hMre11 is responsible for the double-strained DNA break reparation (Carney et al. 1998; Featherstone and Jackson 1998). ATM protein (the product of the *ATM* gene, mutated in Ataxia-Teleangiectasia) is also involved in the regulation of this complex (Savitsky et al. 1995; Gatei et al. 2000). Mutations in NBS1 and consequent nibrin abnormalities lead to impaired control of DNA reparation during the cell cycle. This, in turn, causes one of the main features of NBS—chromosomal instability and increased sensitivity to radiation.

The disease is severe in many cases with a fatal prognosis. In most cases, the reasons of patients' death relate to malignancies or infectious complications of immunodeficiency (Chrzanowska et al. 1995).

To date, more then 100 patients are included in the NBS register; many of them are of Western Slavonic origin (Czech, Polish, and Ukrainian), and most, if not all, Slavonic patients have a similar mutation—657del5 deletion leading to a synthesis of a nonfunctional truncated protein (The International Nijmegen Breakage Syndrome Study Group 2000; Resnick et al. 2002, 2003).

This chapter includes the review of current literature and describes 21 cases of NBS observed in our clinical center.

2. Genetic Basis for NBS

NBS is an autosomal recessive condition. *NBS1* is the only gene known to be associated with NBS and encodes a protein nibrin (Cerosaletti et al. 1998; Featherstone and Jackson 1998). Cells from NBS patients are sensitive to radiation and radiomimetic compounds (Taalman et al. 1983). For instance, peripheral blood lymphocytes display spontaneous and radiation-induced chromosomal instability that frequently involves the immunoglobulin and T cell receptor loci on chromosomes 7 and 14 (Weemaes et al. 1981; Taalman et al. 1989).

DNA damage triggers specific cellular responses that ensure the maintenance of genomic integrity. The induction of DNA strand breakage by ionizing radiation (IR) results in activation of signaling pathways that lead either to elimination of damaged cells by programmed cell death or arrest of cell-cycle progression and repair of the DNA breaks (Elledge 1996; Zhou and Elledge 2000). Among the various proteins that contribute to DNA damage response, (ATM (serine–threonine kinase) protein plays a prominent role. Cells from individuals with ataxia telangiectasia exhibit defects in cell-cycle checkpoints operative in G_1, S, and G_2 phases, as well as radiation hypersensitivity and an increased frequency of chromosome breakage (Kastan and Lim 2000; Kastan et al. 2000). Activated ATM, in turn, triggers the activation of cell-cycle checkpoints and DNA repair through the phosphorylation of various proteins, including nibrin (Petrini 1999; Lim et al. 2000; Wu et al. 2000; D'Amours and Jackson 2002). Mutations in the *NBS1* gene encoding nibrin are responsible for molecular and clinical features of NBS (Varon et al. 1998). IR induces the formation of nuclear foci that contain a complex of NBS1 with MRE11 and RAD50 proteins, and these foci may represent sites of ongoing repair of DNA double-strand breaks (Mirzoeva and Petrini 2001; Howlett et al. 2006).

Genetic studies have provided evidence for a common haplotype of markers present in the families of Eastern European origins, suggesting a common original effect with regard to mutations causing the disorder. After identification of the gene, mutation detection has revealed a truncating 5 bp deletion, 657-661delACAAA, which was proven to be responsible for the disease. However, a few additional truncating mutations were identified in patients with other distinct haplotypes (2000).

3. Clinical Presentation and Therapy

The first NBS patient was described with microcephaly, early growth retardation, specific craniopharyngeal abnormalities, and aberrations of 7 and 14 chromosomes (Weemaes et al. 1981). In 2000, The International Nijmegen Breakage Syndrome Study Group published the analysis of clinical and laboratory data from 55 NBS patients, and at present, the NBS registry includes more then 130 patients.

NBS is characterized by short stature, progressive microcephaly with loss of cognitive skills, premature ovarian failure in females, recurrent sinopulmonary infections, and an increased risk for cancer, particularly lymphoma (van der Burgt et al. 1996; Resnick et al. 2002). There is no sex difference in expansion of NBS. Mental development is normal in 40% of patients, while 50% of patients have the borderline-to-mild retardation and 10% of patients are moderately retarded. No correlation was found between head circumference at birth and mental development (Green et al. 1995). All NBS patients also have a typical distinctive facial appearance, characterized by a receding forehead, prominent mid face with long nose and long philtrum, receding mandible, upward slanting palpebral fissures usually accompanied by epicanthic folds, freckles on the cheeks and nose, large ears with dysplastic helices, and sparse hair (Digweed and Sperling 2004).

Cutaneous manifestations include "Café au lait" spots, vitiligo, sun sensitivity of the eyelids, and pigment deposits in the fundus of the eye. The most common malformations are clinodactyly and/or syndactyly, less common are anal atresia/stenosis, ovarian dysgenesis, hydronephrosis, and hip dysplasia. Other malformations reported are hypoplastic trachea, cavernous angioma, agenesis of phalanges, hypospadias, renal hypoplasia, single kidney, and corpus collosum hypoplasia (Chrzanowska et al. 2002). Infections are common, most frequently of the respiratory tract followed by urinary tract infections. Gastrointestinal infections are reported relatively infrequently. The infections are typically community acquired rather than opportunistic (Chrzanowska 1996). Among the rare clinical features, aplastic anemia, rheumatoid arthritis, and vasculitis were also described (Rosenzweig et al. 2001; New et al. 2005).

Malignancies are developed in approximately 35–40% of NBS patients (Weemaes et al. 1994). Most common are lymphomas and leukemia. Several children have developed solid tumors, such as medulloblastomas, glioma, and rhabdomyosarcoma (Pasic 2002; Niehues et al. 2003; Meyer et al. 2004; Steffen et al. 2006).

Immunodeficiency in NBS patients involves both the humoral and cellular system. The most common defects of humoral immunity are agammaglobulinemia and IgA deficiency, which have been found in 35 and 20% of affected individuals, respectively. Deficiencies in IgG2 and IgG4 are also frequent even when the IgG concentration in serum is normal. The most commonly reported defects in cellular immunity are reduced percentages of total CD3+ and CD4+ T cells. An increased frequency of T cells with a memory phenotype (CD45RO+), a concomitant decrease in naive T cells (CD45RA), and low mitogen responsiveness are also the typical features of NBS patients (Weemaes et al. 1984; Michalkiewicz et al. 2003).

The treatment of NBS includes intravenous immunoglobulin (IVIG) replacement for antibody deficiency, antibiotics, and antifungal and antiviral agents in cases of infections. Although malignancies are treated with common protocols, doses of methotrexate are reduced (20–50%), while alkylating drugs and epipodophyllotoxins were omitted or also reduced by 20–50% according to some authors (Seidemann et al. 2000).

4. Data from Russian Clinical Children's Hospital

We have tested and treated 21 unrelated patients from 21 families (7 males and 14 females). Diagnosis of NBS was suspected in all cases because of coincidence of primary immune deficiency, microcephaly, and chromosomal instability. Nineteen patients were homozygous for 657del5 mutation and one case was heterozygote for 657del5 and 681delT mutations. No specimens were from this patient, but her sister who also had microcephalia, died from pneumonia in early age.

No pregnancy complications were observed. There were no consanguinity relationships in the families also. All patients are of Slavic origin. Family history was positive in three cases. All NBS patients had microcephalia and nine of them had mild or moderate mental retardation. Dysmorphic symptoms, such as receding forehead, prominent midface, receding mandible, and upward slant of palpebral fissures were found in all 21 children in different combinations. Nobody demonstrated ataxia symptoms, and only one patient had cerebral abnormalities: cerebellum agenesia and corpus collosum hypoplasia without ataxia symptoms. Renal dystopia was only seen in one patient, as well as anal atresia. Cafe-au-lait-spots were found in nine children, and one boy had a giant nevus. Growth retardation was identified in 19 NBS patients.

Fifteen and eight patients had infections of respiratory tract and gastrointestinal tract, respectively. Herpes simplex infections with frequent relapses were observed in seven patients while candidiasis superficialis in 12 patients (five of which were after chemotherapy). Lymphoid tissue hypoplasia was observed in one child and hyperplasia of lymph nodes with hepatosplenomegaly was detected in 13 patients; seven of them had lymphoid tissue malignancies. Different cytopenias were identified in 10, bone marrow aplasia in 2, vasculitis in 2, and vasculitis + arthritis in 1 patient. Different malignancies developed in eight cases: embrional rhabdomyosarcoma (1), acute myeloid leukemia (2), Hodgkin disease (1), and B cell leukemia or B lymphoma (4 cases).

Concentrations of serum IgG were <200 mg/dl in eight patients and between 300 and 700 mg/dl in 13 patients. IgM levels were <40 mg/dl in 5 patients, normal in 10, and raised up to 1600–3000 mg/dl in 6 patients. IgA levels were <5 mg/dl in 16 cases, normal in 4 cases, and raised up to 3000 md/dl in 1 patient. Abnormalities in T lymphocyte subset counts including decreased number of CD3+ or CD5+ lymphocytes or inversed CD4/CD8 ratio due to decreased number of CD4+ cells were found in 13 of 13 tested patients; five patients had low or undetectable B cell numbers.

All patients with NBS after verification of diagnosis received the replacement therapy and prophylaxis with trimethoprim/sulfamethoxazole (TMP/SMX): in severe cases of bacterial infections, the combination therapy with phtorchinolones or macrolides was applied. Acyclovir was used for herpes simplex infections and fluconasole or ketoconasole for candidiasis. Autoimmune manifestations were treated with prednisone with good responses. For the treatment of malignancies, we used standard protocols in three patients. In other cases, the doses of methotrexate and alkylating drugs were reduced by 30%; radiotherapy was omitted for a kid with embrional rhabdomyosarcoma. One patient received rituximab (375 mg/m^2). Nine patients died due to different reasons: aplastic anemia, 2; amyloidosis, 1; generalized tuberculosis combined with severe purulent lung infection, 1; septic shock on +1 day after matched-related-donor (MRD) and bone marrow transplantation (BMT), 1; and relapse or resistant course of malignancy, 3.

References

Carney, J.P., Maser, R.S., Olivares, H., Davis, E.M., Le Beau, M., Yates, J.R., 3rd, Hays, L., Morgan, W.F. and Petrini, J.H. (1998) The hMre11/hRad50 protein complex and Nijmegen breakage syndrome: linkage of double-strand break repair to the cellular DNA damage response. Cell 93, 477–486.

Cerosaletti, K.M., Lange, E., Stringham, H.M., Weemaes, C.M., Smeets, D., Solder, B., Belohradsky, B.H., Taylor, A.M., Karnes, P., Elliott, A., Komatsu, K., Gatti, R.A., Boehnke, M. and Concannon, P. (1998) Fine localization of the Nijmegen breakage syndrome gene to 8q21: evidence for a common founder haplotype. Am. J. Hum. Genet. 63, 125–134.

Chrzanowska, K.H. (1996) Microcephaly with chromosomal instability and immunodeficiency–Nijmegen syndrome. Pediatr. Pol. 71, 223–234.

Chrzanowska, K.H., Bekiesinska-Figatowska, M. and Jozwiak, S. (2002) Corpus callosum hypoplasia and associated brain anomalies in Nijmegen breakage syndrome. J. Med. Genet. 39, E25.

Chrzanowska, K.H., Kleijer, W.J., Krajewska-Walasek, M., Bialecka, M., Gutkowska, A., Goryluk-Kozakiewicz, B., Michalkiewicz, J., Stachowski, J., Gregorek, H., Lyson-Wojciechowska, G., et al. (1995) Eleven Polish patients with microcephaly, immunodeficiency, and chromosomal instability: the Nijmegen breakage syndrome. Am. J. Med. Genet. 57, 462–471.

D'Amours, D. and Jackson, S.P. (2002) The Mre11 complex: at the crossroads of DNA repair and checkpoint signalling. Nat. Rev. Mol. Cell. Biol. 3, 317–327.

Digweed, M. and Sperling, K. (2004) Nijmegen breakage syndrome: clinical manifestation of defective response to DNA double-strand breaks. DNA Repair (Amst) 3, 1207–1217.

Elledge, S. (1996) Cell cycle checkpoints: preventing identity crisis. Science 274, 1664–1672.

Featherstone, C. and Jackson, S.P. (1998) DNA repair: the Nijmegen breakage syndrome protein. Curr. Biol. 8, R622–R625.

Gatei, M., Young, D., Cerosaletti, K.M., Desai-Mehta, A., Spring, K., Kozlov, S., Lavin, M.F., Gatti, R.A., Concannon, P. and Khanna, K. (2000) ATM-dependent phosphorylation of nibrin in response to radiation exposure. Nat. Genet. 25, 115–119.

Green, A.J., Yates, J.R., Taylor, A.M., Biggs, P., McGuire, G.M., McConville, C.M., Billing, C.J. and Barnes, N.D. (1995) Severe microcephaly with normal intellectual development: the Nijmegen breakage syndrome. Arch. Dis. Child. 73, 431–434.

Howlett, N.G., Scuric, Z., D'Andrea, A.D. and Schiestl, R.H. (2006) Impaired DNA double strand break repair in cells from Nijmegen breakage syndrome patients. DNA Repair (Amst) 5, 251–257.

Kastan, M.B. and Lim, D.S. (2000) The many substrates and functions of ATM. Nat. Rev. Mol. Cell. Biol. 1, 179–186.

Kastan, M.B., Lim, D.S., Kim, S.T., Xu, B. and Canman, C. (2000) Multiple signaling pathways involving ATM. Cold Spring Harb. Symp. Quant. Biol. 65, 521–526.

Lim, D.S., Kim, S.T., Xu, B., Maser, R.S., Lin, J., Petrini, J.H. and Kastan, M.B. (2000) ATM phosphorylates p95/nbs1 in an S-phase checkpoint pathway. Nature 404, 613–617.

Matsuura, S., Weemaes, C., Smeets, D., Takami, H., Kondo, N., Sakamoto, S., Yano, N., Nakamura, A., Tauchi, H., Endo, S., Oshimura, M. and Komatsu, K. (1997) Genetic mapping using microcell-mediated chromosome transfer suggests a locus for Nijmegen breakage syndrome at chromosome 8q21-24. Am. J. Hum. Genet. 60, 1487–1494.

Meyer, S., Kingston, H., Taylor, A.M., Byrd, P., Last, J., Brennan, B., Trueman, S., Kelsey, A., Taylor, G. and Eden, O. (2004) Rhabdomyosarcoma in Nijmegen breakage syndrome: strong association with perianal primary site. Cancer Genet. Cytogenet. 154, 169–174.

Michalkiewicz, J., Barth, C., Chrzanowska, K., Gregorek, H., Syczewska, M., Weemaes, C.M., Madalinski, K. and Stachowski, J. (2003) Abnormalities in T and NK lymphocyte phenotype in patients with Nijmegen breakage syndrome. Clin. Exp. Immunol. 134, 482–490.

Mirzoeva, O.K. and Petrini, J.H. (2001) DNA damage-dependent nuclear dynamics of the Mre11 complex. Mol. Cell. Biol. 21, 281–288.

New, H.V., Cale, C.M., Tischkowitz, M., Jones, A., Telfer, P., Veys, P., D'Andrea, A., Mathew, C.G. and Hann, I. (2005) Nijmegen breakage syndrome diagnosed as Fanconi anaemia. Pediatr. Blood Cancer 44, 494–499.

Niehues, T., Schellong, G., Dorffel, W., Bucsky, P., Mann, G., Korholz, D. and Gobel, U. (2003) Immunodeficiency and Hodgkin's disease: treatment and outcome in the DAL HD78-90 and GPOH HD95 studies. Klin. Padiatr. 215, 315–320.

Pasic, S. (2002) Aplastic anemia in Nijmegen breakage syndrome. J. Pediatr. 141, 742.

Petrini, J.H. (1999) The mammalian Mre11-Rad50-nbs1 protein complex: integration of functions in the cellular DNA-damage response. Am. J. Hum. Genet. 64, 1264–1269.

Resnick, I.B., Kondratenko, I., Pashanov, E., Maschan, A.A., Karachunsky, A., Togoev, O., Timakov, A., Polyakov, A., Tverskaya, S., Evgrafov, O. and Roumiantsev, A.G. (2003) 657del5 mutation in the gene for Nijmegen breakage syndrome (NBS1) in a cohort of Russian children with lymphoid tissue malignancies and controls. Am. J. Med. Genet. A. 120, 174–179.

Resnick, I.B., Kondratenko, I., Togoev, O., Vasserman, N., Shagina, I., Evgrafov, O., Tverskaya, S., Cerosaletti, K.M., Gatti, R.A. and Concannon, P. (2002) Nijmegen breakage syndrome: clinical characteristics and mutation analysis in eight unrelated Russian families. J. Pediatr. 140, 355–361.

Rosenzweig, S.D., Russo, R.A., Gallego, M. and Zelazko, M. (2001) Juvenile rheumatoid arthritis-like polyarthritis in Nijmegen breakage syndrome. J. Rheumatol 28, 2548–2550.

Savitsky, K., Sfez, S., Tagle, D.A., Ziv, Y., Sartiel, A., Collins, F.S., Shiloh, Y. and Rotman, G. (1995) The complete sequence of the coding region of the ATM gene reveals similarity to cell cycle regulators in different species. Hum. Mol. Genet. 4, 2025–2032.

Seidemann, K., Henze, G., Beck, J.D., Sauerbrey, A., Kuhl, J., Mann, G. and Reiter, A. (2000) Non-Hodgkin's lymphoma in pediatric patients with chromosomal breakage syndromes (AT and NBS): experience from the BFM trials. Ann. Oncol. 11 (Suppl 1), 141–145.

Steffen, J., Maneva, G., Poplawska, L., Varon, R., Mioduszewska, O. and Sperling, K. (2006) Increased risk of gastrointestinal lymphoma in carriers of the 657del5 NBS1 gene mutation. Int. J. Cancer 119, 2970–2973.

Taalman, R.D., Hustinx, T.W., Weemaes, C.M., Seemanova, E., Schmidt, A., Passarge, E. and Scheres, J.M. (1989) Further delineation of the Nijmegen breakage syndrome. Am. J. Med. Genet. 32, 425–431.

Taalman, R.D., Jaspers, N.G., Scheres, J.M., de Wit, J. and Hustinx, T.W. (1983) Hypersensitivity to ionizing radiation, in vitro, in a new chromosomal breakage disorder, the Nijmegen Breakage Syndrome. Mutat. Res. 112, 23–32.

The International Nijmegen Breakage Syndrome Study Group (2000) Nijmegen breakage syndrome. Arch. Dis. Child. 82, 400–406.

van der Burgt, I., Chrzanowska, K.H., Smeets, D. and Weemaes, C. (1996) Nijmegen breakage syndrome. J. Med. Genet. 33, 153–156.

Varon, R., Vissinga, C., Platzer, M., Cerosaletti, K.M., Chrzanowska, K.H., Saar, K., Beckmann, G., Seemanova, E., Cooper, P.R., Nowak, N.J., Stumm, M., Weemaes, C.M., Gatti, R.A., Wilson, R.K., Digweed, M., Rosenthal, A., Sperling, K., Concannon, P. and Reis, A. (1998) Nibrin, a novel DNA double-strand break repair protein, is mutated in Nijmegen breakage syndrome. Cell 93, 467–476.

Weemaes, C.M., Hustinx, T.W., Scheres, J.M., van Munster, P.J., Bakkeren, J.A. and Taalman, R.D. (1981) A new chromosomal instability disorder: the Nijmegen breakage syndrome. Acta Paediatr. Scand. 70, 557–564.

Weemaes, C.M., Smeets, D.F., Horstink, M., Haraldsson, A. and Bakkeren, J.A. (1993) Variants of Nijmegen breakage syndrome and ataxia telangiectasia. Immunodeficiency 4, 109–111.

Weemaes, C.M., Smeets, D.F. and van der Burgt, C.J. (1994) Nijmegen Breakage syndrome: a progress report. Int. J. Radiat. Biol. 66, S185–S188.

Weemaes, C.M., The, T.H., van Munster, P.J. and Bakkeren, J.A. (1984) Antibody responses in vivo in chromosome instability syndromes with immunodeficiency. Clin. Exp. Immunol. 57, 529–534.

Wu, X., Ranganathan, V., Weisman, D.S., Heine, W.F., Ciccone, D.N., O'Neill, T.B., Crick, K.E., Pierce, K.A., Lane, W.S., Rathbun, G., Livingston, D.M. and Weaver, D.T. (2000) ATM phosphorylation of Nijmegen breakage syndrome protein is required in a DNA damage response. Nature 405, 477–482.

Zhou, B.B. and Elledge, S.J. (2000) The DNA damage response: putting checkpoints in perspective. Nature 408, 433–439.

7

Neutrophil Activity in Chronic Granulomatous Disease

Vladimir Pak, Anna Budikhina, Mikhail Pashenkov, and Boris Pinegin

National Research Center Institute of Immunology, FMBA, Moscow, Russia
vgpak@yandex.ru

Abstract. The killing of microorganisms by neutrophils causes degranulation of azurophilic, specific, and gelatinase granules into the formed phagolysosomes. During the degranulation process, increased surface expression of CD63 (localized in the azurophilic granules of resting neutrophils) and CD66b/CD67 (from specific granules) can be detected. This results from the fusion of the granule membrane, containing these markers, with a plasma membrane. Release of granule content into the phagolysosomes or the extracellular environment occurs not only upon proper cell activation but also upon tissue injury. We compared expression of degranulation markers on neutrophils from chronic granulomatous disease (CGD) patients and healthy volunteers. Surface expression of CD63 in non-stimulated and phorbol 12-myristate 13-acetate (PMA)-stimulated neutrophils, bactericidal activity of serum, and alpha-defensins level (HNP 1–3) in plasma of CGD patients were significantly higher in comparison with healthy volunteers. At the same time, the levels of intracellular HNP 1–3 in CGD neutrophils were lower than in normal neutrophils. Thus, our data revealed augmented degranulation of azurophilic neutrophil granules in CGD, which might play a role in tissue destruction observed in this disease.

1. Introduction

Chronic granulomatous disease (CGD) is an uncommon inherited disorder in which phagocytic cells fail to produce antimicrobial oxidants. The disease is characterized by recurrent life-threatening infections with catalase-positive microorganisms and excessive inflammatory reactions with granuloma formation (Orkin 1989).

Although the respiratory burst and its deficiency in CGD have been extensively studied, the process of neutrophil degranulation received much less attention. Different stimuli can activate degranulation, and one of them is triggered by binding of the chemotactic peptide formyl-methionyl-leucyl-phenylalanine (FMLP) to its receptor on cell surface. Phorbol 12-myristate 13-acetate (PMA) activates protein kinase C (PKC) and subsequent phosphorylation of components of the oxidase complex. Treatment of phagocytes with cytochalasin B (CB) and subsequent stimulation with FMLP induce neutrophil degranulation. This process occurs by activation of phospholipase D (PLD) (Mullmann et al. 1993). Another degranulation-promoting agent is PMA, which induces "respiratory burst" and subsequent degranulation through the activation of PLD. Both PMA and FMLP stimulate "respiratory burst" and subsequent degranulation, but FMLP-induced degranulation does not depend on the PKC activation. Degranulation of phagocytes results in increased surface expression of degranulation markers of azurophilic (CD63) and specific (CD66b) granules. Apparently, the release of granule content into the environment may lead to injury of healthy tissue. In fact, biologic fluids from CGD patients possess the high bactericidal activity. Another component of azurophilic granules of neutrophils is α-defensins, and their intracellular content can be used as a marker of degranulation as well. Defensins have multiple host defense functions with the capacity to kill a wide variety of gram-positive and gram-negative bacteria and fungi (Murphy 1994; Gallin 1985; Mullmann et al. 1993; Selvatici et al.2006).

Here, in order to characterize neutrophil degranulation in CGD, we examined expression of degranulation markers CD63 and CD66b before and after stimulation with FMLP and PMA, as well as intracellular defensins in different blood leukocyte populations. We also examined the levels of defensins in plasma and in supernatants of non-stimulated and stimulated neutrophils and bactericidal activity of serum from CGD patients and healthy volunteers.

2. Experimental Procedures

Blood was obtained from eight CGD patients and healthy volunteers. Specimens were collected on 3.8% citrate, centrifuged at 300 *g* 20 min at room temperature, and sedimented in 3% gelatine for 30 min. Cells in the supernatants were treated by red blood cell lysing solution and washed in phosphate-buffered saline (PBS). For the isolation of neutrophils, cells were centrifuged on the Ficoll–Hypaque gradient, washed, and resuspended in RPMI-1640 with 1% fetal calf serum (FCS) (10^7/ml). This procedure yielded 92–98% of neutrophils.

Chemiluminescence assay was done as described by Porter (Porter et al. 1992). Briefly, 10^6 cells were washed in PBS and resuspended in 1 ml prewarmed PBS (pH 7.0) containing either 170 μM lucigenin or 13 μM luminol. Chemiluminescence was monitored using LKB-Wallac 1251 luminometer with sample maintained at 37°C and 1-min reading intervals. Cells were monitored for 5 min prior to the addition of 10 μl of 20 μg/ml PMA.

For the dichlorofluorescin (DCFH) oxidation assay, 5×10^6 neutrophils/ml in PBS were preincubated with 5 mM DCFH-DA for 15 min and then stimulated with PMA (100 ng/ml) for 45 min. The samples were analyzed by flow cytometer FACSCalibur and the CELLQuest software.

Expression of degranulation markers was assessed as described (Nieissen and Verhoeven 1992). Briefly, neutrophils (10^7 cells/ml) in RPMI-1640 with 1% FCS were first stimulated with PMA (100 ng/ml) for 5 min. Then samples were pretreated with CB for 3 min, subsequently stimulated with FMLP (1 μM) for 5 min, and fixed with 1% paraformaldehyde. Next, cells were washed and stained with mouse anti-human CD63 PE-labeled antibody (Caltag) and mouse anti-human CD66b FITC-labeled antibody (BD PharMingen). Expression of the markers was assessed as the mean fluorescence intensities (MFIs) in 5000 events.

Intracellular defensins were also detected by FACScan. Leukocytes were fixed/permeabilized with 2% paraformaldehyde/0.1% saponin solution in PBS for 20 min at 4 degree C, washed, and stained with monoclonal mouse anti-human HNP antibodies (10 μg/ml, 20 min, 4 degree C). The antibodies were kindly provided by Dr. Voitenok (Institute of Hematology and Transfusiology, Minsk, Belorussia). Secondary antibodies were goat anti-mouse PE-labeled IgG1. Cells were analyzed using FACSCalibur.

Levels of defensins in plasma were determined by ELISA (HyCult Biotech) according to the manufacturer's protocol.

Killing of bacteria was evaluated by FITC-conjugated *Staphylococcus aureus* (*S. aureus*). Serum (90 μl) and *S. aureus* (10^6/ml, 90 μl) were incubated for 3 h at 37 degree C. The bacteria were then sedimented at 1000 *g* for 10 min and resuspended in 200 μl of PBS with 2.5 μg/ml propidium iodide (PI) (Sigma). After 10 min, the samples were analyzed by flow cytometry. The percentage of double positive bacteria ($FITC^+PI^+$) among all FITC-labeled bacteria ($FITC^+$) was determined.

3. Results

3.1. Analysis of Oxidative Burst in GCD

The oxidative metabolic burst of stimulated human polymorphonuclear leukocytes (PMNs) has been evaluated by chemiluminescence analysis of oxygen radicals (O_2–, H_2O_2, OH–). PMNs from patients with CGD are known to lack functional NADPH oxidase and oxygen radical generation. However, using flow cytometry and DCFH oxidation by H_2O_2, non-significant DCFH oxidation by PMA-stimulated PMNs from CGD was observed. The chemiluminescence test confirmed lost production of oxidative radicals in cells with or without stimulation with PMA (Table 1). Together with the clinical presentation, these data confirmed the diagnosis of CGD.

TABLE 1. Leukocytes from chronic granulomatous disease patients and healthy volunteers

	Chemiluminescence mV/min			
	Luminol		Lucigenin	
	Control	PMA	control	PMA
CGD patients (M ± SD)	0.9	5.0	0.9	3.0
Volunteers (M ± SD)	12 ± 3	200 ± 46	15 ± 3	252 ± 26
p-value	<0.001	<0.001	<0.001	<0.001

TABLE 2. The up-regulation of CD63 and CD66b on stimulated neutrophils

	CD63			CD66b		
	Control	FMLP	PMA	Control	FMLP	PMA
CGD patients (M ± SD)	97 ± 50	508 ± 348	199 ± 78	107 ± 63	311 ± 135	384 ± 247
Volunteers (M ± SD)	41 ± 29	312 ± 215	100 ± 46	92 ± 43	276 ± 153	247 ± 139
p-value	0.0004	0.3	0.002	0.9	0.9	0.6

PMA, phorbal 12-myristate 13-acetate.

3.2. Expression of Degranulation Markers in CGD

We next evaluated expression of granular markers with CD63 reflecting azurophilic granules and CD66b/CD67 reflecting specific granule membrane. The expression of these markers was measured after stimulation of neutrophils from patients with CGD and healthy individuals. The results revealed significant differences between the surface expression of CD63 in non-stimulated PMN cells from CGD patients and control donors as MFI equal to 97 ± 50 versus 41 ± 29, as well as for PMA-stimulated cells as MFI equal to 199 ± 78 versus.100 ± 46 (Table 2).

3.3. Intracellular and plasma α-defensins in CGD

We revealed that the intracellular level of α-defensins in leukocytes from CGD patients was lower than that in leukocytes from healthy volunteers. However, the level of α-defensins in plasma samples obtained from CGD patients was higher than that from healthy donors (Table 3). Interestingly, both PMA and FMLP induced the release of α-defensins from neutrophils isolated from CGD patients, while cells from healthy volunteers produce the similar levels of α-defensins independently of the stimulation with PMA or FMLP. We also examined the killing of FITC-labeled *S. aureus* by serum received from CGD patients and healthy volunteers. The results revealed that serum from CGD patients possesses stronger bactericidal activity than serum from healthy volunteers. At the same time, defensin levels in plasma from CGD patients were higher than in healthy volunteers (Tables 4 and 5).

TABLE 3. Intracellular defensins in human neutrophils and monocytes

	MFI	
	Neutrophils	Monocytes
CGD patients	458 ± 251	150 ± 45
Volunteers	880 ± 371	191 ± 134
p-value	<0.001	<0.001

MFI, mean fluorescence intensity.

TABLE 4. HNP1-3 levels in plasma and in cell-free medium harvested from non-stimulated and stimulated neutrophils from CGD patients and healthy volunteers

	HNP1-3 contents in plasma, intact and stimulated neutrophils supernatants (ng/ml)			
	Plasma	Supernatants		
		No stimulation	PMA	FMLP
CGD	481 ± 38	16 ± 16	41 ± 16	101 ± 34
Volunteers	7 ± 3	70 ± 17	81 ± 15	98 ± 64
p-value	<0.001	<0.05	<0.01	>0.05

CGD, chronic granulomatous disease; PMA, phorbol 12-myristate 13-acetate.

TABLE 5. Killing FITC-labeled *staphylococus aureus* by serum received from CGD patients and healthy volunteers

	St.aureus ($FITC^+PI^+$), (%)
CGD patients (Mean ± SD) ($n = 8$)	31 ± 6
Volunteers (Mean ± SD) ($n = 30$)	12 ± 5
p-value	<0.001

CGD, chronic granulomatous disease.

4. Discussion

Human neutrophils circulate in the bloodstream ready to be recruited to the site of infection or inflammation where they are expected to protect the host against microorganisms. The neutrophil host defense mechanism includes phagocytosis and killing of microorganisms by the generation of cytotoxic mediators. Human neutrophils could also lyse mammalian targets via oxygen-independent mechanisms (Lehrer et al. 1993). The authors revealed that purified human defensins killed various human and murine target cells in a concentration and time-dependent manner. Defensins are more abundant than myeloperoxidase in the primary granules of human neutrophils (Rice et al. 1987). Considering that human neutrophils contain large amounts of potently cytocidal defensins that can be released externally by various stimuli (Ganz 1987), the participation of defensins in extracellular cell lysis or injury mediated by human neutrophils also seems likely. In fact, cytolysis occurred when extracts from defensin-containing granules or purified defensins (but not elastase and cathepsin-G) were combined with sublytic concentrations of hydrogen peroxide (Lichtenstein 1991).

We hypothesized that defensins may have a leading role in tissue destruction in CGD patients. Our experiments show that CGD patients' plasma/serum contains high levels of α-defensins and displays high level of bactericidal activity in contrast to the values obtained from healthy volunteers. Initial adverse effect of defensins on mammalian cells occurs on cell membrane. It is possible that the second phase of injury is mediated intracellularly by defensins that has been internalized through this leaky membrane (Lichtenstein 1991).

Our data suggest that excess of degranulation process in CGD causes accumulation of high level of defensins in plasma and diminished contents of intracellular defensins in CGD patients. This process may provoke not only host defense against microorganisms but may also result in tissue injury.

References

Gallin, J.I. (1985) Neutrophil specific granule deficiency. Ann. Rev. Med. 36, 263–274.

Ganz, T. (1987) Extracellular release of antimicrobial defensins by human polymorphonuclear leukocytes. Infect. Immunol. 55, 568–571.

Lehrer, R., Lichtenstein, A. and Ganz, T. (1993) Defensins: antimicrobial and cytotoxic peptides of mammalian cells. Ann. Rev. Immunol. 11, 105–128.

Lichtenstein, A. (1991) Mechanism of mammalian cell lysis mediated by peptide defensins. Evidence for an initial alteration of the plasma membrane. J. Clin. Invest. 88, 93–100.

Mullmann, T.J., Cheewatrakoolpong, B., Anthes, J.C., Siegel, M.I., Egan, R.W. and Billah, M.M. (1993) Phospholipase C and phospholipase D are activated independently of each other in chemotactic peptide-stimulated human neutrophils. J. Leukoc. Biol. 53, 630–635.

Murphy P.M. (1994) The molecular biology of leukocyte chemoattractant receptors. Annu. Rev. Immunol. 12, 593–633.

Nieissen, W.M. and Verhoeven A.J. (1992) Differential up-regulation of specific and azurophilic granule membrane markers in electropermeabilized neutrophils. Cell. Signal. 4, 501–509.

Orkin, S.H. (1989) Molecular genetics of chronic granulomatous disease. Ann. Rev. Immunol. 7, 207–307.

Porter, C.D., Parkar, M.G., Collins, M.K.,Levinsky R.J. and Kinnon C. (1992) Superoxide production by normal and chronic granulomatous disease (CGD) patient-derived EBV-transformed B cell lines measured by chemiluminescence-based assays. J. Immunol. Methods 155, 151–157.

Rice, W.G., Ganz, T., Kinkade, J.M., Selsted, M.E., Lehrer R.I. and Parmley R.T. (1987) Defensin-rich dense granules of human neutrophils. Blood 70, 757–765.

Selvatici, R., Mollica, A., Falzarano S., Spisani S. (2006) Signal transduction pathways triggered by selective formylpeptideanalogues in human neutrophils. Eur. J. Pharm. 534, 1–11.

8

Mycobacterial Infections in Primary Immunodeficiency Patients

Elizaveta Galkina, Irina Kondratenko, and Andrey Bologov

Russian Clinical Children's Hospital, Moscow, Russia, elizabethgalklina@yahoo.com

Abstract. Primary immunodeficiencies (PID) are a diverse group of hereditary diseases leading to the impaired immune response that creates high susceptibility to mycobacterium infection. High susceptibility to mycobacterial infections of patients suffering from defects of phagocytosis and combined immunodeficiencies can be explained by predominant participation of macrophages and T lymphocytes in the specific immune response. *Mycobacterium tuberculosis, Bacille Calmette-Guerin*, and non-tuberculosis mycobacterium (NTM) may cause a severe disease in patients with PIDs. We report here our results of the clinical features of mycobacterium infection presentations in 36 patients with various PIDs.

1. Introduction

Primary immunodeficiencies (PIDs) represent a diverse group of hereditary diseases leading to the impaired cell and/or humoral immune response, phagocyte deficiency that causes high susceptibility to infections in particular to mycobacterium. Worldwide, its atypical forms and therapy resistance are reported. While in Russia the growth of tuberculosis is significant because of social instability and population migrations, it is evident that tuberculosis growth influences patients with impaired immunity. For the past 30 years, more than 160 varieties of non-tuberculosis mycobacterium (NTM) were detected, which are poorly virulent and present low risk for immunocompetent people (Rastogi et al. 2001).

Regardless of a high susceptibility, humans are remarkably resistant to disease development. In most scenarios, only 10–20% of exposed individuals develop mycobacterium infection. However, the disease is especially prevalent in crowded urban areas and among patients with AIDS, PIDs, interferon-γ deficiency, and in certain urban groups (American Indians and Eskimos) who have a high genetic susceptibility to mycobacterium infection. Of those infected, only 5–15% developed primary tuberculosis, including children and immunocompromised or malnourished adults. However, the majority of infected individuals develop primary tuberculosis complex

(Ghon complex), which can later be reactivated upon suppression of the immune system (e.g., cancer, and AIDS) (Bloom and Small 1998).

Because effective killing of intracellular bacteria requires interaction of infected macrophages with antigen-specific T cells and natural killer (NK) cells, genetic defects affecting T cells, IFN-γ production and phagocytosis might lead to severe mycobacterium infections (Stenger et al. 1997). Antibodies are produced in response to mycobacterium infection, but they appear to play no beneficial role in host defense (Rolph et al. 2001).

While BCG has been widely administered to newborns successfully preventing complications such as meningitis and miliary tuberculosis, administration of the vaccine has not resulted in a complete control of the disease (Plotkin 1994). BCG vaccination plays a particular role in mycobacterium infection development in immunocompromised patients causing local infection process dissemination of vaccine *Mycobacterium bovis* strain.

2. Mycobacterium Infections and Immunodeficiencies

Mycobacterium tuberculosis, less virulent *Bacille Calmette-Guerin,* and NTM cause severe diseases in patients with primary and acquired immunodeficiencies. Disseminated infection resulted from BCG vaccination, or contact with NTM is usually associated with immunologic defects in a patient (Reichenbach et al. 2001). The inherited impairment of anti-tuberculosis immunity results from genetic defects that lead to IFN-γ, IL-12, or their receptor deficiency playing crucial roles in the immune response to mycobacterium infection. Patients with mycobacterium disease associated with immunodeficiency are susceptible to a wide variety of viruses, bacteria, fungi, and protozoa. In contrast, patients with the so-called idiopathic BCG and NTM infections do not generally have associated infections, apart from salmonellosis, which affects less than half of the cases. Parental consanguinity and familial forms are frequently observed, thus, this syndrome was designated as *Mendelian susceptibility to mycobacterium infection* (Casanova and Abel 2002).

Numerous cases of severe combined immunodeficiency (SCID) patients who developed clinical BCG diseases following vaccination have been described (Minegishi et al. 1985; Gonzalez et al. 1989; Heyderman et al. 1991; Hugosson and Harfi 1991; Abramowsky et al. 1993; Romanus et al. 1993; Fisch et al. 1999). Disseminated disease is observed in most vaccinated patients, and ultimate death occurs unless functions of the immune system can be established by allogeneic BMT or gene therapy. Most children present disseminated disease, but localized BCG infection has also been described (Reichenbach et al. 2001). There are few reports of increased susceptibility to mycobacterial infections in patients with hyper-IgM syndrome (HIGM). Three of 56 HIGM patients in a European survey were exposed to tuberculosis (Levy et al. 1997). In a group of 30 patients with hyper-IgE syndrome (HIES), one adult female patient with cavitary lung disease developed pneumonia due to *Mycobacterium intracellulare* (Grimbacher et al. 1999). Disseminated BCG infection in infancy, with a clinical course similar to that in SCID patients, was described in a girl with HIES who reactivated BCG infection at 7 years (Pasic et al.

1998). A common presentation of mycobacterium infection in chronic granulomatous disease (CGD) patients is BCG lymphadenitis, but disseminated infection may also occur (Urban et al. 1980; Kobayashi et al. 1984; Gonzalez et al. 1989). In contrast to patients with SCID, BCG-infected CGD patients usually succeed in clearing infection following antibiotic therapy. Pulmonary *Mycobacterium avium* infection has been reported in an infant with CGD (Ohga et al. 1997). Three adults have been described as suffering from disseminated *Mycobacterium flavescens* infection, *Mycobacterium fortuitum* pneumonitis, and osteomyelitis, respectively. Six out of seven CGD patients studied in Hong Kong, where tuberculosis is endemic, presented recurrent severe infection with *M. tuberculosis* (Lau et al. 1998). Overall, CGD patients appear to be highly vulnerable to mycobacterium including *M. tuberculosis* in endemic areas (Reichenbach et al. 2001). Among patients with ectodermal dysplasia with immunodeficiency (EDA-ID), one of the tested patients died of miliary tuberculosis at 20 months of age (Doffinger et al. 2001). Other EDA-ID patients with mycobacterium infections generally presented in first 3 years of life with disseminated *M. avium* infections. Disseminated skin infection with *Mycobacterium chelonae* has been observed in a 6-months-old patient, with EDA-ID (Brooks et al. 1994). Two patients had disseminated fatal NTM infection in the first 2 years of life; *Mycobacterium kansasii* was isolated from one of these patients, in the other patient, the NTM was not cultivated (Dupuis-Girod et al. 2002). Four out of 14 patients analyzed in another report developed NTM infection and all died despite aggressive anti-mycobacterium therapy, including IFN-γ therapy in one child (Reichenbach et al. 2001). Primary complement and antibody deficiencies do not appear to result in increased susceptibility to mycobacterium. Patients with human X-linked agammaglobulinemia (XLA), autosomal recessive agammaglobulinemia (ARA), common variable immunodeficiency (CVID), autosomal recessive hyper-IgM syndrome (ARHIGM), and complement factor deficiencies are highly susceptible to extracellular bacterial infections but not to mycobacterium disease (Reichenbach et al. 2001).

3. Results

We have observed 36 patients (among 347 patients with 12 forms of PID which were registered in Russian Children's Clinical Hospital's database from 1992 to 2004) who developed mycobacterium infection. The diagnosis of PID was confirmed according to World Health Organization, European Society of Immunodeficiencies, and Pan-American Immunodeficiency group diagnostic criteria. The diagnoses of IFN-γ receptor (IFNγR) deficiency and IL-12 receptor (IL-12R) deficiency were based on a remarkably severe course of disseminated BCG infection, PID laboratory features absence, and high plasma levels of IFN-γ in one patient and IL-12 in a second patient. Patient age varied from 6 months to 17 years, with the medium being 3.6 ± 0.7 years. Female to male ratio was as following: 6 girls and 30 boys. The boy prevalence is explained by X-linked inheritance of some PIDs: X-linked agammaglobulinemia, X-linked HIGM1, SCID, and CGD.

The diagnosis of mycobacterium infection in PID patients was based on the following diagnostic criteria:

- clinical and family history data (BCG vaccination presence);
- radiological and ultrasound data;
- cytological investigation of infected specimens;
- histological investigation of biopsy material;
- Mbt DNA detection by PCR with Mbt typing.

Incidence of mycobacterium infections in 36 PID patients reached 10.3%, which is statistically significant ($p < 0.05$) in comparison with mycobacteriosis incidence in immunological healthy Russian children (0.0186%). The immunodeficiency manifestation in all patients was presented by various infections. All 36 patients were BCG vaccinated. Twenty-seven kids developed post-vaccination complications, such as local or disseminated BCG-osis. Nine patients developed tuberculosis without preceded BCG-osis. The probability of mycobacetriosis morbidity resulted from BCG vaccination that comprised more than 95% ($p < 0.05$).

The patients with SCID and CGD appeared to be the most susceptible to mycobacterium infection. Among eight BCG-vaccinated SCID patients, seven developed vaccine-associated infection, while among 22 BCG-vaccinated CGD patients, post-vaccination complications took place in 18 children. In both groups, statistically significant difference ($p < 0.05$) of BCG-osis morbidity was observed in comparison with similar morbidity in other PID groups of patients [X-linked HIGM1, Nijmegen breakage syndrome (NBS), and common variable immunodeficiency]. Five of 18 CGD patients with BCG-osis added tuberculosis and developed mixed infection. The probability of mycobacteriosis morbidity, including BCG-osis, among CGD patients reached 81.8% ($p < 0.001$), which makes it evident that BCG vaccination does not protect from mycobacterial infections but presents a factor provoking mycobacteriosis development.

High susceptibility to mycobacterial infections was also observed in patients with combined immunodeficiencies HIGM1 and NBS. Among 13 HIGM1 patients, three kids developed mycobacteriosis, two kids had tuberculosis, and one had BCG-osis. Among 17 NBS patients, two kids developed severe tuberculosis. Two patients out of 17 with CVID had mycobacteriosis. Both patients with IFN-γ/IL-12 defects developed severe disseminated BCG-osis.

The manifestations of mycobacterium infections included local and disseminated BCG-osis, pulmonary tuberculosis, lymph nodes tuberculosis, disseminated tuberculosis, and extra pulmonary (atypical) infection. Fourteen patients had mixed presentations of mycobacteriosis. Among seven SCID patients, post-vaccine complications were presented by local BCG-osis in three kids, and disseminated BCG-osis-affected lungs, lymph nodes, and skin in four patients. Out of 18 CGD patients, mycobacterium infection manifested with local BCG-osis in 13 patients, and in eight of them, it was limited by left axillar calcinate formation without further dissemination of the infection. Three patients developed disseminated BCG-osis with further joint of tuberculosis and in one case complicated by wild *M. Bovis* strain-affected spleen.

In three HIGM1 patients, mycobacterium infection had a severe course. Mycobacteriosis manifested by local BCG-osis with further fatal dissemination of infection-affected lungs. Second patient developed cervical lymph nodes tuberculosis with three relapses of infection in 2, 5, and 7 years. Third patient died of disseminated pulmonary and intestinal tuberculosis. Two NBS patients also died of severe disseminated tuberculosis complications.

Among two CVID patients, one developed disseminated tuberculosis-affected thoracic, mesenteric lymph nodes and spleen and second child presented pulmonary mycobacteriosis influenced by non-tuberculosis *M. chelonae*. Patients with XLA developed sporadic pulmonary tuberculosis with a relapse in 2 years. Both patients with defects in IFN-γ/IL-12-dependent immunity developed severe disseminated mycobacteriosis-affected bones and vertebra as a result of BCG vaccination.

All patients in reported group received two to six specific medications. We used the following schemes of treatment:

- two medications: isoniazid, pyrazinamide;
- three medications: isoniazid, pyrazinamide, ethambutol or amikacin;
- four medications: isoniazid, pyrazinamide, ethambutol, rifampicin or amikacin;
- five medication: izoniazid, pyrazinamide, ethambutol, rifampicin or Amikacin, Levofloxacin;
- six medications: isoniazid, pyrazinamide, ethambutol, rifampicin or amikacin, levofloxacin, azithromycin or clarithromycin.

Therapy was associated with the confirmation of mycobacterium infection diagnosis. The volume and duration of the treatment depended on the course of infection and lasted from several months to years. Two patients with CGD and disseminated mycobacterium infection received granulocyte transfusions that reduced the severity of infection. Three patients with remarkable specific lyphoproliferative process responded to short-course prednisone administration. A patient with IL-12 receptor deficiency was successfully treated by recombinant human IFN-γ administration. When the long-term specific therapy was not efficient enough, we used surgical approaches to eradicate the most clinically significant site of infection (e.g., lymph node, spleen, ribs, and vertebra).

Mycobacterium infection and its complication led to the death of seven out of 36 PID patients. Mortality of mycobacterium infection and its complications comprised 2.0% from all PID patients registered in our database and 19.4% among PID patients with mycobacteriosis.

4. Conclusions

Our data suggest that incidence of mycobacterial infections in children with PID is reliably higher than in immunologically healthy kids. Patients with combined immunodeficiencies (SCID, HIGM1, NBS, and CVID) and phagocytic defects (CGD) appear to be highly susceptible to mycobacterium infection.

References

Abramowsky, C., Gonzalez, B. and Sorensen, R.U. (1993) Disseminated bacillus Calmette-Guerin infections in patients with primary immunodeficiencies. Am. J. Clin. Pathol. 100, 52–56.

Bloom, B.R. and Small, P.M. (1998) The evolving relation between humans and Mycobacterium tuberculosis. N. Engl. J. Med. 338, 677–678.

Brooks, E.G., Klimpel, G.R., Vaidya, S.E., Keeney, S.E., Raimer, S., Goldman, A.S. and Schmalstieg, F.C. (1994) Thymic hypoplasia and T cell deficiency in ectodermal dysplasia: case report and review of the literature. Clin. Immunol. Immunopathol. 71, 44–52.

Casanova, J.L. and Abel, L. (2002) Genetic dissection of immunity to mycobacteria: the human model. Annu. Rev. Immunol. 20, 581–620.

Doffinger, R., Smahi, A., Bessia, C., Geissmann, F., Feinberg, J., Durandy, A., Bodemer, C., Kenwrick, S., Dupuis-Girod, S., Blanche, S., Wood, P., Rabia, S.H., Headon, D.J., Overbeek, P.A., Le Deist, F., Holland, S.M., Belani, K., Kumararatne, D.S., Fischer, A., Shapiro, R., Conley, M.E., Reimund, E., Kalhoff, H., Abinun, M., Munnich, A., Israel, A., Courtois, G. and Casanova, J.L. (2001) X-linked anhidrotic ectodermal dysplasia with immunodeficiency is caused by impaired NF-kappaB signaling. Nat. Genet. 27, 277–285.

Dupuis-Girod, S., Corradini, N., Hadj-Rabia, S., Fournet, J.C., Faivre, L., Le Deist, F., Durand, P., Doffinger, R., Smahi, A., Israel, A., Courtois, G., Brousse, N., Blanche, S., Munnich, A., Fischer, A., Casanova, J.L. and Bodemer, C. (2002) Osteopetrosis, lymphedema, anhidrotic ectodermal dysplasia, and immunodeficiency in a boy and incontinentia pigmenti in his mother. Pediatrics 109, e97.

Fisch, P., Millner, M., Muller, S.M., Wahn, U., Friedrich, W. and Renz, H. (1999) Expansion of gammadelta T cells in an infant with severe combined immunodeficiency syndrome after disseminated BCG infection and bone marrow transplantation. J. Allergy Clin. Immunol. 103, 1218–1219.

Gonzalez, B., Moreno, S., Burdach, R., Valenzuela, M.T., Henriquez, A., Ramos, M.I. and Sorensen, R.U. (1989) Clinical presentation of Bacillus Calmette-Guerin infections in patients with immunodeficiency syndromes. Pediatr. Infect. Dis. J. 8, 201–206.

Grimbacher, B., Holland, S.M., Gallin, J.I., Greenberg, F., Hill, S.C., Malech, H.L., Miller, J.A., O'Connell, A.C. and Puck, J.M. (1999) Hyper-IgE syndrome with recurrent infections–an autosomal dominant multisystem disorder. N. Engl. J. Med. 340, 692–702.

Heyderman, R.S., Morgan, G., Levinsky, R.J. and Strobel, S. (1991) Successful bone marrow transplantation and treatment of BCG infection in two patients with severe combined immunodeficiency. Eur. J. Pediatr. 150, 477–480.

Hugosson, C. and Harfi, H. (1991) Disseminated BCG-osteomyelitis in congenital immunodeficiency. Pediatr. Radiol. 21, 384–385.

Kobayashi, Y., Komazawa, Y., Kobayashi, M., Matsumoto, T., Sakura, N., Ishikawa, K. and Usui, T. (1984) Presumed BCG infection in a boy with chronic granulomatous disease. A report of a case and a review of the literature. Clin. Pediatr. (Phila) 23, 586–589.

Lau, Y.L., Chan, G.C., Ha, S.Y., Hui, Y.F. and Yuen, K.Y. (1998) The role of phagocytic respiratory burst in host defense against Mycobacterium tuberculosis. Clin. Infect. Dis. 26, 226–227.

Levy, J., Espanol-Boren, T., Thomas, C., Fischer, A., Tovo, P., Bordigoni, P., Resnick, I., Fasth, A., Baer, M., Gomez, L., Sanders, E.A., Tabone, M.D., Plantaz, D., Etzioni, A., Monafo, V., Abinun, M., Hammarstrom, L., Abrahamsen, T., Jones, A., Finn, A., Klemola, T., DeVries, E., Sanal, O., Peitsch, M.C. and Notarangelo, L.D. (1997) Clinical spectrum of X-linked hyper-IgM syndrome. J. Pediatr. 131, 47–54.

Minegishi, M., Tsuchiya, S., Imaizumi, M., Yamaguchi, Y., Goto, Y., Tamura, M., Konno, T. and Tada, K. (1985) Successful transplantation of soy bean agglutinin-fractionated, histoincompatible, maternal marrow in a patient with severe combined immunodeficiency and BCG infection. Eur. J. Pediatr. 143, 291–294.

Ohga, S., Ikeuchi, K., Kadoya, R., Okada, K., Miyazaki, C., Suita, S. and Ueda, K. (1997) Intrapulmonary Mycobacterium avium infection as the first manifestation of chronic granulomatous disease. J. Infect. 34, 147–150.

Pasic, S., Lilic, D., Pejnovic, N., Vojvodic, D., Simic, R. and Abinun, M. (1998) Disseminated Bacillus Calmette-Guerin infection in a girl with hyperimmunoglobulin E syndrome. Acta Paediatr. 87, 702–704.

Plotkin, S.A. (1994) Vaccines for varicella-zoster virus and cytomegalovirus: recent progress. Science 265, 1383–1385.

Rastogi, N., Legrand, E. and Sola, C. (2001) The mycobacteria: an introduction to nomenclature and pathogenesis. Rev. Sci. Tech. 20, 21–54.

Reichenbach, J., Rosenzweig, S., Doffinger, R., Dupuis, S., Holland, S.M. and Casanova, J.L. (2001) Mycobacterial diseases in primary immunodeficiencies. Curr. Opin. Allergy Clin. Immunol. 1, 503–511.

Rolph, M.S., Raupach, B., Kobernick, H.H., Collins, H.L., Perarnau, B., Lemonnier, F.A. and Kaufmann, S.H. (2001) MHC class Ia-restricted T cells partially account for beta2-microglobulin-dependent resistance to Mycobacterium tuberculosis. Eur. J. Immunol. 31, 1944–1949.

Romanus, V., Fasth, A., Tordai, P. and Wiholm, B.E. (1993) Adverse reactions in healthy and immunocompromised children under six years of age vaccinated with the Danish BCG vaccine, strain Copenhagen 1331: implications for the vaccination policy in Sweden. Acta Paediatr. 82, 1043–1052.

Stenger, S., Mazzaccaro, R.J., Uyemura, K., Cho, S., Barnes, P.F., Rosat, J.P., Sette, A., Brenner, M.B., Porcelli, S.A., Bloom, B.R. and Modlin, R.L. (1997) Differential effects of cytolytic T cell subsets on intracellular infection. Science 276, 1684–1687.

Urban, C., Becker, H., Mutz, I. and Fritsch, G. (1980) BCG-infection in chronic granulomatous disease (author's translation). Klin. Padiatr. 192, 13–18.

Allergy and Autoimmunity: Prevention or Treatment?

9

SLE 1, 2, 3...Genetic Dissection of Lupus

Jiankun Zhu and Chandra Mohan

Department of Internal Medicine and the Center for Immunology, University of Texas Southwestern Medical School, Dallas, TX, USA, Chandra.mohan@utsouthwestern.edu

Abstract. Systemic lupus erythematosus (SLE) is a chronic and complex autoimmune disease of unknown etiology, characterized by the presence of widespread immunological abnormalities and multiorgan injury. An important advance over the past decade has been our understanding of how different genetic loci (or genes) may dictate specific immune abnormalities in lupus. "Genetic dissection" has unveiled some of the mystery enshrouding lupus pathogenesis. It appears that there are at least two distinct events leading to disease. The first involves a breach in the adaptive immune system and the second involves a dysregulation of innate immunity. Co-ordinate dysregulation of both checkpoints is necessary for full-blown lupus to ensue. The challenge ahead is to understand how these two checkpoints are regulated in human SLE, and to devise therapeutic strategies that target both checkpoints.

1. Introduction

Systemic lupus erythematosus (SLE) is a chronic and complex autoimmune disease of unknown etiology in both humans and animal models, characterized by the presence of widespread immunological abnormalities and multiorgan injury. It is estimated to affect about 1–4 per 2000 people with a striking 9:1 female gender bias and strong ethnic variation. The hallmark of SLE is the production of high titers of autoantibodies directed against nuclear antigens such as double-strand DNA and chromatin, which results in autoantibody–mediated end-organ damage, as reviewed in Wakeland et al. (1999) and Davidson and Aranow (2006).

The etiology of SLE remains unknown. However, it is now apparent that multiple genetic and environmental factors are at play. Over the past decade, several studies have helped uncover genetic susceptibility loci in human (Nath et al. 2004) and murine (Wakeland et al. 2001) lupus. An ample body of evidence also indicates that various immunological abnormalities dysregulate the function of B cells, T cells, and myeloid cells or their interactions, both in SLE patients and lupus mouse models. An important advance over the past decade has been our understanding of how different genetic loci (or genes) may dictate specific immune abnormalities in lupus. The purpose of this chapter is to summarize our current understanding of gene ↔ functions mapping in lupus.

2. Genetic Dissection of Murine Lupus

The NZM2410 mouse strain is a New Zealand Black(NZB)/White(NZW)-derived inbred strain that spontaneously develops highly penetrant lupus with nephritis that is very similar to human SLE (Rudofsky et al. 1993). By performing a genetic analysis (i.e., linkage study) of SLE susceptibility, several different chromosomal intervals, in particular *Sle1*^{z} on chromosome 1, *Sle2*^{z} on chromosome 4, and *Sle3*^{z} on chromosome 7, have been found to confer lupus susceptibility in this mouse model (Morel et al. 1994). By introgressing these different chromosomal intervals individually onto the relatively healthy C57BL/6 (B6) background, congenic strains bearing *Sle*^{z}, *Sle2*^{z}, and *Sle3*^{z} have been generated for further functional analysis (Morel et al. 1997). Hence, for the first time, researchers are able to study "monogenic" models of lupus as opposed to studying "polygenic" strains or patients. Thus, genetic simplification by "congenic dissection" has transformed the study of a polygenic disease into a series of studies of "monogenic" diseases. Over the past decade, these studies have clearly demonstrated that each murine lupus susceptibility locus is responsible for very different component phenotypes of SLE.

The B6.*Sle1*^{z} congenic strain demonstrated a breach of immune tolerance to nuclear antigens, resulting in the production of autoantibodies against chromatin and H2A/H2B/dsDNA subnucleosomes, autoreactive T cells responding to histone epitopes, and an increased expression of cell activation markers on T cells and B cells (Morel et al. 1997; Mohan et al.1998a; Sobel et al. 1999). B6.*Sle2*^{z} mice exhibited B cell hyperactivity and elevated B1-cell formation leading to polyclonal or polyreactive hyper-gammaglobulinemia (Mohan et al. 1997,1998b; Xu et al. 2005). B6 mice bearing the *Sle3*^{z} interval mainly showed phenotypes affecting the T cell compartment, as well as modest levels of antinuclear IgG antibodies (Mohan et al. 1999b; Sobel et al. 2002). An important point to note is that *Sle1*^{z}, *Sle2*^{z}, or *Sle3*^{z} by itself was not sufficient for the development of fatal lupus but only elicited modest serological autoactivity and cellular features of autoreactivity. In contrast, the epistatic interaction of these loci with each other and other loci such as *Fas*^{lpr} and *Yaa* led to highly penetrant fatal glomerulonephritis (Mohan et al. 1999a; Morel et al. 2000; Shi et al. 2002; Subramanian et al. 2006).

The above "congenic dissection" studies illustrate that the development of fatal lupus is the end-result of multiple genes and a multitude of pathways acting in concert. These studies have also revealed that both the innate and adaptive immune systems have to be dysregulated for full-blown lupus to ensue. This review discusses recent evidence indicating that lupus-susceptible genes may impact both arms of the immune system.

3. Aberrant Adaptive Immunity in Lupus

Vertebrates acquire adaptive immunity after birth, which is a response to specific antigens that involves B and T cells of the immune system and frequently leads to a state of immune memory (Iwasaki and Medzhitov 2004). The adaptive immune system produces antibodies and T cells that are highly specific for a particular pathogen (or antigen). The relative specificity of SLE sera to a selected subset of nuclear antigens (as opposed to reacting to the whole universe of antigens) suggests that lupus genes must be impacting adaptive immunity at some level. Our recent genetic dissection studies have indicated that *Sle1*^{z} may be one such locus/gene (Figure. 1).

The *Sle1*^{z} interval, located on distal chromosome 1, is perhaps one of the most extensively studied chromosomal intervals in murine lupus, since it confers disease susceptibility in multiple spontaneous lupus models including the BWF1, SNF1, BXSB, and NZM2410 mice (Morel and Wakeland 2000). The *Sle1*^{z} interval is home to three

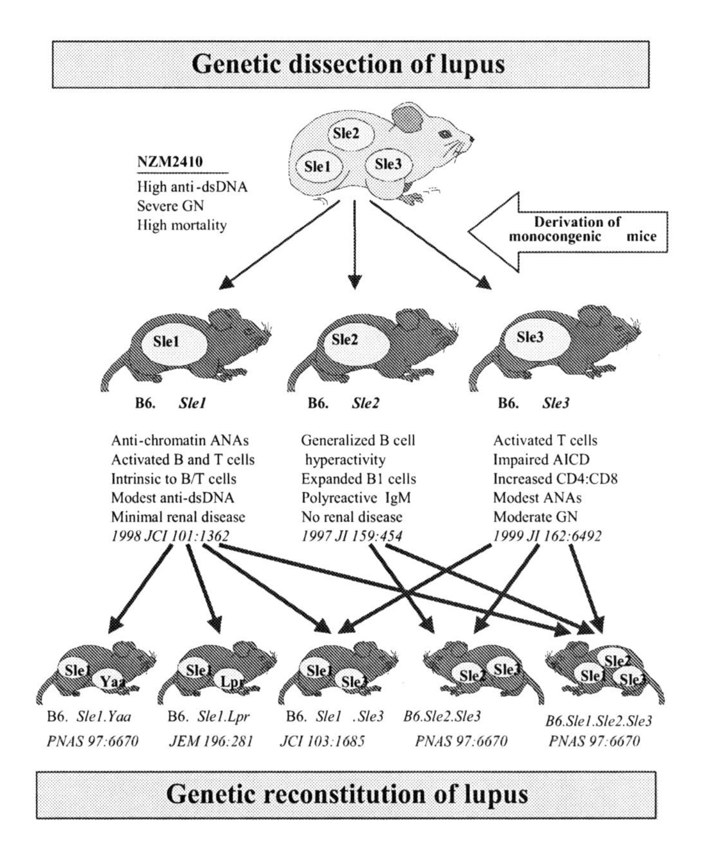

FIGURE 1. Genetic dissection and genetic reconstitution of murine lupus using congenic strains.

Note: The strains exhibit varying degrees IgG anti-dsDNA Abs, renal disease, and Lymphoproliferation. The most severely affected strains are B6*Sle1.lpr* and B6.*Sle1.Sle2.Sle3*

sub-loci: *Sle1a*^{z}, *Sle1b*^{z}, and *Sle1c*^{z}. Among these, the NZM2410/NZW-derived "*z*" allele of *Sle1b*^{z} leads to the highest levels and penetrance of antinuclear autoanti bodies (ANAs) (Morel et al. 2001). We found that *Sle1*^{z} could breach tolerance among B cells with low-avidity but not high-avidity reactivity to self-antigens (Kumar et al. 2006). To understand how *Sle1*^{z} breached B cell tolerance, we examined the immature and mature B cells in these nice further. *Sle1*^{z} did not impact the activation or apoptosis of mature splenic B cells, but caused a significant expansion of transitional immature T1 B cells ($B220^+AA4.1^+CD21^-CD23^-$), alluding to a relative block in the transition from T1 to mature B cells in the presence of *Sle1*^{z}. IL-7-driven immature B cells from the *Sle1*^{z}-and *Sle1b*^{z}-bearing bone marrow (BM) also exhibited a profound reduction in calcium flux and cell death following B-cell receptor (BCR) cross-linking, revealing that the *Sle1b*^{z} lupus susceptibility locus significantly dampens BCR signaling within immature B cells. This observation that is consistent with literature reports that the degree of BCR signaling can dictate B cell tolerance outcomes.

Through a meticulous positional cloning approach, Wakeland and colleagues demonstrated the *SLAM* family of co-stimulatory molecules to be the candidate genes for *Sle1b*^{z}. These molecules, *Ly9, CD84, CD244 (2B4), SLAM (CD150), Ly108*, and *CS1* are expressed in B cells (Wandstrat et al. 2004). Among these genes, *Ly108* has a profound impact on early B cell tolerance (Kumar et al. 2006). The normal "*b*" allele of *Ly108* encodes predominantly the *Ly108.2* isoform (bearing three intracellular ITSM signaling motifs); in contrast, the lupus-associated "*z*" allele of *Ly108* encodes predominantly the *Ly108.1* isoform (bearing two intracellular ITSM motifs) due to a splice-site mutation. Immature B cells represent a critical stage in B cell development, at which point self-reactive B cells are censored by deletion or receptor editing. Importantly, immature B cells transfected with the lupus-associated *Ly108.1* isoform showed impaired calcium flux, apoptotic cell loss, and BCR editing compared to transfectants bearing the normal *Ly108.2* isoform. Thus, the presence of the normal *Ly108.2* isoform may render immature B cells sensitive to BCR cross-linking, effectively facilitating the operation of several tolerance mechanisms including receptor editing and deletion, whereas the lupus-associated isoform, *Ly108.1*, appears to thwart these processes (Kumar et al. 2006).

Collectively, the above studies reveal that a mutant form of the *SLAM* gamily gene, *Ly108,* can profoundly impact key checkpoints in early B cell tolerance, hence leading to the emergence of self-reactive antibodies. Since a similar locus (Tsao et al. 1999) and gene(s) (unpublished observations) are also at play in human SLE, dysregulation of the adaptive immune system, early during B cell development, may constitute a central mechanism leading to lupus, both in mice and in patients.

4. Aberrant Adaptive Immunity in Lupus

The innate immune system is a universal and ancient form of host defense against infection. Before launching an effective adaptive immune response, the host must deal with acute assaults, sense the presence of pathogen, distinguish infectious nonself from non-infectious self, and direct an effective immune response against the invading organism rapidly. Dendritic cells (DCs), interferon-α, and Toll-like receptors all play central roles in innate immune responsiveness (Banchereau and Steinman 1998; Meylan et al. 2006; Stetson and Medzhitov 2006). Unfortunately, when innate immune responses are misdirected to components of self, autoimmunity can ensue. Recent genetic studies in mice have yielded at least two examples of loci/genes that may contribute to lupus by dysregulating innate immunity (Figure 2).

Genetic dissection studies in murine models of lupus have uncovered the existence of a similarly positioned lupus susceptibility locus on mid-chromosome 7 in several strains of mice including the NZM2410, NZB/NZW, and *MRL/lpr* (Morel et al. 1994; Vyse and Kotzin 1998; Mohan et al. 1999; Kono and Theofilopoulos 2000; Kong et al. 2004). This locus has been termed *Sle3*^{z} in the NZM2410 model. B6.*Sle3*^{z} congenics exhibit low levels of ANAs and several lymphocyte phenotypes (Mohan et al. 1999). Importantly, *Sle3*^{z}-bearing T cells were spontaneously activated and exhibited elevated CD4/CD8 ratios and impaired activation-induced cell death (Mohan et al. 1999; Wakeland et al. 1999; Wakui et al. 2004).

To explore the cellular origin of the *Sle3*^{z}-associated phenotypes, Sobel et al.(2002) transferred BM from allotype-marked B6 and B6.*Sle3* congenic mice into B6 hosts. T cells of both origins (i.e., with or without *Sle3*^{z}) exhibited elevated CD4/CD8 ratios, spontaneous T cell activation, and phenotypes that have been attributed to *Sle3*^{z}. These studies demonstrated that the *Sle3*^{z}-associated phenotypes may not be encoded in a T cell-intrinsic fashion although they were BM-transferable. Likewise, the same study

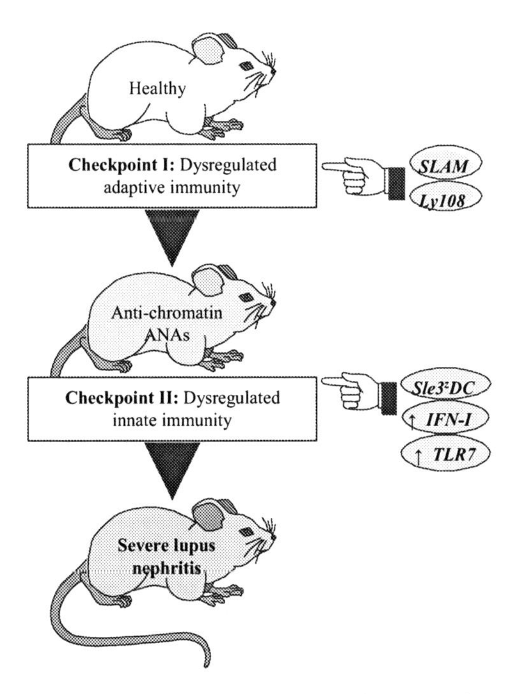

FIGURE 2. Dysregulation in adaptive immunity (checkpoint I) and innate immunity (checkpoint II) can both lead to lupus-like autoimmunity.

also revealed that autoantibody production in the chimeras was also not contingent upon the intrinsic expression of *Sle3*^{z} within B cells (Sobel et al. 2002).

Given that *Sle3*^{z}-associated phenotypes may not be T cell intrinsic, we looked for any quantitative or qualitative differences in *Sle3*^{z}-bearing cells of myeloid origin, including various antigen-presenting cell (APC) subsets. We found that myeloid cells including myeloid DCs or macrophages were more expanded and more activated ex vivo in B6.*Sle3*^{z} congenics (Zhu et al. 2005). DCs and macrophages isolated from B6.*Sle3*^{z} spleens, lymph nodes, and bone marrow exhibited increased surface levels of CD40, CD80, CD86, CD54, CD106 (VCAM-1), and FcR (CD16/32) and increased expression levels of pro-inflammatory cytokines such as IL-12, IL-1ß, and TNF-α. Also, when *Sle3*^{z}-bearing DCs were Ovalbumine (OVA) pulsed and co-cultured with OVA specific T-cell receptor (TCR) Tg T cells, the T cells demonstrated more expansion and reduced apoptosis compared to T cells co-cultured with B6 DCs (Zhu et al. 2005). Most importantly, after adoptive transfer into young B6 hosts, *Sle3*^{z}-bearing DCs elicited two cardinal *Sle3*^{z}-associated phenotypes, including elevated splenic CD4/CD8 ratios and elevated serum autoantibody levels, compared with mice receiving B6-derived DCs. These findings suggested that *Sle3*^{z}-bearing DCs appeared to be sufficient to recreate the *Sle3*^{z}-associated lupus phenotypes (Zhu et al. 2005).

The above congenic dissection studies revealed that hyperactive, pro-inflammatory DCs, arising as a consequence of a dysregulation of the innate immune system, can be a key factor in lupus development. Hyperactive or pro-inflammatory DCs have also shown to contribute to human SLE (Ding et al. 2006). In humans, CD14$^+$ monocytes from SLE blood were reported to be able to act as DCs too to induce proliferation of allogenic CD4 T cells (Blanco et al. 2001). CD14$^+$ monocytes from healthy donor were reported to show DC-like morphology and function in the presence of SLE serum suggesting that components in SLE serum might contain a key initiating factor(s) in driving the differentiation and activation of DCs in human SLE. This serum component has been identified to be IFN-α (Blanco et al. 2001).

As early as 1980s, interferon-α levels have been found to be increased in the serum of SLE patients (Neighbour and Grayzel 1981). Interferon-α treatment for a variety of conditions such as viral infection or tumors frequently resulted in lupus-like manifestations (Gota and Calabrese 2003). Microarray data from peripheral blood mononuclear cells of lupus patients revealed that IFN-α and IFN-γ gene expression signatures correlated with clinical features of SLE (Bennett et al. 2003; Baechler et al. 2004; Crow and Kirou 2004).

Likewise, studies using IFN-α receptor (IFNAR)-deficient New Zealand Black NZB mice, a strain that spontaneously develops mild lupus, have highlighted the potential role of type I IFN in the development of lupus. It was found that lupus was significantly alleviated in the IFNAR- deficient NZB mice, with significantly reduced ANAs, kidney disease, and mortality (Santiago-Raber et al. 2003). These mice also showed reduced numbers and proliferation of most immunocyte subsets including B cells. Likewise, in IFNAR-deficient B6.*Fas*^{lpr} mice, disease reduction was also noted following treatment with poly(I:C), a potent inducer of type I IFN (Braun Geraldes, and Demengeot 2003). However, some conflicting observations have been reported including the findings that suggest that IFN-I may not be a potent disease inducer in the *MRL.FAS*^{lpr} (Hron and Peng 2004) and *B6.Sle2*^{z} murine models of lupus (Li et al. 2005).

A second recent breakthrough in murine lupus genetics that has highlighted the importance of the innate immune system revolves around the *Yaa* lupus susceptibility locus derived from the BXSB lupus strain, which was originally derived from the SB/Le and C57BL/6 (B6) parental strains (Andrews et al. 1978; Murphy and Roths 1979). The male bias in BXSB lupus has clearly been shown to be encoded by the BXSB Y chromosome, and the putative locus on Y chromosome has been termed *Yaa*, or *Y chromosome autoimmunity accelerating locus*, as reviewed in Izui (1990). Interestingly, the *Yaa* locus leads to different degrees of autoimmunity on different genetic backgrounds, in epistasis with other lupus susceptibility loci (Hudgins et al. 1985; Steinberg et al. 1985; Izui et al. 1988). Defined about 30 years ago, it was only recently that the identity of *Yaa* was elucidated —two independent groups identified *Tlr7* as the candidate gene for *Yaa* (Pisitkun et al. 2006; Subramanian et al. 2006), as reviewed below.

Recently, Bolland and colleague have provided data suggesting that *Tlr7* may represent the *Yaa* locus (Pisitkun et al. 2006). Their studies had demonstrated that the locus on the Y chromosome accelerated SLE-like disease on certain strain backgrounds including the lupus-prone Fc_{γ}RIIb–/– model, indicating that *Yaa* is a potent epistatic modifier (Pisitkun et al. 2006). By utilizing a bone marrow chimera strategy, they demonstrated that the B cells in B6.Fc_{γ}RIIb–/– *Yaa* mice produced antibodies preferentially targeting nucleolar antigens, while B cells from B6.Fc_{γ}RIIb–/– mice without *Yaa* recognized chromatin predominantly. This data suggested that *Yaa* was capable of skewing the B cell specificity toward RNA-bearing autoantigens. Further microarray data on *Yaa*-bearing B cells revealed that four out of 26 significantly increased genes were positioned in tandem on the X-chromosome. They hypothesized that there might be a "duplication" and transfer of a segment of the X-chromosome onto Y-chromosome, since *Yaa* is on the Y chromosome. This hypothesis was confirmed by fluorescence in situ hybridization (FISH) experiments using probes specific to a region at the distal end of the X-chromosome (Pisitkun et al. 2006).

Functional analysis focused further on the most interesting molecule hyper-expressed on their microarrays-Toll-like receptor7 (TLR7), which is expressed in B cells. Stimulation of *Yaa* splenocytes with a TLR7 agonist (imiquimod) resulted in enhanced $I_{\kappa}B_{\alpha}$ activity and increased proliferation in vitro. In vivo administration of imiquimod induced Fc_{γ}RIIb–/– mice to produce autoantibodies against nucleolar antigens but not chromatin linking the functional role of TLR7 toward certain nuclear antigen specificities. Subramanian and her colleagues have recently provided confirmatory evidence that dysregulated TLR7 may be the culprit gene for *Yaa* (Subramanian et al. 2006).

Collectively, the above studies suggest that dysregulated expression or function of DCs, the TLRs they bear (e.g., TLR7) and cytokines they express (e.g., IFN-I), can profoundly impact lupus pathogenesis.

5. Conclusions

"Genetic dissection" has unveiled some of the mystery enshrouding lupus pathogenesis. It appears that there are at least two distinct events leading to disease. The first involves a breach in the adaptive immune system as exemplified by *Sle1b*^{z}/*Ly108*^{z} (Figure 2). The second involves a dysregulation of innate immunity as can happen in the context of *Sle3*^{z}-bearing pro-inflammatory DCs, heightened TLR7 activity, and IFN-I production (Figure 2). Co-ordinate dysregulation of both checkpoints is necessary for full-blown lupus to ensue. The challenge ahead is to understand how these two checkpoints are regulated in human SLE and to devise therapeutic strategies that target both checkpoints.

References

Andrews, B.S., Eisenberg, R.A., Theofilopoulos, A.N., Izui, S., Wilson, C.B., McConahey, P.J., Murphy, E.D., Roths, J.B. and Dixon, F.J. (1978) Spontaneous murine lupus-like syndromes. Clinical and immunopathological manifestations in several strains. J. Exp. Med. 148, 1198–1215.

Baechler, E.C., Gregersen, P.K. and Behrens, T.W. (2004) The emerging role of interferon in human systemic lupus erythematosus. Curr. Opin. Immunol. 16, 801–807.

Banchereau, J. and Steinman, R.M. (1998) Dendritic cells and the control of immunity. Nature 392, 245–252.

Bennett, L., Palucka, A.K., Arce, E., Cantrell, V., Borvak, J., Banchereau, J. and Pascual, V. (2003) Interferon and granulopoiesis signatures in systemic lupus erythematosus blood. J. Exp. Med. 197, 711–723.

Blanco, P., Palucka, A.K., Gill, M., Pascual, V. and Banchereau, J. (2001) Induction of dendritic cell differentiation by IFN-alpha in systemic lupus erythematosus. Science 294, 1540–1543.

Braun, D., Geraldes, P. and Demengeot, J. (2003) Type I interferon controls the onset and severity of autoimmune manifestations in lpr mice. J. Autoimmun. 20, 15–25.

Crow, M.K. and Kirou, K.A. (2004) Interferon-alpha in systemic lupus erythematosus. Curr. Opin. Rheumatol. 16, 541–547.

Davidson, A. and Aranow, C. (2006) Pathogenesis and treatment of systemic lupus erythematosus nephritis. Curr. Opin. Rheumatol. 18, 468–475.

Ding, D., Mehta, H., McCune, W.J. and Kaplan, M.J. (2006) Aberrant phenotype and function of myeloid dendritic cells in systemic lupus erythematosus. J. Immunol. 177, 5878–5889.

Gota, C. and Calabrese, L. (2003) Induction of clinical autoimmune disease by therapeutic interferon-alpha. Autoimmunity 36, 511–518.

Hron, J.D. and Peng, S.L. (2004) Type I IFN protects against murine lupus. J. Immunol. 173, 2134–2142.

Hudgins, C.C., Steinberg, R.T., Klinman, D.M., Reeves, M.J. and Steinberg, A.D. (1985) Studies of consomic mice bearing the Y chromosome of the BXSB mouse. J. Immunol. 134, 3849–3854.

Iwasaki, A. and Medzhitov, R. (2004) Toll-like receptor control of the adaptive immune responses. Nat. Immunol. 5, 987–995.

Izui, S. (1990) Autoimmune accelerating genes, lpr and Yaa, in murine systemic lupus erythematosus. Autoimmunity 6, 113–129.

Izui, S., Higaki, M., Morrow, D. and Merino, R. (1988) The Y chromosome from autoimmune BXSB/MpJ mice induces a lupus-like syndrome in (NZW x C57BL/6)F1 male mice, but not in C57BL/6 male mice. Eur. J. Immunol. 18, 911–915.

Kong, P.L., Morel, L., Croker, B.P. and Craft, J. (2004) The centromeric region of chromosome 7 from MRL mice (Lmb3) is an epistatic modifier of Fas for autoimmune disease expression. J. Immunol. 172, 2785–2794.

Kono, D.H. and Theofilopoulos, A.N. (2000) Genetics of systemic autoimmunity in mouse models of lupus. Int. Rev. Immunol. 19, 367–387.

Kumar, K.R., Li, L., Yan, M., Bhaskarabhatla, M., Mobley, A.B., Nguyen, C., Mooney, J.M., Schatzle, J.D., Wakeland, E.K. and Mohan, C. (2006) Regulation of B cell tolerance by the lupus susceptibility gene Ly108. Science 312, 1665–1669.

Li, J., Liu, Y., Xie, C., Zhu, J., Kreska, D., Morel, L. and Mohan, C. (2005) Deficiency of type I interferon contributes to Sle2-associated component lupus phenotypes. Arthritis Rheum. 52, 3063–3072.

Meylan, E., Tschopp, J. and Karin, M. (2006) Intracellular pattern recognition receptors in the host response. Nature 442, 39–44.

Mohan, C., Alas, E., Morel, L., Yang, P. and Wakeland, E.K. (1998a) Genetic dissection of SLE pathogenesis. Sle1 on murine chromosome 1 leads to a selective loss of tolerance to H2A/H2B/DNA subnucleosomes. J. Clin. Invest. 101, 1362–1372.

Mohan, C., Morel, L., Yang, P. and Wakeland, E.K. (1998b) Accumulation of splenic B1a cells with potent antigen-presenting capability in NZM2410 lupus-prone mice. Arthritis Rheum. 41, 1652–1662.

Mohan, C., Morel, L., Yang, P. and Wakeland, E.K. (1997) Genetic dissection of systemic lupus erythematosus pathogenesis: Sle2 on murine chromosome 4 leads to B cell hyperactivity. J. Immunol. 159, 454–465.

Mohan, C., Morel, L., Yang, P., Watanabe, H., Croker, B., Gilkeson, G. and Wakeland, E.K. (1999a) Genetic dissection of lupus pathogenesis: a recipe for nephrophilic autoantibodies. J. Clin. Invest. 103, 1685–1695.

Mohan, C., Yu, Y., Morel, L., Yang, P. and Wakeland, E.K. (1999b) Genetic dissection of Sle pathogenesis: Sle3 on murine chromosome 7 impacts T cell activation, differentiation, and cell death. J. Immunol. 162, 6492–6502.

Morel, L., Blenman, K.R., Croker, B.P. and Wakeland, E.K. (2001) The major murine systemic lupus erythematosus susceptibility locus, Sle1, is a cluster of functionally related genes. Proc. Natl. Acad. Sci. U.S.A. 98, 1787–1792.

Morel, L., Croker, B.P., Blenman, K.R., Mohan, C., Huang, G., Gilkeson, G. and Wakeland, E.K. (2000) Genetic reconstitution of systemic lupus erythematosus immunopathology with polycongenic murine strains. Proc. Natl. Acad. Sci. U.S.A. 97, 6670–6675.

Morel, L., Mohan, C., Yu, Y., Croker, B.P., Tian, N., Deng, A. and Wakeland, E.K. (1997) Functional dissection of systemic lupus erythematosus using congenic mouse strains. J. Immunol. 158, 6019–6028.

Morel, L., Rudofsky, U.H., Longmate, J.A., Schiffenbauer, J. and Wakeland, E.K. (1994) Polygenic control of susceptibility to murine systemic lupus erythematosus. Immunity 1, 219–229.

Morel, L. and Wakeland, E.K. (2000) Lessons from the NZM2410 model and related strains. Int. Rev. Immunol. 19, 423–446.

Morel, L., Yu, Y., Blenman, K.R., Caldwell, R.A. and Wakeland, E.K. (1996) Production of congenic mouse strains carrying genomic intervals containing SLE-susceptibility genes derived from the SLE-prone NZM2410 strain. Mamm. Genome 7, 335–339.

Murphy, E.D. and Roths, J.B. (1979) A Y chromosome associated factor in strain BXSB producing accelerated autoimmunity and lymphoproliferation. Arthritis Rheum. 22, 1188–1194.

Nath, S.K., Kilpatrick, J. and Harley, J.B. (2004) Genetics of human systemic lupus erythematosus: the emerging picture. Curr. Opin. Immunol. 16, 794–800.

Neighbour, P.A. and Grayzel, A.I. (1981) Interferon production of vitro by leucocytes from patients with systemic lupus erythematosus and rheumatoid arthritis. Clin. Exp. Immunol. 45, 576–582.

Pisitkun, P., Deane, J.A., Difilippantonio, M.J., Tarasenko, T., Satterthwaite, A.B. and Bolland, S. (2006) Autoreactive B cell responses to RNA-related antigens due to TLR7 gene duplication. Science 312, 1669–1672.

Rudofsky, U.H., Evans, B.D., Balaban, S.L., Mottironi, V.D. and Gabrielsen, A.E. (1993) Differences in expression of lupus nephritis in New Zealand mixed H-2z homozygous inbred strains of mice derived from New Zealand black and New Zealand white mice. Origins and initial characterization. Lab. Invest. 68, 419–426.

Santiago-Raber, M.L., Baccala, R., Haraldsson, K.M., Choubey, D., Stewart, T.A., Kono, D.H. and Theofilopoulos, A.N. (2003) Type-I interferon receptor deficiency reduces lupus-like disease in NZB mice. J. Exp. Med. 197, 777–788.

Shi, X., Xie, C., Kreska, D., Richardson, J.A. and Mohan, C. (2002) Genetic dissection of SLE: SLE1 and FAS impact alternate pathways leading to lymphoproliferative autoimmunity. J. Exp. Med. 196, 281–292.

Sobel, E.S., Mohan, C., Morel, L., Schiffenbauer, J. and Wakeland, E.K. (1999) Genetic dissection of SLE pathogenesis: adoptive transfer of Sle1 mediates the loss of tolerance by bone marrow-derived B cells. J. Immunol. 162, 2415–2421.

Sobel, E.S., Morel, L., Baert, R., Mohan, C., Schiffenbauer, J. and Wakeland, E.K. (2002) Genetic dissection of systemic lupus erythematosus pathogenesis: evidence for functional expression of Sle3/5 by non-T cells. J. Immunol. 169, 4025–4032.

Steinberg, R.T., Miller, M.L. and Steinberg, A.D. (1985) Effect of the BXSB Y chromosome accelerating gene on autoantibody production. Clin. Immunol. Immunopathol. 35, 67–72.

Stetson, D.B. and Medzhitov, R. (2006) Type I interferons in host defense. Immunity 25, 373–381.

Subramanian, S., Tus, K., Li, Q.Z., Wang, A., Tian, X.H., Zhou, J., Liang, C., Bartov, G., McDaniel, L.D., Zhou, X.J., Schultz, R.A. and Wakeland, E.K. (2006) A Tlr7 translocation accelerates systemic autoimmunity in murine lupus. Proc. Natl. Acad. Sci. U.S.A. 103, 9970–9975.

Tsao, B.P., Cantor, R.M., Grossman, J.M., Shen, N., Teophilov, N.T., Wallace, D.J., Arnett, F.C., Hartung, K., Goldstein, R., Kalunian, K.C., Hahn, B.H. and Rotter, J.I. (1999) PARP alleles within the linked chromosomal region are associated with systemic lupus erythematosus. J. Clin. Invest. 103, 1135–1140.

Vyse, T.J. and Kotzin, B.L. (1998) Genetic susceptibility to systemic lupus erythematosus. Ann. Rev. Immunol. 16, 261–292.

Wakeland, E.K., Liu, K., Graham, R.R. and Behrens, T.W. (2001) Delineating the genetic basis of systemic lupus erythematosus. Immunity 15, 397–408.

Wakeland, E.K., Wandstrat, A.E., Liu, K. and Morel, L. (1999) Genetic dissection of systemic lupus erythematosus. Curr. Opin. Immunol. 11, 701–707.

Wakui, M., Kim, J., Butfiloski, E.J., Morel, L. and Sobel, E.S. (2004) Genetic dissection of lupus pathogenesis: Sle3/5 impacts IgH CDR3 sequences, somatic mutations, and receptor editing. J. Immunol. 173, 7368–7376.

Wandstrat, A.E., Nguyen, C., Limaye, N., Chan, A.Y., Subramanian, S., Tian, X.H., Yim, Y.S., Pertsemlidis, A., Garner, H.R., Jr., Morel, L. and Wakeland, E.K. (2004) Association of extensive polymorphisms in the SLAM/CD2 gene cluster with murine lupus. Immunity 21, 769–780.

Xu, Z., Duan, B., Croker, B.P., Wakeland, E.K. and Morel, L. (2005) Genetic dissection of the murine lupus susceptibility locus Sle2: contributions to increased peritoneal B-1a cells and lupus nephritis map to different loci. J. Immunol. 175, 936–943.

Zhu, J., Liu, X., Xie, C., Yan, M., Yu, Y., Sobel, E.S., Wakeland, E.K. and Mohan, C. (2005) T cell hyperactivity in lupus as a consequence of hyperstimulatory antigen-presenting cells. J. Clin. Invest. 115, 1869–1878.

10

Network of Myeloid and Plasmacytoid Dendritic Cells in Atopic Dermatitis

Natalija Novak, Wenming Peng, and Chunfeng Yu

Department of Dermatology, University of Bonn, Bonn, Germany, Natalija.Novak@ukb.uni-bonn.de

Abstract. Atopic dermatitis (AD) presents as a chronic relapsing skin disease with high prevalence in children. The typical distributed skin lesions make the clinical diagnosis of AD very simple and clear-cut in most of the cases. In contrast, the underlying mechanisms leading to the manifestation of AD are more than complex and consist of genetic components combined with various deficiencies on the level of innate and adaptive immune mechanisms. Challenged by this puzzle, scientific approaches of the last years have made considerable progress in gaining insights into the mechanisms, which cause AD. AD is a biphasic inflammatory skin disease characterized by an initial phase predominated by Th2 cytokines which switches into a second, more chronic Th1-dominated eczematous phase. Two different dendritic cell (DC) subtypes bearing the high-affinity receptor for IgE (FcεRI) have been identified in the epidermal skin of AD patients: FcεRI^{high} Langerhans cells (LCs) and FcεRI^{high} inflammatory dendritic epidermal cells (IDECs). These two DC subtypes are believed to contribute distinctly to the biphasic nature and the outcome of T cell responses in AD. In contrast, plasmacytoid DCs, which play an important role in the defence against viral infections, have been shown to bear the high-affinity receptor for IgE too but are nearly absent from the epidermal skin lesions of AD patients. In light of recent developments, the picture emerges that different IgE-receptor bearing DC subtypes in the blood and skin of AD patients play a pivotal role in the complex network of DCs, which is highlighted in this review.

1. Introduction

Atopic dermatitis (AD) is a chronic inflammatory skin disease, which currently affects 10–30% of children and 1–3% of adults worldwide (Leung and Bieber 2003). The pathophysiology of AD is complex, and several lines of evidence indicate that multifaceted genetic and environmental factors contribute to the clinical manifestation of the disease (Novak et al. 2003a). Food allergens, aeroallergens in addition to microbial components, and stress are regarded as important trigger factors and impact on both the severity and the duration of the disease. Besides that, chronic tissue damage and frequent allergen challenges seem to predispose subgroups of patients to the manifestation of immunoglobulin E (IgE) hyperreactivity, which together with numerous other factors paves the way for the chronification of AD and the concomitant development of other atopic disorders such as allergic rhinoconjunctivitis and asthma (Mothes et al. 2005). Although there is incontrovertible evidence for a capability of environmental allergens to initiate severe flare-ups of eczema or contribute to the impairment of the characteristic, xerotic and itchy skin lesions which predominate the clinical picture of AD, the main mediators of these reactions on the cellular level, were elusive for a long time. This changed over a decade ago, when IgE-bearing dendritic cells (DCs) have been discovered in the epidermal skin lesions of AD for the first time. Later on, the high-affinity receptor for IgE (FcεRI) has been identified as main IgE-binding structure on these cells (Bieber et al. 1992; Wang et al. 1992). In contrast to the classical tetrameric FcεRI on the surface of effector cells of allergic reactions such as mast cells and basophils, FcεRI on DC has a trimeric structure and consists of the IgE-binding α-chain and the γ-chain dimer which are responsible for downstream-signal propagation, while the β-chain is absent. Further on, it has been shown that the FcεRI surface expression is regulated distinctly in antigen-presenting cells of atopic and non-atopic donors. In DCs of non-atopic individuals, most of the IgE-binding α-chain remains in the intracellular space, while only few γ-chain dimers are present. In contrast, in DCs of atopic individuals, a second variant of the α-chain can be detected, which is capable of associating with the γ-chains present in high levels in these cells. This is the basis for the transportation of the complete trimeric FcεRI complex to the cell surface and accounts for the high capability of DCs in the skin of AD to bind IgE molecules (Novak et al. 2004b; 2003b; 2003c).

2. Network of Myeloid DCs in AD

DCs as antigen-presenting cells are outposts of the immune system, which are located at the border zones of the body to the environment. Since they have been discovered over a century ago in 1868 by Paul Langerhans, they remained enigmatic and fascinated researchers all over the world (Jolles 2002). With the help of their dendrites, they form a sophisticated network within the epidermis and encounter foreign antigens. Subsequently, they internalize these antigens, secret cytokines, process them, migrate to the lymph nodes, and present the processed antigens to T cells. Obviously, their function of antigen uptake and presentation as well as T cell priming is of particular immunologic importance in diseases with an impaired epidermal skin barrier such as AD (Cork et al. 2006). This impaired skin barrier has been shown to result partially from genetic modifications in genes encoding important proteins of the epidermal differentiation complex such as Filaggrin or S100 (Weidinger et al. 2006; Palmer et al. 2006; Cookson 2004). Together with an increased water loss and shift of the pH, this contributes to the fatal "loss" of the skin barrier as a protective shield in AD patients, which allows foreign antigens to invade easily into the skin (Strid and Strobel 2005). As a basic principle, two different types of DCs are known to regulate the immune haemostasis in our body: First, the so-called myeloid DCs, which express the DC marker CD1a. Myeloid DCs are

CD11c+CD123– and suspected to encounter a high number of immune reactions, which lead to immunity (Bonasio and von Andrian 2006).

The most prominent members of this class of DCs are the classical Langerhans cells (LCs), which are characterized by the Birbeck granules, electron-microscopically visible as tennis racket-shaped organelles originating from the accumulation of the C-type lectin Langerin (Villadangos and Heath 2005). LCs reside in both, healthy and inflamed skin and are constantly renewed under steady-state conditions. As a characteristic feature of AD, LCs are equipped with the high-affinity receptor for IgE (FcεRI) on their cell surface, which enables them to take up allergens penetrating into the skin (Villadangos and Heath 2005). In vitro studies of LCs combined with Atopy-patch test results in which type I allergens applied to the skin induce an eczematous reaction within 24–48 h in sensitized individuals, provide evidence that LCs are in the foreground in the initial phase of AD (Novak et al. 2004b). LC are capable of taking up invading allergens and presenting these allergens to T cells (Kerschenlohr, Decard, Przybilla, and Wollenberg 2003). In this step, primarily T cells of the Th2 type are primed by LCs in vitro, which are characterized by the production of IL-4, IL-5, and IL-13, typical for the initial phase of AD (Novak et al., 2004b). Other factors, such as the release of thymic stromal lymphopoietin (TSLP) by keratinocytes, are suspected to aggravate this DC-mediated Th2-immune response (Wang et al. 2006). Further on, allergen challenge and concomitant IgE receptor cross linking on LCs or activation by microbial products such as staphylococcal enterotoxins lead to the release of different chemotactic mediators, which together with soluble factors contribute to the recruitment of other cell types such as inflammatory dendritic epidermal cells (IDECs) from their precursor cells from the blood into the skin (Gunther et al. 2005; Pivarcsi et al. 2004; Homey et al. 2006).

IDECs are only present at inflammatory epidermal sites and bear significantly high numbers of FcεRI in combination with CD11b molecules and the mannose receptor (CD206) on their cell surface but are Langerin (CD207)-negative DCs (Wollenberg et al. 1995; Stary, Bangert et al. 2005). In time, kinetics acquired with the help of atopy patch tests, invasion of IDECs into the epidermis within 24–48 h after allergen application, and concomitant up-regulation of FcεRI expression on LCs and IDECs in the developing skin lesions has been observed (Kerschenlohr et al. 2003), supporting the view of a two-step model directed by DC subtypes in AD (Figure 1) Due to their high ability to release distinct pro-inflammatory mediators after IgE receptor-mediated allergen challenge, it is assumed that IDECs are the main dendritic amplifiers of the epidermal allergic-inflammatory reaction in AD. Allergen challenge of IDECs in vitro leads, beside the release of pro-inflammatory cytokines and chemokines, to the production of IL-12 and IL-18, which might contribute to the alteration of the initial Th2-immune micromilieu to the predominance of interferon (IFN)-γ-producing T cells in the skin, which seem to be a crucial step for the chronification of the skin lesions (Novak et al. 2004b; Grewe et al. 1995).

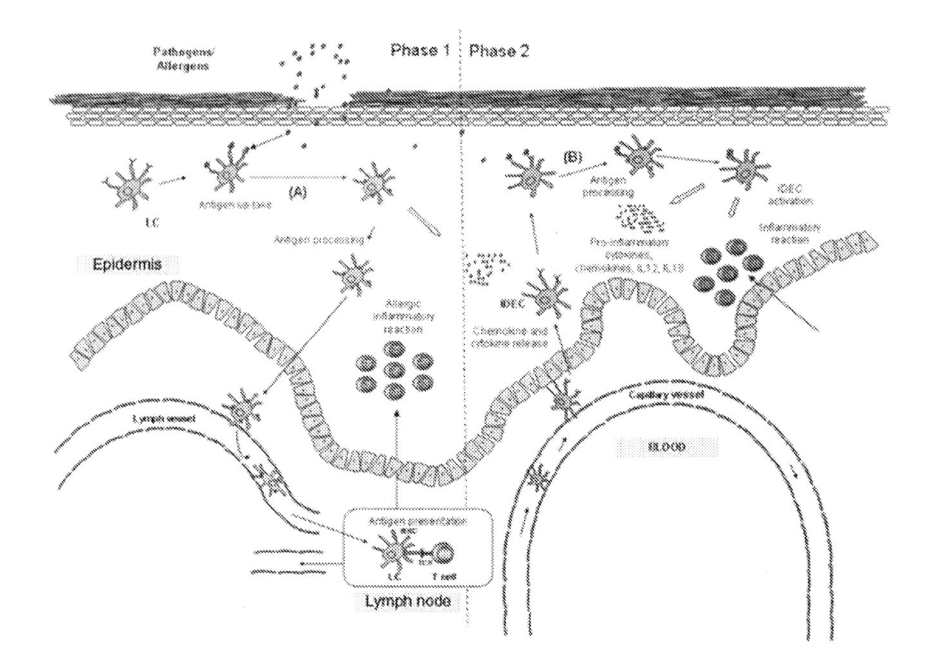

FIGURE 1. Network of Langerhans cells (LCs) and inflammatory dendritic epidermal cells (IDECs) in the epidermis of atopic dermatitis (AD) patients. (A) LC residing in the epidermis capture pathogens, secret cytokines/chemokines, process antigens, migrate to the lymph nodes, and present them to T cells. In the lymph node, LC prime large amounts of Th2 cells, which produce high amounts of Th2 cytokines such as IL-4, IL-5, and IL-13. (B) Allergen challenge and concomitant IgE receptor cross linking on LC or activation by microbial products lead to the release of different chemotactic mediators, which together with soluble factors contribute to the recruitment of IDECs into the skin. IDEC present at inflammatory epidermal sites and bear significantly high numbers of FcεRI on cell surface. Allergen challenge of IDEC in vitro leads to the production of IL-12 and IL-18 and release of pro-inflammatory cytokines and chemokines, which might contribute to the amplification of the allergic-inflammatory reaction and alteration of the initial Th2-immune micromilieu to the predominance of IFN-γ-producing Th1 T cells in the skin, which seem to be a crucial step for the chronification of the skin lesions. MHC, major histocompatibility class; TCR, T cell receptor.

3. Balance of LCs and IDECs Predicts the Severity of AD

It is assumed that LCs are capable of maintaining, to some degree, a state of tolerance toward invading allergens and pathogens within the skin. Most likely, the intensity and number of danger signals and other yet-unidentified signals might lead to the break-down of this state of tolerance under certain conditions and an overbalance of inflammatory reactions accompanied by the invasion of cell types with pro-inflammatory characteristics into the skin such as IDECs. After therapy with topical immunmodulators, for instance, the surface expression of the IgE receptors is reduced on both LCs and IDECs, and the number of IDECs within the epidermis decreases below the detectable level (Schuller et al. 2004; Novak et al. 2005; Simon et al. 2004). These findings support the view that IDEC should be the main DC subtypes targeted and eliminated by therapeutic approaches, while the rather anti-inflammatory and pro-tolerogenic functions of LCs should be sustained or even strengthened by therapeutic intervention.

4. Deficiency of Plasmacytoid DCs in AD

Plasmacytoid DCs are CD1a negative as well as CD11c negative but positive for the α-chain of the IL-3 receptor (CD123) and the blood-dendritic cell antigen (BDCA)-2. They are equipped with specific pattern recognition receptors of the innate immune system which enable them to sense microbial pathogens and thereby defend our organism against bacterial and viral infections (Soumelis and Liu 2006). Plasmacytoid DCs in the peripheral blood of AD patients have been shown to bear the FcεRI receptor on their surface which is densely occupied with IgE molecules (Novak et al. 2004a). IgE receptor expression of pDC correlates with the IgE serum levels, indicating that IgE in the micromilieu might be necessary to stabilize this structure on the cell surface of pDC. Further on, activation of FcεRI on pDC counter-regulates the toll-like receptor (TLR)-9 pathway involved in the regulation of the type I IFN production, which is required for the defence against virus infections (Schroeder et al. 2005). Aggregation of FcεRI on pDC induces the release of IL-10 and increases in an endogenous loop IL-10-mediated apoptosis of pDC in vitro. Further on, the pre-activation of pDC via allergen challenge significantly reduces the capacity of pDC to produce IFN-α and IFN-β in response to subsequent stimulation with viral DNA motifs (Novak et al., 2004a). The reduced capacity of pDC to produce type I IFN after allergen challenge might be one of the reasons for the high susceptibility of AD patients to viral infections. There is also evidence for a reduced amount of pDC in the epidermis of AD patients in comparison to other chronic inflammatory skin diseases such as psoriasis, allergic contact dermatitis, or lupus erythematodes which might be based on a lower recruitment of these cells into the skin due to reduced expression of skin homing molecules on pDC of atopic donors or a higher rate of apoptosis of pDC in the Th2-prone micromilieu of AD skin (Wollenberg et al. 2002). The Th2-predominated immune state (Wollenberg et al. 2003; Novak and Peng 2005), modifications on the level of pDC together with other deficiencies of the innate immune system, for instance, reduced amounts of antimicrobial peptides such as cathelicidin in the skin of AD patients (Ong et al. 2002; Howell et al. 2006) have been identified as important risk factors for the manifestation of bacterial and viral infections of the skin, which significantly aggravate the course of the disease in a subgroup of AD patients.

5. Conclusions

Distinct DC subtypes in the blood and in the skin of AD patients play a pivotal role in the pathophysiology of this chronic inflammatory skin disease. Targeting DC subtypes by boosting their anti-inflammatory, pro-tolerogenic capacities while impairing their pro-inflammatory, pro-allergic attributes would represent a promising therapeutic strategy to take advantage of DCs and the complex network of DCs in AD.

Acknowledgments

This work was supported by grants from the DFG (NO454/1-3, NO454/2-3), the University of Bonn (BONFOR), and in part, by the NIH/NIAID contract N01 AI40029. N.N. was supported by a Heisenberg-Fellowship of the DFG NO454/3-1. We are thankful to Andreas Neubauer for critical reading of this manuscript.

References

Bieber, T., de la Salle, H., Wollenberg, A., Hakimi, J., Chizzonite, R., Ring, J., Hanau, D. and de la Salle, C. (1992) Human epidermal Langerhans cells express the high affinity receptor for immunoglobulin E (Fc epsilon RI). J. Exp. Med. 175, 1285–1290.

Bonasio, R. and von Andrian, U.H. (2006) Generation, migration and function of circulating dendritic cells. Curr. Opin. Immunol. 18, 503–511.

Cookson, W. (2004) The immunogenetics of asthma and eczema: a new focus on the epithelium. Nat. Rev. Immunol. 4, 978–988.

Cork, M.J., Robinson, D.A., Vasilopoulos, Y., Ferguson, A., Moustafa, M., MacGowan, A., Duff, G.W., Ward, S.J. and Tazi-Ahnini, R. (2006) New perspectives on epidermal barrier dysfunction in atopic dermatitis: gene-environment interactions. J. Allergy Clin. Immunol. 118, 3–21.

Grewe, M., Walther, S., Gyufko, K., Czech, W., Schopf, E. and Krutmann, J. (1995) Analysis of the cytokine pattern expressed in situ in inhalant allergen patch test reactions of atopic dermatitis patients. J. Invest. Dermatol. 105, 407–410.

Gunther, C., Bello-Fernandez, C., Kopp, T., Kund, J., Carballido-Perrig, N., Hinteregger, S., Fassl, S., Schwarzler, C., Lametschwandtner, G., Stingl, G., Biedermann, T. and Carballido, J.M. (2005) CCL18 is expressed in atopic dermatitis and mediates skin homing of human memory T cells. J. Immunol. 174, 1723–1728.

Homey, B., Steinhoff, M., Ruzicka, T. and Leung, D.Y. (2006) Cytokines and chemokines orchestrate atopic skin inflammation. J. Allergy Clin. Immunol. 118, 178–189.

Howell, M.D., Wollenberg, A., Gallo, R.L., Flaig, M., Streib, J.E., Wong, C., Pavicic, T., Boguniewicz, M. and Leung, D.Y. (2006) Cathelicidin deficiency predisposes to eczema herpeticum. J. Allergy Clin. Immunol. 117, 836–841.

Jolles, S. (2002) Paul Langerhans. J. Clin. Pathol. 55, 243.

Kerschenlohr, K., Decard, S., Przybilla, B. and Wollenberg, A. (2003) Atopy patch test reactions show a rapid influx of inflammatory dendritic epidermal cells in patients with extrinsic atopic dermatitis and patients with intrinsic atopic dermatitis. J. Allergy Clin. Immunol. 111, 869–874.

Leung, D.Y. and Bieber, T. (2003) Atopic dermatitis. Lancet 361, 151–160.

Mothes, N., Niggemann, B., Jenneck, C., Hagemann, T., Weidinger, S., Bieber, T., Valenta, R. and Novak, N. (2005) The cradle of IgE autoreactivity in atopic eczema lies in early infancy. J. Allergy Clin. Immunol. 116, 706–709.

Novak, N., Allam, J.P., Hagemann, T., Jenneck, C., Laffer, S., Valenta, R., Kochan, J. and Bieber, T. (2004a) Characterization of FcepsilonRI-bearing CD123 blood dendritic cell antigen-2 plasmacytoid dendritic cells in atopic dermatitis. J. Allergy Clin. Immunol. 114, 364–370.

Novak, N., Valenta, R., Bohle, B., Laffer, S., Haberstok, J., Kraft, S. and Bieber, T. (2004b) FcepsilonRI engagement of Langerhans cell-like dendritic cells and inflammatory dendritic epidermal cell-like dendritic cells induces chemotactic signals and different T cell phenotypes *in vitro*. J. Allergy Clin. Immunol. 113, 949–957.

Novak, N., Bieber, T. and Leung, D.Y. (2003a) Immune mechanisms leading to atopic dermatitis. J. Allergy Clin. Immunol. 112, S128–S139.

Novak, N., Kraft, S. and Bieber,T. (2003b) Unraveling the mission of FcepsilonRI on antigen-presenting cells. J. Allergy. Clin. Immunol. 111, 38–44.

Novak, N., Tepel, C., Koch, S., Brix, K., Bieber, T.,and Kraft, S. (2003c) Evidence for a differential expression of the FcepsilonRIgamma chain in dendritic cells of atopic and nonatopic donors. J. Clin. Invest. 111, 1047–1056.

Novak, N., Kwiek, B. and Bieber, T. (2005) The mode of topical immunomodulators in the immunological network of atopic dermatitis. Clin. Exp. Dermatol. 30, 160–164.

Novak, N. and Peng, W. (2005) Dancing with the enemy: the interplay of herpes simplex virus with dendritic cells. Clin. Exp. Immunol. 142, 405–410.

Ong, P.Y., Ohtake, T., Brandt, C., Strickland, I., Boguniewicz, M., Ganz, T., Gallo, R.L. and Leung, D.Y. (2002) Endogenous antimicrobial peptides and skin infections in atopic dermatitis. N. Engl. J. Med. 347, 1151–1160.

Palmer, C.N., Irvine, A.D., Terron-Kwiatkowski, A., Zhao, Y., Liao, H., Lee, S.P., Goudie, D.R., Sandilands, A., Campbell, L.E., Smith, F.J., O'Regan, G.M., Watson, R.M., Cecil, J.E., Bale, S.J., Compton, J.G., DiGiovanna, J.J., Fleckman, P., Lewis-Jones, S., Arseculeratne, G., Sergeant, A., Munro, C.S., El, H.B., McElreavey, K., Halkjaer, L.B., Bisgaard, H., Mukhopadhyay, S. and McLean, W.H. (2006) Common loss-of-function variants of the epidermal barrier protein filaggrin are a major predisposing factor for atopic dermatitis. Nat. Genet. 38, 441–446.

Pivarcsi, A., Gombert, M., Dieu-Nosjean, M.C., Lauerma, A., Kubitza, R., Meller, S., Rieker, J., Muller, A., Da, C.L., Haahtela, A., Sonkoly, E., Fridman, W.H., Alenius, H., Kemeny, L., Ruzicka, T., Zlotnik, A. and Homey, B. (2004) CC chemokine ligand 18, an atopic dermatitis-associated and dendritic cell-derived chemokine, is regulated by staphylococcal products and allergen exposure. J. Immunol. 173, 5810–5817.

Schroeder, J.T., Bieneman, A.P., Xiao, H., Chichester, K.L., Vasagar, K., Saini, S. and Liu, M.C. (2005) TLR9- and FcepsilonRI-mediated responses oppose one another in plasmacytoid dendritic cells by down-regulating receptor expression. J. Immunol. 175, 5724–5731.

Schuller, E., Oppel, T., Bornhovd, E., Wetzel, C. and Wollenberg, A. (2004) Tacrolimus ointment causes inflammatory dendritic epidermal cell depletion but no Langerhans cell apoptosis in patients with atopic dermatitis. J. Allergy Clin. Immunol. 114, 137–143.

Simon, D., Vassina, E., Yousefi, S., Kozlowski, E., Braathen, L.R. and Simon, H.U. (2004) Reduced dermal infiltration of cytokine-expressing inflammatory cells in atopic dermatitis after short-term topical tacrolimus treatment. J. Allergy Clin. Immunol. 114, 887–895.

Soumelis, V. and Liu, Y.J. (2006) From plasmacytoid to dendritic cell: morphological and functional switches during plasmacytoid pre-dendritic cell differentiation. Eur. J. Immunol. 36, 2286–2292.

Stary, G., Bangert, C., Stingl, G. and Kopp, T. (2005) Dendritic cells in atopic dermatitis: expression of FcepsilonRI on two distinct inflammation-associated subsets. Int. Arch. Allergy Immunol. 138, 278–290.

Strid, J. and Strobel, S. (2005) Skin barrier dysfunction and systemic sensitization to allergens through the skin. Cur. Drug Targets Inflamm. Allergy 4, 531–541.

Villadangos, J.A. and Heath, W.R. (2005) Life cycle, migration and antigen presenting functions of spleen and lymph node dendritic cells: limitations of the Langerhans cells paradig. Semin. Immunol. 17, 262–272.

Wang, B., Rieger, A., Kilgus, O., Ochiai, K., Maurer, D., Fodinger, D., Kinet, J.P. and Stingl, G. (1992) Epidermal Langerhans cells from normal human skin bind monomeric IgE via Fc epsilon receptor. J. Exp. Med. 175, 1353–1365.

Wang, Y.H., Ito, T., Wang, Y.H., Homey, B., Watanabe, N., Martin, R., Barnes, C.J., McIntyre, B.W., Gilliet, M., Kumar, R., Yao, Z. and Liu, Y.J. (2006) Maintenance and polarization of human TH2 central memory T cells by thymic stromal lymphopoietin-activated dendritic cells. Immunity 24, 827–838.

Weidinger, S., Illig, T., Baurecht, H., Irvine, A.D., Rodriguez, E., Diaz-Lacava, A., Klopp, N., Wagenpfeil, S., Zhao, Y., Liao, H., Lee, S.P., Palmer, C.N., Jenneck, C., Maintz, L., Hagemann, T., Behrendt, H., Ring, J., Nothen, M.M., McLean, W.H. and Novak, N. (2006) Loss-of-function variations within the filaggrin gene predispose for atopic dermatitis with allergic sensitizations. J. Allergy Clin. Immunol. 118, 214–219.

Wollenberg, A., Wagner, M., Gunther, S., Towarowski, A., Tuma, E., Moderer, M., Rothenfusser, S., Wetzel, S., Endres, S. and Hartmann, G. (2002) Plasmacytoid dendritic cells: a new cutaneous dendritic cell subset with distinct role in inflammatory skin diseases. J. Invest. Dermatol. 119, 1096–1102.

Wollenberg, A., Wen, S. and Bieber, T. (1995) Langerhans cell phenotyping: a new tool for differential diagnosis of inflammatory skin diseases [letter]. Lancet 346, 1626–1627.

Wollenberg, A., Zoch, C., Wetzel, S., Plewig, G. and Przybilla, B. (2003) Predisposing factors and clinical features of eczema herpeticum: a retrospective analysis of 100 cases. J. Am. Acad. Dermatol. 49, 198–205.

11

Peptide-Based Therapy in Lupus: Promising Data

Fanny Monneaux and Sylviane Muller

Institut de Biologie Moléculaire et Cellulaire, 67000 Strasbourg, France,
S.Muller@ibmc.u-strasbg.fr

Abstract. Systemic lupus erythematosus (SLE) is a multisystem chronic inflammatory disease of multifactorial aetiology, characterized by inflammation and damage of various tissues and organs. Current treatments of the disease are mainly based on immunosuppressive drugs such as corticosteroids and cyclophosphamide. Although these treatments have reduced mortality and morbidity, they cause a non-specific immune suppression. To avoid these side effects, our efforts should focus on the development of alternative therapeutic strategies, which consist, for example in specific T cell targeting using autoantigen-derived peptides identified as sequences encompassing major epitopes.

1. Introduction

Systemic lupus erythematosus (SLE) is a multisystem chronic inflammatory disease of multifactorial aetiology, characterized by inflammation and damage of various tissues, including the joints, skin, kidneys, heart, lungs, blood vessels and brain (Kotzin 1996). At the immunological and biological levels, SLE is characterized by complement deficiencies, modification of cytokine secretion, hypergammaglobulinemia, production of autoantibodies and formation of immune complexes, which play a crucial role in associated glomerulonephritis and cutaneous lesions (Toskos 1999). At least 100 different antigens targeted by specific antibodies have been characterized in SLE. These antibodies often target components of the nucleus and more specifically macromolecular complexes such as the nucleosome, spliceosome, and Ro/La particle (Mattioli and Reichlin 1973; Pinnas et al. 1973). The cause of the illness is unknown, but it is clear that it depends on three types of risk factors, namely a genetic part, a hormonal influence with a female prevalence (9:1) and a part of acquisition that remains extremely difficult to evaluate. Environmental factors cannot only exacerbate existing lupus conditions but can also trigger the initial onset. They include extreme stress, exposure to sunlight and infections, for example. In its worst form, lupus can be fatal: 10% of patients die from kidney disease, cardiovascular disease or infections. Current treatments of the disease are mainly based on immunosuppressive drugs such as corticosteroids and cyclophosphamide, which are often administered at high doses in acute exacerbation phases. Although these treatments have reduced mortality and significantly lengthened patients' life expectancies, they cause a non-specific immune suppression leading to side effects, which are sometimes worse than the disease itself. Certain adverse effects such as obesity, diabetes mellitus, hyperlipidemia and hypertension are reversible and generally improve after reducing the corticoid dosage. However, they may contribute to late irreversible complications. To avoid such side effects, our efforts should focus on the development of alternative therapeutic strategies, which consist, for example, in specific T cell targeting using autoantigen-derived peptides identified as sequences encompassing major epitopes.

2. Are Peptides Suitable Agents for Therapeutics in Autoimmune Diseases?

Synthetic peptides have long been recognized as potential candidates for therapeutic vaccines in autoimmune diseases (Fairchild 1997; Anderton 2001; Sela and Mozes 2004; Larché and Wraith 2005, Root-Bernstein 2006). One important advantage of short peptides is that in the absence of adjuvant, their immunogenicity is very weak, making them very attractive immunomodulatory agents. In general, however, peptides are short-lived molecules, which degrade rapidly in biological fluids, and this poor stability has been used to deny their potential usefulness. It should be mentioned that in contrast to what is generally believed, all peptides are not immediately cleaved in serum; certain sequences possess a sufficient intrinsic stability allowing their target, such as major histocompatibility complex (MHC) molecules or albumin, which serves at the periphery as a carrier "sponge." However, to overcome this instability drawback, several attempts have been made, replacing standard peptides by more stable peptidomimetics. Thus, some pseudopeptides, which contain modified peptide bonds, possess favourable structural and antigenic properties, which can have a considerable interest for therapeutic applications (Jameson et al. 1994; Allen et al. 2005). For example, B cell epitopes (and autoepitopes) can be very efficiently mimicked by means of peptide analogues, which contain changes in the natural CO – NH peptide bond, such as retro-inverso and reduced peptide bond analogues (Guichard et al. 1994; Briand et al. 1995; Benkirane et al. 1996). Such analogues can have mean half-life increased by ten to several hundred in serum (Stemmer et al. 1999; Ben-Yedidia et al. 2002). All D-peptides, which are also very resistant to proteases, can be efficiently recognized by anticarbohydrate antibodies (Pinilla et al. 1998).

Some pseudopeptide analogues readily bind class I and II MHC molecules and can also induce differential effects on T cell responsiveness, similar to those described with altered peptide ligands, which contain single amino acid replacements. For example, retro-inverso analogues corresponding to two $CD4^+$ T cell epitopes (one corresponded to the third constant region of mouse heavy chain IgG2a allotype $\gamma 2a^b$, which mimics a corneal antigen implicated in autoimmune keratitis) were shown to retain their binding capacity to murine I-E^d and I-A^d MHC class II molecules and induce in vivo T cell responses equivalent to those obtained with the wild-type peptides (Mézière et al. 1997). Recently, using the peptides 88–99 of histone H4, which contains a supradominant epitope recognized by Th cells induced to nucleosomes (Decker et al. 2000), we identified several analogues containing aza-β^3-amino acid residue substitutions Ψ[CONHNRCH$_2$], which were recognized by T cells generated to the cognate sequence (Dali et al. in press). Most interestingly, T cells primed to certain of these peptide analogues and recalled ex vivo with the nominal peptide secreting a distinct cytokine pattern (with IL-2 without IFN-γ or IFN-γ without IL-2, according to the considered analogue, or IL-2 and IFN-γ in the presence of the nominal peptide) (Dali et al. in press). Altogether, these results emphasize the considerable advantages that peptides can have for developing a new generation of therapeutic agents and immunomodulators.

3. Potential of Peptide Therapeutics for Treating Lupus Patients

Relatively recently has emerged the idea of using peptides described as sequences able to modulate the autoimmune response, as these peptides will only target specific autoreactive B and T cells (Singh 2000). Several successful attempts of peptide-based therapy have been described in murine model of lupus (Monneaux and Muller 2004). Some peptides corresponding to antibody idiotypes—for example the pCONS peptide, a consensus peptide derived from the variable heavy chain (VH) region of NZBxNZW (BW) IgG antibodies to DNA and predicted to possess T cell stimulatory activity (Hahn et al. 2001; Hahn et al. 2005), or peptides derived from the sequence of the complementary-determining regions 1 and 3 (pCDR1 and pCDR3) of a human anti-DNA monoclonal antibody that bears the so-called 16/6 Id (Zinger et al. 2003)—have been used with remarkable efficacy in BW lupus mice. An impressive protective effect was also observed in MRL/lpr mice with a peptide identified using combinatorial chemistry approaches and able to interfere with Fcγ-receptor recognition (Marino et al. 2000). For therapeutic application, this immunoglobulin-binding peptide (called TG19320) was used as a protease-resistant tetrameric tripeptide containing D-amino acid residues.

Regarding peptides from nuclear autoantigens, Datta and colleagues showed that repeated intravenous (i.v.) or intraperitoneal (i.p.) administration into (SWRxNZB)F1 (SNF1) lupus mice with established glomerulonephritis of a single peptide of histone H4 (sequences 16–39), which behaves as a “promiscuous” T cell epitope, prolonged survival of treated animals and halted progression of renal disease (Kaliyaperumal et al. 1999). The protective properties of another peptide of histone H4 (sequences 71–93) accompanied by an increased level of IL-10 and suppression of IFN-γ secreted by lymph node (LN) cells were described by Staines and collaborators who administrated the peptide to SNF1 mice by the intranasal (i.n.) route (Wu et al. 2002). Wu and Staines (2004) showed further that following i.n. (but not intradermal) administration of H4 peptides 71–93, the number of CD4/CD25$^+$ regulatory T cells, which is low in BW and SNF1 mice as compared with normal mice, was restored in both strains. Very low-dose therapy of SNF1 mice with H4 peptides 71–94 was also found to induce CD8$^+$ and CD4/CD25$^+$ regulatory T cells containing autoantigen-specific cells to decrease IFN-γ levels secreted by pathogenic T cells and to decrease the antibody levels by 90–100% (Kang et al. 2005). The histone H3 peptides 111–130 encompassing a T cell epitope in BW mice was also used with success when administrated intradermally (4 X 100 μg in Freund's adjuvant) into BW mice (Suen et al. 2004). Treatment (5 X 80 μg i.p. in saline weekly for 260 days) of MRL/lpr mice with a 21-mer peptide of laminin α-chain targeted by lupus antibodies also prevented antibody deposition in the kidneys, ameliorated renal disease, decreased the weight gain caused by accumulating ascitic fluid and markedly improved longevity of treated mice (Amital et al. 2005).

4. A New Peptide Analogue for Treating Lupus Patients? The Potential of Peptide P140

By testing a series of overlapping peptides, we identified an epitope present in residues 131–151 of the spliceosomal U1-70K small nuclear ribonucleoprotein (snRNP), recognized very early by IgG antibodies and $CD4^+$ LN T cells from both $H\text{-}2^k$ MRL/lpr and $H\text{-}2^{d/z}$ BW lupus-prone mice (Monneaux et al. 2000, 2001). Fibroblasts transfected with MHC class II molecules were used to demonstrate that peptides 131–151 readily binds $I\text{-}A^k$, $I\text{-}E^k$, $I\text{-}A^d$ and $I\text{-}E^d$ murine MHC molecules (Monneaux et al. 2000, 2001). We further showed that an analogue of this sequence phosphorylated on Ser^{140} (named peptide P140) was strongly recognized by LN and peripheral $CD4^+$ T cells and by IgG antibodies from MRL/lpr mice (Monneaux, 2003, 2004). This analogue and the cognate peptides 131–151 were used in therapeutic trials in lupus-prone mice to investigate their ability to restore tolerance. Young MRL/lpr mice were given the peptides i.v. in saline (4 X 100 μg), and we found that P140 peptide, but not the non-phosphorylated peptide 131-151, reduced proteinuria and dsDNA IgG antibody levels and significantly enhanced the survival of treated mice (Monneaux et al. 2003). When administrated s.c. in Freund's adjuvant, P140 peptide accelerated lupus nephritis.

Our studies revealed that P140 peptide and the non-phosphorylated corresponding sequence behave as promiscuous epitopes (Monneaux et al. 2005). As in the mouse model, we found that peptides 131–151 of the U1-70K protein induces ex vivo proliferation of $CD4^+$ T cells from lupus patients. Interestingly, however, we observed that phosphorylation of Ser^{140} prevented proliferation while favouring secretion of high levels of regulatory cytokines, which are produced specifically when lupus patients' peripheral CD4 T cells are incubated in the presence of P140 analogue (Monneaux et al. 2005).

The identification of a tolerogenic $CD4^+$ T cell epitope within P140 peptide is remarkable, because this sequence, which is completely conserved in the mouse and human U1-70K protein, contains an RNA-binding motif called RNP1, also present in other sn/hnRNPs and often targeted by antibodies from lupus patients and mice. The 131–151 sequence of the spliceosomal U1-70K protein is located within an 80–90 amino acid-long RNA-binding domain. It encompasses a conserved sequence, called RNP1 motif, which is also present in other RNA-binding proteins, such as snRNP (e.g. U1-A) and heterogeneous nuclear (hn)RNP (e.g. hnRNP-A2/B1) proteins.

Starting from the observation that sequences containing this RNP1 motif are often targeted by antibodies from lupus patients and mice, we hypothesized that the RNP1 motif could be involved in the earliest stages of the T–B intramolecular diversification process to other regions of one of the spliceosomal proteins containing this unique motif and might promote intermolecular spreading to epitopes of other proteins present within the same spliceosomal particle and containing or not an RNP1 motif (Monneaux and Muller 2001, 2002). We experimentally demonstrated that an intramolecular T and B cell spreading effectively occurs in MRL/lpr mice tested at different ages (Monneaux et al. 2004). Moreover, we showed that repeated administration of phosphorylated analogue P140 in saline into pre-autoimmune MRL/lpr mice transiently abolishes both T cell intramolecular spreading to other regions of the U1-70K protein (Monneaux et al. 2004) and T cell intermolecular spreading to regions of other spliceosomal proteins, suggesting that the P140 analogue might originate a mechanism of so-called "tolerance spreading."

These results are extremely promising. From a conceptual point of view, however, we have to determine how in such a polymorphic and multifactorial pathology, administration of a single peptide (P140 peptide, histone peptides or peptides from other self-antigens) can be sufficient to suppress a complex autoimmune response to multiple cell components. Several possible pathways by which peptide P140 might exert a modulating effect are currently envisaged. P140 peptide could act as a partial agonist of the autoreactive T cell receptor, as suggested by our results obtained with lupus patients' T cells (Monneaux et al. 2005). It may also expand the regulatory T cell pool and/or restore defective regulatory T cell function as seen in the case of therapeutic histone peptides (see above). These and other possible mechanisms are presently under evaluation.

5. Final Remarks

Peptides encompassing T cell epitopes represent promising tools for manipulating immune regulation in autoimmune diseases, such as lupus. Examples in other experimental models of autoimmunity (e.g. in experimental autoimmune encephalomyelitis, experimental myasthenia gravis or diabetic NOD mice) also show spectacular protective effects. It is possible that for long-term therapeutic applications it will be necessary to combine several independent strategies, which can be introduced simultaneously or sequentially. These strategies could target T cells but also B cells, $CD8^+$ suppressor cells and regulatory T and B cells. The results obtained in different laboratories are quite encouraging and a number of T cell-specific agents with clinical potential have recently emerged (Isenberg and Rahman 2006; Kaul et al. 2006; Liu et al. 2007). The P140 peptide—which interferes with intra and intermolecular epitope spreading in MRL/lpr mice and consequently interrupts, at least transiently, the spiral of events leading to antibody production and tissue inflammation—represents a potential candidate that deserves most attention. Since in addition we have shown that P140 therapy does not alter immune response against a viral challenge (Monneaux et al. submitted), it might constitute the basis of a specific and safe immunointervention strategy in lupus.

Acknowledgements

This work is supported by CNRS, the Fondation pour la Recherche Médicale, the Région Alsace and ImmuPharma-France.

References

Allen, S.D., Rawale, S.V., Whitacre, C.C. and Kaumaya, P.T. (2005) Therapeutic peptidomimetic strategies for autoimmune diseases: costimulation blockade. J. Pept. Res. 65, 591–604.
Amital, H., Heilweil, M., Ulmansky, R., Szafer, F., Bar-Tana, R., Morel, L., Foster, M.H., Mostoslavsky, G., Eilat, D., Pizov, G. and Naparstek, Y. (2005) Treatment with a laminin-derived peptide suppresses lupus nephritis. J. Immunol. 175, 5516–5523.
Anderton, S.M. (2001) Peptide-based immunotherapy of autoimmunity: a path of puzzles, paradoxes and possibilities. Immunology 104, 367–376.
Benkirane, N., Guichard, G., Briand, J.P. and Muller, S. (1996) Exploration of requirements for peptidomimetic immune recognition. Antigenic and immunogenic properties of reduced peptide bond pseudopeptide analogues of a histone hexapeptide. J. Biol. Chem. 271, 33218–33224.
Ben-Yedidia, T., Beignon, A.S., Partidos, C.D., Muller, S. and Arnon, R. (2002) A retro-inverso peptide analogue of influenza virus hemagglutinin B-cell epitope 91-108 induces a strong mucosal and systemic immune response and confers protection in mice after intranasal immunization. Mol. Immunol. 39, 323–331.
Briand, J.P., Guichard, G., Dumortier, H. and Muller, S. (1995) Retro-inverso peptidomimetics as new immunological probes. Validation and application to the detection of antibodies in rheumatic diseases. J. Biol. Chem. 270, 20686–20691.
Dali, H., Busnel, O., Hoebeke, J., Bi, L., Decker, P., Briand, J.P., Baudy-Floc'h, M. and Muler, S. Heteroclitic properties of mixed-α-and aza-β3-peptides mimicking a supradominant CD4 T cell epitope presented by nucleosome. Mol. Immunol. 44, 3024–3036.
Decker, P., Le Moal, A., Briand, J.P. and Muller, S. (2000) Identification of a minimal T cell epitope recognized by antinucleosome Th cells in the C-terminal region of histone H4. J. Immunol. 165, 654–662.
Fairchild, P.J. (1997) Altered peptide ligands: prospects for immune intervention in autoimmune disease. Eur. J. Immunogenet. 24, 155–167.
Guichard, G., Benkirane, N., Zeder-Lutz, G., van Regenmortel, M.H., Briand, J.P. and Muller, S. (1994) Antigenic mimicry of natural L-peptides with retro-inverso-peptidomimetics. Proc. Natl. Acad. Sci. U.S.A. 91, 9765–9769.
Hahn, B.H., Singh, R.P., La Cava, A. and Ebling, F.M. (2005) Tolerogenic treatment of lupus mice with consensus peptide induces Foxp3-expressing, apoptosis-resistant, TGFbeta-secreting CD8+ T cell suppressors. J. Immunol. 175, 7728–7737.
Hahn, B.H., Singh, R.R., Wong, W.K., Tsao, B.P., Bulpitt, K. and Ebling, F.M. (2001) Treatment with a consensus peptide based on amino acid sequences in autoantibodies prevents T cell activation by autoantigens and delays disease onset in murine lupus. Arthritis Rheum. 44, 432–441.
Isenberg, D. and Rahman, A. (2006) Systemic lupus erythematosus--2005 annus mirabilis? Nat. Clin. Pract. Rheumatol. 2, 145–152.
Jameson, B.A., McDonnell, J.M., Marini, J.C. and Korngold, R. (1994) A rationally designed CD4 analogue inhibits experimental allergic encephalomyelitis. Nature 368, 744–746.

Kaliyaperumal, A., Michaels, M.A. and Datta, S.K. (1999) Antigen-specific therapy of murine lupus nephritis using nucleosomal peptides: tolerance spreading impairs pathogenic function of autoimmune T and B cells. J. Immunol. 162, 5775–5783.

Kang, H.K., Michaels, M.A., Berner, B.R. and Datta, S.K. (2005) Very low-dose tolerance with nucleosomal peptides controls lupus and induces potent regulatory T cell subsets. J. Immunol. 174, 3247–3255.

Kaul, A., D'Cruz, D. and Hughes, G.R.V. (2006) New therapies for systemic lupus erythematosus: has the future arrived? Future Rheumatol. 1, 235–247.

Kotzin, B.L. (1996) Systemic lupus erythematosus. Cell 85, 303–306.

Larché, M. and Wraith, D.C. (2005) Peptide-based therapeutic vaccines for allergic and autoimmune diseases. Nat. Med. 11, S69–S76.

Liu, E.H., Siegel, R.M., Harlan, D.M. and O'Shea, J.J. (2007) T cell-directed therapies: lessons learned and future prospects. Nat. Immunol. 8, 25–30.

Marino, M., Ruvo, M., De Falco, S. and Fassina, G. (2000) Prevention of systemic lupus erythematosus in MRL/lpr mice by administration of an immunoglobulin-binding peptide. Nat. Biotechnol. 18, 735–739.

Mattioli, M. and Reichlin, M. (1973) Physical association of two nuclear antigens and mutual occurrence of their antibodies: the relationship of the SM and RNAprotein (MO) systems in SLE sera. J. Immunol. 110, 1318–1324.

Mézière, C., Viguier, M., Dumortier, H., Lo-Man, R., Leclerc, C., Guillet, J.G., Briand, J.P. and Muller, S. (1997) In vivo T helper cell response to retro-inverso peptidomimetics. J. Immunol. 159, 3230–3237.

Monneaux, F., Briand, J.P. and Muller, S. (2000) B and T cell immune response to small nuclear ribonucleoprotein particles in lupus mice: autoreactive CD4(+) T cells recognize a T cell epitope located within the RNP80 motif of the 70K protein. Eur. J. Immunol. 30, 2191–2200.

Monneaux, F., Dumortier, H., Steiner, G., Briand, J.P. and Muller, S. (2001) Murine models of systemic lupus erythematosus: B and T cell responses to spliceosomal ribonucleoproteins in MRL/Fas(lpr) and (NZB x NZW)F(1) lupus mice. Int. Immunol. 13, 1155–1163.

Monneaux, F., Hoebeke, J., Sordet, C., Nonn, C., Briand, J.P., Maillère, B., Sibillia, J. and Muller, S. (2005) Selective modulation of CD4+ T cells from lupus patients by a promiscuous, protective peptide analog. J. Immunol. 175, 5839–5847.

Monneaux, F., Lozano, J.M., Patarroyo, M.E., Briand, J.P. and Muller, S. (2003) T cell recognition and therapeutic effect of a phosphorylated synthetic peptide of the 70K snRNP protein administered in MR/lpr mice. Eur. J. Immunol. 33, 287–296.

Monneaux, F. and Muller, S. (2001) Key sequences involved in the spreading of the systemic autoimmune response to spliceosomal proteins. Scand. J. Immunol. 54, 45–54.

Monneaux, F. and Muller, S. (2002) Epitope spreading in systemic lupus erythematosus: identification of triggering peptide sequences. Arthritis Rheum. 46, 1430–1438.

Monneaux, F. and Muller, S. (2004) Peptide-based immunotherapy of systemic lupus erythematosus. Autoimmun. Rev. 3, 16–24.

Monneaux, F., Parietti, V., Briand, J.P. and Muller, S. (2004) Intramolecular T cell spreading in unprimed MRL/lpr mice: importance of the U1-70k protein sequence 131–151. Arthritis Rheum. 50, 3232–3238.

Pinilla, C., Appel, J.R., Campbell, G.D., Buencamino, J., Benkirane, N., Muller, S. and Greenspan, N.S. (1998) All-D peptides recognized by an anti-carbohydrate antibody identified from a positional scanning library. J. Mol. Biol. 283, 1013–1025.

Pinnas, J.L., Northway, J.D. and Tan, E.M. (1973) Antinucleolar antibodies in human sera. J. Immunol. 111, 996–1004.

Root-Bernstein, R. (2006) Peptides vaccines against arthritis. Future Rheumatol. 1, 339–344.

Sela, M. and Mozes, E. (2004) Therapeutic vaccines in autoimmunity. Proc. Natl. Acad. Sci. U.S.A. 101 Suppl 2, 14586–14592.
Singh, R.R. (2000) The potential use of peptides and vaccination to treat systemic lupus erythematosus. Curr. Opin. Rheumatol. 12, 399–406.
Stemmer, C., Quesnel, A., Prévost-Blondel, A., Zimmermann, C., Muller, S., Briand, J.P. and Pircher, H. (1999) Protection against lymphocytic choriomeningitis virus infection induced by a reduced peptide bond analogue of the H-2Db-restricted CD8(+) T cell epitope GP33. J. Biol. Chem. 274, 5550–5556.
Suen, J.L., Chuang, Y.H., Tsai, B.Y., Yau, P.M. and Chiang, B.L. (2004) Treatment of murine lupus using nucleosomal T cell epitopes identified by bone marrow-derived dendritic cells. Arthritis Rheum. 50, 3250–3259.
Toskos, G.C. (1999) Overview of cellular function in systemic lupus erytheatosus. In: Lahita R. (Ed), *Systemic Lupus Erythematosus*, 3rd ed., Academic Press, San Dego, pp. 17–54.
Wu, H.Y. and Staines, N.A. (2004) A deficiency of CD4+CD25+ T cells permits the development of spontaneous lupus-like disease in mice, and can be reversed by induction of mucosal tolerance to histone peptide autoantigen. Lupus 13, 192–200.
Wu, H.Y., Ward, F.J. and Staines, N.A. (2002) Histone peptide-induced nasal tolerance: suppression of murine lupus. J. Immunol. 169, 1126–1134.
Zinger, H., Eilat, E., Meshorer, A. and Mozes, E. (2003) Peptides based on the complementarity-determining regions of a pathogenic autoantibody mitigate lupus manifestations of (NZB x NZW)F1 mice via active suppression. Int. Immunol. 15, 205–214.

12

Reduced Number and Function of $CD4^{+}CD25^{high}FoxP3^{+}$ Regulatory T Cells in Patients with Systemic Lupus Erythematosus

Elena Yu. Lyssuk, Anna V. Torgashina, Sergey K. Soloviev, Evgeny L. Nassonov, and Svetlana N. Bykovskaia

Laboratory of Cell Monitoring, Institution of Rheumatology, Russian Academy of Medical Sciences, Moscow, Russia, sbykovk@mailcity.com

Abstract. $CD4^{+}CD25^{+}$ regulatory T cells (Tregs) play an important role in maintaining tolerance to self-antigens controlling occurrence of autoimmune diseases. Recently, it has been shown that the transcription factor forkhead box P3 (FoxP3) is specifically expressed on $CD4^{+}CD25^{+}$ T cells. FoxP3 has been described as the master control gene for the development and function of Tregs. We characterized $CD4^{+}CD25^{+}CTLA\text{-}4^{+}FoxP3^{+}$ T cells in 43 patients with systemic lupus erythematosus (SLE). Twenty of them comprised a group of newly admitted patients with the first manifestations of the disease, and the second group included patients that were treated with cytostatics and steroids. The results revealed a significant decrease in $CD4^{+}CD25^{+}$ and $CD4^{+}CD25^{high}$ T cells numbers in patients from group I compared with control and group II patients. Coexpression of FoxP3 on $CD4^{+}CD25^{+}$ T cells was significantly reduced in both groups regardless the therapy. The ability of Tregs to suppress proliferation of autologous $CD8^{+}$ and $CD4^{+}$ T cells was significantly reduced in both groups of patients compared to healthy donors. Our data revealed impaired production of Tregs in SLE patients that can be partly restored by conventional treatments.

1. Introduction

Regulatory $CD4^{+}CD25^{+}$ T cells (Tregs) represent a small subset of $CD4^{+}$ T cells that constitutively express the IL-2 receptor-α chain, CTL-associated antigen 4 (CTLA-4), glucocorticoid-induced TNF receptor (GITR), and class II MHC markers (Beacher-Allan et al. 2001). Unique lineage of immunoregulatory $CD4^{+}CD25^{+}$ T cells comprises ~5–10% of $CD4^{+}$ T cell population. Tregs do not proliferate in response to T cell antigens but can inhibit activation of other T cells by the contact-dependent or cytokine-mediated mechanisms (Shevach 2001). The Treg-specific gene, forkhead box protein P3 (FoxP3), encodes a transcription factor, which is explicitly expressed in Tregs (Hori et al. 2003). FoxP3 acts as a negative regulator of cytokine production by CD4 T cells and repress transcription of IL-2 and other cytokine genes including IL-4 and IFN-γ (Schubert et al. 2001).

FoxP3 can be induced upon TCR-mediated activation (Allan et al. 2005), and the function of FoxP3 is not restricted to Tregs (Chen et al. 2005). Small numbers of human $CD4^+$ and $CD8^+$ T cells transiently upregulated FoxP3 upon in vitro stimulation (Gavin et al. 2006). However, at present, high levels of expression of CD25 and FoxP3 are the most valued markers of $CD4^+CD25^{high}$ Tregs.

In human, autoimmune diseases, including diabetes (Lindley et al. 2005), multiple sclerosis (Viglietta et al. 2004), rheumatoid arthritis (van Amelsfort et al. 2004), psoriasis (Sugiyama et al. 2005), and type II autoimmune polyendocrinopathy (Kriegel et al. 2004), are characterized by low numbers and/or defective function of $CD4^+CD25^{high}$ T cells. However, reports on Tregs in patients with systemic lupus erythematosus (SLE) are controversial. Recent data indicate diminished numbers of $CD4^+CD25^+$ T cells that mainly associated with the disease's flares but not in remission (Crispin et al. 2003; Liu et al. 2004; Lee et al. 2006; Miyara et al. 2005). However, Alvarado-Sanchez et al. (2006) did not show any significant differences in the levels of regulatory T cells in SLE patients. The explanation of these differences might be in the variety in treatments and stages of the disease.

We aimed to estimate the total number and function of $CD4^+CD25^+FoxP3^+CTLA\text{-}4^+$ T cells and $CD4^+CD25^{high}FoxP3^+$ cells in a group of newly admitted patients with early SLE manifestations prior to any therapy. Healthy volunteers and SLE patients after therapy served as controls.

2. Experimental Design

Forty-three patients with the diagnosis of active SLE according to the American College of Rheumatology criteria were enrolled in the study. The group included 35 women and 8 men, 14–49 (median 29.4) years old. Twenty newly admitted patients and 23 patients, who were recently treated with 3 g methylprednisolone/1 g cyclophosphamide pulse therapy, were studied prior to receiving corticosteroids. Seventeen matched healthy donors were included as controls.

Peripheral blood mononuclear cells (PBMCs) were isolated by Ficoll gradient centrifugation. Antibodies used for flow cytometry were anti-CD3-FITC, anti-CD25-FITC, anti-CD4-PE, anti-CD8-PE, anti-CD19-PC5 (Beckman Coulter), anti-CD19-PC5, anti-CD152-PC5, anti-CD25-FITC (BD Pharmingen), and appropriate isotype controls. For the detection of intracellular markers, PBMCs were stained with surface membrane antibodies (anti-CD25-FITC and anti-CD4-PE), fixed, permeabilized with 0.1% saponin, and stained with anti-Foxp3-PC5 or isotype control antibody (eBioscience). Flow cytometry analysis was performed on FACSCalubur (Becton Dickinson) using CellQuest Pro software. The results are presented as the percentage of positive cells.

For $CD4^+CD25^+$ cell isolation, PBMCs were labeled with Biotin-Antibody Cocktail and anti-Biotin MicroBeads (Miltenyi Biotec) followed by a negative selection of $CD4^+$ T cells. Cells from the negative fraction were labeled with CD25 MicroBeads and positive $CD4^+CD25^+$ cells were harvested. To increase the purity of cell populations, the positive fraction was isolated on magnetic columns two times, and the purity of $CD4^{+}CD25^{+}$ cells was >80–90%. For the isolation of $CD4^{+}$ and $CD8^{+}$ T cells, PBMCs were labeled with anti-CD4 or anti-CD8 MicroBeads, and the purity of these cells was >95–98%.

For the proliferation/suppression assay, CD3-depleated mitomycin C-treated PBMCs were used as accessory cells. All cultures were performed in 96-well plates in a final volume of 200 μl. 5 X 10^4 $CD4^{+}CD25^{-}$ and $CD4^{+}CD25^{+}$ cells were incubated in the presence of 5 X 10^4 accessory cells and anti-CD3 antibodies (5 μg/ml, ICO-90, MedBioSpectr). $CD4^{+}CD25^{+}$ cells were added to CFSE (carboxifluorescein succinimidylester, 5 μM, 10 min, 37° C Fluka)-labeled $CD4^{+}CD25^{-}$ cells at 1:1 ratio, and 6 days later, cell proliferation was estimated by flow cytometry as a reduction of CFSE intensity. The results were presented as the proliferative index (PI) calculated as a ratio between the sum of the events in each generation and the number of original parent cells (obtained by dividing the number of the events in each generation by 2 raised to the power of the generation number) (Lyons 2000). Statistical analysis was performed with Statistica 6.0 software. To compare differences between groups, the Mann–Whitney U-test or Kruskal–Wallis ANOVA test was used. The Wilcoxon-matched pair test was used to compare the difference before and after the treatment.

3. Results

3.1. $CD4^{+}CD25^{+}$ and $CD4^{+}CD25^{high}FoxP3^{+}$ T Cell Numbers are Reduced in Patients with SLE

The phenotypic analysis of the frequency of $CD4^{+}CD25^{+}$ Tregs in SLE patients demonstrated that the percentage of Tregs was almost twice higher in healthy donors (10.3 ± 3.9%) than in patients (6.1 ± 3.8%). The levels of $CD4^{+}CD25^{+}FoxP3^{+}$ cells in SLE patients were lower than in healthy donors: 1.8 ± 0.8% versus 4.9 ± 1.4%, $p<0.05$. Coexpression of CTLA-4 (CD152) on $CD4^{+}CD25^{+}$ T cells was estimated as 2.5 ± 1.2% in patients and 4.1 ± 2.4% in donors, but the numbers were not statistically significant.

Characterization of $CD4^{+}CD25^{+}$ and $CD4^{+}CD25^{high}FoxP3^{+}$ cells in SLE patients and healthy donors is shown in Figure 1. Analysis of Tregs in newly admitted untreated patients (group I) and treated SLE patients (group II) demonstrated significantly decreased numbers of $CD4^{+}CD25^{high}$ T cells in group I. In contrast, differences in $CD4^{+}CD25^{+}$ T cell numbers between group II and healthy volunteers were not statistically significant (Figure 1A). Furthermore, evaluation of Tregs coexpressing FoxP3 and $CD4/CD25^{high}$ molecules revealed that coexpression of FoxP3 and $CD4^{+}CD25^{high}$ on Tregs in group I was significantly lower when compared with group II or healthy donors ($p < 0.05$). Importantly, although the expression of FoxP3 was almost equal in both groups of patients, the numbers of $CD4^{+}CD25^{+}$ cells coexpressing FoxP3 were twice as low in patients than in healthy donors (Figure 1B).

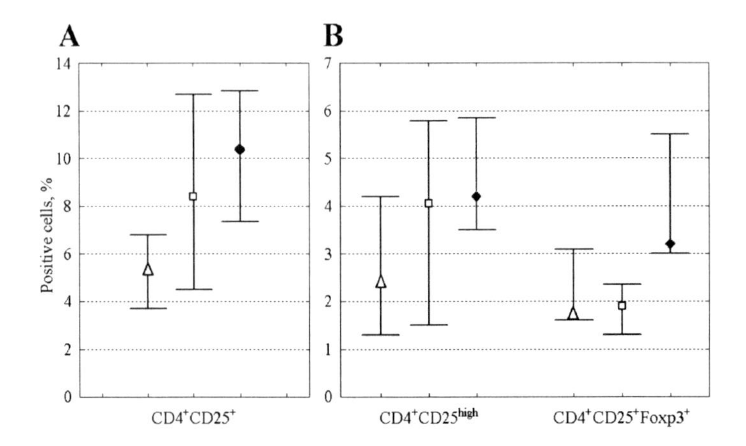

FIGURE 1. PBMC from SLE patients contain low numbers of $CD4^+CD25^+$ and $CD4^+CD25^{high}$ $FoxP3^+$ T cells. (A) The number of $CD4^+25^+$ T cells and (B) coexpression of $CD4^+CD25^{high}$ and FoxP3 in newly admitted non-treated patients, recently treated patients, and healthy volunteers (transparent triangulars, transparent squares, and filled rhombs, respectively). Data are shown as a box-plot, where the points represent the median, and the lines represent the 25–75th percentiles.

3.2. Patients with SLE Display Impaired Function of $CD4^+CD25^+$ T Cells

We next tested the suppressive activity of $CD4^+CD25^+$ cells toward autologous responder T cells. First, $CD4^+CD25^+$ T cells from SLE patients were anergic and did not proliferate upon activation with anti-CD3 antibodies in the presence of allogeneic antigen-presenting cells (data not shown). Proliferation of donors' $CD8^+$ T cells mixed (1:1) with autologous $CD4^+CD25^+$ Tregs was completely inhibited, while patients' Tregs were able to suppress proliferation of only 35.4–55.2% of $CD8^+$ T cells (Figure 2 C–D). The response of $CD4^+$ T cells mixed with autologous $CD4^+CD25^+$ T cells was significantly lower then the response of donors' $CD4^+$ cells. Obviously, suppressive function of patients' $CD4^+CD25^+$ Tregs was impaired (Figure 2 II.E–F). We, however, revealed no significant differences in the suppressive function of Tregs between patients from group I and group II.

4. Discussion

Data regarding the functional activity and the numbers of $CD4^+CD25^+$ cell population in patients with SLE are conflicting (Crispin et al. 2003; Liu et al. 2004; Miyara et al. 2005; Alvarado-Sanchez et al. 2006; Lee et al. 2006). These conflicting results might be due to specific therapies. Patients with SLE, as usual, are on the treatments that include cytostatics and steroids. For example,

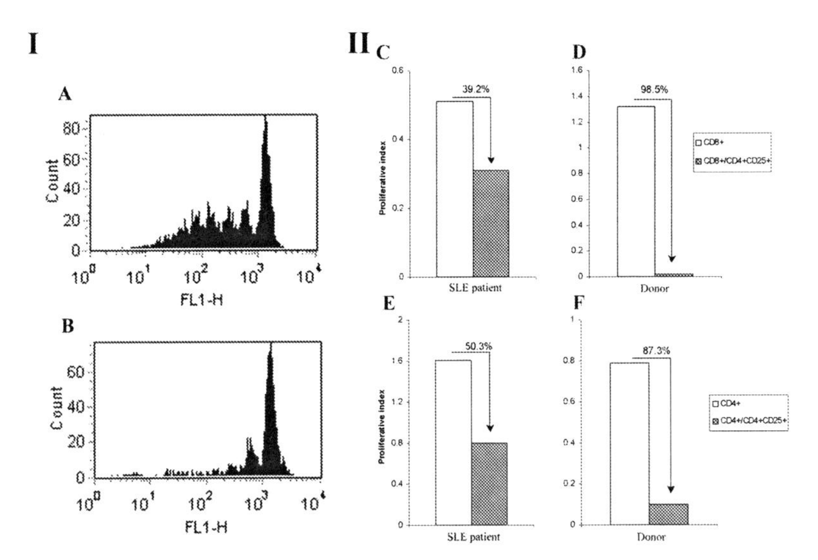

FIGURE 2. Reduced suppressive function of $CD4^+CD25^+$ T cells in SLE patients. Suppressive activity of Tregs was assessed in cocultures with autologous $CD8^+$ and $CD4^+$ responder T cells in the presence of CD3-depleted pheripheral blood mononuclear cells (PBMCs). I. Proliferation of CFSE-labeled $CD4^+$ T cells cultured with medium (A) or Tregs (B). II. Inhibition of $CD8^+$ (C and D) and $CD4^+$ (E and F) T cell proliferation in the presence of $CD4^+CD25^+$ T cells. C and E, SLE patients' T cells; D and F, donors' T cells. Arrows show the percentage of inhibition of T cell proliferation. The results of one representative out of seven experiments are shown.

Suarez et al. (2006) have reported that patients treated with steroids have elevated levels of $CD4^+CD25^{high}$ cells, whereas untreated patients have values similar to those in healthy controls. Estimating the total number of $CD4^+CD25^+$ T cells in SLE patients in newly diagnosed non-treated and recently treated patients with SLE, we revealed reduced numbers of $CD4^+CD25^+$ T cells in newly admitted patients compared to treated patients and healthy donors.

Cell surface molecule CTLA-4 (CD152) is often considered as a marker of Tregs and might be involved in the suppressive activity of Tregs. Indeed, it has been recently reported that anti-CTLA-4 therapy completely inhibits the function of Tregs (Read et al. 2006). However, we did not find significant differences in the expression of CTLA-4 on Tregs between two groups of patients and donors.

FoxP3 has been described as the gene controlling development and function of Tregs. FoxP3 regulates expression of the IL-2 and CD25 genes in Tregs (Fontenot et al. 2003; Hori et al. 2003), and FoxP3 deficiency induces severe autoimmune diseases in humans and animals (Bennett et al. 2001). Mutations in the X-linked FoxP3 are responsible for human autoimmune diseases, immune dysregulation, polyendocrinopathy, enteropathy, and X-linked syndrome (IPEX) (Hori et al. 2003). Our data show that expression of FoxP3 in $CD4^+CD25^{high}$ T cells in SLE patients was significantly lower than in control group and did not differ between two groups of SLE patients. Noticeably, treatment of SLE patients did not affect a potential genetic impairment in FoxP3 expression in T cells. Analyzing the regulatory capacity of T cells obtained from SLE patients, we revealed significant deficiency in suppressive function of Treg cells assessed by their coculture with $CD4^+$ or $CD8^+$ responder T cells. However, again we did not observe any difference in the functional capability of Tregs between two groups of patients.

In conclusion, we have reported here that the conventional therapy results in increased numbers of Tregs in SLE patients, which were lower in patients prior to the treatment. However, expression of FoxP3 in Tregs was equally lower in both treated and non-treated patients, which was associated with reduced immunosuppressive function of Tregs. It is possible that therapeutic manipulations with ex vivo-generated Tregs might be used for the treatment of SLE patients.

References

Allan, S.E., Passerini, L., Bacchetta, R., Crellin, N., Dai, M., Orban, P.C., Ziegler, S.F., Roncarolo, M.G. and Levings, M.K. (2005) The role of 2 FOXP3 isoforms in the generation of human CD4+ Tregs. J. Clin. Invest. 115, 3276–3284.

Alvarado-Sanchez, B., Hernandez-Castro, B., Portales-Perez, D., Baranda, L., Layseca-Espinosa, E., Abud-Mendoza, C., Cubillas-Tejeda, A.C. and Gonzalez-Amaro, R. (2006) Regulatory T cells in patients with systemic lupus erythematosus. J. Autoimmun. 27, 110–118.

Beacher-Allan, C., Brown, J.A., Freeman, G.J. and Hafler, D.A. (2001) CD4+CD25high regulatory cells in human peripheral blood. J. Immunol. 167, 1245–1253.

Bennett, C.L., Christie, J., Ramsdell, F., Brunkow, M.E., Ferguson, P.J., Whitesell, L., Kelly, T.E., Saulsbury, F.T., Chance, P.F. and Ochs, H.D. (2001) The immune dysregulation, polyendocrinopathy, enteropathy, X-linked syndrome (IPEX) is caused by mutations of FOXP3. Nat. Genet. 27, 20–21.

Chen, Z., Herman, A.E., Matos, M., Mathis, D. and Benoist, C. (2005) Where CD4+CD25+ T reg cells impinge on autoimmune diabetes. J. Exp. Med. 202, 1387–1397.

Crispin, J.C., Martinez, A. and Alcocer-Varela, J. (2003) Quantification of regulatory T cells in patients with systemic lupus erythematosus. J. Autoimmun. 21, 273–276.

Fontenot, J.D., Gavin, M.A. and Rudensky, A.Y. (2003) Foxp3 programs the development and function of CD4+CD25+ regulatory T cells. Nat. Immunol. 4, 330–336.

Gavin, M.A., Taogerson, T.R., Houston, E., DeRoos, P., Ho, W.Y., Stray-Pedersen, A., Ocheltree, E.L., Greenberg, P.D., Ochs, H.D. and Rudensky, A.Y. (2006) Single-cell analysis of normal and FOXP3-mutant human T cells: FOXP3 expression without regulatory T cell development. Proc. Natl. Acad. Sci. U.S.A. 103, 6659–6664.

Hori, S., Takahashi, T. and Sakaguchi, S. (2003) Control of autoimmunity by naturally arising regulatory CD4+ T cells. Adv. Immunol. 81, 331–371.

Kriegel, M.A., Lohmann, T., Gabler, C., Blank, N., Kalden, J.R. and Lorenz, H.M. (2004) Defective suppressor function of human CD4+ CD25+ regulatory T cells in autoimmune polyglandular syndrome type II. J. Exp. Med. 199, 1285–1291.

Lee, J.H., Wang, L.C., Lin, Y.T., Yang, Y.H., Lin, D.T. and Bhiang, B.L. (2006) Inverse correlation between CD4+ regulatory T cell population and autoantibody levels in paediatric patients with systemic lupus erythematosus. Immunology 117, 280–286.

Lindley, S., Dayan, C.M., Bishop, A., Roep, B.O., Peakman, M. and Treem T.I. (2005) Defective suppressor function in CD4(+)CD25(+) T cells from patients with type 1 diabetes. Diabetes 54, 92–99.

Liu, M.F., Wang, C.R., Fung, L.L. and Wu, C.R. (2004) Decreased CD4+CD25+ T cells in peripheral blood of patients with systemic lupus erythematosus. Scand. J. Immunol. 59, 198–202.

Lyons, A.B. (2000) Analyzing cell division *in vivo* and *in vitro* using flow cytometric measurement of CFSE dye dilution. J. Immunol. Methods 243, 147–154.

Miyara, M., Amoura, Z., Parizot, C., Badoual, C., Dorgham, K., Trad, S., Nochy, D., Debre, P., Piette, J.C. and Gorochov, G. (2005) Global natural regulatory T cell depletion in active systemic lupus erythematosus. J. Immunol. 175, 8392–8400.

Read, S., Greenwald, R., Izcue, A., Robinson, N., Mandelbrot, D., Francisco, L., Sharpe, A.H. and Powrie, F. (2006) Blockade of CTLA-4 on CD4+CD25+ regulatory T cells abrogates their function in vivo. J. Immunol. 177, 4376–4383.

Schubert, L.A., Jeffery, E., Zang, Y., Ramsdell, F. and Ziegler, S.F. (2001) Scurfin (FOXP3) acts as a repressor of transcription and regulates T cell activation. J. Biol. Chem. 276, 37672–37679.

Shevach, E.M. (2001) Certified professionals: CD4(+)CD25(+) suppressor T cells. J. Exp. Med. 193, F41–F46.

Suarez, A., Lopez, P., Gomez, J. and Gutierrez, C. (2006) Enrichment of CD4+ CD25high T cell population in patients with systemic lupus erythematosus treated with glucocorticoids. Ann. Rheum. Dis. 65, 1512–1517.

Sugiyama, H., Gyulai, R., Toichi, E., Garaczi, E., Shimada, S., Stevens, S.R., McCormick, T.S. and Cooper, K.D. (2005) Dysfunctional blood and target tissue CD4+CD25high regulatory T cells in psoriasis: mechanism underlying unrestrained pathogenic effector T cell proliferation. J. Immunol. 174, 164–173.

van Amelsfort, J.M., Jaclbs, K.M., Bijlsma, J.W., Lafeber, F.P. and Taams, L.S. (2004) CD4(+)CD25(+) regulatory T cells in rheumatoid arthritis: differences in the presence, phenotype, and function between peripheral blood and synovial fluid. Arthritis Rheum. 50 2775–2785.

Viglietta, V., Baecher-Allan, C., Wiener, H.L. and Hafler, D.A. (2004) Loss of functional suppression by CD4+CD25+ regulatory T cells in patients with multiple sclerosis. J. Exp. Med. 199, 971–979.

Tumor Immunity in New Dimensions

13

Role of Altered Expression of HLA Class I Molecules in Cancer Progression

Natalia Aptsiauri, Teresa Cabrera, Rosa Mendez, Angel Garcia-Lora, Francisco Ruiz-Cabello, and Federico Garrido

Departmiento de Analisis Clinicos, Hospital Universitario Virgen de las Nieves, Granada, Spain, federico.garrido.sspa@juntadeandalucia.es

Abstract. HLA class I antigens play a key role in immune recognition of transformed and virally infected cells via binding to the peptides of "non-self" or aberrantly expressed proteins and subsequent presentation of the newly formed "HLA-I-peptide" complex to T lymphocytes. Consequently, a chain of immune reactions is initiated leading to tumor cell elimination by cytotoxic T cells. Altered tumor expression of HLA class I is frequently observed in various types of malignancies. It represents one of the main mechanisms used by cancer cells to evade immunosurveillance. Because of immune selection, HLA class I-negative variants escape and lead to tumor growth and metastatic colonization. Loss or downregulation of HLA class I antigens on tumor cell surface is a factor that limits clinical outcome of peptide-based cancer vaccines aimed to increasing specific anti-tumor activity of cytotoxic T lymphocytes. Thus, gaining more knowledge regarding frequency of HLA class I defect, its tissue specificity, and underlying molecular mechanisms may help designing appropriate therapeutic strategies in cancer treatment. Here, we describe various types of HLA class I alterations found in different malignancies and molecular mechanisms that underlie these defects. We also discuss a correlation between HLA class I defects cancer progression in melanoma patients with poor clinical response to autologous vaccination.

1. Introduction

Malignant transformation of normal cells and cancer progression results from an accumulation of mutational and epigenetic changes that alter normal cell growth and survival pathways. High genomic instability and diversity produces a great degree of heterogeneity seen in human tumors (Figure 1A). Tumor cells often express new antigens because of multiple genetic alterations that are associated with malignant transformation. These antigens are recognized by T cells in association with HLA class I molecules. These activated T cells consequently eliminate tumor cells. The process that involves the recognition and destruction of cancer cells by numerous innate and adaptive immune effector cells and molecules is known as immunosurveillance (Garrido et al. 1997; Zitvogel et al. 2006; Smyth et al. 2006).

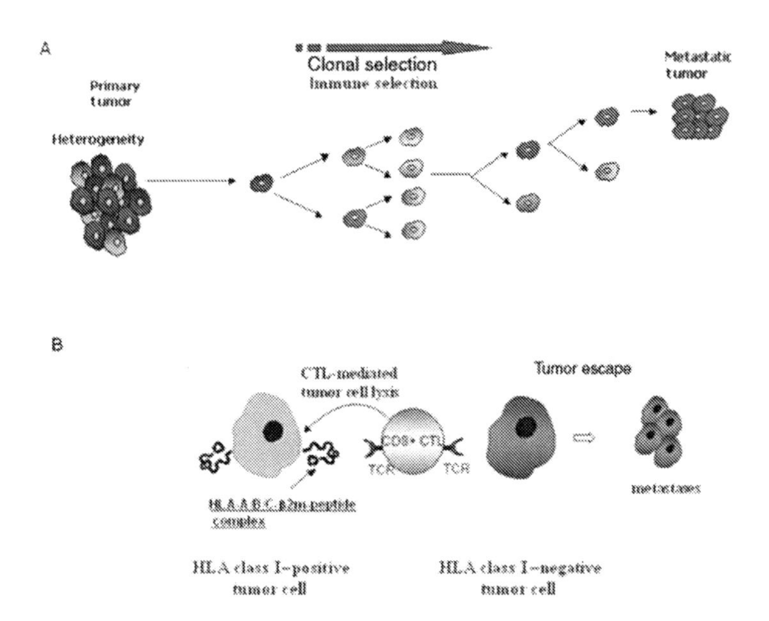

FIGURE 1. Schematic representation of a somatic and immunological selection of tumor escapes variants leading to metastatic progression (A) and CTL-mediated selection and progression of HLA class I-negative tumor cells (B).

Nevertheless, cancer cells escape from immunosurveillance through the outgrowth of poorly immunogenic tumor cell variants, which emerge due to growth advantages created by a combination of certain features of tumor cells and the factors of tumor environment (Garrido, Cabrera, Concha, Glew, Ruiz-Cabello, and Stern 1993). Therefore, cancer progression may be considered because of a balance between tumor immunosurveillance and tumor escape.

Several mechanisms might contribute to the failure of the immune control of tumor growth. Lack of expression of costimulatory molecules (ligands for T cell activation) can induce tolerance to neoplastic cells. Tumor cells also produce immune suppressive factors (VEGF, TGF-β, IL-10, and PGE_2) that have systemic effects on immune cell function. Disabled dendritic cell differentiation, maturation, migration, and function are of special importance in immune control failure, because they are the most potent antigen-presenting cells (APCs) of the immune system, interacting with T and B lymphocytes and natural killer (NK) cells to induce and modulate immune responses (Shurin et al. 2006). Tumors also alter host hematopoiesis and produce large numbers of immature dendritic cells with a direct immune suppressive activity (Zou 2005). Finally, tumor cells have a very low level of MHC class I and, in many cases, a low expression of tumor-specific antigens (Garrido et al. 1993; Khong and Restifo 2002).

Major contributor to the appearance of MHC class I-negative tumor clones is T cell immunoselection. Cells that are highly immunogenic and express high levels of MHC class I are eliminated by cytotoxic T lymphocytes (CTLs) (Figure 1B).

Malignant cells with total MHC class I loss are susceptible to NK cell lysis because of inactivation of inhibitory receptors on NK cells. Another immunoselection route is provided by the partial loss of HLA class I antigens that allows tumor cells to escape both CTL and NK attack. For instance, a recent study of colorectal cancer showed that a high level of HLA I expression or total loss of HLA class I was associated with similar disease-specific survival times, possibly due to T cell reactivity or NK cell-mediated clearance of class I-positive and class I-negative tumor cells, respectively. However, tumors with intermediate HLA class I expression were reported to be associated with a poor prognosis, suggesting that these tumors may avoid both NK- and T cell-mediated immune surveillance (Watson, Ramage, Madjd, Splendlove, Ellis, Scholefeld, and Durrant 2006).

2. Altered Phenotypes of HLA Class I in Various Types of Cancer

Different altered HLA class I tumor phenotypes are produced because of immune selection (Garrido and Algarra 2001; Seliger et al. 2002). Total or selective losses of HLA class I antigens have been reported in different human tumor samples. The loss of an MHC antigen associated with H-2K^k class I molecule was first described in 1976 in a mouse lymphoma, and the detection of HLA losses in human tumors followed in 1977 (Garrido et al. 1976). In subsequent years, an increasing proportion of tumors were found to have these alterations supporting the theory that altered HLA expression phenotypes represent a major mechanism of tumor escape from T cell recognition.

We have classified all known HLA-altered phenotypes in seven groups: (i) phenotype I, HLA class I total loss (a) or downregulation (b); (ii) phenotype II, haplotype loss; (iii) phenotype III, locus loss; (iv) phenotype IV, allelic losses; (v) phenotype V, compound phenotype; (vi) phenotype VI, unresponsiveness to IFN-γ; and (vii) phenotype VII, downregulation of classical HLA-A, -B, and -C molecules and appearance of non-classical HLA-E molecules (Figure 2) (Garrido et al. 1997; Aptsiauri et al. 2007).

All of these phenotypes can be found in various types of tumor, regardless of the tissue origin or of the carcinogen inducing the tumor. Differences are observed in the distribution of the phenotypes and in the combination of molecular mechanisms leading to each phenotype. An example of such distribution in various types of solid tumors and melanoma cell lines (ESTDAB project, http://www.ebi.ac.uk/ipd/estdab/) (Pawelec and Marsh 2006) is presented in Table 1.

3. Molecular Mechanisms of HLA Class I Alterations in Tumors

Importantly, the same phenotype of HLA alteration seen in different tumors can be produced by a combination of different mechanisms. Two or more mechanisms responsible for HLA alteration can frequently be observed within the same phenotype. The mechanism leading to total or partial HLA alterations can occur at any step required for HLA synthesis, assembly, transport, or expression on cell surface.

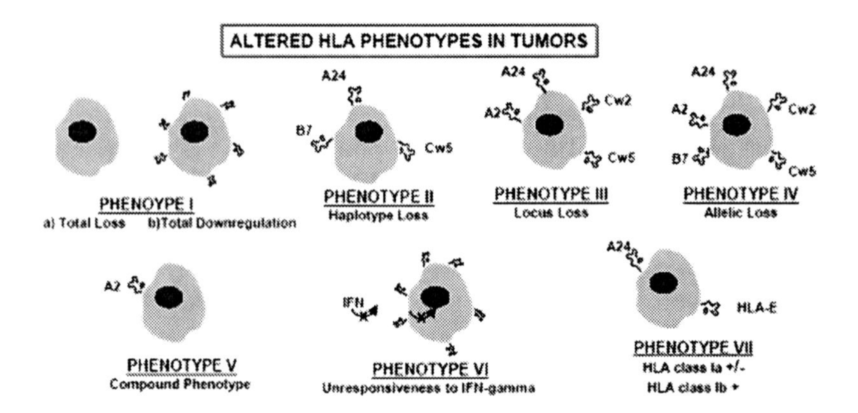

FIGURE 2. Classification of HLA class I-altered phenotypes identified in various types of tumor.

The defects can represent a structural defect, such as gene mutation or regulatory, on transcriptional level, which can be corrected with cytokine treatment.

Normally, two structural defects are necessary to produce the total loss of HLA class I on malignant cells: mutation in one copy of β2m gene and loss of heterozygosity (LOH) in the second allele (Paschen et al. 2003). The coordinated downregulation of several antigen-processing machinery (APM) components can also produce this phenotype. We have recently observed coordinated transcriptional downregulation of APM components (LMP2, LMP7, TAP1, TAP2, and tapasin) and of HLA-ABC in bladder carcinoma (Romero et al. 2005).

Phenotype II, or loss of HLA haplotype, is associated with LOH due to loss of one copy of chromosome 6 or loss of a DNA fragment containing the HLA-A, HLA-B, and HLA-C genes. It has been described in practically all types of tumors analyzed to date (Ramal et al. 2000; Koene et al. 2004). LOH in chromosome 6 is a very frequent mechanism underlying HLA class I defect. Mechanisms underlying this phenotype also vary. Deletion of a large genomic region of chromosome 6 (that includes the HLA genes), total chromosome 6 deletion, chromosomal non-dysjunction, or mitotic recombination have been described as contributing factors to HLA class I haplotype loss. LOH in chromosome 15 has also been reported in the literature. We determined that about 40% of studied laryngeal carcinomas have LOH for at least one short tandem repeats (STR) marker in chromosome 15 (Maleno et al. 2002). LOH in chromosome 15 was reported to be more frequent than mutations in β2m gene in melanoma cell lines and could be found in cells with normal or low expression of HLA class I (Paschen et al. 2006). LOH in chromosome 15 is also frequently observed in head and neck cancer. However, microsatellite LOH analysis does not always represent the β2m gene copy number, since in some cases the loss of one of chromosome 15 is accompanied by duplication of the second one according to the results obtained from comparative genomic hybridization (CGH) studies (Koene et al. 2004). Thus, this defect could be one of the early HLA class I alterations in malignant cells which could play an important role in predisposition to further generation of

TABLE 1. Distribution of altered HLA class I phenotypes (Ph I–PhV) in various types of solid tumors and melanoma cell lines

Tumor \ Phenotype	Normal HLA expression	Ph I, total loss or downreg.	Ph II, haplot. loss	Ph III, locus loss	Ph IV, allelic loss	Ph V, comp.
Colorectal cancer[1]	22%	18%	28%	8.5%	15%	8.5%
Laryngeal cancer[2]	23%	11%	36%	20%	10%	-
Cervical cancer[3]	10%	10%	33%	13%	17%	17%
Bladder cancer[4]	23%	25%	17%	7%	14%	13%
Melanoma[5]	31%	18%	-	18%	33%	-
Melanoma cell lines[6]	44%	11% (9%-total loss and 2% - downregulation	15%	22%	0%	6%

[1]Garrido et al. 1995; 1997.
[2]Maleno et al. 2002.
[3]Koopman et al. 1998; 2000.
[4]Romero et al. 2005.
[5] Kageshita et al. 2005.
[6]Rodriguez et al. manuscript in preparation (ESTDAB project, http://www.ebi.ac.uk/ipd/estdab/).

tumor escape variants. Identification of LOH in chromosome 15 may be unnoticed if tumor cells have normal HLA class I surface expression or when methods used to analyze chromosome 15 are not adequate.

HLA class I loss can be selective. Unlike total HLA class I loss, which requires two mutational events in the β2m gene, selective HLA loss variants result from a single mutational change in a heterozygous allelic background (Cabrera et al. 2003).

4. Association of HLA Class I Alterations with Tumor Escape and Cancer Progression

The identification of the genes encoding tumor-associated antigens (TAAs) and the development of means for immunizing against these antigens have opened new avenues for the development of an effective anti-cancer immunotherapy. However, current protocols of cancer immunotherapy aimed at enhancing anti-tumor T cell activity cause cancer regression only in a small number of patients (Rosenberg et al. 2004). Immune selection of CTL- and NK-cell-resistant tumor cells might explain the rapid progression and poor prognosis of cancers that exhibit HLA class I downregulation. Thus, HLA class I downregulation represents a significant challenge for the successful application of cancer immunotherapy.

Recently, we have observed an interesting association of HLA class I expression and metastatic progression in a melanoma patient, whose subcutaneous metastases responded differently to autologous tumor cell vaccination (Cabrera et al. 2007). We analyzed three progressing and three regressing metastatic lesions for HLA class I expression using immunohistochemistry, tissue microdissection, LOH analysis, and other molecular techniques. Interestingly, the progressors had low HLA class I expression and higher frequency of LOH in chromosomes 6 and 15. In addition, all of the progressing metastases were HLA-B negative. Expression of class I molecules was normal in the regressing metastases. LOH at chromosome 6 was detected in all six studied metastases, suggesting that this defect may also be found in the primary tumor, contributing to the mechanism of tumor escape and metastatic progression. Real-time quantitative PCR of the samples obtained from microdissected tumor showed lower mRNA levels of HLA-ABC heavy chain and β2m in progressing metastases than in regressing ones, confirming the immunohistological findings. Sequence analysis of the β2m-coding region was performed in progressing metastasis, and no alterations of β2m region were found. The vaccination may have led to specific CTL immune responses against these tumor variants (with LOH in chromosome 6), leading to the expansion of new tumor variants with other alterations in HLA class I expression, including HLA locus B downregulation and LOH in chromosome 15. Reduced levels of HLA class I expression may be due to deletions in both HLA class I (LOH, chromosome 6) and β2m (LOH, chromosome 15) genes.

Similar correlations were previously described by our group in a murine model. MHC class I phenotype of metastatic lung colonies produced by a mouse fibrosarcoma tumor clone (B9) was MHC class I negative in immunocompetent mice and MHC class I positive in immunodeficient athymic nude/nude mice. In addition, TAP-1, TAP2, LMP2, LMP7, LMP10, tapasin, and calnexin mRNA were absent in metastases produced in immunocompetent mice. Interestingly, the MHC class I-positive or class I-negative phenotypes of the metastatic colonies correlated with in vivo immunogenicity (Garcia-Lora et al. 2003).

5. Implication of HLA Class I Defects for Cancer Immunotherapy

Tumor peptide-based immunotherapy is an established approach of cancer treatment, with the aim of boosting anti-tumor T cell reactivity by stimulation with tumor-specific peptides. However, the overall clinical outcome of this type of treatment is poor. In many cases, the failure of this therapy and the progression of the cancer are associated with total loss of HLA class I tumor expression. Normal expression of HLA class I molecules on tumor cell surface is crucial for the successful outcome of peptide-based cancer therapy, since cytotoxic T cells can only recognize tumor-derived peptides in a complex with self-MHC class I molecules.

When the mechanism underlying total HLA class I loss is at a transcriptional level, the expression of surface HLA class I antigens can be reversed by cytokine treatment, and T cell-based therapy can be successfully applied.

Patients with HLA class I loss may not benefit from peptide-based immunotherapy, since peptide-mediated activation of cytotoxic T cells can occur only in complex with HLA class I molecules. In some cases, HLA class I downregulation can be restored by cytokine treatment. In cases of structural irreversible alterations, such as mutations in β2-microglobulin or LOH in chromosome 15, gene therapy aimed at reconstituting wild-type β2m gene would be more appropriate in order to restore normal HLA class I expression and consequent tumor recognition and elimination by cytotoxic T cells.

For this purpose, in an ongoing study, we generated in our laboratory a "gutless" adenoviral vector with β2m gene (AdCMVβ2m) using a cre-lox recombination system. Infection of two melanoma cell lines and one cell line derived from a Burkitt lymphoma with this vector produced a significant recovery of surface expression of HLA class I molecules.

6. Conclusions

HLA class I expression on tumor cells is important for initiation of an immune response against malignant cells. Downregulation or total loss of HLA class I antigens is frequently seen in tumors providing an escape mechanism for cancer progression. Evaluation of HLA class I expression in primary tumors and metastases of cancer patients should be considered before selection of cancer immunotherapy. Comprehensive information on the tumor and the immune status of an individual could provide a precise picture of the ongoing evolution of the tumor, as well as to yield invaluable information about which strategy will result in optimal therapeutic outcome.

Acknowledgements

This work was supported by grants from the Fondo de Investigaciones Sanitarias (FIS), Red Genomica del Cancer (C03/10), Plan Andaluz de Investigacion, Servicio Andaluz de Salud (SAS) in Spain, the ESTDAB project (contract no. QLRI-CT-2001-01325), the European Network for the identification and validation of antigens and biomarkers in Cancer and their application in clinical tumor immunology (ENACT, contract No. 503306), and the European Cancer Immunotherapy project (Comunidad Economica Europea OJ2004/C158, 518234).

References

Aptsiauri, N., Cabrera, T., Garcia-Lora, A., Lopez-Nevot, M.A., Ruiz-Cabello, F. and Garrido, F. (2007) MHC class I antigens and immune surveillance in transformed cells. Int. Rev. Cytol. 256, 139–189.

Cabrera, T., Lopez-Nevot, M.A., Gaforio, J.J., Ruiz-Cabello, F. and Garrido, F. (2003) Analysis of HLA expression in human tumor tissues. Cancer Immunol. Immunother. 52, 1–9.

Cabrera, T., Lara E., Romero, J.M., Maleno, I., Real, L.M., Ruiz-Cabello, F., Valero, P., Camacho, F.M. and Garrido, F. (2007) HLA class I expression in metastatic melanoma correlates with tumor development during autologous vaccination. Cancer Immunol. Immunother. 56, 709–717.

Garcia-Lora, A., Algarra, I. and Garrido, F. (2003) MHC class I antigens, immune surveillance, and tumor escape. J. Cell. Physiol. 195, 346–55.

Garrido, F., Festenstein, H. and Schirrmacher, V. (1976) Further evidence for depression of H-2 and Ia-like specificities of foreign haplotypes in mouse tumour cell lines. Nature 261, 705–707.

Garrido, F., Cabrera, T., Concha, A., Glew, S., Ruiz-Cabello, F. and Stern, P.L. (1993) Natural history of HLA expression during tumour development. Immunol. Today 14, 491–499.

Garrido, F., Cabrera, T., Lopez-Nevot, M.A. and Ruiz-Cabello, F. (1995) HLA class I antigens in human tumors. Adv. Cancer Res. 67, 155–195.

Garrido, F., Ruiz-Cabello, F., Cabrera, T., Perez-Villar, J.J., Lopez-Botet, M., Duggan-Keen, M. and Stern, P.L. (1997) Implications for immunosurveilance of altered HLA class I phenotypes in human. Immunol. Today 18, 89–95.

Garrido, F. and Algarra, I. (2001) MHC antigens and tumor escape from immune surveillance. Adv. Cancer Res. 83, 117–158.

Kageshita, T., Ishihara, T., Campoli, M. and Ferrone, S. (2005) Selective monomorphic and polymorphic HLA class I antigenic determinant loss in surgically removed melanoma lesions. Tissue Antigens 65, 419–428.

Khong, H.T. and Restifo, N.P. (2002) Natural selection of tumor variants in the generation of "tumor escape" phenotypes. Nat. Immunol. 3, 999–1005.

Koene, G.J., Arts-Hilkes, Y.H., van der Ven, K.J., Rozemuller, E.H., Slootweg, P.J., de Weger, R.A. and Tilanus, M.G. (2004) High level of chromosome 15 aneuploidy in head and neck squamous cell carcinoma lesions identified by FISH analysis: limited value of beta2-microglobulin LOH analysis. Tissue Antigens 64, 452–461.

Koopman, L.A., Mulder, A., Corver, W.E., Anholts, J.D., Giphart, M.J., Claas, F.H. and Fleuren, G.J. (1998) HLA class I phenotype and genotype alterations in cervical carcinomas and derivative cell lines. Tissue Antigens 51, 623–636.

Koopman, L.A., Corver, W.E., van der Slik, A.R., Giphart, M.J. and Fleuren, G.J. (2000) Multiple genetic alterations cause frequent and heterogeneous human histocompatibility leukocyte antigen class I loss in cervical cancer. J. Exp. Med. 191, 961–976.

Maleno, I., Lopez-Nevot, M.A., Cabrera, T., Salinero, J. and Garrido, F. (2002) Multiple mechanisms generate HLA altered phenotypes in laryngeal carcinomas: high frequency of HLA haplotype loss associated with loss of heterozygosity in chromosome region 6p21. Cancer Immunol. Immunother. 51, 389–396.

Paschen, A., Mendez, R.M., Jimenez, P., Sucker, A., Ruiz-Cabello, F., Song, M., Garrido, F. and Schadendorf, D. (2003) Complete loss of HLA class I antigen expression on melanoma cells: a result of successive mutational events. Int. J. Cancer 103, 759–767.

Paschen, A., Arens, N., Sucker, A., Greulich-Bode, K.M., Fonsatti, E., Gloghini, A., Striegel, S, Schwinn, N., Carbone, A., Hildenbrand, R., Cerwenka, A., Maio, M. and Schadendorf, D. (2006) The coincidence of chromosome 15 aberrations and beta2-microglobulin gene mutations is causative for the total loss of human leukocyte antigen class I expression in melanoma. Clin. Cancer Res. 12, 3297–3305.

Pawelec, G. and Marsh, S.G. (2006) ESTDAB: a collection of immunologically characterised melanoma cell lines and searchable databank. Cancer Immunol. Immunother. 55, 623–627.

Ramal, L.M., Feenstra, M., van der Zwan, A.W., Collado, A., Lopez-Nevot, M.A., Tilanus, M. and Garrido, F. (2000) Criteria to define HLA haplotype loss in human solid tumors. Tissue Antigens 55, 443–448.

Romero, J.M., Jimenez, P., Cabrera, T., Pedrinaci, S., Tallada, M., Garrido, F. and Ruiz-Cabello, F. (2005) Coordinated downregulation of the antigen presentation machinery and HLA class I/beta2-microglobulin complex is responsible for HLA-ABC loss in bladder cancer. Int. J. Cancer 113, 605–610.

Rosenberg, S.A., Yang, J.C. and Restifo, N.P. (2004) Cancer immunotherapy: moving beyond current vaccines. Nature Med. 10, 909–915.

Seliger, B., Cabrera, T., Garrido, F. and Ferrone, F. (2002) HLA class I antigen abnormalities and immune escape by malignant cells. Semin. Cancer Biol. 12, 3–13.

Shurin, M.R., Shruin, G., Lokshin, A., Yurkovetsky, Z.R., Gutkin, D.W., Chatta, G., Zhong, H., Han, B. and Ferris, R.L. (2006) Intratumoral cytokines/chemokines/growth factors and tumor infiltrating dendritic cells: friends or enemies? Cancer Met. Rev. 25, 333–356.

Smyth, M.J., Dunn, G.P. and Schreiber, R.D. (2006) Cancer immunosurveillance and immunoediting: the roles of immunity in suppressing tumor development and shaping tumor immunogenicity. Adv. Immunol. 90, 1–50.

Watson, N.F., Ramage, J.M., Madjd, Z., Splendlove, I., Ellis, I.O., Scholefeld, J.H. and Schreiber, R.D. Durrant, L.G. (2006) Immunosurveillance is active in colorectal cancer as downregulation but not complete loss of MHC class I expression correlates with a poor prognosis. Int. J. Cancer 118, 6–10.

Zitvogel, L., Tesniere, A. and Schreiber, R.D. Kroemer, G. (2006) Cancer despite immunosurveillance: immunoselection and immunosubversion. Nature Rev. Immunol. 6, 715–727.

Zou, W. (2005) Immunosupressive networks in the tumor environment and their therapeutic relevance. Nature Rev. Cancer 5, 263–274.

14

Intrathymic Selection: New Insight into Tumor Immunology

Dmitry B. Kazansky

Institute of Carcinogenesis, Blokhin Cancer Research Center, Moscow, Russia, kazansky@dataforce.net

Abstract. Central tolerance to self-antigens is formed in the thymus where deletion of clones with high affinity to "self" takes place. Expression of peripheral antigens in the thymus has been implicated in T cell tolerance and autoimmunity. During the last years, it has been shown that medullary thymic epithelial cells (mTECs) are the unique cell type expressing a diverse range of tissue-specific antigens. Promiscuous gene expression is a cell autonomous property of thymic epithelial cells and is maintained during the entire period of thymic T cell output. The array of promiscuously expressed self-antigens was random and included well-known targets for cancer immunotherapy, such as α-fetoprotein, P1A, tyrosinase, and gp100. Gene expression in normal tissues may result in tolerance of high-avidity cytotoxic T lymphocyte (CTL), leaving behind low-avidity CTL that cannot provide effective immunity against tumors expressing the relevant target antigens. Thus, it may be evident that tumor vaccines that targeted the tumor-associated antigens should be inefficient due to the loss of high-avidity T cell clones capable to be stimulated. Stauss with colleagues have described a strategy to circumvent immunological tolerance that can be used to generate high-avidity CTL against self-proteins, including human tumor-associated antigens. In this strategy, the allorestricted repertoire of T cells from allogenic donor is used as a source of T cell clones with high avidity to tumor antigens of recipient for adoptive immunotherapy. Then, the T cell receptor (TCR) genes isolated from antigen-specific T cells can be exploited as generic therapeutic molecules for antigen-specific immunotherapy.

1. Background

The T lymphocyte repertoire is formed in the thymus as a result of random rearrangement of germinal sequences of the T cell receptor (TCR) gene fragments and other processes that bring about the diversity of TCRs. Immunologists used to consider the traits of the immune system in the context of its coevolution with pathogenic microorganisms. Apparently, the most important problem for the immune system is to avoid the transplantation conflict within the organism. Indeed, the receptors of adaptive immunity "see" self-antigens much more frequently than pathogenic bacteria. In any case, an existence of potentially dangerous system in the organism could be not less important factor for the evolution of the immune system than pathogenic microorganisms.

Therefore, major histocompatibility comple (MHC)-restricted recognition could develop not as much in the struggle with pathogenic microorganisms but in inhibiting the reactions with "self." On this way, restriction of immune reactions by several types of recognized molecules would be helpful, because other biological macromolecules are rescued from danger. MHC molecules have allowed focusing these reactions on short peptides containing amino acid substitutions not presented in the responding organism. On the other hand, they have allowed an efficient formation of specific central tolerance to self not involving in this process a huge diversity of other protein molecules. Thus, thymic selection can be considered as a "first and last" life-long immune reaction of preselected repertoire to self-transplantation antigens. This process solves two problems: (i) to delete autoimmune and cross-reactive (promiscuous) clones that react with self transplantation antigens and (ii) to keep the repertoire of specificities to all conceivable pathogens as broad as possible. The first problem is resolved by negative selection in the thymus and deletion of autoimmune and cross-reactive clones. The second task is accomplished via preservation of a portion of T cell repertoire through a weak degenerated interaction with self-transplantation antigens. Since the repertoire is specific to the whole spectrum of the species' MHC molecules, the remaining portion of the repertoire would contain the clones that might be "autoimmune" or "cross-reactive" in different MHC environment (Zerrahn et al. 1997; Logunova et al. 2005). This provides explanation for frequent cross-reactivity of clones raised in the allogeneic response and provides the reason for degenerated mode of recognition of foreign antigens by TCR. Definitely, the degenerative fashion of recognition is necessary for preserving a broad spectrum of specificity of the repertoire to yet non-encountered pathogens.

As a part of the original demonstration that syngeneic anticancer immunity is possible, it has been shown that, among sarcomas induced in mice by a hydrocarbon, each tumor, when transplanted, could arouse an inhibitory immune reaction against itself (Prehn and Main 1957). However, it also became clear that each tumor was antigenically unique, even if each had been induced by identical means in one and the same animal; although cross-reactions were reported, these were the exceptions. Since it is probable that the immunogenicity was caused by the mutations induced by the carcinogen, obviously these chemicals produced different spectra of mutations in each tumor with very little overlap. Consequently, one had to conclude that any of a vast array of possible mutations could be found in phenotypically similar cancers, a not impossible idea. However, it was also clear that none of the carcinogen-induced mutations, at least among those identified by their resulting antigenicity, could be considered essential or causative for the induction of the cancer (Prehn 2005). Moreover, the vast majority of tumor antigens were identified as non-mutated "self" proteins. The identification of many tumor-associated epitopes as non-mutated "self" antigens led to the hypothesis that the induction of large numbers of self/tumor antigen-specific T cells would be prevented because of central and peripheral tolerance (Rosenberg et al. 2005).

2. Tumor-Specific and Tumor-Associated Antigens as the Consequences of Genetic and Epigenetic Alterations

Accumulating evidence shows that tumor formation is accompanied by both genetic and epigenetic alterations of the genome (Hahn and Weinberg 2002; Felsher 2003; Egger et al. 2004). Correspondingly, cancers in mouse and man express multiple tumor-specific as well as tumor-associated antigens encoded by mutant and normal cellular genes.

Most chemical or physical carcinogens are mutagens. Therefore, it is generally assumed that tumor-specific antigens on tumors induced by these carcinogens are products of mutated genes, possibly single genes with "hot spots" for mutations. Some tumor-specific antigens are retained during tumor progression possibly because they are essential for survival of the malignant phenotype. Since 1995, the genetic origins of several T cell-recognized unique antigens from murine and human cancers have been identified, and in every case, the antigen was caused by a somatic mutation (i.e., by a genetic change absent from autologous normal DNA) and thus found to be truly tumor specific (Monach et al. 1995; Coulie et al. 1995; Wolfel et al. 1995; Robbins et al. 1996; Brandle et al. 1996; Dubey et al. 1997).

Unlike genetic changes, epigenetic changes do not alter the primary DNA sequence and are therefore reversible. Examples of epigenetic modifications are the methylation of DNA and histones, the acetylation/deacetylation of histones, and the packing of chromatin into euchromatic and heterochromatic regions (Li 2002). Epigenetic modifications play an important role during normal development by regulating gene expression through stable activation or silencing of differentiation-associated genes. Similarly, epigenetic changes can promote cell proliferation, inhibit apoptosis, and induce angiogenesis during tumorigenesis by activating oncogenes and silencing tumor suppressor genes (Felsher 2003). Treatment of tumor cells with methylation- and histone-modifying drugs can inhibit malignancy, and this inhibition correlates with the reactivation of important tumor suppressor loci (Egger et al. 2004).

Despite differences in their tissue of origin, in many tumors, certain tumor-associated proteins are highly expressed. Many studies have been focused on the possibility of utilizing antigenic components of these proteins as a focus for T cell immunotherapy of cancer. The advantage of targeting such commonly expressed proteins is founded on the fact that such therapy could be of value in eliminating many different tumor types. A potential barrier in the identification of T cell epitopes derived from these proteins and presented by tumor cells is that these proteins are also expressed at low levels in normal tissues, and therefore, self-tolerance may eliminate T cells capable of recognizing these epitopes with high avidity (Sherman et al. 1998).

3. Does Immunological Surveillance Make Tumor-Specific Antigens Undetectable?

The idea of immunological surveillance against cancer has existed for nearly 100 years. However, the importance of the cellular immune defense in the detection and removal of incipient or existing tumors is still a hotly debated subject. In order to select a relevant immunotherapeutic strategy in cancer treatment, a fundamental understanding of the basic immunological conditions under which tumor is developed and exists is a prerequisite. Therefore, several murine models were set up that would enable to confirm or decline the theory of immunological surveillance.

The initial studies have shown that the incidence of spontaneous tumors in immunodeficient *nude* mice was similar to that reported for the thymus-bearing background strain arguing against the thymus dependency of the putative immunological surveillance mechanisms (Pelleitier and Montplaisir 1975; Sharkey and Fogh 1979). Possibly, many spontaneous tumors are induced by oncogenic viruses requiring the host immune system for propagation. For example, MMTV utilizes cells of the immune system in its infection pathway. Therefore, subsequent carcinogenesis was highly dependent on T cells (Golovkina et al. 1992; Pobezinskaya et al. 2004). It has been argued that the apparent general lack of tumor immunogenicity may be an artifact caused by immune selection for non-immunogenic tumor variants. Perhaps most tumors, according to the immunosurveillance hypothesis, are really highly immunogenic and what we see is actually a small surviving, relatively non-immunogenic, highly selected subpopulation. This popular concept can account for the paucity of tumor immunogenicity.

First, cloned cell lines of chemically induced murine fibrosarcomas maintained in tissue culture usually fail to grow when transplanted to normal syngeneic mice. They grow, however, in various categories of T cell-deficient mice, and after such passages grow readily in normal mice. Both cultured and mouse-passaged lines possess strong tumor transplantation antigens (Woodruff and Hodson 1985).

Second, methylcholanthrene-induced tumors originating from the immunodeficient *nude* mice turned out to be far more immunogenic than tumors from normal mice, resulting in a high rejection rate after transplantation back to normal histocompatible congenic mice. *Nude* mice developed tumors most quickly and with the highest incidence, leading to the conclusion that in this model the immune system constituted a "tumor-suppressive factor" delaying and sometimes abrogating tumor growth, that is, performing immune surveillance. Cytotoxic $CD8^+$ T cells were found to be indispensable for this rejection, leading to the conclusion that the cytotoxic T cells perform immune selection in normal mice, eliminating immunogenic tumor cell variants in the incipient tumor (Svane et al. 1996; Svane et al. 1999).

Third, the rejection of murine UV-induced skin cancers by normal mice is a striking example of powerful immune surveillance of the normal host against malignant cells. UV-induced regressor tumors grew progressively and killed mice that were depleted of $CD8^+$ T cells. Depletion of $CD4^+$ T cells had no effect, suggesting that $CD8^+$ but not $CD4^+$ T cells were required for this immune surveillance. There was no correlation between the ability of a tumor to grow progressively in a normal immunocompetent host and the level of constitutive class I expression or the level of expression induced in vitro by γ-interferon (Ward et al. 1990).

Thus, tumor antigenicity can be detected during carcinogenesis in immunodeficient animals but gradually lost in normal ones. An important point here is that immunological surveillance provides a borderline between immunity and tolerance in response to tumor cells.

4. Gaze into Cells: MHC-Binding Motifs

To "see" mutant or inappropriately expressed proteins in transformed cells, the immune system needs to "look" into these cells. This capability is provided by expression of MHC class I molecules associated with endogenous peptides. The mechanism of how antigen presenting cell(s) (APC) determines the restriction of effector T cells was uncovered by the Rammensee's group. As the author mentioned in his review on MHC-binding motifs in antigens, immunology owes two students, Olaf Rotzschke and Kirsten Folk, who were interested in the structure of peptides that interact with class I MHC molecules. To isolate these molecules, proteins of cell membranes were adsorbed on an immune affine column followed by the acidic elution and dissociation of MHC/β_2-microglobulin/peptide complexes. The Edman sequencing of isolated peptides showed invariant amino acid residues near the C and N ends. Most importantly, the peptides bound by different allelic forms of MHC class I molecules had similar length but different allele-specific motifs (Falk et al. 1990, 1991a; Rotzschke, et al. 1990). The authors realized that they possess a powerful method for identification of the allele-specific sequences among a huge variety of peptides that could be derived from one antigenic protein (Falk et al. 1991b).

The fact that these motifs have been formed by "anchoring" amino acid residues necessary for high-affinity binding of the peptide with respective MHC molecule also indicated that APC expressing different MHC haplotypes can present various peptides of the same antigen (Rammensee et al. 1993; 1995). For example, the nucleoprotein of influenza virus contains the epitope that binds with H-2K^d in positions AA147–155 (T**Y**QRTRAL**V**) and with H-2D^b in positions AA366–374 (MTEM**N**ENS**A**). For human MHC molecules, the epitope for binding with HLA-A2 is within AA85–94 (K**L**GEFYNQ**M**), and with HLA-A3 within AA265–273 (I**L**RGSVAH**K**), with HLA-B8 within AA380–388 (IAWY**R**SR**L**E), and with HLA-B27 in AA383–391 (S**R**YWAIRT**R**) (underlined are anchoring, i.e., motif forming, residues). The highest binding affinity of influenza virus nucleoprotein-derived peptides with H-2D^b is achieved if the peptide has the canonical motif (Cerundolo et al. 1991). To see whether the expression of certain MHC molecule influenced the efficacy of peptide epitope presentation, the amounts of correctly processed K^b-restricted epitope of the minor histocompatibility antigen H-4b in H-2K^b-positive and-negative cells were estimated. The difference was 3000-fold, indicating an instructive role of MHC molecules in peptide-processing machinery (Wallny et al. 1992a). The mechanisms of peptide/MHC molecule association allowed predicting the structure of T cell peptide epitopes including tumor antigens (Rotzschke et al. 1991; Wallny et al. 1992b). Furthermore, these studies set rational molecular basis for an association between autoimmune diseases and certain MHC haplotypes (Vartdal et al. 1996; Kalbus et al. 2001; Munz et al. 2002). Finally, the ability of individual allelic products of MHC molecules to bind particular peptides of the pathogen directly links MHC with genetically determined immune response to pathogens.

The experience gained in the studies of MHC-binding motifs in MHC-associated peptides is currently used for constructing combinatorial peptide libraries. These tools make possible the positional screening for high-affinity ligands and cross-reacting ones for TCRs. Furthermore, making MHC tetramers can help to directly visualize the antigen-specific T lymphocytes (Wilson et al. 2004; Xu and Screaton 2002). There are a number of on-line services making possible the search for T cell epitopes among different protein sequences. One of them, RANKPEP service is a powerful mean for search of potential T cell epitopes presented by allelic forms of mouse and human MHC classes I and II molecules, taking into account protein degradation by proteasome (Reche et al. 2004). Obviously, the analysis of a specific protein reveals restricted number of epitopes capable to be presented by the MHC molecules of an individual. It assumes that some mutations in cancer-related genes can be invisible to the immune system making possible tumor escape from immunological surveillance.

5. Aire and Thymic Selection

Aire (autoimmune regulator), the gene responsible for the clinical disorder autoimmune polyendocrinopathy syndrome type I, has recently been identified as an important mediator of central tolerance. Structural characteristics and biochemical data suggest that Aire might play a direct role in transcription and function as an ubiquitin ligase. Aire up-regulates the transcription of certain organ-specific self-antigens in medullary thymic epithelial cells (mTECs) and has a role in the negative selection of organ-specific thymocytes (Su and Anderson 2004). Aire promotes the tolerance of thymocytes by inducing the expression of a battery of peripheral-tissue antigens in mTECs. The mechanism whereby Aire exerts its tolerance-promoting function is not primarily positive selection of regulatory T cells but rather negative selection of T effector cells. Surprisingly, supplementing its influence on the transcription of genes encoding peripheral-tissue antigens, Aire somehow enhances the antigen-presentation capability of mTECs. Thus, this transcriptional control element promotes central tolerance both by furnishing a specific thymic stromal cell type with a repertoire of self-antigens and by better arming such cells to present these antigens to differentiating thymocytes. In Aire's absence, autoimmunity and ultimately overt autoimmune disease develops (Anderson et al. 2005).

At certain stages of male gametogenesis, a broad range of genes is expressed. The underlying mechanism and the biological significance of this phenomenon remain elusive. Derbinsky et al. inquired whether the spectrum of ectopically expressed genes is distinct or shared between mTECs and testis. They analyzed gene expression in cDNA libraries prepared either from whole testis or highly enriched immature gametocytes. Of 19 tissue-specific genes expressed in the thymus, 14 were also expressed in the testis. The transcription factors Whn and Aire, both of which have been ascribed specific roles in thymus biology, were also found in male gametocytes. The authors suggested that this mechanism may facilitate tolerance induction to self-antigens that would otherwise be temporally or spatially secluded from the immune system.

Such spatially secluded antigens as cancer/testis (CT) antigens, of which more than 40 have now been identified, are encoded by genes normally expressed in the human germ line but are also expressed in melanoma and carcinomas of the bladder, lung, and liver. These immunogenic proteins are being vigorously pursued as targets for therapeutic cancer vaccines (Van Der Bruggen, Zhang, Chaux, Stroobant, Panichelli, Schultz, Chapiro, Van Den Eynde, Brasseur, and Boon 2002; Simpson, Caballero, Jungbluth, Chen, and Old 2005). It is very important that the array of promiscuously expressed self-antigens in mTECs includes well-known targets for cancer immunotherapy, such as α-fetoprotein, tyrosinase, and gp100. Therefore, intrathymic selection makes the immune system tolerant to tumor-associated antigens. Gene expression in normal tissues may result in tolerance of high-avidity CTL, leaving behind low-avidity CTL that cannot provide effective immunity against tumors expressing the relevant target antigens. Evidently, any antitumor vaccine targeted to tumor-associated antigens should be inefficient due to the loss of high-avidity T cell clones capable to be stimulated.

6. Transduction with Allorestricted TCR as a Mean to Overcome Central Tolerance

Normally negative selection ablates high-avidity T cell clones that can react with self antigens of individual 1 in the context of self MHC molecules (H-2^x-P1). But the clones specific to H-2^x-P1 can be presented in allogeneic individual 2 because negative selection in this organism deletes the clones specific to tumor-associated antigens in the context of "another self" (i.e., allogeneic to P1) MHC (H-2^y-P2). Therefore, allorestricted recognition can supposedly provide the basis for obtaining clones of individual 2 specific to the combination of MHC molecule with H-2^x-P1 peptide of allogeneic recipient for adoptive immunotherapy (Figure 1).

One example of successful use of allorestricted recognition of tumor-associated antigens is the work of Elena Sadovnikova and Hans Stauss who obtained allorestricted CTL clones of H-2^d mice. These clones were specific to the complex of H-$2K^b$ plus mdm-2-derived peptide; it is noteworthy that mdm-2 is frequently overexpressed in tumor cells. In culture, these clones selectively reacted with tumor versus normal cells and killed melanoma and lymphoma cells but not normal H-$2K^b$- expressing dendritic cells. In vivo, the allorestricted clones caused retardation of growth of melanoma and lymphoma in syngeneic (H-2^b) recipients (Sadovnikova and Stauss 1996). The same authors attempted to obtain allorestricted clones specific to a cyclin D1 peptide in the context of human HLA-A2. The clones lysed cyclin D1 overexpressing breast carcinoma cells but not Epstein–Barr-transformed lymphoblastoid cells (Sadovnikova et al. 1998). Allorestricted recognition became an efficient means of breaking tolerance to tumor-associated antigens and to get responses to leukemia-associated markers such as WT1, CD68, and CD45 (Gao et al. 2000; Sadovnikova et al. 2002; Amrolia et al. 2003).

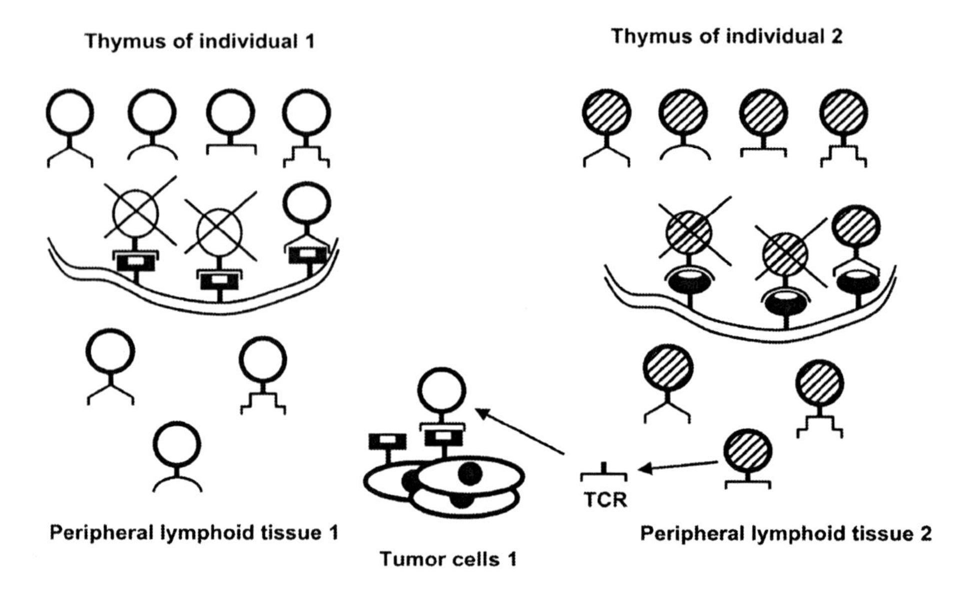

FIGURE 1. Negative selection ablates high-avidity clones specific to tumor-associated antigens of individual 1. This process makes antitumor immune response and vaccination of individual 1 inefficient. Clones with required specificity may be isolated from allogeneic individual 2 for subsequent cloning of T cell receptors(TCRs) and retroviral transduction of T lymphocytes of individual 1.

Then, the TCR genes isolated from antigen-specific T cells can be exploited as generic therapeutic molecules for antigen-specific immunotherapy. Retroviral TCR gene transfer into patient T cells can readily produce populations of antigen-specific lymphocytes after a single round of polyclonal T cell stimulation. The TCR gene-modified lymphocytes are functionally competent in vitro and can have therapeutic efficacy in murine models in vivo. The TCR gene expression is stable, and the modified lymphocytes can develop into the memory T cells. Introduction of TCR genes into $CD8^+$ and $CD4^+$ lymphocytes provides an opportunity to use the same TCR specificity to produce antigen-specific killer and helper T lymphocytes. Thus, TCR gene therapy provides an attractive strategy to develop antigen-specific immunotherapy with autologous lymphocytes as a generic treatment option (Morris et al. 2005; Xue et al. 2005).

The ultimate goal of cancer immunotherapy is to utilize the immune system to eliminate malignant cells. Recently published research has mainly focused on the generation of effective antigen-specific T cell responses because of the general belief that T cell immunity is essential in controlling tumor growth and protection against viral infections. However, isolation of antigen-specific T cells for therapeutic application is a laborious task. Therefore, strategies were developed to genetically transfer tumor-specific immune-receptors into patients' T cells. To this end, the chimeric receptors were constructed that comprise the antibody fragments specific for tumor associated antigens linked to genes encoding signaling domains of TCR or Fc receptor. T cells with such chimeric antibody receptors recapitulate the immune-specific responses mediated by the receptor (Willemsen et al. 2003).

It is noteworthy that murine TCRs are highly functional when expressed in human lymphocytes. Recently published work compared human and mouse TCR function and expression to delineate the molecular basis for the apparent superior biological activity of murine receptors in human T lymphocytes. To this end, authors created hybrid TCRs where they swapped the original constant regions with either human or mouse ones. Murine or "murinized" receptors were overexpressed on the surface of human lymphocytes compared with their human/humanized counterparts and were able to mediate higher levels of cytokine secretion when cocultured with peptide-pulsed antigen-presenting cells. Preferential pairing of murine constant regions and improved CD3 stability seemed to be responsible for these observations. This approach allowed circumventing the natural low avidity of class I MHC TCR in $CD4^+$ cells by introducing the murinized TCR into $CD4^+$ lymphocytes, giving them the ability to recognize melanoma. These findings have implications for human TCR gene therapy (Sommermeyer et al. 2006).

In another work, the TCR gene transfer was investigated as a convenient method to produce antigen-specific T cells for adoptive therapy. The authors focused on the expression of two TCRs in T cells, which could impair their function or cause unwanted effects of mixed TCR heterodimers. With five different TCRs and four different T cells, either mouse or human, they have shown that some TCRs were strong—in terms of cell surface expression—and replaced weak TCRs on the cell surface, resulting in exchange of antigen specificity. Two strong TCRs were coexpressed. The mouse TCR replaced human TCR on human T cells. Even though it is still poorly understood why some TCRαβ combinations are preferentially expressed on T cells, the data suggest that T cells with exclusive tumor reactivity can potentially be generated by T cell engineering (Cohen et al. 2006).

References

Amrolia, P.J., Reid, S.D., Gao, L., Schultheis, B., Dotti, G., Brenner, M.K., Melo, J.V., Goldman, J.M. and Stauss, H.J. (2003) Allorestricted cytotoxic T cells specific for human CD45 show potent antileukemic activity. Blood 101, 1007–1014.

Anderson, M.S., Venanzi, E.S., Chen, Z., Berzins, S.P., Benoist, C. and Mathis, D. (2005) The cellular mechanism of Aire control of T cell tolerance. Immunity 23, 227–239.

Brandle, D., Brasseur, F., Weynants, P., Boon, T. and Van den Eynde, B. (1996) A mutated HLA-A2 molecule recognized by autologous cytotoxic T lymphocytes on a human renal cell carcinoma. J. Exp. Med. 183, 2501–2508.

Cerundolo, V., Elliott, T., Elvin, J., Bastin, J., Rammensee, H.G. and Townsend, A. (1991) The binding affinity and dissociation rates of peptides for class I major histocompatibility complex molecules. Eur. J. Immunol. 21, 2069–2075.

Cohen, C.J., Zhao, Y., Zheng, Z., Rosenberg, S.A. and Morgan, R.A. (2006) Enhanced antitumor activity of murine-human hybrid T cell receptor (TCR) in human lymphocytes is associated with improved pairing and TCR/CD3 stability. Cancer Res. 66, 8878–8886.

Coulie, P.G., Lehmann, F., Lethe, B., Herman, J., Lurquin, C., Andrawiss, M. and Boon, T. (1995) A mutated intron sequence codes for an antigenic peptide recognized by cytolytic T lymphocytes on a human melanoma. Proc. Natl. Acad. Sci. U.S.A. 92, 7976–7980.
Dubey, P., Hendrickson, R.C., Meredith, S.C., Siegel, C.T., Shabanowitz, J., Skipper, J.C., Engelhard, V.H., Hunt, D.F. and Schreiber, H. (1997) The immunodominant antigen of an ultraviolet-induced regressor tumor is generated by a somatic point mutation in the DEAD box helicase p68. J. Exp. Med. 185, 695–705.
Egger, G., Liang, G., Aparicio, A. and Jones, P.A. (2004) Epigenetics in human disease and prospects for epigenetic therapy. Nature 429, 457–463.
Falk, K., Rotzschke, O. Stevanovic, S., Jung, G. and Rammensee, H.G. (1991a) Allele-specific motifs revealed by sequencing of self-peptides eluted from MHC molecules. Nature 351, 290–296.
Falk, K., Rotzschke, O., Deres, K., Metzger, J., Jung, G. and Rammensee, H.G. (1991b) Identification of naturally processed viral nonapeptides allows their quantification in infected cells and suggests an allele-specific T cell epitope forecast. J. Exp. Med. 174, 425–434.
Falk, K., Rotzschke, O. and Rammensee, H.G. (1990) Cellular peptide composition governed by major histocompatibility complex class I molecules. Nature 348, 248–251.
Felsher, D.W. (2003) Cancer revoked: oncogenes as therapeutic targets. Nat. Rev. Cancer 3, 375–380.
Gao, L., Bellantuono, I., Elsasser, A., Marley, S.B., Gordon, M.Y., Goldman, J.M. and Stauss, H.J. (2000) Selective elimination of leukemic $CD34^+$ progenitor cells by cytotoxic T lymphocytes specific for WT1. Blood 95, 2198–2203.
Golovkina, T.V., Chervonsky, A., Dudley, J.P. and Ross, S.R. (1992) Transgenic mouse mammary tumor virus superantigen expression prevents viral infection. Cell 69, 637–645.
Hahn, W.C. and Weinberg, R.A. (2002) Rules for making human tumor cells. N. Engl. J. Med. 347, 1593–1603.
Kalbus, M., Fleckenstein, B.T., Offenhausser, M., Bluggel, M., Melms, A., Meyer, H.E., Rammensee, H.G., Martin, R., Jung, G. and Sommer, N. (2001) Ligand motif of the autoimmune disease-associated mouse MHC class II molecule $H2\text{-}A^s$. Eur. J. Immunol. 31, 551–562.
Li, E. (2002) Chromatin modification and epigenetic reprogramming in mammalian development. Nat. Rev. Genet. 3, 662–673.
Logunova, N.N., Viret, C., Pobezinsky, L.A., Miller, S.A., Kazansky, D.B., Sundberg, J.P. and Chervonsky, A.V. (2005) Restricted MHC-peptide repertoire predisposes to autoimmunity. J. Exp. Med. 202, 73–84.
Monach, P.A., Meredith, S.C., Siegel, C.T. and Schreiber, H. (1995) A unique tumor antigen produced by a single amino acid substitution. Immunity 2, 45–59.
Morris, E.C., Tsallios, A., Bendle, G.M., Xue, S.A. and Stauss, H.J. (2005) A critical role of T cell antigen receptor-transduced MHC class I-restricted helper T cells in tumor protection. Proc. Natl. Acad. Sci. U.S.A. 102, 7934–7939.
Munz, C., Hofmann, M., Yoshida, K., Moustakas, A.K., Kikutani, H., Stevanovic, S., Papadopoulos, G.K. and Rammensee, H.G. (2002) Peptide analysis, stability studies, and structural modeling explain contradictory peptide motifs and unique properties of the NOD mouse MHC class II molecule $H2\text{-}A^{g7}$. Eur. J. Immunol. 32, 2105–2116.
Pelleitier, M. and Montplaisir, S. (1975) The nude mouse: a model of deficient T cell function. Methods Achiev. Exp. Pathol. 7, 149–166.
Pobezinskaya, Y., Chervonsky, A.V. and Golovkina, T.V. (2004) Initial stages of mammary tumor virus infection are superantigen independent. J. Immunol. 172, 5582–5587.
Prehn, R.T. (2005) On the nature of cancer and why anticancer vaccines don't work. Cancer Cell Int. 5, 25.

Prehn, R.T. and Main, J.M. (1957) Immunity to methylcholanthrene-induced sarcomas. J. Natl. Cancer Inst. 18, 769–778.
Rammensee, H.G. (1995) Chemistry of peptides associated with MHC class I and class II molecules. Curr. Opin. Immunol. 7, 85–96.
Rammensee, H.G., Falk, K. and Rotzschke, O. (1993) Peptides naturally presented by MHC class I molecules. Annu. Rev. Immunol. 11, 213–244.
Reche, P.A., Glutting, J.P., Zhang, H. and Reinherz, E.L. (2004) Enhancement to the RANKPEP resource for the prediction of peptide binding to MHC molecules using profiles. Immunogenetics 56, 405–419.
Robbins, P.F., El-Gamil, M., Li, Y.F., Kawakami, Y., Loftus, D., Appella, E. and Rosenberg, S.A. (1996) A mutated beta-catenin gene encodes a melanoma-specific antigen recognized by tumor infiltrating lymphocytes. J. Exp. Med. 183, 1185–1192.
Rosenberg, S.A., Sherry, R.M., Morton, K.E., Scharfman, W.J., Yang, J.C., Topalian, S.L., Royal, R.E., Kammula, U., Restifo, N.P., Hughes, M.S., Schwartzentruber, D., Berman, D.M., Schwarz, S.L., Ngo, L.T., Mavroukakis, S.A., White, D.E. and Steinberg, S.M. (2005) Tumor progression can occur despite the induction of very high levels of self/tumor antigen-specific $CD8^+$ T cells in patients with melanoma. J. Immunol. 175, 6169–6176.
Rotzschke, O., Falk, K., Deres, K., Schild, H., Norda, M., Metzger, J., Jung, G. and Rammensee, H.G. (1990) Isolation and analysis of naturally processed viral peptides as recognized by cytotoxic T cells. Nature 348, 252–254.
Rotzschke, O., Falk, K., Stevanovic, S., Jung, G., Walden, P. and Rammensee, H.G. (1991) Exact prediction of a natural T cell epitope. Eur. J. Immunol. 21, 2891–2894.
Sadovnikova, E., Jopling, L.A., Soo, K.S. and Stauss, H.J. (1998) Generation of human tumor-reactive cytotoxic T cells against peptides presented by non-self HLA class I molecules. Eur. J. Immunol. 28, 193–200.
Sadovnikova, E. and Stauss, H.J. (1996) Peptide-specific cytotoxic T lymphocytes restricted by nonself major histocompatibility complex class I molecules: reagents for tumor immunotherapy. Proc. Natl. Acad. Sci. U.S.A. 93, 13114–13118.
Sadovnikova, E., Parovichnikova, E.N., Savchenko, V.G., Zabotina, T. and Stauss, H.J. (2002) The CD68 protein as a potential target for leukaemia-reactive CTL. Leukemia 16, 2019–2026.
Sharkey, F.E. and Fogh, J. (1979) Incidence and pathological features of spontaneous tumors in athymic nude mice. Cancer Res. 39, 833–839.
Sherman, L.A., Theobald, M., Morgan, D., Hernandez, J., Bacik, I., Yewdell, J., Bennink, J. and Biggs, J. (1998) Strategies for tumor elimination by cytotoxic T lymphocytes. Crit. Rev. Immunol. 18, 47–54.
Simpson, A.J., Caballero, O.L., Jungbluth, A., Chen, Y.T. and Old, L.J. (2005) Cancer/testis antigens, gametogenesis and cancer. Nat. Rev. Cancer 5, 615–625.
Sommermeyer, D., Neudorfer, J., Weinhold, M., Leisegang, M., Engels, B., Noessner, E., Heemskerk, M.H., Charo, J., Schendel, D.J., Blankenstein, T., Bernhard, H. and Uckert, W. (2006) Designer T cells by T cell receptor replacement. Eur. J. Immunol. 36, 3052–3059.
Su, M.A. and Anderson, M.S. (2004) Aire: an update. Curr. Opin. Immunol. 16, 746–752.
Svane, I.M., Boesen, M. and Engel, A.M. (1999) The role of cytotoxic T lymphocytes in the prevention and immune surveillance of tumors - lessons from normal and immunodeficient mice. Med. Oncol. 16, 223–238.
Svane, I.M., Engel, A.M., Nielsen, M.B., Ljunggren, H.G., Rygaard, J. and Werdelin, O. (1996) Chemically induced sarcomas from nude mice are more immunogenic than similar sarcomas from congenic normal mice. Eur. J. Immunol. 26, 1844–1850.
Van Der Bruggen, P., Zhang, Y., Chaux, P., Stroobant, V., Panichelli, C., Schultz, E.S., Chapiro, J., Van Den Eynde, B.J., Brasseur, F. and Boon, T. (2002) Tumor-specific shared antigenic peptides recognized by human T cells. Immunol. Rev. 188, 51–64.

Vartdal, F., Johansen, B.H., Friede, T., Thorpe, C.J., Stevanovic, S., Eriksen, J.E., Sletten, K., Thorsby, E., Rammensee, H.G. and Sollid, L.M. (1996) The peptide binding motif of the disease associated HLA-DQ (alpha 1* 0501, beta 1* 0201) molecule. Eur. J. Immunol. 26, 2764–2772.

Wallny, H.J., Deres, K., Faath, S., Jung, G., Van Pel, A., Boon, T. and Rammensee, H.G. (1992b) Identification and quantification of a naturally presented peptide as recognized by cytotoxic T lymphocytes specific for an immunogenic tumor variant. Int. Immunol. 4, 1085–1090.

Wallny, H.J., Rotzschke, O., Falk, K., Hammerling, G. and Rammensee, H.G. (1992a) Gene transfer experiments imply instructive role of major histocompatibility complex class I molecules in cellular peptide processing. Eur. J. Immunol. 22, 655–659.

Ward, P.L., Koeppen, H.K., Hurteau, T., Rowley, D.A. and Schreiber, H. (1990) Major histocompatibility complex class I and unique antigen expression by murine tumors that escaped from $CD8^+$ T cell-dependent surveillance. Cancer Res. 50, 3851–3858.

Willemsen, R.A., Debets, R., Chames, P. and Bolhuis, R.L. (2003) Genetic engineering of T cell specificity for immunotherapy of cancer. Hum. Immunol. 64, 56–68.

Wilson, D.B., Wilson, D.H., Schroder, K., Pinilla, C., Blondelle, S., Houghten, R.A. and Garcia, K.C. (2004) Specificity and degeneracy of T cells. Mol. Immunol. 40, 1047–1055.

Wolfel, T., Hauer, M., Schneider, J., Serrano, M., Wolfel, C., Klehmann-Hieb, E., De Plaen, E., Hankeln, T., Meyer zum Buschenfelde, K.H. and Beach, D. (1995) A p16INK4a-insensitive CDK4 mutant targeted by cytolytic T lymphocytes in a human melanoma. Science 269, 1281–1284.

Woodruff, M.F. and Hodson, B.A. (1985) The effect of passage *in vitro* and *in vivo* on the properties of murine fibrosarcomas I. Tumorigenicity and immunogenicity. Br. J. Cancer 51, 161–169.

Xu, X.N. and Screaton, G.R. (2002) MHC/peptide tetramer-based studies of T cell function. J. Immunol. Methods 268, 21–28.

Xue, S., Gillmore, R., Downs, A., Tsallios, A., Holler, A., Gao, L., Wong, V., Morris, E. and Stauss, H.J. (2005) Exploiting T cell receptor genes for cancer immunotherapy. Clin. Exp. Immunol. 139, 167–172.

Zerrahn, J., Held, W. and Raulet, D.H. (1997) The MHC reactivity of the T cell repertoire prior to positive and negative selection. Cell 88, 627–636.

15

Crosstalk Between Apoptosis and Antioxidants in Melanoma Vasculogenic Mimicry

Amalia Vartanian and Anatoly Yu. Baryshnikov

Department of Experimental Diagnosis and Biotherapy of Tumors, Russian Cancer Research Center, Moscow, Russia, amaliavartanian@hotmail.com

Abstract. The concept of "vasculogenic mimicry" (VM) was introduced to describe the unique ability of highly aggressive tumor cells to form capillary-like structure (CLS) and matrix-rich patterned network in three-dimensional cultures that mimic embryonic vasculogenic network. Here, we provide the experimental evidence that CLS structure formation requires apoptotic cell death through activation of caspase-dependent mechanism. Our results indicate that the formation of CLS is also related to the reactive oxygen species (ROS) levels.

1. Introduction

It is generally assumed that tumors require a blood supply for growth and metastasis, and much attention has been focused on angiogenesis, that is, recruitment of new vessels into a tumor from the pre-existing vessels (Folkman 2003). Angiogenesis allows the tumor cells to express their critical growth advantage and permits the establishment of continuity with the existing vasculature of the host. In addition, increased angiogenesis coincides with increased tumor entry into the circulation and thereby facilitates hematogenous dissemination and metastasis. Angiogenesis is, however, not the only mechanism by which tumors acquire a microcirculation. Recently, it was demonstrated that highly aggressive and metastatic melanoma cells in the absence of endothelial cells (ECs) and fibroblasts are capable of forming highly patterned vascular channels in vitro that are composed of a basement membrane that stains positive with periodic-acid-Schiff reagent; in contrast, less aggressive melanoma cell lines did not generate these patterns (Maniotis et al. 1999). The channels, formed in vitro, were morphologically identical to PAS-positive channels in histological preparations from highly aggressive uveal and cutaneous melanomas found in patients. The generation of microvascular channels by aggressive tumor cells was termed "vasculogenic mimicry" (VM) to emphasize their de novo generation without participation by ECs and independent of angiogenesis. Recent observations using intravenous tracer have confirmed the presence of PAS-positive fluid-conducting channels containing red blood cells and plasma and, possibly, providing a perfusion mechanism within the tumor compartment that functions either independent of or simultaneously with angiogenic vessels (Maniotis et al. 2002). A strong statistical correlation was established between the presence of PAS-positive patterns and outcome in primary uveal and cutaneous melanoma: tumor-exhibiting VM had poor prognosis (Warso et al. 2001).

The analysis of differential gene expression of highly aggressive melanoma cell lines revealed the co-expression of multiple phenotype-specific genes [VE-cadherin, EphA2, CD31, VEGF and receptors for VEGF, von Willebrand Factor (vWF), endothelial protein receptor TIE-1, lymphocyte-specific protein 1 (LSP1), major his to compatibility class (MHC) class II, KIT, and desmin] (Hendrix et al. 2003). These findings suggested that aggressive melanoma cells were able to engage in VM owing to their genetic reversion to an undifferentiated, embryonic-like phenotype and adopted EC-like properties through an apparent transdifferentiation process.

Melanoma accounts for only 4% of skin cancer cases but most of skin cancer-related death (Forguson 2005). The rising incidence of melanoma makes this tumor an important public health problem. Standard systemic therapies have not been adequately effective in the management of melanoma (Mandara et al. 2006). Chemoprevention by naturally occurring agents has shown benefits in many types of cancer in animal models, including melanoma (Demierre and Natanson 2003; Mukhtar and Ahmad 1999; Jang et al. 1997), and antioxidation, antilipogenesis, and antiinflammation have been proposed as possible mechanisms of these preventive effects (Fujiki 1999; Soleas et al. 1997).

In this study, we used six human cell lines derived from patients with disseminated melanoma to explore parameters driving aggressive melanoma cells to express a latent "angiogenic program" that recapitulates the early events of capillary-like structure (CLS) formation.

2. Early CLS Formation Requires Activation of Apoptotic Pathways

To determine the molecular mechanisms underlying the phenomenon of VM in the tumor, we prepared six cell lines from the surgical species of patients with disseminated melanoma (Mikhailova et al. 2005). Melanoma status was confirmed by immunocytochemistry for the expression of four melanoma-associated markers—MelanA, S100, tyrosinase, and high-molecular weight antigen (HMW). Furthermore, using the short-term Matrigel assay, we showed that five cell lines with high invasive potential were engaged in CLS formation.

In the presence of growth factors, melanoma cells were induced to align and form CLS in cultures. A gradual decrease in cell number during the CLS formation could readily be observed in in vitro cultures. Cell counts revealed that only 65–70% of cells were viable 24 h later, suggesting that apoptosis might be an essential part of the VM phenomenon. Both morphological data and the PI staining suggested that apoptosis may be responsible for the decrease in cell numbers (Figure 1A and B). To confirm the role for apoptosis in CLS formation, we demonstrated that the addition of a caspase

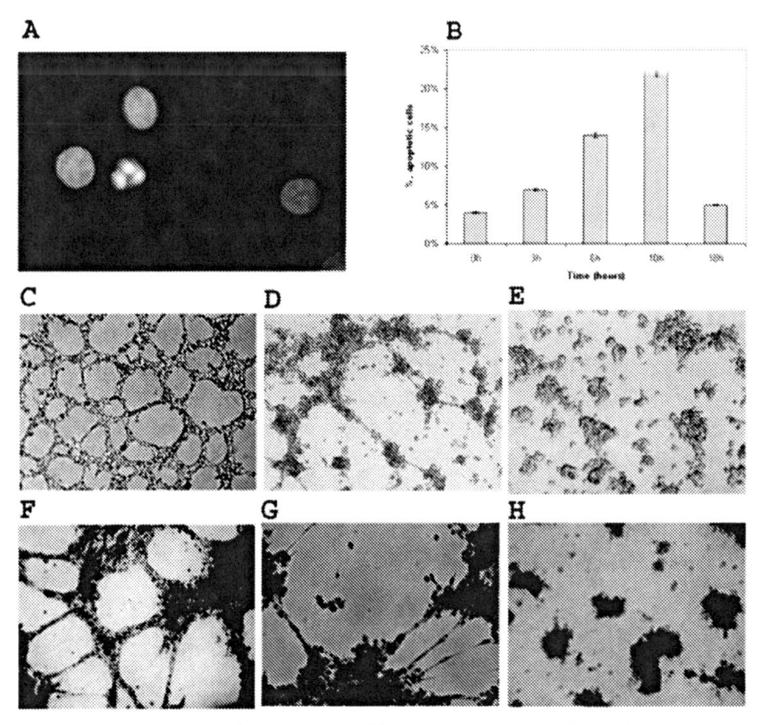

FIGURE 1. Apoptosis occurs during capillary-like structure (CLS) formation on Matrigel. (A) Apoptotic melanoma cells were visualized by Hoecht 33342. (B) Time course analysis of apoptotic cells by flow cytometry. (C) Control, non-treated cells. (D) Cells treated with 0.1mM zVAD-fmk. (E) Cells treated with 0.2 mM DEVD. MTT staining of cells during CLS formation. (F) Control. (G) zVAD-fmk, 0.1 mM. (H) DEVD, 0.2 mM. Original magnifications: ×100 (A), ×40 (C–H).

inhibitor zVAD-fmk simultaneously inhibited CLS formation: 95% of cells remaining in the cultures were alive and no CLS was seen (Figure 1C and D). Interestingly, the metabolic activity of melanoma cells assessed in dimetylthiazol-diphenyltetrazolium-bromide (MTT) assay and cell motility tested in the Transwell inserts, that are both involved in CLS formation, were unaffected by zVAD-fmk (Figure 1F and G). These results were further confirmed by DEVD, a caspase-3 inhibitor, which also induced malformation in CLS similar to that induced by zVAD-fmk (Figure 1E and H). Thus, our results indicate that apoptosis may be a general mechanism of elimination of tumor cells, which failed to turn on the gene program that leads to the differentiation of melanoma cells into endothelial-like cells.

The involvement of caspases in differentiation was initially thought to be associated with enucleation process. For instance, caspase activation is required for maturation of erythroid progenitor (Zermati et al. 2001). However, platelet formation from megakaryocytes, skeletal muscle differentiation, monocytes differentiation into macrophages, and osteoblastic differentiation also require caspase activation (De Botton et al. 2002; Fernando et al. 2002; Sordet et al. 2002; Mogi and Togari 2003). Further studies are required to reveal the mechanisms protecting cells from apoptosis during intrinsic caspase activation.

2.1. Time- and Dose-Dependent Effects of Antioxidants on Melanoma Cells

Based on a large body of evidence, it is clear now that Resveratrol (RT, trans-3,4', 5-trihydroxystilbene), a polyphenolic phytoalexin found in grapes, fruits, and root extracts of the weed *Polygonum cuspidatum*, is an effective chemopreventive agent for many types of cancer. We hypothesized that this naturally occurring compound could also be involved in vessel formation. In order to exclude the participation of cytotoxic and antiproliferative effects of RT in VM, we examined the influence of RT on melanoma cell viability and proliferation. As assessed by MTT assay, the metabolic activity of cells treated with RT for 24 h was similar to the control values (data not shown). The Trypan blue exclusion assay did not reveal any cytotoxicity of RT (up to 100 μM for 24 h). The treatment of melanoma cells with RT (1, 2, 5, 10, 20, and 50 μM) for 48 h resulted in a concentration-dependent decrease of cell viability and growth (Figure 2A), which could be attributed to its antiproliferative effect. Since Ki-67, a cell proliferation marker, is rapidly degraded as cells enter the non-proliferative stage (Endl and Gerdes 2000), we evaluated the effect of RT on

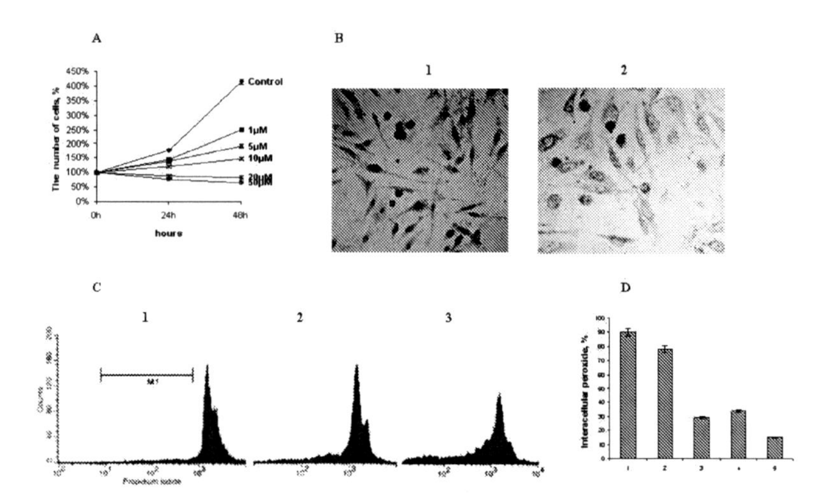

FIGURE 2. Effect of Resveratrol (RT) on melanoma cell growth. (A) Inhibitory effect of RT on kinetic growth of melanoma cells. After 24 h in cultures, cells were treated with RT or medium and harvested 24 and 48 h later. The number of growing cells was estimated by Trypan blue exclusion. The data are shown as the percentage of cell viability (the mean from three experiments, $p < 0.05$). (B) Immunocytochemical analysis of melanoma cells treated with RT or medium. (A1) Control, non-treated cells. (A2) Cells were treated with 50 μM RT. Ki-67 staining was strongly inhibited in RT-treated cells. Magnifications, ×100. (C) Flow cytometry analysis of RT-treated cells. Cells were treated with 10, 20, and 50 μM RT: C1, C2, and C3, respectively. Apoptotic cells appear as a peak to the left of the G1 peak. (D) Effect of antioxidants on intracellular ROS levels. (D1, D2, and D3) Cells incubated with 1, 5, and 10 μM RT, respectively. (D4) Cells incubated with 50 μM EGCG. (D5) Cells incubated with 20 mM NAC. The values are the percentage of increase of the mean cell fluorescence.

Ki-67 expression and observed a down-regulation of Ki-67 protein levels after 48 h of treatment with 50 μM RT (Figure 2B). These data suggested that RT-mediated decrease of cell growth was a result of decreased cellular proliferation. To evaluate whether the antiproliferative effect of RT is mediated by changing the rate of apoptosis, we performed a flow cytometry analysis of DNA fragmentation in the presence of RT at different concentrations (Figure 2C). RT at 50 μM induced the highest level of apoptosis (up to 32%). We have also examined the effect of RT on cell migration and revealed that RT at 1–20 μM did not influence the melanoma cells motility (data not shown). We also revealed that RT at 1, 5, and 10 μM caused 10, 24, and 71% reduction of intracellular peroxide, respectively, as was determined by flow cytometry using fluorescent dye 2′,7′-dichlorofluorescein-diacetate (DCF-DA) (Figure 2D). Control experiments were carried out with epigallocatechin-3-gallate (EGCG, the major constituent of green tea) and a thiolic antioxidant *N*-acetylcysteine (NAC) known to have no effect on growth or metabolic activity of tumor cells. As can be seen, 50 μM EGCG and 20 mM NAC did not reduce the proliferative rate or migration of melanoma cells and caused 65 and 86% reduction of intracellular peroxide, respectively (Figure 2D).

2.2. Antioxidants Inhibit the Formation of CLS

Recently, RT and EGCG have been found to be effective in inhibiting VEGF-induced angiogenesis (Brakenhielm et al. 2002; Lamy et al. 2002). We, thus, tested whether they can also affect CLS formation. RT had no effect on CLS formation at 1–5 μM but completely abolished the generation of CLS at 10 μM. In contrast to other studies utilizing high concentrations of RT (>100 μM) (Garcia-Garcia et al. 1999; Manna et al. 2000), we observed that low concentrations of RT were sufficient to significantly inhibit CLS formation. Furthermore, EGCG (50 μM, 24 h) also resulted in 65% reduction of intracellular peroxide and blocked CLS formation. Similarly, an antioxidant NAC completely abolished the generation of peroxide and blocked CLS formation. These observations suggest that CLS formation might be related to the reactive oxygen species (ROS) levels.

At present, there is no evidence that tumor cell redox status might be associated with cell ability to form CLS although numerous studies have pointed that ROS can act as important mediators of melanoma growth and VEGF-induced angiogenic signaling including EC-tube formation (Yasuda et al. 1999; Stone and Collins 2002; van der Schaft et al. 2004; Forguson 2005; Meykens et al. 2001). In agreement with these studies, we found that CLS formation also correlated with the ROS levels. More importantly, antioxidants resulted in a concentration-dependent inhibition of CLS formation. These findings support the hypothesis that ROS levels might be a sensitive indicator of CLS formation. Many intracellular signaling molecules, including AP-1, NF-κB, MAPK, and PKC have been shown to be affected by RT (Manna et al. 2000); however, the effective inhibiting doses of RT were always >100 μM, that is, significantly higher than in our studies. Although additional studies are needed to define the nature of RT inhibition of vasculogenesis, our data raise an exciting possibility of using RT for both prevention and therapy of disseminated melanoma.

3. CLS Formation Requires Activation of Apoptotic Pathways Mediated by Mitochondrial Cytochrome c Release and Caspase-3 Activation

There are two major mechanisms mediating apoptosis: intrinsic and extrinsic. Cell surface Fas-receptor signaling upon FasL binding triggers caspase-8 activity and caspase cascade (Schulze-Osthoff et al. 1998). Involvement of this pathway in CLS formation by melanoma cells was ruled out in our studies, since antagonistic anti-Fas antibody did not block the in vitro tube-like organization process and CLS formation (Figure 3A). Fas and FasL expression on melanoma cells was also analyzed by flow cytometry and neither Fas nor FasL was seen during CLS formation (data not shown). These findings suggest that extrinsic apoptotic pathways did not participate in CLS formation.

Alternatively, under the stress conditions, mitochondrial cytochrome c is released into the cytosol. On release, cytochrome c interacts with Apaf-1 and pro-caspase-9 to form apoptosomes activating downstream effector caspase-3 responsible for

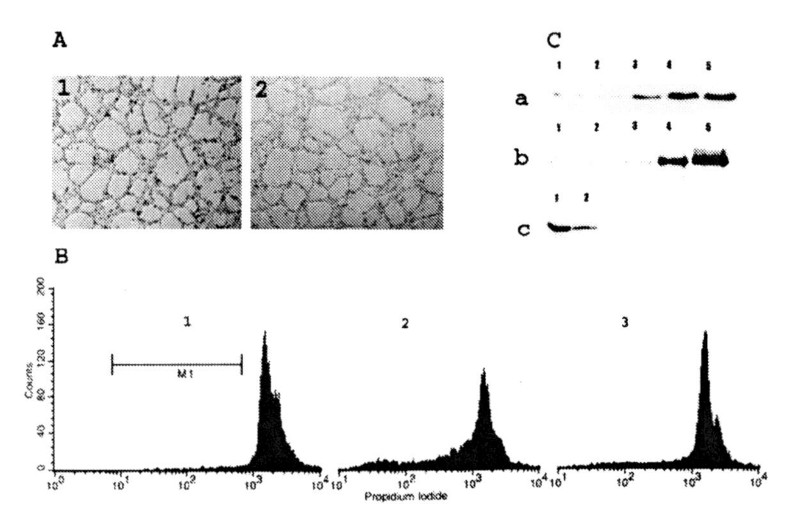

FIGURE 3. Pathways of capillary-like structure (CLS) formation. (A) Fas–FasL croslinkage does not modify CLS formation. (A1) Cells treated with anti-Fas blocking antibody; (A2) Control, non-treated cells. (B) Flow cytometry analysis of melanoma cells stained with PI. (B1) Control: cells were detached from plastic 10 h after seeding; (B2) cells were detached from Matrigel 10 h after seeding; (B3) cells were incubated with 10 μM RT for 24 h, detached from plastic, seeded on Matrigel, then detached from Matrigel and analyzed. (C) CLS formation requires activation of apoptotic pathways mediated by mitochondrial cytochrome c release and caspase-3 activation. (Ca) Time-dependent cytochrome c release into cytosol. Lane 1, cells grown on plastic for 10 h, lanes 2–5, cells grown on Matrigel for 2, 4, 6, and 8 h. (Cb) Active caspase-3 in cytosol. Lane 1, cells grown on plastic; lanes 2–5, cells grown on Matrigel for 2, 4, 8, and 10 h . (Cc) Effect of antioxidants on active caspase-3. Lane 1, active caspase-3 in control cells; lane 2, active caspase-3 in cells pre-incubated with 10 μM RT for 24 h.

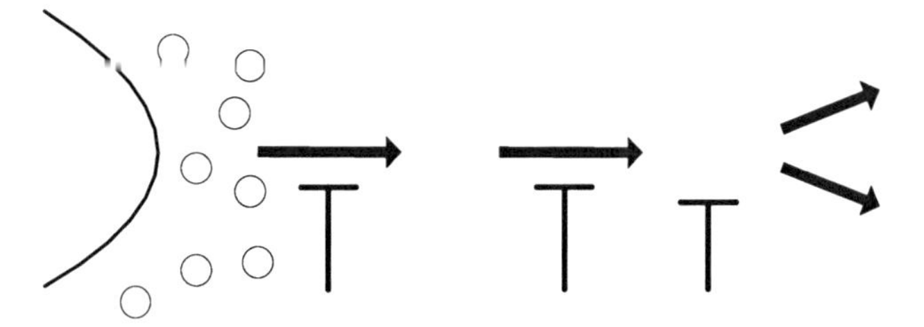

Figure 4. Experimental model for melanoma VM cells highlighting the involvement of apoptosis and ROS in the formation of CLS.

apoptotic destruction of cells (Meier et al. 2000). In our studies, cytochrome c was found in the cytosol as early as 4 h after seeding cells on Matrigel (Figure 3C). Initially, undetectable, active caspase-3 appeared 7 h later and progressively accumulated during CLS formation. This suggests activation of intrinsic apoptotic pathways at the first step of CLS formation. FACScan analysis of apoptotic cells has shown an increase in PI-positive cells (10 h, 22–25%, Figure 3B) compared with control values (2–4%, Figure 3B). Interestingly, pretreatment of cells with antioxidants for 24 h prior to seeding them on Matrigel decreases the level of apoptosis (10 h, 8–10%, Figure 3). To verify whether the antiapoptotic effect of antioxidants reflects their ability to alter activation of caspase-3, cells were incubated with antioxidants for 24 h and examined for caspase-3 activation by Western blot. Figure 3C shows that RT decreased the level of active caspase-3 by threefold. These results suggest that deregulation of non-apoptotic pathways of caspases might contribute to tumor vascularization.

Based on our data, we suggest a scenario of signal transduction pathways involved in the process of melanoma VM (Figure 4). In this model, VEGF can trigger the tube formation in VEGF-R-expressing tumor. The binding of VEGF to VEGF-R results in accumulation of ROS and activation of the gene program leading to differentiation of melanoma cell to endothelial-like cells. Apoptosis is relevant for vascular tissue rearrangement. Inhibition of apoptosis by caspase inhibitors or antioxidants would block CLS formation. This scenario deserves further experimental verifications.

4. Conclusions

Despite impressive results in a variety of pre-clinical models, antiangiogenic agents have met only limited success in the clinic. The heterogeneity of vascular supply might be one of the key factors required for tumor survival. In fact, in addition to traditional angiogenesis, vascular channels, cords, sinuses (without EC lining), and mosaic vessels (EC and tumor cell lined) are also involved in the vascularization of tumors. The presence of tumor cell-lined channels provides a potentially new route by which metastasis may spread to the distant sites. The data discussed above, thus, should help understanding the mechanisms of VM phenomenon and may be of potential importance for the development of novel chemotherapeutic approaches.

Acknowledgments

This work was supported by the Foundation of Russian Ministry of Education and Science (to M.R. Lichinitser) and ZAO "PROTEC" (to E.V. Stepanova).

References

Brakenhielm, E., Cao, R. and Cao, Y. (2002) Suppression of angiogenesis, tumor growth and wound healing by resveratrol, a natural compound in red wine and grapes. FASEB J. 15, 1798–1800.

De Botton, S., Sabri, S., Daugas, E., Zermati, Y., Gudorti, J.E., Hermine, O. and Debili, N. (2002) Platelet formation is the consenquence of caspase activation within megakariocytes. Blood 100, 310–317.

Demierre, M.F. and Natanson, L. (2003) Chemoprevention of melanoma: an unexpected strategy. J. Clin. Oncol. 21, 158–165.

Endl, E. and Gerdes J. (2000) Ki-67 protein: fascinating forms and unknown functions. Exp. Cell. Res. 257, 231–237.

Fernando, P., Kelly, J.F., Balazsi, K., Slack, R.S. and Megeney, L.A. (2002) Caspase-3 activity is required for skeletal muscle differentiation. Proc. Natl. Acad. Sci. U.S.A. 99, 11025–11030.

Folkman, J. (2003) Fundamental concepts of the angiogenic process. Curr. Mol. Med. 3, 643–651.

Forguson, K (2005) Melanoma. J. Cont. Educ. Nurs. 36, 242–243.

Fujiki, H. (1999) Two stages of cancer prevention with green tea. J. Cancer Res. Clin. Oncol. 125, 589–597.

Garcia-Garcia, J., Micol, V., de Godes, A. and Gomez-Fernandes, J.C. (1999) The cancer chemopreventive agent resveratrol is incorporated into model membranes and inhibits protein kinase C alpha activity. Arch. Biochem. Biophys. 372, 382–388.

Hendrix, M.J., Seftor, E.A., Hess, A.R. and Seftor, R.E. (2003) Molecular plasticity of human melanoma cells. Oncogene 22, 3070–3075.

Jang, M., Cai, L., Udeani, G.O., Slowing, K.V., Thomas, C.F., Beecher, C.W. and Pezzuto, J.M. (1997) Cancer chemopreventive activity of resveratrol, a natural product derived from grapes. Science 275, 218–220.

Lamy, S., Gingras, D. and Beliveau, R. (2002) Green tea catechins inhibit vascular endothelial growth factor receptor phosphorylation. Cancer Res. 62, 381–385.

Mandara, M., Nortilli, R., Sava, T. and Cetto, G.L. (2006) Chemotherapy for metastatic melanoma. Expert Rev. Anticancer Ther. 6, 121–130.

Maniotis, A., Folberg, R., Hess, A., Seftor, E.A., Gardner, L.M., Pe'er, J. and Hendrix, M.J. (1999) Vascular channel formation by human melanoma cells *in vivo* and *in vitro*: vasculogenic mimicry. Am. J. Pathol. 55, 739–752.

Maniotis, A.J., Chen, X., Garcia, C., deChristopher, P.J., Wu, D. and Folberg, R. (2002) Control of melanoma morphogenesis, endothelial survival and perfusion by extracellular matrix. Lab. Invest. 82, 1031–1043.

Manna, S.K., Mukhopadhyay, A. and Aggarwal, B.B. (2000) Resveratrol suppresses TNF-induced activation of NF-kappa B, activator protein-1 and apoptosis: potential role of reactive oxygen intermediates and lipid peroxidation. J. Immunol. 164, 6509–6519.

Meier, P., Finch, A. and Evan, G. (2000) Apoptosis in development. Nature 407, 796–801.

Meykens, F.L., Farmer, P. and Fruehanf, J.P. (2001) Redox regulation in human melanocytes and melanoma. Pigment Cell Res. 14, 148–152.
Mikhailova, I.N., Lukashina, M.I., Baryshnikov, A.Yu., Morozova, L.F., Burova, O.S., Kiselev, S. and Georgiev, G.P. (2005) Melanoma cell line as a basis for antitumor vaccine preparation. Vest. Ross. Akad. Med. Nauk [Rus] 7, 37–40.
Mogi, M. and Togari, A. (2003) Activation of caspases is required for osteoblastic differentiation. J. Biol. Chem. 278, 47477–47482.
Mukhtar, H. and Ahmad, N. (1999) Green tea in chemoprevention of cancer. Toxicol. Sci. 52, 111–117.
Schulze-Osthoff, K., Ferrari, D. and Loss, M. (1998) Apoptosis signaling by death receptors. Eur. J. Biochem. 17, 1675–1687.
Soleas, G.J., Diamandis, E.P. and Goldbery, D.M. (1997) Resveratrol: a molecule whose time has come? And gone? Clin. Biochem. 30, 91–113.
Sordet, O., Rebe, C., Plenchette, S., Hermine, O., Vanchenker, W. and Dubrez-Daloz, L. (2002) Specific involvement of caspases in the differentiation of monocytes into macrophages. Blood 100, 4446–4453.
Stone, J.R. and Collins, T. (2002) The role of hydrogen peroxide in endothelial proliferative response. Endothelium 9, 231–238.
Van der Schaft, D.W., Seftor, R.B., Seftor, E.A., Hess, A.R., Gruman, L.M., Kirschmann, D.A. and Hendrix, M.J. (2004) Effects of angiogenesis inhibitors on vascular network formation by human endothelial and melanoma cells. J. Natl. Cancer Inst. 96, 1473–1476.
Warso, M.A., Maniotis, A.J., Chen, X., Majumdar, D., Patel, M.K., Shilkaitis, A. and Folberg, R. (2001) Prognostic significance of Periodic acid-Schiff-positive patterns in primary cutaneous melanoma. Clin. Cancer Res. 7, 473–377.
Yasuda, M., Ohzeki, Y., Shimizu, S., Naito, S., Ohtsuru, A., Yamamoto, T. and Kuroiwa, Y. (1999) Stimulation of *in vitro* angiogenesis by hydrogen peroxide and the relation with ETS-1 in endothelial cells. Life Sci. 64, 249–258.
Zermati, Y., Garrido, C., Amsellem, S., Fashelson, S., Bouscary, D., Valenci, F. and Hermine, O. (2001) Caspase activation is required for terminal erytroid differentiation. J. Exp. Med. 193, 247–254.

16

Immunological Role of Dendritic Cells in Cervical Cancer

Alagar Manickam[1], Muthukumaran Sivanandham[2], and Irina L. Tourkova[3]

[1]Department of Biotechnology, Government College of Technology, Tamil Nadu, India
[2]Department of Biotechnology, Sri Venkateswara College of Engineering, Tamil Nadu, India
[3]University of Pittsburgh, Department of Pathology, Pittsburgh, PA, USA, turkovai@upmc.edu

Abstract. Cervical cancer is the second most frequent gynecological malignancy in the world. Human papillomavirus (HPV) infection is the primary etiologic agent of cervical cancer. However, HPV alone is not sufficient for tumor progression. The clinical manifestation of HPV infection depends also on the host's immune status. Both innate and adaptive immunity play a role in controlling HPV infection. In untransformed HPV-infected keratinocytes, the innate immunity is induced to eliminate the invading HPV pathogen through sensitization to HPV-related proteins by epithelial-residing Langerhans cells (LCs), macrophages, and other immune cells. Once the HPV infection escapes from initial patrolling by innate immunity, cellular immunity becomes in charge of killing the HPV-infected keratinocytes of the uterine cervix through systemic immune response developing by dendritic cells (DCs) in the regional lymphoid organs or through local immune response developing by LCs in the cervix. Thereby, DC/LC plays a critical role in eliciting innate and adaptive cellular immune responses against HPV infection. HPV-associated cervical malignancies might be prevented or treated by induction of the appropriate virus-specific immune responses in patients. Encouraging results from experimental vaccination systems in animal models have led to several prophylactic and therapeutic vaccine clinical trials.

1. Human Papillomavirus and Cervical Cancer

Although the recent advances in cervical cancer diagnosis and treatment reduce the mortality of women with cervical cancer from the second leading cause of death to fifth place worldwide, still about 500,000 new cases are reported annually, and one-third of them are under the high risk of death (Sasagawa et al. 2005). Clinically, cervical cancer is an advanced stage of cervical intraepithelial lesions manifested from pre-invasive cervical intraepithelial neoplasia (CIN) due to the progressive infection of human papillomavirus (HPV) in keratinocytes of the uterine cervix (Walboomers et al. 1999). HPV belongs to a family of 120 double-stranded DNA viruses that have been linked to a number of epithelial cancers, and more often with the uterine cervix, where more than 90% of tumors contain HPV DNA (Tindle 2002). The abnormal uterine cervical clinical features as a result of HPV infection have a wide spectrum of diseases including warts (skin disease), low-grade dysplasia [cervical intraepithelial neoplasia 1 (CIN1)], high-grade dysplasia (CIN2/3), and cervical cancer. Low-risk HPV, such as HPV-6 and HPV-11, causes benign genital warts, whereas high-risk types, such as HPV-16 (about 60%) and HPV-18 (about 10–20%), are associated with cervical cancer. Other high-risk types include types 31, 33, 35, 39, 45, 51, 52, 56, 58, 59, 68, and 73 (Mahdavi and Monk 2005).

The clinical manifestation of HPV infection depends not only on the type of HPV but also on the location of the epithelial lesion and the host's immune status (Lee et al. 2006). It was shown that people with history of deficit immunological response against infectious agents (HIV) have more probability to have cervical carcinoma (Del Mistro et al. 1994; Eckert et al. 1999). Evasion of the immune response is shown to contribute to the survival and propagation of HPV-infected cells. The HPV genome is organized into three regions: early region encoding proteins (E1, E2, E4, E5, E6, and E7) which are responsible for viral replication and transformation of host cells, late region which codes the viral capsid proteins (L1 and L2), and a noncoding regulatory region (Park et al. 1995). Expression of HPV E6 and E7 genes in differentiating keratinocytes directly alters the expression of genes that influence host resistance to infection and immune function. The squamous epithelium of the cervix is composed not only of keratinocytes but also of a type of dendritic cell (DC) [an epithelial-residing Langerhan cell (LC)], which are important for immunosurveillance of the epithelium. The following section will highlight the LCs in the cervix and their function in cervical carcinoma.

2. Langerhans Cells in the Uterine Cervix

DCs are antigen-presenting cells (APC) with a unique ability to induce primary immune responses. Many subsets of DCs exist, characterized by differences in developmental pathways, surface marker expression, and anatomical localization. The level of heterogeneity reflected by anatomical localization includes epidermal LCs, dermal (interstitial) DCs (intDC), splenic marginal DCs, T-zone interdigitating cells, germinal-center DCs, thymic DCs, liver DCs, and blood DCs (Banchereau, Briere, Caux, Davoust, Lebecque, Liu, Pulendran, and Palucka 2000). Since the first evidence for the presence of LCs in the cervix in 1968 (Younes et al. 1968), more attention has been focused on cervical LCs in the normal cervix and in diseased state (Caorsi and Figueroa 1986; Bonilla-Musoles et al. 1987; Hawthorn and MacLean 1987).

Immature LCs, efficient in the uptake of antigens, continually monitor the epidermal microenvironment for damage and infection. Following antigen processing, LCs mature and migrate to dermal lymphatic or peripheral lymphoid organs to present antigen and provide signals for T cell activation (Aiba and Katz 1990; Groves et al. 1995; Cumberbatch et al. 2005; Griffiths et al. 2005). The migration of LCs is known to be under the control of adhesion molecules, cytokines and chemokines, expressed by keratinocytes and by LCs themselves. LC migration is associated with a down-regulation of the membrane expression of E cadherin, an adhesion molecule expressed by both LCs and keratinocytes and which is believed to retain resident LCs within the tissue matrix (Borkowski et al. 1994; Griffiths et al. 2005). Particularly important cytokines known to be involved in LC migration include IL-1α and TNF-α (mainly produced by keratinocytes) and IL-1β (mainly produced by LCs), all of which up-regulated LC migration, and IL-10, which down-regulated LC migration (Wang et al. 1999; Scott et al. 2001). Chemokine MIP-3α/CCL20, secreted also by keratinocytes, has been reported to recruit LCs through the CCR6 receptor, which is highly expressed in immature LCs. The LC maturation process is associated with a coreceptor switch (CCR6 to CCR7) allowing the migration of LCs from epithelial structure to the lymph nodes (Caux et al. 2002; Cremel, Hamzeh Cognasse et al. 2006).

LCs, which capture antigens for transport to the local draining lymph nodes and presentation to naïve T cells, were demonstrated to be essential for the initiation of an adaptive immune response against viral antigens, particularly HPV antigens, encountered within the epidermis (Scott et al. 2001; Matthews et al. 2003). In un-transformed HPV-infected keratinocytes, the innate immunity is induced to eliminate the invading HPV pathogen through sensitization to HPV-related proteins by epithelial-residing LCs, macrophages, and other immune cells. Once the HPV infection escapes from initial patrolling by innate immunity, cellular immunity becomes in charge of eliminating the HPV-infected keratinocytes of the uterine cervix through systemic immune response developed by DCs in the regional lymphoid organs or local immune response developed by LCs in the cervix (Lee et al. 2006). Thereby, LCs/DCs are involved in controlling HPV infection through both local and systemic immune responses.

3. Langerhans Cells in Cervical Neoplasia

Many studies have addressed the potential role for LCs in HPV infection or associated malignancy. A number of observations have shown that LCs in HPV lesions may be quantitatively reduced and functionally impaired, which may contribute to the persistence of infection (Viac et al. 1990; Scott et al. 2001; Hubert et al. 2005; Walker et al. 2005). Analysis of cervical biopsies revealed that the number of LCs was significantly reduced in intraepithelial neoplasia (Morris et al. 1983; Tay et al. 1987; Hawthorn et al. 1988; Barberis et al. 1998; Connor et al. 1999; Guess and McCance 2005). Immunolabeling of cervical epithelial sheets, prepared from fresh cervical biopsies with antibodies to the three LC markers (MHC-II, CD1a, and Langerin), revealed reduced (to approximately 50% for the three markers) frequency of LCs in samples from HPV-infected patients compared with samples from controls (Jimenez-Flores et al. 2006). Immunostaining of formalin-fixed, paraffin-embedded tissue sections of cervical epithelium from punch biopsies of low-grade or atypical lesions from 96 patients provides evidence that active HPV16 infection in cervical epithelium causes a depletion of LCs, which closely correlates with reduced and atypical E-cadherin expression (Matthews et al. 2003). A 2.5-fold reduction in the number of LCs in HPV16-infected regions of epidermis has been found compared with adjacent, uninfected regions of the same tissue samples, clearly indicating that the reduction in LC density in the skin was a direct result of the active viral infection.

It has also been confirmed, that the LCs in HPV-infected tissues were generally morphologically less dendritic than LCs in normal tissue, which has previously been observed in several studies on LC density in CIN (Morris et al. 1983; Morelli et al. 1993; Uchimura et al. 2004). Significant reduction of cytoplasmic profiles of LCs was found in viral lesions when compared with normal tissue of the epithelium and has been postulated as a mechanism by which HPV could be involved in the genesis of neoplasia (Uchimura et al. 2004). Recently, a correlation between E-cadherin immunostaining and the density of CD1a+ LC within the normal and (pre-) neoplastic squamous epithelium was reported. In that study, HPV-transformed keratinocytes were used to demonstrate that the E-cadherin-mediated contact between keratinocytes and LCs is potentially important for initiating or maintaining the immune response during chronic HPV infection (Hubert et al. 2005).

Several studies have suggested that LCs in HPV lesions may be functionally impaired, contributing to the persisting infection (Scott et al. 2001; Fausch et al. 2005). The phosphoinositide-3-kinase (PI3-K) activation in LCs was proposed as a possible escape mechanism used by HPV to lack the ability of LCs to induce an anti-HPV immune response, and inhibition of PI3-K was suggested as an effective clinical target to enhance HPV immunity (Fausch et al. 2005). Since cell-mediated immune responses are suggested to be important in controlling both HPV infections and HPV-associated neoplasia, the way of generating powerful anticancer immune response may be to generate large numbers of autologous antigen-loaded DCs for vaccination.

4. DC-Based Vaccines for Cervical Cancer

The unique ability of DCs to induce and sustain primary immune responses makes them optimal candidates for vaccination protocols in cancer. DCs derived from cultured hematopoietic progenitors appear to have an antigen-presenting function similar to that of purified mature DCs. Ex vivo generation of DCs, therefore, provides a source of professional APCs for the use in experimental immunotherapy. There are several vaccine strategies utilizing DCs prepared with HPV-16 E6/E7. Vaccine strategies using DC-generated ex vivo can be classified as follows: (i) DCs pulsed with tumor peptides/proteins; (ii) DCs transduced with tumor antigen and cytokine genes or tumor RNA; (iii) DCs that have engulfed HPV virus capsids or virus-like particles; (iv) DCs that are loaded with killed allogeneic tumor cells; and (v) DCs that have fused with tumor cells (Biragyn and Kwak 2000; Ling et al. 2000; Tindle 2002).

Phase 1 trial using immature DCs pulsed with the autologous or allogeneic tumor lysates has been conducted in patients with advanced recurrent cervical cancer (Adams et al. 2001, 2003). It was possible to induce HPV-specific CTL and HPV-specific T-helper responses in the peripheral blood in a few patients after the vaccination with 10^8 DCs pulsed with a tumor lysate. Another pilot clinical study performed with HPV E7 antigen-loaded autologous DCs demonstrated the induction of T cell responses in a portion of late-stage cervical cancer patients (Ferrara et al. 2003). Since the use of immature DCs for antitumor vaccination may not be an optimum, vaccine of DCs primed with tumor lysates and activated with maturation agent poli IC was reported to be currently tested in a clinical trial setting (Adams et al. 2003). A recent study of tumor antigen-pulsed DCs vaccination in cervical cancer patients who have failed all conventional treatment modalities has shown that autologous mature DCs pulsed with HPV16/18 E7 proteins can induce systemic B and T cell responses in late-stage cervical cancer patients (Santin et al. 2006). However, treatment-induced and/or disease-induced immunosuppression may limit the efficacy of vaccination strategies in cervical cancer patients harboring recurrent/metastatic disease refractory to the standard treatments. DC-based vaccination trials in immunocompetent cervical cancer patients with early-stage disease have been suggested. The clinical data such as dose, route of administration, toxicity, efficacy, immune response, and biological response of HPV-pulsed DCs from the past, and future clinical studies of HPV type 16-and HPV type 18-positive patients would contribute to the development of therapeutic as well as prophylactic protein-based DCs vaccine against cervical cancer in nearest future.

5. Conclusions

In the management or treatment of cervical cancer, chemotherapy alone is generally ineffective against this relatively slow-growing disease. Immunotherapy may be the most promising additions to the current therapeutic inventory. Since LCs/DCs play important role in eliciting cellular immunity, it is essential to study the function of LCs and DCs in the HPV-infected patients with or without other infectious disease. The understanding of antigen processing and presentation in this disease would be also helpful to develop a strategy for DC-based immunotherapy. The messages obtained from clinical studies will open new avenues to establish how the functional properties of LCs/DCs could be exploited to fight immunosuppression and get desired clinical responses in cervical cancer patients.

References

Adams, M., Borysiewicz, L., Fiander, A., Man, S., Jasani, B., Navabi, H., Lipetz, C., Evans, A. S. and Mason, M. (2001) Clinical studies of human papilloma vaccines in pre-invasive and invasive cancer. Vaccine 19, 2549–2556.

Adams, M., Navabi, H., Jasani, B., Man, S., Fiander, A., Evans, A. S., Donninger, C. and Mason, M. (2003) Dendritic cell (DC) based therapy for cervical cancer: use of DC pulsed with tumour lysate and matured with a novel synthetic clinically non-toxic double stranded RNA analogue poly [I]:poly [C(12)U] (Ampligen R). Vaccine 21, 787–790.

Aiba, S. and Katz, S. I. (1990) Phenotypic and functional characteristics of *in vivo*-activated Langerhans cells. J. Immunol. 145, 2791–2796.

Banchereau, J., Briere, F., Caux, C., Davoust, J., Lebecque, S., Liu, Y. J., Pulendran, B. and Palucka, K. (2000) Immunobiology of dendritic cells. Annu. Rev. Immunol. 18, 767–811.

Barberis, M. C., Vago, L., Cecchini, G., Bramerio, M., Banfi, G., D'Amico, M. and Cannone, M. (1998) Local impairment of immunoreactivity in HIV-infected women with HPV-related squamous intraepithelial lesions of the cervix. Tumori 84, 489–492.

Biragyn, A. and Kwak, L. W. (2000) Designer cancer vaccines are still in fashion. Nat. Med. 6, 966–968.

Bonilla-Musoles, F., Castells, A., Simon, C., Serra, V., Pellicer, A., Ramirez, A. and Pardo, G. (1987) Importance of Langerhans cells in the immune origin of carcinoma of the uterine cervix. Eur. J. Gynaecol. Oncol. 8, 44–60.

Borkowski, T. A., Van Dyke, B. J., Schwarzenberger, K., McFarland, V. W., Farr, A. G. and Udey, M. C. (1994) Expression of E-cadherin by murine dendritic cells: E-cadherin as a dendritic cell differentiation antigen characteristic of epidermal Langerhans cells and related cells. Eur. J. Immunol. 24, 2767–2774.

Caorsi, I. and Figueroa, C. D. (1986) Langerhans' cell density in the normal exocervical epithelium and in the cervical intraepithelial neoplasia. Br. J. Obstet. Gynaecol. 93, 993–998.

Caux, C., Vanbervliet, B., Massacrier, C., Ait-Yahia, S., Vaure, C., Chemin, K., Dieu, N., Mc and Vicari, A. (2002) Regulation of dendritic cell recruitment by chemokines. Transplantation 73, S7–S11.

Connor, J. P., Ferrer, K., Kane, J. P. and Goldberg, J. M. (1999) Evaluation of Langerhans' cells in the cervical epithelium of women with cervical intraepithelial neoplasia. Gynecol. Oncol. 75, 130–135.

Cremel, M., Hamzeh-Cognasse, H., Genin, C. and Delezay, O. (2006) Female genital tract immunization: evaluation of candidate immunoadjuvants on epithelial cell secretion of CCL20 and dendritic/Langerhans cell maturation. Vaccine 24, 5744–5754.

Cumberbatch, M., Clelland, K., Dearman, R. J. and Kimber, I. (2005) Impact of cutaneous IL-10 on resident epidermal Langerhans' cells and the development of polarized immune responses. J. Immunol. 175, 43–50.

Del Mistro, A., Insacco, E., Cinel, A., Bonaldi, L., Minucci, D. and Chieco-Bianchi, L. (1994) Human papillomavirus infections of the genital region in human immunodeficiency virus seropositive women: integration of type 16 correlates with rapid progression. Eur. J. Gynaecol. Oncol. 15, 50–58.

Eckert, L. O., Watts, D. H., Koutsky, L. A., Hawes, S. E., Stevens, C. E., Kuypers, J. and Kiviat, N. B. (1999) A matched prospective study of human immunodeficiency virus serostatus, human papillomavirus DNA, and cervical lesions detected by cytology and colposcopy. Infect. Dis. Obstet. Gynecol. 7, 158–164.

Fausch, S. C., Fahey, L. M., Da Silva, D. M. and Kast, W. M. (2005) Human papillomavirus can escape immune recognition through Langerhans cell phosphoinositide 3-kinase activation. J. Immunol. 174, 7172–7178.

Ferrara, A., Nonn, M., Sehr, P., Schreckenberger, C., Pawlita, M., Durst, M., Schneider, A. and Kaufmann, A. M. (2003) Dendritic cell-based tumor vaccine for cervical cancer II: results of a clinical pilot study in 15 individual patients. J. Cancer Res. Clin. Oncol. 129, 521–530.

Griffiths, C. E., Dearman, R. J., Cumberbatch, M. and Kimber, I. (2005) Cytokines and Langerhans cell mobilisation in mouse and man. Cytokine 32, 67–70.

Groves, R. W., Allen, M. H., Ross, E. L., Barker, J. N. and MacDonald, D. M. (1995) Tumour necrosis factor alpha is pro-inflammatory in normal human skin and modulates cutaneous adhesion molecule expression. Br. J. Dermatol. 132, 345–352.

Guess, J. C. and McCance, D. J. (2005) Decreased migration of Langerhans precursor-like cells in response to human keratinocytes expressing human papillomavirus type 16 E6/E7 is related to reduced macrophage inflammatory protein-3alpha production. J. Virol. 79, 14852–14862.

Hawthorn, R. J. and MacLean, A. B. (1987) Langerhans' cell density in the normal exocervical epithelium and in the cervical intraepithelial neoplasia. Br. J. Obstet. Gynaecol. 94, 815–818.

Hawthorn, R. J., Murdoch, J. B., MacLean, A. B. and MacKie, R. M. (1988) Langerhans' cells and subtypes of human papillomavirus in cervical intraepithelial neoplasia. BMJ 297, 643–646.

Hubert, P., Caberg, J. H., Gilles, C., Bousarghin, L., Franzen-Detrooz, E., Boniver, J. and Delvenne, P. (2005) E-cadherin-dependent adhesion of dendritic and Langerhans cells to keratinocytes is defective in cervical human papillomavirus-associated (pre)neoplastic lesions. J. Pathol. 206, 346–355.

Jimenez-Flores, R., Mendez-Cruz, R., Ojeda-Ortiz, J., Munoz-Molina, R., Balderas-Carrillo, O., de la Luz Diaz-Soberanes, M., Lebecque, S., Saeland, S., Daneri-Navarro, A., Garcia-Carranca, A., Ullrich, S. E. and Flores-Romo, L. (2006) High-risk human papilloma virus infection decreases the frequency of dendritic Langerhans' cells in the human female genital tract. Immunology 117, 220–228.

Lee, B. N., Follen, M., Rodriquez, G., Shen, D. Y., Malpica, A., Shearer, W. T. and Reuben, J. M. (2006) Deficiencies in myeloid antigen-presenting cells in women with cervical squamous intraepithelial lesions. Cancer 107, 999–1007.

Ling, M., Kanayama, M., Roden, R. and Wu, T. C. (2000) Preventive and therapeutic vaccines for human papillomavirus-associated cervical cancers. J. Biomed. Sci. 7, 341–356.

Mahdavi, A. and Monk, B. J. (2005) Vaccines against human papillomavirus and cervical cancer: promises and challenges. Oncologist 10, 528–538.

Matthews, K., Leong, C. M., Baxter, L., Inglis, E., Yun, K., Backstrom, B. T., Doorbar, J. and Hibma, M. (2003) Depletion of Langerhans cells in human papillomavirus type 16-infected skin is associated with E6-mediated down regulation of E-cadherin. J. Virol. 77, 8378–8385.

Morelli, A. E., Sananes, C., Di Paola, G., Paredes, A. and Fainboim, L. (1993) Relationship between types of human papillomavirus and Langerhans' cells in cervical condyloma and intraepithelial neoplasia. Am. J. Clin. Pathol. 99, 200–206.

Morris, H. H., Gatter, K. C., Sykes, G., Casemore, V. and Mason, D. Y. (1983) Langerhans' cells in human cervical epithelium: effects of wart virus infection and intraepithelial neoplasia. Br. J. Obstet. Gynaecol. 90, 412–420.

Park, T. W., Fujiwara, H. and Wright, T. C. (1995) Molecular biology of cervical cancer and its precursors. Cancer 76, 1902–1913.

Santin, A. D., Bellone, S., Palmieri, M., Ravaggi, A., Romani, C., Tassi, R., Roman, J. J., Burnett, A., Pecorelli, S. and Cannon, M. J. (2006) HPV16/18 E7-pulsed dendritic cell vaccination in cervical cancer patients with recurrent disease refractory to standard treatment modalities. Gynecol. Oncol. 100, 469–478.

Sasagawa, T., Tani, M., Basha, W., Rose, R. C., Tohda, H., Giga-Hama, Y., Azar, K. K., Yasuda, H., Sakai, A. and Inoue, M. (2005) A human papillomavirus type 16 vaccine by oral delivery of L1 protein. Virus Res. 110, 81–90.

Scott, M., Nakagawa, M. and Moscicki, A. B. (2001) Cell-mediated immune response to human papillomavirus infection. Clin. Diagn. Lab. Immunol. 8, 209–220.

Tay, S. K., Jenkins, D., Maddox, P., Campion, M. and Singer, A. (1987) Subpopulations of Langerhans' cells in cervical neoplasia. Br. J. Obstet. Gynaecol. 94, 10–15.

Tindle, R. W. (2002) Immune evasion in human papillomavirus-associated cervical cancer. Nat. Rev. Cancer 2, 59–65.

Uchimura, N. S., Ribalta, J. C., Focchi, J., Simoes, M. J., Uchimura, T. T. and Silva, E. S. (2004) Evaluation of Langerhans' cells in human papillomavirus-associated squamous intraepithelial lesions of the uterine cervix. Clin. Exp. Obstet. Gynecol. 31, 260–262.

Viac, J., Guerin-Reverchon, I., Chardonnet, Y. and Bremond, A. (1990) Langerhans cells and epithelial cell modifications in cervical intraepithelial neoplasia: correlation with human papillomavirus infection. Immunobiology 180, 328–338.

Walboomers, J. M., Jacobs, M. V., Manos, M. M., Bosch, F. X., Kummer, J. A., Shah, K. V., Snijders, P. J., Peto, J., Meijer, C. J. and Munoz, N. (1999) Human papillomavirus is a necessary cause of invasive cervical cancer worldwide. J. Pathol. 189, 12–19.

Walker, F., Adle-Biassette, H., Madelenat, P., Henin, D. and Lehy, T. (2005) Increased apoptosis in cervical intraepithelial neoplasia associated with HIV infection: implication of oncogenic human papillomavirus, caspases, and Langerhans cells. Clin. Cancer Res. 11, 2451–2458.

Wang, B., Amerio, P. and Sauder, D. N. (1999) Role of cytokines in epidermal Langerhans cell migration. J. Leukoc. Biol. 66, 33–39.

Younes, M. S., Robertson, E. M. and Bencosme, S. A. (1968) Electron microscope observations on Langerhans cells in the cervix. Am. J. Obstet. Gynecol. 102, 397–403.

17

Sensitizing Tumor Cells to Immune-Mediated Cytotoxicity

Anil Shanker and Thomas Sayers

Cancer and Inflammation Program, SAIC-Frederick Inc., NCI-Frederick, Frederick, MD, USA, sayerst@mail.nih.gov

Abstract. The molecular basis underlying tumor destruction in vivo by specific antitumor $CD8^+$ T cells remains unclear. We propose that the local production of certain tumor necrosis factor (TNF)-family members (death ligands) may be more important for tumor destruction in vivo than previously thought. Also, the apoptotic response of some tumor cells to the TNF-family member TRAIL can be augmented by the proteasome inhibitor bortezomib (Velcade). Thus, bortezomib may sensitize tumor cells to T cell-mediated cytotoxicity and could potentially improve the beneficial effects of immunotherapy.

1. Introduction

For a number of years, there has been great enthusiasm for the idea of triggering the immune system to eradicate tumors. However, in contrast to bacteria or viruses, tumor cells closely resemble the normal tissue from which they were originally derived. Therefore, in order for an immune response to tumor cells to occur, normal self-tolerance must be overcome and tumor antigens must be shown to exist. Over the last decade, great progress has been made in tumor immunology, particularly in the identification of numerous tumor antigens. These antigens fall into five main categories (Coulie et al. 2001). (i) Cancer-germline or activation antigens: Tumor antigens encoded by genes that are completely silent in most normal tissues but are activated in a number of tumors of various histological types such as melanomas and carcinomas of the lung, head and neck, bladder, and esophagus. Prototype antigens of this group are those encoded by gene *P1A* in the mouse and by the *MAGE* genes in the humans. Such antigens ought to be strictly tumor specific and constitute very promising targets for anticancer vaccines. (ii) Mutation antigens: The tumor specificity of the antigens resulting from mutations in the DNA of a cell is absolute but unique to individual tumors. (iii) Tissue-specific differentiation antigens: These antigens are not tumor-specific but attracted tumor immunologists because the corresponding cytotoxic T lymphocytes (CTLs) were found in cancer patients. (iv) Overexpressed tumor antigens: A number of CTLs raised against tumor cells have been found to recognize antigens by genes expressed both in normal and tumoral tissues. (v) Viral antigens: Antigens encoded by oncogenic viruses have and are recognized by tumor-specific CTLs.

Encouragingly, in some tumor models, the T cell-mediated immune response efficiently eradicated tumor. Functional effector T cells have been shown to develop against immunogenic as well as self-tumor antigens in some melanoma patients (Rosenberg et al. 2004), pancreatic cancer patients (Ryschich et al. 2005), Epstein–Barr virus-associated malignancies (Gottschalk et al. 2005), and some breast cancer patients (Gillmore et al. 2005).

From studies in animals and man, it has become clear that cellular immune responses can have an important role in the immunologic rejection of leukemias and lymphomas as well as certain highly vascularized solid tumors (Rosenberg et al. 2004). Therefore, in mouse tumor models, the transfer of immune T lymphocytes, but not antibodies, protects mice from tumor challenge. Furthermore, elimination of $CD8^+$ T cells often abrogates both protective and therapeutic antitumor effects. In addition, extensive T cell infiltrates are commonly seen in tumors or allografts undergoing rejection. However, three criteria are assumed to be crucial for the immunologic destruction of established tumors: (i) sufficient numbers of immune cells with high-avidity recognition of tumor antigens must be generated in vivo (ii) these cells must traffic and infiltrate the tumor stroma, and (iii) the immune cells must be activated at the tumor site to manifest appropriate effector mechanisms capable of causing tumor destruction.

Although immune cells can be generated by vaccination, this procedure only seems to provide therapeutic benefit in treating tumors a few days after transplantation when they are not yet vascularized. Indeed, even in mice transgenic for T cell receptors that recognize tumor antigens, where virtually all T lymphocytes can recognize tumor, tumor growth and lethality are often unaffected. Inadequate numbers or avidity of immune cells, the inability of tumor to activate quiescent or precursor lymphocytes, tolerance mechanisms including anergy, and suppressor influences produced by the tumor or the immune system itself are among the mechanisms that can prevent tumor destruction by immune cells. However, some of these problems can be circumvented by the adoptive transfer ofantitumor T cells. Indeed, large B16 melanomas can be rejected in mice following host lymphocyte immunodepletion when antigen-specific antitumor cells are transferred with antigen-specific vaccination and IL-2. The success of this approach has its counterpart in recent human clinical trials (Gattinoni et al. 2006). Since fully activated $CD8^+$ T cells have the ability to directly lyse tumor cells in vitro, it has been assumed that this cell-mediated cytotoxicity is crucial to their antitumor effector function.

2. Direct Cell–Mediated Cytotoxicity

$CD8^+$ T cells, in common with natural killer (NK) cells, can directly lyse various tumor cells in vitro following appropriate recognition of distinct structures on the tumor cell surface. From a large number of studies, two main molecular pathways used by these cytolytic lymphocytes to effect tumor cell lysis in vitro have been defined. One pathway involves the directed release of cytotoxic granules found in both activated T cells and NK cells. These granules contain the pore-forming protein perforin as well as various other constituents including a unique family of serine proteases known as granzymes. Expression of both perforin and the granzymes seems mostly confined to activated $CD8^+$ T cells and NK cells. Cytolysis of tumor target cells by granule exocytosis is completely dependent on perforin, since T cells or NK cells isolated from mice where the perforin gene is eliminated by gene-targeting (pfp–/–) are unable to perform granule-mediated cytolysis. How perforin functions in this lytic event is still somewhat unclear. It was initially thought that perforin created membrane pores in a similar manner to complement. More recently, it has been proposed that perforin promotes the access and trafficking of the granzymes into target cells, subsequently triggering the apoptotic signaling cascade (Voskoboinik and Trapani 2006).

The second lytic mechanism used by cytotoxic T cells and NK cells involves the production by these effector lymphocytes of various proapoptotic members of the tumor necrosis factor (TNF) family such as TNF-α, FasL, or TNF-related apoptosis-inducing ligand (TRAIL or Apo2L). These so-called "death ligands" on binding to the appropriate cell surface receptors cause aggregation of these receptors. This in turn induces intracellular motifs (death domains) in these receptors to bind various adaptor molecules such as FADD and TRADD into a complex. Then, apical procaspases can also be recruited and activated in this death-inducing signaling complex (DISC). Subsequently, this leads to activation of downstream caspases, the activation of proapototic Bcl-2 family members, and ultimately apoptotic death of the cell. Interestingly in contrast to granzymes and perforin, expression of these death ligands is not restricted to activated $CD8^+$ T cells and NK cells. They are also produced by a variety of cells of the innate immune system such as dendritic cells and macrophages. The intracellular signaling pathways in the cytotoxic cells required to engage these two lytic pathways differ substantially. Once appropriate recognition and triggering of granule release from the cytotoxic cell has occurred, it seems unlikely that the tumor target cell can influence its own demise. By contrast, many cancer cells are resistant to the apoptotic effects following engagement of death receptors with their ligands. This is presumably due to the fact that the target cell itself must participate in the transmission of the apoptotic signal. Indeed, the resistance of tumor cells to apoptotic signaling is now thought to be one of the prerequisites of cancer development. This loss of apoptotic signaling potential can occur at multiple points along the apoptotic signaling pathway. In the response to death ligands, a loss of the appropriate receptor or any of the intracellular components of the apoptotic signaling pathway can prevent cellular apoptosis. Nonetheless, it has been known for many years that agents such as cycloheximide or actinomycin D can sensitize some tumor cells to apoptosis triggered by death ligands. Thus, it may be that the apoptotic signaling pathway in response to death ligands still exists in many tumors, but this suicide pathway is blocked at some particular stage resulting in tumor cell survival.

3. Immune-Mediated Tumor Destruction In vivo

Since activated $CD8^+$ T cells possess direct cytotoxic capacities and are responsible for most immune-mediated tumor regression, it has been assumed that their cyotoxic potential is a vital component for the promotion of tumor regression in vivo. However, the recent availability of a variety of gene-targeted or mutant mice has allowed for this hypothesis to be tested in well-defined tumor models. Surprisingly, in a variety of mouse tumor models including the B16 melanoma, MCA-101 sarcoma (Winter et al. 1999), MCA-205 sarcoma (Peng et al. 2000), and the mouse renal cancer Renca (Seki et al. 2002), it has been demonstrated that tumor rejection by the transfer of activated T cells could occur in the total absence of perforin. This despite the fact that the perforin-dependent granule-mediated lytic pathway is usually the predominant cytotoxic pathway employed by activated $CD8^+$ T cells in vitro. This suggests that granule-mediated cytotoxic effects may not be as crucial for tumor rejection in vivo as was originally thought. In the absence of a major role for granule-mediated tumor cell lysis, it is reasonable to assume that death ligands produced by $CD8^+$ T cells or NK cells may be major mediators of tumor destruction in vivo. Interestingly, in the B16 model, the rejection of lung micrometastases can occur in the absence of perforin, FasL, or interferon-γ. Under such circumstances, a role for other members of the TNF family such as TNF-α or lymphotoxin in tumor destruction is revealed (Poehlein et al. 2003). By contrast, rejection of Renca tumors by activated T cells seems to be very dependent on FasL production by the antitumor T cells. Since tumors do not have a uniform response to death ligands, it is possible that their apoptotic responses in vivo are dependent on characteristics of the individual tumor. Therefore, for tumors sensitive to the apoptotic effects of TNF-α, yet resistant to FasL or TRAIL, local production of TNF-α (from whatever cell source) would be crucial for any antitumor effects. As such, prior knowledge of the apoptotic signaling of the individual tumor in response to specific death ligands could be helpful in determining the likelihood of success of different immunotherapeutic approaches. Since activated T cells are a major source of FasL, their direct cytotoxic effects could be a major mediator for the destruction of tumor cells uniquely sensitive to this death ligand. However, solid tumors are often infiltrated with other cells of the innate immune system such as macrophages and dendritic cells. In this case, the local production of TNF-α or TRAIL by these cells may be crucial for the destruction of tumors particularly sensitive to these death ligands. In such cases, the major antitumor effector role of the activated $CD8^+$ T cells may be in the local production of proinflammatory cytokines, thus increasing levels of death ligands produced by cells in the tumor stroma. Interestingly, the production of IFN-γ is often essential for tumor rejection by activated T cells in a number of mouse tumor models. Since the tumor stoma can play a major role in the maintenance of tumor cell growth, it must be borne in mind that $CD8^+$ T cells may also exhibit antitumor effects indirectly due to their influence on stromal elements such as endothelial cells (Qin and Blankenstein 2000; Spiotto et al. 2004).

Over the last few years, immunologists have performed a great deal of work to determine the optimal requirements necessary for transferred T cells to promote clinically relevant antitumor responses in vivo (Gattinoni et al. 2006). Nonetheless, surprisingly little is known of the influence of the tumor in determining the outcome of this response. Therefore, although the transfer of highly activated T cells can be optimized, there are still a significant number of patients in whom no antitumor response is observed. Our laboratory has therefore focused its efforts on attempting to sensitize tumors to activated immune effector cells. From a practical standpoint, it is difficult to imagine how the sensitivity of tumor cells to perforin-dependent granule-mediated cytotoxicity can be substantially enhanced. Granule-mediated cytotoxicity is not tumor cell specific, and components of the lytic granules are relatively unstable. As such, it is difficult to deliver this antitumor killing mechanism by any other means than by transfer of T cells with a high complement of lytic granules. By contrast, the antitumor effects of death ligands may be more amenable to pharmacological modification at the level of the tumor itself.

4. Augmenting Immune-Mediated Cytotoxicity

Considering TNF-α and FasL, there are major problems of toxicity associated with the direct administration of these death ligands to animals. By contrast, TRAIL is cytotoxic to a number of tumor cells in vitro and in vivo in the absence of major associated toxicities (Almasan and Ashkenazi 2003). Therefore, we have used TRAIL as a model to test our hypothesis that the response of tumor cells to death ligand-mediated apoptosis can be augmented by the administration of sensitizing agents. The proapoptotic activities of TRAIL on cancer cells are known to be enhanced by both radiation and chemotherapy. However, it is hard to envisage how radiation or chemotherapy could be effectively used in combination with immunotherapy, as both radiation and chemotherapy are quite immunosuppressive. We have therefore chosen to study the efficacy of combining the proteasome inhibitor bortezomib (Velcade) and TRAIL in promoting tumor cell apoptosis. We and others have shown that bortezomib can sensitize a number of mouse and human tumor cells to TRAIL-mediated apoptosis (Sayers and Murphy 2006). The effects of the combination of bortezomib and TRAIL on various human renal cell carcinomas (RCCs) were tested. These RCCs were resistant to the apoptotic effects of low dose of TRAIL. However, treatment with bortezomib dramatically sensitized some of these carcinoma cells such as A498, ACHN, and UO-31 to the antitumor effects of TRAIL (Figure 1). This dramatic reduction in cell number was due to apoptotic death of the cancer cells (data not shown). We also tested a panel of 60 human cancer cells of different origin and found essentially the same pattern. That is, bortezomib at 5–20 nM concentrations sensitized about 25–30% of the panel of cancer cells to the apoptotic effects of TRAIL. None of the human RCCs we tested were intrinsically resistant to TRAIL-mediated apoptosis, since treatment with cycloheximide dramatically sensitized all seven cell lines to TRAIL. Thus, the apoptotic pathways in response to TRAIL could potentially be triggered in all the RCCs. Therefore, differences between the sensitization of ACHN and Caki-1 were specifically manifest in response to bortezomib (Figure 1).

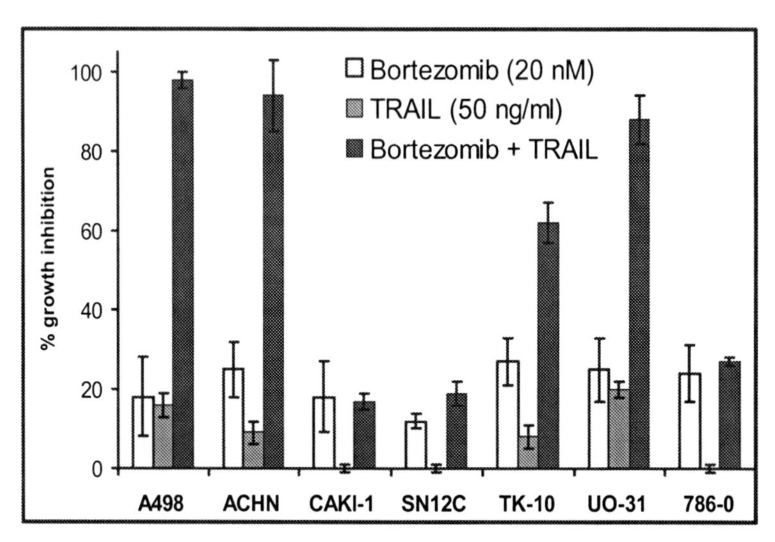

FIGURE 1. Effects of bortezomib and TRAIL on human renal carcinoma cells. Viable cell counts were estimated 18 h after TRAIL addition by MTS staining, and percent growth inhibition was then calculated.

The molecular mechanism underlying this selective sensitization of tumor cells to TRAIL-mediated apoptosis remains unclear. It was initially thought that the ability of bortezomib to block activation of the transcription factor NF-κB would be crucial for sensitizing tumor cells to TRAIL. However, this activity of bortezomib does not seem to be crucial for the sensitization of a variety of tumor cells to TRAIL. Other possibilities include increases in the levels of TRAIL death receptors, increases in proapototic Bcl-2 family members, as well as decreases in cellular levels of antiapoptotic proteins such as cellular FADD-like IL-1-converting enzyme-inhibitory protein (cFLIP) or inhibitors of apoptosis proteins (IAPs). It was recently demonstrated that bortezomib treatment of tumor cells resulted in an enhanced activation of caspase-8 following TRAIL binding. Therefore, it is possible that the major effects of bortezomib in sensitizing tumor cells to TRAIL may occur at a proximal level of the apoptotic signaling cascade (Ganten et al. 2005). Further work is required to identify the precise molecular mechanism(s), whereby bortezomib sensitizes tumor cells to TRAIL apoptosis. As yet, there is very little information available on whether the combination of bortezomib and TRAIL can display therapeutic benefit in an in vivo setting. We first reported that this combination was superior to either agent alone in purging C1498 murine acute myeloid leukemia cells from a bone marrow inoculum (Sayers et al. 2003). Furthermore, more recent studies have shown that bortezomib can be combined with allogeneic bone marrow transfer allowing for a reduction of graft-versus-host disease while maintaining beneficial graft-versus-tumor effects (Sun et al. 2004). Current studies from our laboratory also indicate that the combination of bortezomib with an agonist antibody to TRAIL death receptor 5 can provide therapeutic benefit in tumor-bearing mice in the absence of detectable toxicity (Shanker and Sayers, in preparation). It seems likely that bortezomib could also be effectively combined with agents that induce TRAIL production in vivo, particularly if this induction could be localized to the vicinity of the tumor or its metastases. Human genome sciences have recently commenced with a clinical trial investigating any therapeutic benefit of combining bortezomib and agonist antibodies to TRAIL death receptors in patients with multiple myeloma.

5. Future Directions and Concluding Remarks

The molecular basis underlying bortezomib sensitization of tumor cells to TRAIL apoptosis remains unclear. Bortezomib could be affecting multiple components of the apoptotic signaling pathway. Identification of the molecular mechanism underlying bortezomib sensitization of tumor cells to TRAIL could be important in two respects: (1) It may allow for molecular profiling of tumor cells or tissue samples in advance, thus allowing for selection of tumors based on characteristics associated with the "sensitive" phenotype and (2) more information on the critical components of the apoptotic pathway could result in the development of more specific novel agents to augment death ligand-mediated apoptosis of tumors.

As previously mentioned, some cancer cells may be resistant to TRAIL, yet sensitive to FasL or TNF-α. We have evidence that bortezomib can also sensitize some tumor cells to the apoptotic effects of FasL and TNF-α. However, there remains the dilemma as to how to locally deliver these death ligands to the tumor. The adoptive transfer of specific antitumor $CD8^+$ T cells may be beneficial in this regard. These T cells could be a source of multiple death ligands as well as other proinflammatory cytokines. Furthermore, the death ligands should only be produced locally following contact between the transferred $CD8^+$ T cells and tumor antigens. We are currently addressing whether bortezomib treatment of mice can be successfully combined with adoptive transfer of T cells in a mouse tumor model. One concern is that bortezomib could block the trafficking or effector functions of the transferred T cells. However, preliminary results from our laboratory suggest that antitumor effector functions of activated T cells are not overtly influenced by bortezomib either in vitro or in vivo. To address this, more detailed studies are required, and are currently underway. In addition, even if bortezomib does inhibit some functions of activated T cells, this may not be a major practical concern. The effects of bortezomib are short-lived in animals. Therefore, appropriate scheduling of agents may allow for bortezomib to be administered for tumor sensitization prior to a later transfer of activated T cells, at a time when bortezomib is no longer present in the circulation.

In conclusion, it is hoped that these strategies to sensitize tumor cells to the apoptotic effects of death ligands could in the future be coupled to the recent advances in adoptive T cell transfer protocols, to improve the beneficial effects of immunotherapy in cancer patients.

Acknowledgments

This project has been funded in whole or in part with federal funds from the National Cancer Institute, National Institutes of Health, under contract N01-CO-12400. The content of this publication does not necessarily reflect the views or policies of the Department of Health and Human Services nor does mention of trade names, commercial products, or organizations, imply endorsement by the US Government. This Research was supported, in part, by the Intramural Research Program of the NIH, National Cancer Institute, Center for Cancer Research.

References

Almasan, A. and Ashkenazi, A. (2003) Apo2L/TRAIL: apoptosis signaling, biology, and potential for cancer therapy. Cytokine Growth Factor Rev. 14, 337–348.

Coulie, P.G., Hanagiri, T. and Takenoyama, M. (2001) From tumor antigens to immunotherapy. Int. J. Clin. Oncol. 6, 163–170.

Ganten, T.M., Koschny, R., Haas, T.L., Sykora, J., Li-Weber, M., Herzer, K. and Walczak, H. (2005) Proteasome inhibition sensitizes hepatocellular carcinoma cells, but not human hepatocytes, to TRAIL. Hepatology 42, 588–597.

Gattinoni, L., Powell, D.J., Rosenberg, S.A. and Restifo, N.P. (2006) Adoptove immunotherapy for cancer: building on success. Nature 6, 383–393.

Gillmore, R., Xue, S-A., Holler, A., Kaeda, J., Hadjiminas, D., Healy, V., Dina, R., Parry, S.C., Bellantuono, I., Ghani, Y., Coombes, R.C., Waxman, J. and Strauss, H.J. (2006) Detection of Wilms' tumor antigen-specific CTL in tumor-draining lymph nodes of patients with early breast cancer. Human Cancer Biol. 12, 34–42.

Gottschalk, S., Hislop, H.E. and Rooney, C.M. (2005) Adoptive immunotherapy of EBV-associated malignancies. Leuk. Lymphoma 46, 1–10.

Peng, L., Krauss, J.C., Plautz, G.E., Mukai, S., Shu, S. and Cohen, P.A. (2000) T cell-mediated tumor rejection displays diverse dependence upon perforin and IFN-γ mechanisms that cannot be predicted from *in vitro* T cell characteristics. J. Immunol. 165, 7116–7124.

Poehlein, C.H., Hu, H-M., Yamada, J., Assmann, I., Alvord, W.G., Urba, W.J. and Fox, B.A. (2003) TNF plays an essential role in tumor regression after adoptive transfer of perforin /IFN-γ double knockout effector T cells. J. Immunol. 170, 2004–2013.

Qin, Z. and Blankenstein, T. (2000) $CD4^+$ T cell-mediated tumor rejection involves inhibition of angiogenesis that is dependent on IFN-γ receptor expression by nonhematopoetic cells. Immunity 12, 677–686.

Rosenberg, S.A., Yang, J.C. and Restifo,N.P. (2004) Cancer immunotherapy: moving beyond current vaccines. Nat. Med. 10, 909–915.

Ryschich, E., Notzel, T., Hinz, U, Autschbach, F., Ferguson, J., Simon, I., Weitz, J., Frohlich, B., Klar, E., Buchler, M.W. and Schmidt, J. (2005) Control of T cell-mediated immune response by HLA class I in human pancreatic cancer. Clin. Cancer Res. 11, 498–504.

Sayers, T.J., Brooks, A.D., Koh, C.Y., Ma, W., Seki, N., Raziuddin, A. Blazar, B.R., Zhang, X., Elliott, P.J. and Murphy, W.J. (2003) The proteasome inhibitor PS-341 sensitizes neoplastic cells to TRAIL-mediated apoptosis by reducing levels of c-FLIP. Blood 102, 303–310.

Sayers, T.J. and Murphy, W.J. (2006) Combining proteasome inhibition with TNF-related apoptosis-inducing ligand (Apo2L/TRAIL) for cancer therapy. Cancer Immunol. Immunother. 55, 76–84.

Seki, N., Brooks, A.D., Carter, C.R.D., Back,T.C., Parsoneault, E.M., Smyth, M.J., Wiltrout, R.H. and Sayers, T.J. (2002) Tumor-specific CTL kill murine renal cancer cells using both perforin and Fas ligand-mediated lysis *in vitro*, but cause tumor regression *in vivo* in the absence of perforin. J. Immunol. 168, 3484–3492.

Spiotto, M.T., Rowley, D.A. and Schrieber, H. (2004) Bystander elimination of antigen loss Variants in established tumors. Nat. Med. 10, 294–298.

Sun, K., Welniak, L., Panoskaltsis-Mortari, A., O' Shaughnessy, M.J., Liu, H., Barao, I., Riordan, W., Sitcheran, R., Wysocki, C., Serody, J.S., Blazar, B.R., Sayers, T.J. and Murphy, W.J. (2004) Inhibition of acute graft-versus-host disease with retention of graft -versus-tumor effects by the proteasome inhibitor bortezomib. Proc. Natl. Acad. Sci. U.S.A. 101,8120–8125.

Voskoboinik, I. and Trapani, J.A. (2006) Addressing the mysteries of perforin function. Immunol. and Cell Biol. 84, 66–71.

Winter, H., Hu, H-M., Urba, W.J. and Fox, B.A. (1999) Tumor regression after adoptive transfer of effector T cells is independent of perforin or Fas ligand (Apo-1L/CD95L). J. Immunol. 163, 4462–4472.

18

Inhibition of Dendritic Cell Generation and Function by Serum from Prostate Cancer Patients: Correlation with Serum-Free PSA

Maryam Aalamian-Matheis[1], Gurkamal S. Chatta[2], Michael R. Shurin[3], Edith Huland[4], Hartwig Huland[4], and Galina V. Shurin[5]

[1]Department of Medicine, Johns Hopkins University, Baltimore, MD, USA
[2]Department of Medicine, University of Pittsburgh, Pittsburgh, PA, USA
[3]Department of Pathology and Immunology, University of Pittsburgh, Pittsburgh, PA, USA
[4]Department of Urology, University Clinics Hamburg-Eppendorf, Hamburg, Germany
[5]Department of Pathology, Pittsburgh, University of Pittsburgh, PA, USA, shuringv@upmc.edu

Abstract. Tumor produces a number of immunosuppressive factors that block maturation of dendritic cells (DCs). Here, we demonstrated that endogenous factors presented in the serum of patients with prostate cancer (CaP) inhibited the generation of functionally active DCs from CD14+ monocytes in vitro. We have shown a significant inhibitory potential of serum obtained from patients with CaP and benign prostate hyperplasia benign prostatic hyperplasia (BPH) when compared with serum from healthy volunteers. As assessed by flow cytometry, expression of CD83, CD86, and CD40 molecules was strongly inhibited by CaP and BPH serum. In addition, these DCs were weak stimulators of allogeneic T cell proliferation when compared with DCs produced in the presence of healthy volunteer serum. Statistical analysis of the results revealed a positive relationship between the inhibition of expression of DC markers CD83 and CD80 and the levels of serum-free prostate-specific antigen (PSA). These data suggest that the DC system may be impaired in CaP patients.

1. Introduction

Prostate Cancer (CaP) is responsible for more than 40,000 deaths every year in the United States; more than 180,000 new cases are diagnosed yearly (Landis et al. 1999). It is the leading malignancy in incidence and the second in cancer mortality among North American men. Despite the apparent magnitude of this medical problem, fundamental questions regarding the biology of CaP remain unanswered. Due to the unique biologic features of this disease, no curative treatment exists when the disease spreads beyond the prostate. The absence of effective conventional therapy for advanced CaP justifies the application of novel, experimental approaches (Comuzzi and Sadar 2006). Immunotherapeutic approaches for cancer treatment have shown great promise in experimental studies, and a number of immunotherapies are now being tested in clinical settings (Colaco 1999a). This includes intraprostatic injections of cytokines, growth factors, vaccines consisting of liposome-encapsulated recombinant prostate-specific antigen (PSA) and lipid A, genetically modified tumor or immune cell vaccines, and other approaches (Slovin et al. 1998; Harris et al. 1999; Simons et al. 1999; Li et al. 2006; Mohamedali et al. 2006). Although most of human tumor immunotherapies have reported only limited success, recent insights into the role of dendritic cells (DCs) may provide the basis for generating more effective antitumor immune response.

Initiation of specific antitumor immune responses requires the involvement of professional antigen-presenting cells. DCs are the most potent antigen-presenting cells and have been shown to be capable of recognizing, processing, and presenting tumor antigens, in turn initiating a specific antitumor immune response in both animals and humans (Shurin 1996). It has been recently postulated that the transfer of antigens from tumor to professional antigen-presenting cells, or cross priming, is mandatory for tumor antigens to be processed and presented to naïve T lymphocytes (Huang et al. 1994; Speiser et al. 1997; Armstrong et al. 1998). This fact additionally underlines the key role of DCs in the induction of tumor-specific immune responses and suggests that adequate functioning of the DC system is essential for the development of antitumor immunity. During the last several years, a number of clinical trials of DC-based immunotherapies for various tumors, including CaP, have been initiated (Colaco 1999b; Ragde et al. 2004). However, to date, DC therapy for CaP patients has met with only limited success. One reason of the limited efficacy of DC immunotherapy in CaP patients is associated with the local or systemic suppression of the DC system by the CaP-derived factors, resulting in inhibition of immune responsiveness. In fact, the immunological condition in CaP patients has been evaluated, and moderate immunosuppression was already noted in relatively confined prostate tumors. Several studies demonstrated significant depression of the immune system at advanced stages in CaP patients (Herr 1980; Ivshina et al. 1995; Healy et al. 1998; Salgaller et al. 1998). We have recently demonstrated that CaP-derived factors inhibit the generation of human DCs from monocytes and hematopoietic progenitor cells (Aalamian et al. 2001; Shurin et al. 2001). DCs co-incubated with CaP cells expressed lower levels of CD80 and CD86, produced less IL-12, and induced weak T cell proliferation. These data were confirmed by others who demonstrated that the supernatants from human CaP cell-line cultures inhibit CD80 and CD86 expression on DCs (Troy et al. 1999). Our recent data showed that CaP cell-derived PSA, as well as purified PSA, is one of the immunosuppressive factors responsible for inhibition of DC generation and maturation (Aalamian et al. 2003).

PSA is a protein originated primarily from the prostatic epithelium and has emerged as the most important tumor marker for CaP (Polascik et al. 1999). PSA, as known, circulates in the serum in free (unbound) and complexe (bound to protease inhibitors) forms (Lilja 1993). Approximately 65–95% of the PSA is bound to the serine proteinase inhibitor (serpin) α1-antichymotrypsin (ACT); free PSA (fPSA) represents, on average, only 5–35% of the total PSA (tPSA) concentration. The relative amount of fPSA tends to be increased in benign disease compared with CaP (Lilja 1993; Diamandis and Yu 1997). The fPSA/tPSA ratio is now routinely used to increase the specificity for CaP and reduce unnecessary biopsies. Other PSA complexes with alpha$_2$-macroglobulin and α1-protease inhibitor are now measurable in serum as well. PSA–ACT has been used as markers for the diagnosis of CaP ever since Stamey et al. found that the total serum PSA concentration was elevated in CaP patients. Although increase in serum PSA does not distinguish patients with CaP from those with benign prostatic hyperplasia (BPH), the ratio of free to total PSA is lower for CaP than for BPH (Catalona et al. 1998), and is useful for the diagnosis of early stage CaP.

Based on these facts and our previous data, we designed the study to determine whether an endogenous CaP-derived factor(s) might inhibit dendropoiesis. At present study, serum obtained from patients with CaP and BPH and healthy volunteers (10–15 per group) has been tested in vitro for its ability to suppress dendropoiesis. We demonstrated that the addition of CaP or BPH patients' serum to CD14+ monocytes cultures resulted in significant inhibition of expression of DC-related markers (CD83, CD80, and CD40) when compared with control serum. Furthermore, monocyte-derived DCs generated in the presence of patients' serum were weaker stimulators of allogeneic T cell proliferation when compared with DCs produced in the presence of serum obtained from healthy volunteers. Furthermore, statistical analysis of samples revealed a positive relationship between the inhibition of expression of CD83 and CD80 molecules and the percentage of serum-free PSA.

2. Experimental Procedures

The serum specimens from 40 patients and 10 healthy control donors were taken in a sterile fashion and stored at –80°C. The current study was undertaken of serum from 20 CaP, 20 BPH patients, and 10 healthy volunteers. The serum level of total and free PSA was detected by Abbott AXSYM System. The levels of total PSA in serum of CaP patients were 30.1 ± 16.1 ng/ml in BPH serum and were 10.5 ± 1.2 and 0.5 ± 0.2 ng/ml in control serum. The percentage of free PSA relative to total PSA, or the f/t PSA ratio, was calculated as [fPSA (ng/ml)/tPSA (ng/ml)] × 100%. For CaP patients, this ratio was 15.0 ± 2.4 ng/ml, for BPH patients 20.0 ± 1.8 ng/ml. All serum specimens were obtained from the Department of Urology, University Clinic Hamburg-Eppendorf (Hamburg, Germany).

Human DC cultures were initiated from CD14+-adherent peripheral blood mononuclear cells (PBMCs). PBMCs, resuspended in a complete AIM V medium (Gibco) supplemented with heat-inactivated 1% human AB serum (Sigma) and 50 μg/ml gentamicin sulfate (Gibco), were plated in 75 cm^2 cell-culture flasks (Costar) and incubated at 37°C for 2 h. Non-adherent cells were removed and adherent monocytes were gently washed with a warm (37°C) medium followed by culture in AIM V medium supplemented with human granulocyte-macrophage colony stimulating factor (GM-CSF) (1000 U/ml) and IL-4 (1000 U/ml) (PeproTech) for 7 days.

Evaluation of expression of surface molecules on DCs was performed with fluorescein isothiocyanate (FITC) - or phycoerythrin (PE) - conjugated antibodies recognizing CD83, CD86, CD80, CD40, CD1a, and human leukocyte antigen-DR molecules (PharMingen). Flow cytometric analysis of stained cells was performed on Becton-Dickinson FACScan cytometer by Cell Quest software.

Functional activity of DC was determined in one-way allogeneic-mixed leukocyte reaction (MLR) using T lymphocytes as responders. Allogeneic T cells were isolated from PBMCs by purification on a nylon wool column. Cultured human DCs, serving as stimulators, were irradiated at 3000 rads and added in triplicate in graded doses to 3×10^5 allogeneic T cells in 96-well plates in total volume of 200 μl. Ninety-six hours later, cell cultures were pulsed with 1 μCi of [^{3}H] thymidine/well (DuPont-NEN) for the last 16–18 h and harvested onto GF/C glass fiber filter paper (Whatman) using a MACH III M harvester (Tomtec). The 1450 MicroBeta TRILUX liquid scintillation counter (Wallac) was used to detect the incorporation of [^{3}H]-thymidine. The counts were expressed as stimulatory index (SI), which was calculated by division of counts per minute (cpm) of each well onto mean of cpm of spontaneous T cell proliferation and presented as the mean SI ± standard error of the mean (SEM).

3. Results

To determine whether an endogenous CaP-derived factor(s) might inhibit dendropoiesis, serum specimens obtained from CaP patients, BPH patients, CaP patients after prostatectomy, and healthy volunteers have been tested in vitro. To exclude the influence of the abnormal levels of any major serum proteins, a high-resolution total protein electrophoresis (TPE) was performed with all specimens (Helena Labs), and suspicious samples were excluded from the further analysis. Hundred microliters of serum was added daily to DC cultures (10 ml) initiated from normal CD14+-adherent PBMCs (equivalent to 50 ml blood/specimen). Media and cytokines were replaced in all cultures on day 3, and cells were harvested and analyzed on day 7. Specimens were randomized and tested in several experiments. In addition, the same samples were tested several times on different days. Total of eight experiments (with 10–12 samples) were conducted. FACScan analysis of cultured cells revealed that the presence of CaP or BPH serum in DC cultures markedly altered the expression of DC-related markers when compared with control serum (Figure 1A). For instance, expression of CD83, CD1a, and CD40 was strongly suppressed, whereas expression of CD14 (monocyte/macrophage marker) was increased. Furthermore, the results of the MLR assay on Figure 1B revealed that DC cultures treated with serum obtained from patients with non-treated CaP displayed a significantly reduced ability to induce T cell proliferation ($p < 0.05$, two-way ANOVA). Thus, these data suggest that there are not yet identified factors in CaP and BPH serum, which could inhibit dendropoiesis in vitro.

Statistical analysis of these data revealed a positive relationship between the inhibition of expression of CD83 and CD80 on DC and serum levels of free PSA. In fact, Figure 2A demonstrates the mean values of inhibition of expression of DC surface molecules (reflecting the decrease in numbers of marker-positive cells in cultures) and serum-free PSA in three groups of patients: CaP patients after prostatectomy (no PSA), CaP patients, and BPH patients.

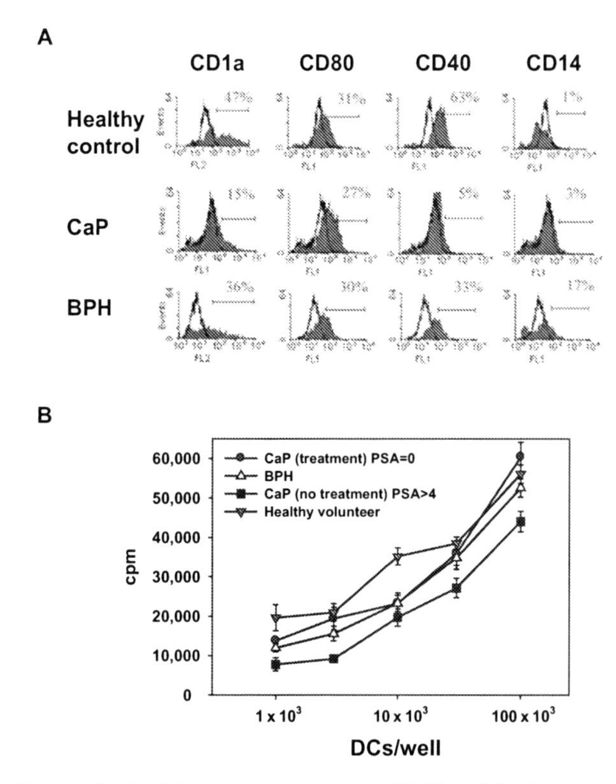

FIGURE 1. Serum obtained from prostate cancer (CaP) and benign prostatic hyperplasia (BPH) patients inhibit dendropoiesis in vitro. Addition of CaP or BPH patient serum to CD14-derived DC cultures resulted in significant inhibition of expression of DC-related markers when compared with control serum (A). Furthermore, DCs generated in the presence of patients' serum were weaker stimulators of T cell proliferation in the MLR assay when compared with DCs produced with serum obtained from healthy volunteers (B).

As can be seen, high levels of inhibition of dendropoiesis was associated with high levels of free PSA, which was confirmed by a correlation analysis of combined samples (Figure 2B). Interestingly, no correlation between dendropoiesis and concentrations of total PSA was demonstrated. Thus, it is possible that serum-free PSA might be involved in inhibition of DC function and thus, suppression of specific immune responsiveness in CaP patients.

4. Discussion

CaP is the most common malignancy in men in the United States and the second most frequent cause of cancer-related death in men. It is variable in its natural history with some patients having slowly progressive low-grade disease while others with poorly differentiated cancers having rapidly declining course (Landis et al. 1999).

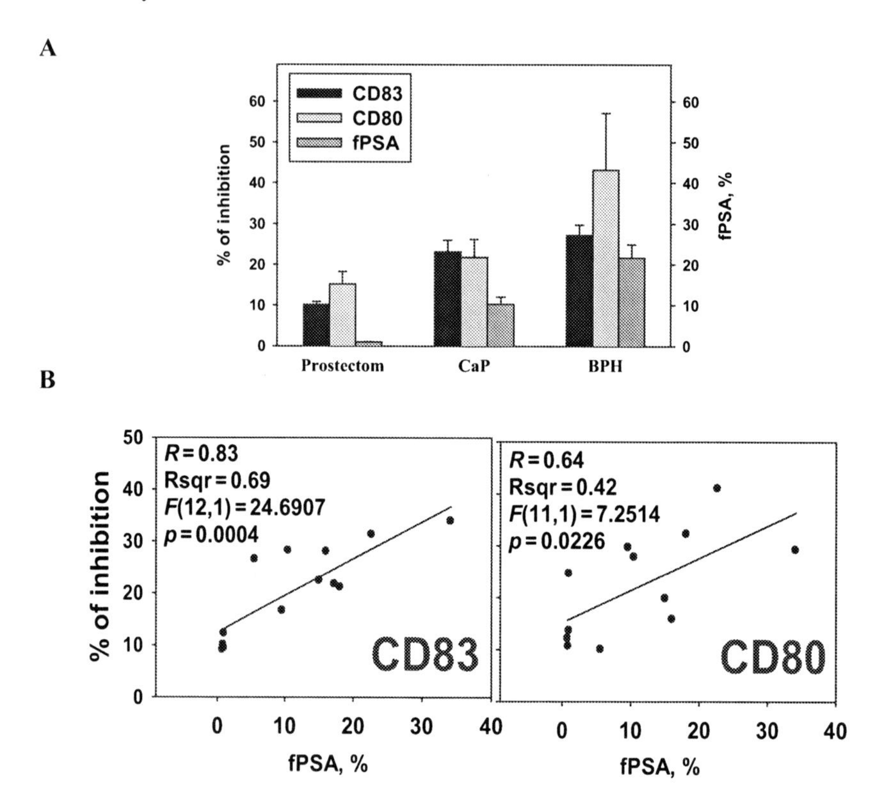

FIGURE 2. Serum-induced inhibition of dendropoiesis, assessed as a percentage of CD83+ CD80+ expression on dendritic cells (DCs), was associated with increased levels of free prostate-specific antigen (PSA) in serum obtained from Prostate cancer (CaP) and benign prostatic hyperplasia (BPH) patients (A). Correlation analysis of combined data revealed a significant correlation between inhibition of CD83 and CD80 expression and the levels of serum-free PSA (B).

The extensive mortality and morbidity associated with CaP is caused by the high prevalence of metastatic disease at the time of diagnosis (Arya et al. 2006). One of the features of CaP is the bone destruction during the metastatic bone disease and disturbed hemopoiesis (Frassica et al. 1992; Hall et al. 2005). The dysfunction of hemopoiesis is one of the major results of the bone destruction during the metastatic bone disease (Frassica et al. 1992). In fact, disturbed hemopoiesis characterized by the presence of immature leukocytes in CaP invading the bone marrow has been repeatedly reported (McGregor et al. 1990; Shamdas et al. 1993). In addition, we have demonstrated the alteration of dendropoiesis by prostate carcinoma cells in vitro. Prostate carcinoma cell line LNCaP-derived factors affected early hematopoietic precursor cells and markedly inhibited the generation of functionally active DCs in vitro. LNCaP cells added to DC cultures inhibited the generation of functionally active DCs from CD34+ precursor cells as well as from CD14+ cells in vitro (Aalamian et al. 2001).

At present study, we determined whether endogenous CaP-derived factors in patients' serum might inhibit dendropoiesis in a similar fashion. We showed a significant DC-inhibitory potential of serum obtained from CaP and BPH patients but not from healthy volunteers. As assessed by flow cytometry, expression of CD83, CD86, and CD40 molecules on DCs was strongly inhibited in the presence of serum received from CaP and BPH patients when compared with control serum. The ability of DCs to stimulate proliferation of naïve allogeneic T cells was also significantly decreased, if DCs were cultured in the presence of serum received from CaP patients. These data suggest that the DC system might be impaired in CaP patients. These results also suggest that malfunction of dendropoiesis in CaP patients is associated with a progression of the disease. In fact, a correlation between the number of DCs infiltrating tumor and longer patient survival or reduced frequency of metastatic disease has been described for a number of cancers (Becker 1992; Lotze 1997). In general, progressive tumor growth seems to be accompanied by a decreased density or loss of DCs and, conversely, regression is often associated with infiltration of these cells into tumor. It has been shown that the number of DCs in prostatic carcinomas is inversely correlated to the histopathological grade of the tumor, with those having no DC infiltrate representing the highest grade of carcinoma (Troy et al. 1998). Since DCs are known to play a key role in the initiation of the antitumor immune responses, impairment in their number or activity might result in deficient expansion of specific T lymphocytes. Thus, the resulting immunosuppression might play an important role in further tumor progression and resistance to immunotherapy.

For the protection of DCs from tumor-induced immunosuppression, it is important to investigate the mechanisms of their deregulation. We have recently demonstrated that purified and LNCaP-derived PSA suppressed DC generation and maturation in vitro, resulting in the low expression of costimulatory molecules on DCs and decreased ability of DCs to induce T cell proliferation (Aalamian et al. 2003). Here, we observed significant positive correlation between the inhibition of CD83 and CD80 expression on DCs treated with Cap and BPH serum and the percentage of free PSA relative to total PSA in serum of these patients. This correlation was detected in 60% of cases. Thus, the percentage of free PSA may be an important marker of DC function inhibition and thus, suppression of specific immune responsiveness in CaP patients.

PSA was the first tumor marker approved by FDA for use in early detection of CaP in 1994 (Stamey et al. 1987). Elevation in this marker in serum has been associated with CaP progression (Oesterling 1991; Williams et al. 2006). PSA complexed to α1- ACT is the predominant form of immunoreactive PSA in serum (~85%), and the free (uncomplexed) form makes up the rest (Lilja 1993). The concentration of these forms of PSA varies according to the disease state, and the analysis of these molecular forms may improve the clinical use of PSA (Christensson et al. 1990; Elabbady and Khedr 2006). For instance, it has been reported that the f/t PSA ratio was significantly lower in men with untreated CaP than in men with benign disease (Christensson et al. 1993; Elabbady and Khedr 2006; Jung et al. 2000; Stephan et al. 2000).

Our data suggest that the percentage of free PSA might reflect the inhibition of DC function and, thus, suppression of specific immune responsiveness in cancer patients. Further investigation of the role of free PSA for the prognosis of CaP, induction of immunosuppression, and understanding the mechanisms of immune dysbalance in CaP may lead to the discovery of new therapeutic agents that will be directed to decrease tumorigenicity and increase immunogenicity of CaP cells.

Acknowledgments

This work was supported by the NCI NIH (RO1 CA84270 to MRS) and DoD (PC050252 to MRS).

References

Aalamian, M., Pirtskhalaishvili, G., Nunez, A., Esche, C., Shurin, G.V., Huland, E., Huland, H. and Shurin, M.R. (2001) Human prostate cancer regulates generation and maturation of monocyte-derived dendritic cells. Prostate 46, 68–75.

Aalamian, M., Tourkova, I.L., Chatta, G.S., Lilja, H., Huland, E., Huland, H., Shurin, G.V. and Shurin, M.R. (2003) Inhibition of dendropoiesis by tumor derived and purified prostate specific antigen. J. Urol. 170, 2026–2030.

Armstrong, T.D., Pulaski, B.A. and Ostrand-Rosenberg, S. (1998) Tumor antigen presentation: changing the rules. Cancer Immunol. Immunother. 46, 70–74.

Arya, M., Bott, S.R., Shergill, I.S., Ahmed, H.U., Williamson, M. and Patel, H.R. (2006) The metastatic cascade in prostate cancer. Surg. Oncol. 23,77–83.

Becker, Y. (1992) Anticancer role of dendritic cells (DC) in human and experimental cancers -- a review. Anticancer Res. 12, 511–520.

Catalona, W.J., Partin, A.W., Slawin, K.M., Brawer, M.K., Flanigan, R.C., Patel, A., Richie, J.P., deKernion, J.B., Walsh, P.C., Scardino, P.T., Lange, P.H., Subong, E.N., Parson, R.E., Gasior, G.H., Loveland, K.G. and Southwick, P.C. (1998) Use of the percentage of free prostate-specific antigen to enhance differentiation of prostate cancer from benign prostatic disease: a prospective multicenter clinical trial. JAMA 279, 1542–1547.

Christensson, A., Bjork, T., Nilsson, O., Dahlen, U., Matikainen, M.T., Cockett, A.T., Abrahamsson, P.A. and Lilja, H. (1993) Serum prostate specific antigen complexed to alpha 1-antichymotrypsin as an indicator of prostate cancer. J. Urol. 150, 100–105.

Christensson, A., Laurell, C.B. and Lilja, H. (1990) Enzymatic activity of prostate-specific antigen and its reactions with extracellular serine proteinase inhibitors. Eur. J. Biochem. 194, 755–763.

Colaco, C.A. (1999a) DC-based cancer immunotherapy: the sequel. Immunol. Today 20, 197–198.

Colaco, C.A. (1999b) Why are dendritic cells central to cancer immunotherapy? Mol. Med. Today 5, 14–17.

Comuzzi, B. and Sadar, M.D. (2006) Proteomic analyses to identify novel therapeutic targets for the treatment of advanced prostate cancer. Cell Sci. 3, 61–81.

Diamandis, E.P. and Yu, H. (1997) Nonprostatic sources of prostate-specific antigen. Urol. Clin. North. Am. 24, 275–282.

Elabbady, A.A. and Khedr, M.M. (2006) Free/total PSA ratio can help in the prediction of high gleason score prostate cancer in men with total serum prostate specific antigen (PSA) of 3-10 ng/ml. Int. Urol. Nephrol. 38, 553–557.
Frassica, F.J., Gitelis, S. and Sim, F.H. (1992) Metastatic bone disease: general principles, pathophysiology, evaluation, and biopsy. Instr. Course Lect. 41, 293–300.
Hall, C.L., Bafico, A., Dai, J., Aaronson, S.A. and Keller, E.T. (2005) Prostate cancer cells promote osteoblastic bone metastases through Wnts. Cancer Res. 65, 7554–7560.
Harris, D.T., Matyas, G.R., Gomella, L.G., Talor, E., Winship, M.D., Spitler, L.E. and Mastrangelo, M.J. (1999) Immunologic approaches to the treatment of prostate cancer. Semin. Oncol. 26, 439–447.
Healy, C.G., Simons, J.W., Carducci, M.A., DeWeese, T.L., Bartkowski, M., Tong, K.P. and Bolton, W.E. (1998) Impaired expression and function of signal-transducing zeta chains in peripheral T cells and natural killer cells in patients with prostate cancer. Cytometry 32, 109–119.
Herr, H.W. (1980) Suppressor cells in immunodepressed bladder and prostate cancer patients. J. Urol. 123, 635–639.
Huang, A.Y.C., Golumbek, P., Ahmadzadeh, M., Jaffee, E., Pardoll, D. and Levitsky, H. (1994) Role of bone marrow-derived cells in presenting MHC class I-restricted tumor antigens. Science 264, 961–965.
Ivshina, A.V., Zhumagazin Zh, D., Zabotina, T.N., Matveev, B.P. and Kadagidze, Z.G. (1995) Effect of the spread of the process and treatment on the phenotype of peripheral blood lymphocytes in patients with prostatic cancer. Urologiia. i. Nefrologiia 9, 36–38.
Jung, K., Brux, B., Lein, M., Rudolph, B., Kristiansen, G., Hauptmann, S., Schnorr, D., Loening, S.A. and Sinha, P. (2000) Molecular forms of prostate-specific antigen in malignant and benign prostatic tissue: biochemical and diagnostic implications. Clin. Chem. 46, 47–54.
Landis, S.H., Murray, T., Bolden, S. and Wingo, P.A. (1999) Cancer statistics, 1999. CA Cancer J. Clin. 49, 8–31.
Li, N., Qin, H., Li, X., Zhou, C., Wang, D., Ma, W., Lin, C., Zhang, Y., Wang, S. and Zhang, S. (2006) Potent systemic antitumor immunity induced by vaccination with chemotactic-prostate tumor associated antigen gene-modified tumor cell and blockade of B7 H1. J. Clin. Immunol. 27, 117–130.
Lilja, H. (1993) Structure, function, and regulation of the enzyme activity of prostate- specific antigen. World J. Urol. 11, 188–191.
Lotze, M.T. (1997) Getting to the source: dendritic cells as therapeutic reagents for the treatment of patients with cancer [editorial; comment]. Ann. Surg. 226, 1–5.
McGregor, A.R., Moore, M.L., Bailey, R.R., Robson, R.A. and Lynn, K.L. (1990) Disseminated prostatic carcinoma presenting with acute interstitial nephritis and microangiopathic haemolytic anaemia. Aust. N. Z. J. Med. 20, 170–172.
Mohamedali, K.A., Poblenz, A.T., Sikes, C.R., Navone, N.M., Thorpe, P.E., Darnay, B.G. and Rosenblum, M.G. (2006) Inhibition of prostate tumor growth and bone remodeling by the vascular targeting agent VEGF121/rGel. Cancer Res. 66, 10919–10928.
Oesterling, J.E. (1991) Prostate specific antigen: a critical assessment of the most useful tumor marker for adenocarcinoma of the prostate. J. Urol. 145, 907–923.
Polascik, T.J., Oesterling, J.E. and Partin, A.W. (1999) Prostate specific antigen: a decade of discovery--what we have learned and where we are going. J. Urol. 162, 293–306.
Ragde, H., Cavanagh, W.A. and Tjoa, B.A. (2004) Dendritic cell based vaccines: progress in immunotherapy studies for prostate cancer. J. Urol. 172, 2532–2538.
Salgaller, M.L., Lodge, P.A., McLean, J.G., Tjoa, B.A., Loftus, D.J., Ragde, H., Kenny, G.M., Rogers, M., Boynton, A.L. and Murphy, G.P. (1998) Report of immune monitoring of

prostate cancer patients undergoing T cell therapy using dendritic cells pulsed with HLA-A2-specific peptides from prostate-specific membrane antigen. Prostate 35, 144–151.

Shamdas, G.J., Ahmann, F.R., Matzner, M.B. and Ritchie, J.M. (1993) Leukoerythroblastic anemia in metastatic prostate cancer. Clinical and prognostic significance in patients with hormone-refractory disease. Cancer 71, 3594–3600.

Shurin, G.V., Aalamian, M., Pirtskhalaishvili, G., Bykovskaia, S., Huland, E., Huland, H. and Shurin, M.R. (2001) Human prostate cancer blocks the generation of endritic cells from CD34+ hematopoietic progenitors. Eur. Urol. 39 (suppl), 37–40.

Shurin, M.R. (1996) Dendritic cells presenting tumor antigen. Cancer Immunol. Immunother. 43, 158–164.

Simons, J.W., Mikhak, B., Chang, J.F., DeMarzo, A.M., Carducci, M.A., Lim, M., Weber, C.E., Baccala, A.A., Goemann, M.A., Clift, S.M., Ando, D.G., Levitsky, H.I., Cohen, L.K., Sanda, M.G., Mulligan, R.C., Partin, A.W., Carter, H.B., Piantadosi, S., Marshall, F.F. and Nelson, W.G. (1999) Induction of immunity to prostate cancer antigens: results of a clinical trial of vaccination with irradiated autologous prostate tumor cells engineered to secrete granulocyte-macrophage colony-stimulating factor using ex vivo gene transfer. Cancer Res. 59, 5160–5168.

Slovin, S.F., Kelly, W.K. and Scher, H.I. (1998) Immunological approaches for the treatment of prostate cancer. Semin. Urol. Oncol. 16, 53–59.

Speiser, D.E., Miranda, R., Zakarian, A., Bachmann, M.F., McKall-Faienza, K., Odermatt, B., Hanahan, D., Zinkernagel, R.M. and Ohashi, P.S. (1997) Self antigens expressed by soild tumors do not efficiently stimulate naïve or activated T cells: implications for immunotherapy. J. Exp. Med. 186, 645–653.

Stamey, T.A., Yang, N., Hay, A.R., McNeal, J.E., Freiha, F.S. and Redwine, E. (1987) Prostate-specific antigen as a serum marker for adenocarcinoma of the prostate. N. Engl. J. Med. 317, 909–916.

Stephan, C., Jung, K., Lein, M., Sinha, P., Schnorr, D. and Loening, S.A. (2000) Molecular forms of prostate-specific antigen and human kallikrein 2 as promising tools for early diagnosis of prostate cancer. Cancer Epidemiol. Biomarkers Prev. 9, 1133–1147.

Troy, A., Davidson, P., Atkinson, C. and Hart, D. (1998) Phenotypic characterization of the dendritic cell infiltrate in prostate cancer. J. Urol. 160, 214–219.

Troy, A., Davidson, P., Atkinson, C. and Hart, D. (1999) Renal cell carcinoma and prostate cancer inhibit dendritic cell activation. Aust. N. Z. J. Surg. 69, A111–A112.

Williams, S.A., Singh, P., Isaacs, J.T. and Denmeade, S.R. (2006) Does PSA play a role as a promoting agent during the initiation and/or progression of prostate cancer? Prostate 67, 312–329.

Immunomodulators

19

Alarmins Initiate Host Defense

Joost J. Oppenheim[1], Poonam Tewary[1], Gonzalo de la Rosa[1], and De Yang[2]

[1]Laboratory of Molecular Immunoregulation, Center for Cancer Research, NCI, Frederic, MD, USA, Oppenhei@ncifcrf.gov
[2]Basic Research Program, SAIC-Frederick, Inc., Frederic, MD, USA

Abstract. In response to infection and/or tissue injury, cells of the host innate immune system rapidly produce a variety of structurally distinct mediators (we elect to call alarmins) that not only function as potent effectors of innate defense but also act to alarm the immune system by promoting the recruitment and activation of host leukocytes through interaction with distinct receptors. Alarmins are capable of activating antigen-presenting cells (APCs) and enhancing the development of antigen-specific immune responses. Here, we discuss the characteristics of several alarmins, a variety of potential alarmin candidates and potential implications of alarmins.

1. Introduction: Properties of Alarmins

There are a number of structurally unrelated endogenous proteins that are very rapidly released or induced to be produced in response to microbial invasion or tissue injury. Although these readily available proteins could be considered a subset of danger signals or cytokines, they are already known to have antimicrobial, enzymatic, or chromatin-binding activities. However, we have chosen to call them "alarmins" because they also have the capacity to interact with receptors on host cells, which enables them to induce chemotactic migration and to activate receptor-expressing cells. Alarmins interact with receptors on antigen-presenting cells (APCs) such as dendritic cells (DCs) and as a result are able to rapidly initiate innate as well as lymphocyte-dependent adaptive immune responses to components of damaged tissues or invading pathogens.

In our laboratory, we detect alarmins based on their capacity to induce directional migration (chemotaxis) of APCs such as DCs in a pertussis toxin (PTx)-sensitive manner. Since PTx is selectively inhibitory for Gαi protein-coupled receptors (GiPCRs) utilized by chemokines and a wide variety of other chemotactic ligands, this provides evidence that alarmins also interact with such receptors on host APCs/DCs. Furthermore, some of the ligands that interact with GiPCR have been

TABLE 1. Characteristics of alarmins

Ligand	Chemotactic receptor	Activation of DC to mature ($CCR7^+$, $Comigen^+$, $IL12p70^+$)	TLR	Constitutive (necrotic cell release)	Inducible	In vivo Adjuvant activity
α-Defensins	GiPCR	Yes	?	Yes (Degranulates)	No	Yes
β-Defensins	CCR6	Yes	Mu βD2/ TLR4	HBD1	All others in epithelium	Yes
Cathelicidin LL37/ CRAMP	FPRL-1/ FPR-2	No (acts on pre-DCs)	?	Yes (Degranulates)	Yes (monos and epithelial)	Yes
EDN/ EARM	GiPCR	Yes	TLR2	Yes (Degranulates)	Yes (monos)	Yes
HMGB1	GiPCR/ RAGE	Yes	TLR2 TLR4	Yes	Yes (monos)	Yes

DC, dendritic cell; EDN, eosinophil-derived neurotoxin; HBD, human beta defensin; HMGBI, high-mobility group box 1; ?, unknown.

shown to be internalized by GiPCRs on DCs and to co-localize intracellularly with MHC class II proteins (Biragyn et al. 2004). Thus, alarmins have the potential to be internalized into the endosomal and Golgi compartments of APCs favoring antigen processing and presentation. Since all the above properties of alarmins are based on in vitro assays, to ensure their physiological relevance, an alarmin also has to have demonstrable in vivo immunoadjuvant effects. As summarized in Table 1, we will briefly review the various endogenous proteins that have been identified as having alarmin functions. Additional information can be obtained in a number reviews (Yang et al. 2004a; Oppenheim and Yang 2005; Kornbluth and Stone 2006; Zedler and Faist 2006; Bianchi 2007).

2. Alpha Defensins

Defensins consist of a family of small (3–6 kDa) antimicrobial peptides. Human alpha defensins (HAD) 1–4, also known as human neutrophil peptides (HNP) (1–4), are produced by neutrophils and monocytes, while some lymphocytes and natural killer (NK) cells produce HNP1. HAD 5–6 are generated by Paneth cells of the small intestine and epithelial cells in reproductive tissues, placenta, and nasal and bronchial lining. Mice do not have neutrophil alpha defensins but instead possess multiple Paneth cell-derived alpha defensins known as cryptidins. Alpha defensins are constitutively expressed and released from intracellular granules. HD5 has potent anti-human papillovirus (HPV) activity, and their production by cells lining the vaginal mucosa is inversely correlated to the incidence of HPV infection and cervical cancer incidence (Buck et al. 2006). HNP1–4 have antiviral effects against many viruses including HIV1 (Klotman and Chang 2006). Furthermore, deficiencies in these alpha defensins are also associated with an increased incidence of inflammatory bowel disease (IBD) (Wehkamp and Stange 2006). We have found alpha defensins to be chemotactic for the subset of $CD4^+$ $CD45^+$ RA^+-naïve T cells, some $CD8^+$ T cells, and immature DCs (Chertov et al. 1996; Yang et al. 2000, 2004a). Although this chemotactic effect is PTx inhibitable, the specific HNP receptor has not been identified. Alpha defensins also activate immature DCs (iDCs) to mature into mDCs. This is characterized by phenotypic changes such as increased expression of CD80, CD83, CD86, CD40, and MHC class II, increased production of inflammatory cytokines, and loss of receptor expression such as CCR1 and 5 but an increase in the expression of CCR7. Mature DCs also develop the capacity to induce allogeneic mixed leukocyte reaction (MLR) in vitro.

Administration of HAD to mice together with antigens shows them to have immunoenhancing adjuvant effects and to augment the production of both TH1 and TH2 cytokines (Lillard et al. 1999; Yang et al. 2004a). Consequently, HAD have the capacity to interact with receptors on both human and mouse cells. However, we have not been able to identify the receptor responsible for the activating effects or its relationship to the chemotactic receptor.

About eight beta defensins have been identified in mice and man, but genome sequences suggest that there may be close to 30 distinct beta-defensin entities in both species. Beta defensins are produced by keratinocytes of the skin and epithelial cells in the gastrointestinal (GI), genitourinary (GU) and tracheobronchial lining. In fact, human beta defensin 1 (HBD1) is constitutively produced, stored and released by these cells at the interface with our environment. Furthermore, "dangerous" stimuli such as microbial products (e.g. LPS) and inflammatory cytokines (e.g. IL-1, IL-17, TNF and IFN), in addition to inducing the release of HBD1, also upregulate the expression of HBD2–4. Beta defensins are also essential for host defense, since MBD1 null mice exhibit reduced resistance to challenge with *Hemophylus Influenza* (Moser et al. 2002). HBD1–4 are chemotactic for the subset of $CD4^+$ $CD45^+$ RO^+-memory T cells and iDCs by interacting with the CCR6 receptor, which is also the cognate receptor for the MIP3α chemokine. Although the amino acid sequences of the HBD2 and MIP3α are unrelated, their tertiary structure shows some similarity, which may account for their capacity to share CCR6. In addition, HBD3 and HBD4 are chemotactic for human monocytes which do not express CCR6 (Garcia et al. 2001a; Wu et al. 2003). Consequently, HBD3 and HBD4 use an additional PTx-inhibitable GiPCR, which has not been identified as yet (Garcia et al. 2001a,b; Wu, et al. 2003). The beta defensins also have the capacity to activate iDCs to differentiate into mDCs. TLR4 has been identified as the receptor signalling pathway responsible for this activating effect only in the case of mDB2 (Biragyn et al. 2002). The receptor(s) responsible for the activating effects of the other defensins remain unknown, but they are likely to be distinct from TLR4, since the beta defensins have varied immune effects. For example, MBD2 has been shown to be an effective enhancer of TH1-dependent cellular immune antitumor responses (Biragyn et al. 2001, 2002). Recently, MBD2 has also been shown to promote antileukemia immunity (Ma et al.

2006). In contrast, MBD3 favors TH2-dependent humoral immune responses and does not promote antitumor immunity (Biragyn et al. 2001). Nevertheless, both these beta defensins have considerable albeit distinct immunoadjuvant effects.

3. Cathelicidin (LL37)/CRAMP

Human cathelicidin (LL37) is the alpha helical C-terminal 37 amino acid peptide domain cleaved from the hCAP18 intact protein. LL37 has broad antimicrobial activities and just like the defensins is stored in the granules of neutrophils, monocytes, T cells and keratinocytes. In addition, LL37 and its mouse homologue (CRAMP) can be induced by "danger" signals and inflammatory cytokines secreted by the usual Golgi apparatus and ER pathway. LL37 is chemotactic and interacts with the FPRL1 GiPCR present on neutrophils, monocytes, mast cells and lymphocytes (Yang et al. 2000, 2004a). Unlike the other alarmins, LL37/CRAMP cannot attract DCs but activates "DC precursors" (monocytes) to develop into APCs. Once again, the receptor(s) used by LL37 to activate APCs is unknown. However, LL37 clearly has considerable immunoadjuvant activity equivalent to that of the standard aluminum hydroxide adjuvant used in human vaccines (Kurosaka et al. 2005). Consequently, LL37/CRAMP contributes to host defense based on its antimicrobial and immunostimulating activities. Furthermore, cathelicidin, like the defensins, based on the overall negative charge of its molecule can also directly bind and neutralize endotoxin. These observations are supported by data showing that CRAMP null mice have reduced resistance to challenge by pathogens (Nizet et al. 2001; Rosenberger et al. 2004). Furthermore, mice transfected to overexpress cathelicidin have increased resistance to endotoxin challenge, but greater inflammatory reactions with increased secretion of mucous and IL-8 by the tracheobronchial lining cells (Bals and Wilson 2003). However, these studies cannot determine which of the multiple activities of cathelicidin is more important in host defense.

4. Eosinophil-Derived Neurotoxin

Eosinophil-Derived Neurotoxin (EDN) is an 18-kDa member of the large RNAse family. EDN is stored in eosinophil granules and to a lesser extent in neutrophil granules but can also be induced to be expressed in monocytes/macrophages in response to danger signals and cytokines. EDN's name derives from the identification of eosinophils as a major cell source and its cytotoxic effects on cultured Purkinje cells. EDN has potent anti-HIV-1 activities (Rugeles et al. 2003). EDN and its mouse homologue mEAR2 are the only RNases with GiPCR-dependent chemotactic effects on both iDCs and mDCs (Yang et al. 2003. 2004a). The specific chemotactic receptor for EDN remains unknown. EDN and mEARM2 also activate iDCs in a TLR2-specific manner (Yang et al. 2004b). They activate only cells transfected with TLR2, independent of TLR1 and TLR6 (Yang et al. submitted). Immunization of mice with ovalbumin (OVA) plus EDN enhances the OVA-specific proliferative responses of splenocytes of wild type but not of TLR2 null mice. EDN also promotes antigen-specific anti-OVA responses but favors the production of IgG1 anti-OVA. In comparison, human angiogenin, another RNAse family member, did not activate iDCs and had no immunoadjuvant effects (Yang et al. 2004b, submitted) Furthermore, TLR2–/– mice also failed to show enhanced anti-OVA responses. Splenocytes from TLR2+/+ mice immunized with EDN and OVA, when stimulated with OVA, produced elevated levels of IL-5, IL-10 and IL-13 but did not augment IFN-γ and IL-4 production (Yang et al. submitted). Thus, the immunoadjuvant effects of EDN, unlike those of the defensins and LL37, favor polarization of TH2 responses.

5. High-Mobility Group Box 1

High-Mobility Group Box 1 (HMGB1), a prominent nonhistone chromosomal structural protein with DNA-binding activity is also an alarmin. HMGB1 located in the nucleosome by bending DNA promotes gene transcription. HMGB1 is a 215 amino acid antimicrobial peptide consisting of two binding A and B box domains. In 1999, HMGB1 was discovered to be released by necrotic cells and to have potent "cytokine-like" activities (Wang et al. 1999). HMGB1, to everyone's surprise, was reported to be a "late mediator" of endotoxin shock. Since then, HMGB1 has been shown to transduce signals by interacting with ubiquitously expressed RAGE receptors. HMGB1 also uses RAGE receptor for the regulation of neurite outgrowth (Hori et al. 1995). In addition to being released by necrotic cells and being constitutively present in the nucleus, HMGB1 can also be induced to be expressed in the cytoplasmic compartment of monocytes, DCs and NK cells by inflammatory cytokines and "danger" signals. The prepro-HMGB1 is enzymatically cleaved, similarly to IL-1, for release from cells. HMGB1 is chemotactic for muscle cells, monocytes and iDCs in a PTx-sensitive manner (Rouhiainen et al. 2004; Yang et al. 2007). However, anti-RAGE antibody also reduces the chemotactic activity, suggesting that both RAGE and an unidentified GiPCR are somehow involved.

The chemokine-like and cytokine-inducing properties have been attributed to the B box of HMGB1. Purified HMGB1 is also a potent activator of iDCs, and this may be RAGE dependent (Kokkola et al. 2005). However, HMGB1 has a great propensity to complex DNA, RNA and components of LPS and has been reported to activate both TLR2 and TLR4 (Park et al. 2006). Since purified HMGB1 fails to act on these receptors, it may do so only in such a complex form in vitro and in vivo. In fact, serum from most rheumatoid patients contains antibodies directed against nucleosomes including the histone and HMGB1 components (Uesugi et al. 1998). Thus, HMBG1 may be present in vivo in pathophysiological conditions in the form of complexes containing DNA, RNA and lipoproteins. HMGB1 null mice die in utero, thus it is essential for development as well. HMGB1 has immunoadjuvant effects with a TH1-polarizing capacity (Rovere-Querini et al. 2004). A series of papers have been published in the January 2007 issue of *Journal of Leukocyte Biology* presenting and reviewing data on the many biological repair and immunological activities of HMGB1.

6. Candidate Alarmins and Pseudo-Alarmins

As summarized in Table 2, we and others have investigated a number of molecules that have some but not all of the properties of alarmins. A number of cytokines such as granulocyte-macrophage colony stimulating factor (GM-CSF), interferon αβ, TNF, IL-1, IL-12 and IL-15 have been shown to have modest immunoadjuvant activities, with GM-CSF being the most potent inducer of antitumor activities (Kornbluth and Stone 2006). Yet, GM-CSF is only weakly chemotactic and does not activate iDCs; it merely primes them. None of the other cytokines are chemotactic, and only type I IFN, TNF and IL-1 activate DCs. Granulysin, a cytolytic protein present in the granules of CD8+ T lymphocytes, has been reported to have both chemotactic and activating effects on iDCs (Deng et al. 2005).

In addition to being released by degranulation of CD8+ T cells, granulysin can also be induced by inflammatory stimulants. The host cell receptors and immunoadjuvant effects of granulysin have not been identified.

There have been a few reports that selected chemokines such as CCL21 and CCL20 can promote T lymphocyte activation (Marsland et al. 2005; Molon et al. 2005; Kornbluth and Stone 2006). However, most chemokines, although chemotactic, fail to induce full maturation of iDCs in vitro. This could be due to suboptimal folding of recombinant chemokines used for such studies. They may also be capable of priming immune responses and have been shown to have immunoadjuvant activities when covalently linked to antigens (Biragyn et al. 2001). This may be based on their capacity to deliver fused-protein antigens more effectively via their receptors for antigen processing and presentation (Biragyn et al. 2004). The chemotactic ef-

TABLE 2. Candidate Alarmins

Ligand	Chemotactic Receptor	Activation of DC to Mature $CCR7^{+}$,Comitogen^{+},IL12p70^{+}	TLR	Constitutive (Necrotic cell release)	Inducible	In vivo Adjuvant Activity
GM-CSF	GM-CSFR	No	No	No	Yes	Priming
IFNαβ, IL12, IL15	No	Yes/No/No	No	No	Yes	Yes
Chemokines	GiPCRs	±	No	No (only in plts)	Yes	Priming
Granulysin	GiPCR	Yes	?	Yes	Yes	?
ATP	GiPCR	Yes	No	Yes	No	No
Uric acid	No	Yes	No	Yes	No	Yes
HSP60,70 96	No	Yes	TLR2 TLR4	Yes	Yes	Yes
Lactoferrin	Yes (Monos)	±	No	Yes	Yes	Yes
S100A7/ psoriasin	RAGE	No	No	No	Yes	?

?, unknown

fects of ATP are based on interactions with GiPCR (Idzko et al. 2002). ATP can also activate iDCs in vitro, however, ATP does not have potent immunoadjuvant effects, presumably because it is rapidly degraded. Uric acid, although not chemotactic, in crystallized form activates iDCs, is exceedingly inflammatory and has immunoadjuvant effects (Shi et al. 2003). The likelihood that it probably induces alarmins needs to be investigated.

The heat shock proteins (HSP 60, 70 and 96) with the exception of chemotactic activities have all the other properties of alarmins. They activate iDCs by stimulating TLR2 and TLR4 and have adjuvant effects (Manjili et al. 2004). Perhaps based on their ability to bind proteins, they also are able to deliver antigens to receptors on APCs favoring antigen processing and presentation (Manjili et al. 2004; Kornbluth and Stone 2006). Thus, they provide an example of an "alarmin-like" molecule that has the ability to deliver antigens to APCs and therefore does not need to be able to attract such cells. We are in the process of evaluating a number of other candidate endogenous proteins for their alarmin activities including lactoferrin, cyclophillin and the S100 A7 family member known as psoriasin. However, the studies are ongoing and incomplete, and these proteins are still lacking one or more of the properties of alarmins.

In conclusion, we have identified a number of endogenous proteins and peptides as having the characteristics of alarmins. They include defensins, cathelicidin, EDN and HMGB1. They all have chemotactic effects on host defense cells expressing GiPCR. They induce the maturation of iDCs and have immunoadjuvant effects. One of the aims of these studies is to identify endogenous proteins that could be more potent and effective nontoxic vaccine adjuvants. Perhaps, alarmins as endogenous proteins may be less toxic and more effective than currently available adjuvants.

Acknowledgements

This research was supported by the Intramural Research Program of the NIH, NCI.

References

Bals, R. and Wilson, J.M. (2003) Cathelicidins-a family of multifunctional antimicrobial peptides. Cell Mol. Life Sci. 60, 711–720.

Bianchi, M.E. (2007) DAMPs, PAMPs and alarmins: all we need to know about danger. J. Leukoc. Biol. 81, 1–5.

Biragyn, A., Ruffini, P.A., Coscia, M., Harvey, L.K., Neelapu, S.S., Baskar, S., Wang, J.M. and Kwak, L.W. (2004) Chemokine receptor-mediated delivery directs self-tumor antigen efficiently into the class II processing pathway *in vitro* and induces protective immunity *in vivo*. Blood 104, 1961–1969.

Biragyn, A., Ruffini, P.A., Leifer, C.A., Klyushnenkova, E., Shakhov, A., Chertov, O., Shirakawa, A.K., Farber, J.M., Segal, D.M., Oppenheim, J.J. and Kwak, L.W. (2002) Toll-like receptor 4-dependent activation of dendritic cells by β-defensin 2. Science 298, 1025–1029.

Biragyn, A., Surenhu, M., Yang, D., Ruffini, P.A., Haines, B.A., Klyushnenkova, E., Oppenheim, J.J. and Kwak, L.W. (2001) Mediators of innate immunity that target immature, but not mature, dendritic cells induce antitumor immunity when genetically fused with nonimmunogenic tumor antigens. J. Immunol. 167, 6644–6653.

Buck, C.B., Day, P.M., Thompson, C.D., Lubkowski, J., Lu, W., Lowy, D.R. and Schiller, J.T. (2006) Human alpha-defensins block papillomavirus infection. Proc. Natl. Acad. Sci. U.S.A. 103, 1516–21.

Chertov, O., Michiel, D.F., Xu, L., Wang, J.M., Tani, K., Murphy, W.J., Longo, D.L., Taub, D.D. and Oppenheim, J.J. (1996) Identification of defensin-1, defensin-2, and CAP37/azurocidin as T cell chemoattractant proteins released from interleukin-8-stimulated neutrophils. J. Biol. Chem. 271, 2935–2940.

Deng, A., Chen, S., Li, Q., Lyu, S.C., Clayberger, C. and Krensky, A.M. (2005) Granulysin, a cytolytic molecule, is also a chemoattractant and proinflammatory activator. J. Immunol. 174, 5243–8.

Garcia, J.R., Jaumann, F., Schulz, S., Krause, A., Rodriguez-Jimenez, J., Forssmann, U., Adermann, K., E, E.K., Vogelmeier, C., Becker, D., Hedrich, R., Forssmann, W.G. and Bals, R. (2001a) Identification of a novel, multifunctional β-defensin (human β-defensin 3) with specific antimicrobial activity: its interaction with plasma membranes of Xenopus oocytes and the induction of macrophage chemoattraction. Cell Tissue Res. 306, 257–264.

Garcia, J.R., Krause, A., Schulz, S., Rodriguez-Jimenez, F.J., Kluver, E., Adermann, K., Forssmann, U., Frimpong-Boateng, A., Bals, R. and Forssmann, W.G. (2001b) Human β-defensin 4: a novel inducible peptide with a specific salt-sensitive spectrum of antimicrobial activity. FASEB J. 15, 1819–1821.

Hori, O., Brett, J., Slattery, T., Cao, R., Zhang, J., Chen, J.X., Nagashima, M., Lundh, E.R., Vijay, S., Nitecki, D. Morser, J., Stern, D. and Schmidt, A.M. (1995) The receptor for advanced glycation end products (RAGE) is a cellular binding site for amphoterin. Mediation of neurite outgrowth and co-expression of rage and amphoterin in the developing nervous system. J. Biol. Chem. 270, 25752–25761.

Idzko, M., Dichmann, S., Ferrari, D., Di Virgilio, F., la Sala, A., Girolomoni, G., Panther, E. and Norgauer, J. (2002) Nucleotides induce chemotaxis and actin polymerization in immature but not mature human dendritic cells via activation of pertussis toxin-sensitive P2y receptors. Blood 100, 925–32.

Klotman, M.E. and Chang, T.L. (2006) Defensins in innate antiviral immunity. Nat. Rev. Immunol. 6, 447–456.

Kokkola, R., Andersson, A., Mullins, G., Ostberg, T., Treutiger, C.J., Arnold, B., Nawroth, P., Andersson, U., Harris, R.A. and Harris, H.E. (2005) RAGE is the major receptor for the proinflammatory activity of HMGB1 in rodent macrophages. Scand. J. Immunol. 61, 1–9.

Kornbluth, R.S. and Stone, G.W. (2006) Immunostimulatory combinations: designing the next generation of vaccine adjuvants. J. Leukoc. Biol. 80, 1084–102.

Kurosaka, K., Chen, Q., Yarovinsky, F., Oppenheim, J.J. and Yang, D. (2005) Mouse cathelin-related antimicrobial peptide chemoattracts leukocytes using formyl peptide receptor-like 1/mouse formyl peptide receptor-like 2 as the receptor and acts as an immune adjuvant. J. Immunol. 174, 6257–6265.

Lillard Jr., J.W., Boyaka, P.N., Chertov, O., Oppenheim, J.J. and McGhee, J.R. (1999) Mechanisms for induction of acquired host immunity by neutrophil peptide defensins. Proc. Natl. Acad. Sci. U.S.A. 96, 651–656.

Ma, X.T., Xu, B., An, L.L., Dong, C.Y., Lin, Y.M., Shi, Y. and Wu, K.F. (2006) Vaccine with beta-defensin 2-transduced leukemic cells activates innate and adaptive immunity to elicit potent antileukemia responses. Cancer Res. 66, 1169–1176.

Manjili, M.H., Wang, X.Y., MacDonald, I.J., Arnouk, H., Yang, G.Y., Pritchard, M.T. and Subjeck, J.R. (2004) Cancer immunotherapy and heat-shock proteins: promises and challenges. Expert Opin. Biol. Ther. 4, 363–373.

Marsland, B.J., Battig, P., Bauer, M., Ruedl, C., Lassing, U., Beerli, R.R., Dietmeier, K., Ivanova, L., Pfister, T., Vogt, L., Nakano, H., Nembrini, C., Saudan, P., Kopf, M. and Bachmann, M.F. (2005) CCL19 and CCL21 induce a potent proinflammatory differentiation program in licensed dendritic cells. Immunity 22, 493–505.

Molon, B., Gri, G., Bettella, M., Gomez-Mouton, C., Lanzavecchia, A., Martinez, A.C., Manes, S. and Viola, A. (2005) T cell costimulation by chemokine receptors. Nat. Immunol. 6, 465–471.

Moser, C., Weiner, D.J., Lysenko, E., Bals, R., Weiser, J.N. and Wilson, J.M. (2002) β-Defensin 1 contributes to pulmonary innate immunity in mice. Infect. Immun. 70, 3068–3072.

Nizet, V., Ohtake, T., Lauth, X., Trowbridge, J., Rudisill, J., Dorschner, R.A., Pestonjamasp, V., Piraino, J., Huttner, K. and Gallo, R.L. (2001) Innate antimicrobial peptide protects the skin from invasive bacterial infection. Nature 414, 454–457.

Oppenheim, J.J. and Yang, D. (2005) Alarmins: chemotactic activators of immune responses. Curr. Opin. Immunol. 17, 359–365.

Park, J.S., Gamboni-Robertson, F., He, Q., Svetkauskaite, D., Kim, J.Y., Strassheim, D., Sohn, J.W., Yamada, S., Maruyama, I., Banerjee, A., Ishizaka, A. and Abraham, E. (2006) High mobility group box 1 protein interacts with multiple Toll-like receptors. Am. J. Physiol. Cell. Physiol. 290, C917–C924.

Rosenberger, C.M., Gallo, R.L. and Finlay, B.B. (2004) Interplay between antibacterial effectors: a macrophage antimicrobial peptide impairs intracellular Salmonella replication. Proc. Natl. Acad. Sci. U.S.A. 101, 2422–2427.

Rouhiainen, A., Kuja-Panula, J., Wilkman, E., Pakkanen, J., Stenfors, J., Tuominen, R.K., Lepantalo, M., Carpen, O., Parkkinen, J. and Rauvala, H. (2004) Regulation of monocyte migration by amphoterin (HMGB1). Blood 104, 1174–1182.

Rovere-Querini, P., Capobianco, A., Scaffidi, P., Valentinis, B., Catalanotti, F., Giazzon, M., Dumitriu, I.E., Muller, S., Iannacone, M., Traversari, C., Bianchi, M.E. and Manfredi, A.A. (2004) HMGB1 is an endogenous immune adjuvant released by necrotic cells. EMBO Rep. 5, 825–830.

Rugeles, M.T., Trubey, C.M., Bedoya, V.I., Pinto, L.A., Oppenheim, J.J., Rybak, S.M. and Shearer, G.M. (2003) Ribonuclease is partly responsible for the HIV-1 inhibitory effect activated by HLA alloantigen recognition. AIDS 17, 481–486.

Shi, Y., Evans, J.E. and Rock, K.L. (2003) Molecular identification of a danger signal that alerts the immune system to dying cells. Nature 425, 516–521.

Uesugi, H., Ozaki, S., Sobajima, J., Osakada, F., Shirakawa, H., Yoshida, M. and Nakao, K. (1998) Prevalence and characterization of novel pANCA, antibodies to the high mobility group non-histone chromosomal proteins HMG1 and HMG2, in systemic rheumatic diseases. J. Rheumatol. 25, 703–709.

Wang, H., Bloom, O., Zhang, M., Vishnubhakat, J.M., Ombrellino, M., Che, J., Frazier, A., Yang, H., Ivanova, S., Borovikova, L., Manogue, K.R., Faist, E., Abraham, E., Andersson, J., Andersson, U., Molina, P.E., Abumrad, N.N., Sama, A. and Tracey, K.J. (1999) HMG-1 as a late mediator of endotoxin lethality in mice. Science 285, 248–251.

Wehkamp, J. and Stange, E.F. (2006) A new look at Crohn's disease: breakdown of the mucosal antibacterial defense. Ann. N. Y. Acad. Sci. 1072, 321–331.

Wu, Z., Hoover, D.M., Yang, D., Boulegue, C., Santamaria, F., Oppenheim, J.J., Lubkowski, J. and Lu, W. (2003) Engineering disulfide bridges to dissect antimicrobial and chemotactic activities of human β-defensin 3. Proc. Natl. Acad. Sci. U.S.A. 100, 8880–8885.

Yang, D., Biragyn, A., Hoover, D.M., Lubkowski, J. and Oppenheim, J.J. (2004a) Multiple roles of antimicrobial defensins, cathelicidins, and eosinophil-derived neurotoxin in host defense. Annu. Rev. Immunol. 22, 181–315.

Yang, D., Chen, Q., Rosenberg, H.F., Rybak, S.M., Newton, D.L., Wang, Z.Y., Fu, Q., Tchernev, V.T., Wang, M., Schweitzer, B., Kingsmore, S.F., Patel, D.D., Oppenheim, J.J. and Howard, O.M. (2004b) Human ribonuclease A superfamily members, eosinophil-derived neurotoxin and pancreatic ribonuclease, induce dendritic cell maturation and activation. J. Immunol. 173, 6134–6142.

Yang, D., Chen, Q., Chertov, O. and Oppenheim, J.J. (2000) Human neutrophil defensins selectively chemoattract naïve T and immature dendritic cells. J. Leukoc. Biol. 68, 9–14.

Yang, D., Chen, Q., Schmidt, A.P., Anderson, G.M., Wang, J.M., Wooters, J., Oppenheim, J.J. and Chertov, O. (2000) LL-37, the neutrophil granule- and epithelial cell-derived cathelicidin, utilizes formyl peptide receptor-like 1 (FPRL1) as a receptor to chemoattract human peripheral blood neutrophils, monocytes, and T cells. J. Exp. Med. 192, 1069–1074.

Yang, D., Chen, Q., Su, S.B., Zhang, P., Kurosaka, K., Caspi, R.R., Michalek, S.M., Rosenberg, H.F. and Oppenheim, J.J. (submitted) Eosinophil-derived neurotoxin acts as an alarmin to enhance Th2 immune responses by activating TLR2-MyD88 signal pathway in dendritic cells. (submitted).

Yang, D., Chen, Q., Yang, H., Tracey, K.J., Bustin, M. and Oppenheim, J.J. (2007b) High mobility group box-1 (HMGB1) protein induces the migration and activation of human dendritic cells and acts as an alarmin. J. Leukoc. Biol. 81, 59–66.

Yang, D., Rosenberg, H.F., Chen, Q., Dyer, K.D., Kurosaka, K. and Oppenheim, J.J. (2003) Eosinophil-derived neurotoxin (EDN), an antimicrobial protein with chemotactic activities for dendritic cells. Blood 102, 3396–3403.

Zedler, S. and Faist, E. (2006) The impact of endogenous triggers on trauma-associated inflammation. Curr. Opin. Crit. Care 12, 595–601.

20

Gangliosides as Immunomodulators

Miroslava Potapenko[1], Galina V. Shurin[1], and Joel de León[2]

[1]Department of Pathology, Division of Clinical Immunopathology, University of Pittsburgh, Pittsburgh, PA, USA
[2]Center of Molecular Immunology, Havana, Cuba, yoel@cim.sld.cu

Abstract. Gangliosides are glycosphingolipids expressed at the outer leaflet of the plasmatic membrane of cells from vertebrate organisms. These molecules exert diverse biological functions including modulation of the immune system responses. Aberrant expression of gangliosides has been demonstrated on malignant cells. Besides expression on tumor cell membranes, gangliosides are also shed in the tumor microenvironment and eventually circulate in patients blood. Gangliosides derived from tumors posses the capability to affect the immune system responses by altering the function of lymphocytes and antigen-presenting cells and promoting tumor growth. These molecules can be considered as tumor weapons directed to attack and destroy immunosurveillance mechanisms devoted to control cancer progression.

1. Gangliosides structure

Gangliosides are glycosphingolipid molecules containing two well-defined moieties: a hydrophilic lineal oligosaccharide with different carbohydrate residues containing at least one sialic acid molecule and a hydrophobic ceramide portion containing primarily saturated fatty acids with 16–22 carbon atoms (Figure 1).

Carbohydrate components are mainly glucose, galactose, and hexosamines (usually *N*-acetylgalactosamine). Up to 40 different oligosaccharide structures have been detected in this lipid class to date, and when variation in sialic acid type is taken into account, the number increases to approximately 50 (Ledeen and Yu 1982; Angata and Varki 2002).

In 1941, Klenk for the first time determined the content of gangliosides in white and gray matters of brain (Klenk 1941). Subsequently, gangliosides of varying complexity have been found in virtually every studied vertebrate tissue. In 1942, Klenk educed neuraminic acid from ganglioside (Klenk 1942), but the chemical structure of neuraminic acid had to wait until 1955 to be completely described (Gottschalk 1955). The major types are *N*-acetylneuraminic acid (Neu5Ac) and *N*-glycolylneuraminic acid (Neu5Gc) (Figure 1). The dominant sialic acid form in humans is *N*-acetylneuraminic

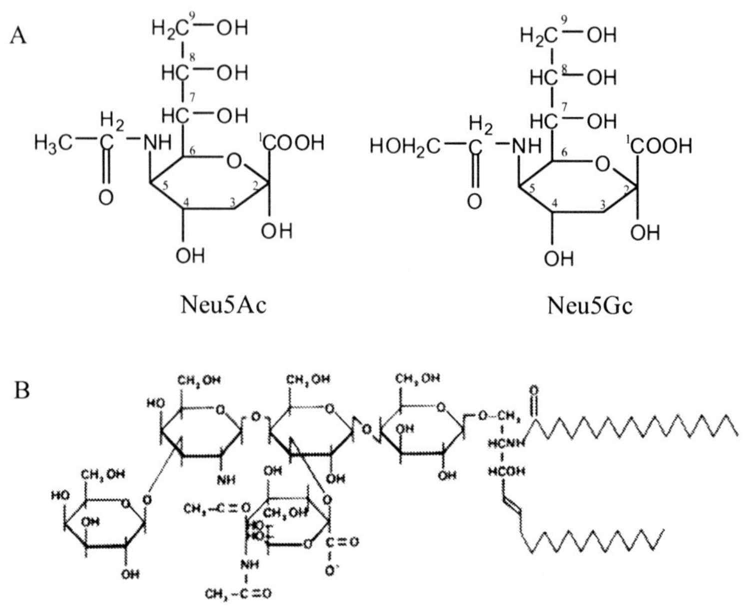

FIGURE 1. Chemical structure of gangliosides. (A) N-acetylated (Neu5Ac) and N-glycolylated (Neu5Gc) neuraminic acid. (B) Chemical structure of the N-acetylated GM1 ganglioside.

acid. *N*-acetylneuraminic acid has been considered as the major type of sialic acid present in human brain. Neu5Gc, a modified form of *N*-acetylneuraminic acid bearing a hydroxyl group at the *N*-acyl position, was found to be a minor species in brain gangliosides of some mammals but is widely distributed in peripheral neural tissues. Some of di- and oligosialogangliosides contain two different types of sialic acid within the same molecule. These acidic units are attached to the oligosaccharide backbone or to each other through an α-ketosidic linkage (Yu and Ledeen 1969).

2. Gangliosides Biosynthesis

The first stages of ganglioside biosynthesis leading to the formation of ceramide are catalyzed by membrane-bound enzymes located in cytosolic fraction of endoplasmic reticulum (Mandon et al. 1992). Ceramides (*N*-acylsphingosins) are synthesized from sphingosine interacting with acyl-CoA in the membranes of endoplasmic reticulum. The next stages of ganglioside formation take place in the Golgi apparatus and endoplasmatic reticulum, including stepwise glycosylation of ceramide, catalyzed by glycosyltransferases and sialyltransferases enzymes (Keenan et al. 1974). Gangliosides are synthesized from glucosylceramide by a sequential addition of nucleotide derivative-activated galactose, sialic acid and *N*-acetylgalactosamine. Ganglioside biosynthesis occurs via four main pathways designated "o" (GA2, GA1, GM1b, and GD1c), "a" (GM3, GM2, GM1a, GD1a, and GT1a), "b" (GD3, GD2, GD1b, GT1b, GQ1b), and "c" series (GT3, GT2, GT1c, GQ1c, and GP1c) (Iber et al. 1992).

3. Functions

Gangliosides are present in the outer leaflet of the cell plasma membrane (Sorice et al. 1997). The hydrophobic ceramide moiety of gangliosides is embedded into the lipid bilayer and the oligosaccharide chain with one or more sialic acids facing the extracellular environment (Steck and Dawson 1974).

Gangliosides are involved in various cellular functions, including signal transduction (Fujitani et al. 2005), regulation of cell proliferation and differentiation (Yogeeswaran and Hakomori 1975), cell–cell recognition, (Feizi 1985), adhesion (Cheresh et al. 1986), and cell death (De Maria et al. 1997; Collel et al. 2001).

Biological functions of membrane-bond gangliosides are associated with the role of these molecules in specialized membrane regions called lipid rafts. These domains are considered as a platform for the formation of multicomponent transduction complexes containing lipid-modified signaling proteins (Pizzo and Viola 2003).

Gangliosides have been considered as the Ca^{2+}-binding co-factor in synaptic transmission (Svennerholm 1980). They might be a co-factor of membrane adenylate cyclase (Partington and Daly 1979). Gangliosides are also present in non-cell-associated forms in plasma and other body fluids. There is constant exchange between cell-associated and non-cell-associated gangliosides (Bergelson 1995). In plasma, most gangliosides are bound to low-density lipoproteins (Senn et al. 1989). The ganglioside spectra in plasma from healthy donors are remarkably stable and exhibit only minor variations. The major ganglioside there is GM3, followed by GD3, GD1a, GM2, GT1b, GD1b, and GQ1b (Senn et al. 1989). In pathological conditions, such as cancer and atherosclerosis, the plasma spectrum of gangliosides is characterized by their elevated levels and another qualitative distribution. For example, elevated levels of GD3 ganglioside were determined in serum of patients with cancer of mammary gland and stomach (Somova et al. 1991). The plasma concentration of disialoganglioside GD2 in patients with neuroblastoma was found to be 50 times higher than in healthy volunteers. The increase in the level of GM3 ganglioside was observed in serum of head and neck carcinoma patients when compared to healthy donors. Furthermore, this increase was correlated with the tumor size (Portoukalian et al. 1989).

4. Changing Ganglioside Profile in Transformed Cells

Each differentiated cell line presents a unique ganglioside pattern. The dramatic changes in glycolipid composition and metabolism, associated with oncogenic transformation, implicate a specific role for membrane glycolipids in regulation of cell growth and interaction. Composition changes results from deletion of complex glycolipids due to a block in synthesis, which frequently leads to accumulation of precursor structures (Hakomori and Kannagi 1983) and synthesis of new glycolipid due to activation of normally unexpressed glycotransferases. Both changes can produce tumor-distinctive glycolipids, some of which may be tumor-associated antigens or markers (Hakomori 1981). Gangliosides, actively shed from tumors, might be inserted into the plasmatic membrane of surrounding cells and inhibit their functions. It has been demonstrated that tumor-derived gangliosides inhibited function of lymphocytes, (Ladisch et al. 1983; Gonwa et al. 1984), monocytes (Ladisch et al. 1984), natural killer cells (Diatlovitskaia et al. 1985), and antigen-presenting cells (Caldwell et al. 2003; Bennaceur et al. 2006).

Tumorigenicity and malignancy of tumor cells depend on their ganglioside content. Inhibition of glucosylceramide synthetase, affecting ganglioside synthesis and reducing the cellular ganglioside content in the ganglioside-rich MEB4 cells, results in markedly reduced tumorigenicity of this cell line (Deng et al. 2000). On the other hand, characterization of murine lymphoma cell lines with markedly different tumorigenic potential demonstrated a strong correlation between levels of ganglioside shedding and tumorigenicity (Ladisch et al. 1987). Changes in ganglioside composition were shown to be specific for different tumor cell types. There is strong link between expression of certain ganglioside species and behavior of corresponding tumors (Hettmer et al. 2005). For example, malignant melanomas overexpress GM3, GM2, GD3, and GD2 gangliosides (Ritter and Livingston 1991). Neuroblastoma is characterized by accumulation of GM2 and GD2 gangliosides (Ladisch et al. 1987; Li and Ladisch 1991; Hettmer et al. 2005). Renal cell carcinomas (RCCs) have elevated levels of GD1a, GM1, and GM2 gangliosides (Hoon et al. 1993).

5. Immunomodulatory Activities of Gangliosides

There is evidence suggesting initiation and maintenance of antitumor immune responses in tumor-bearing hosts (Boon et al. 1997; Wang et al. 2002). However, it is well known that tumor cells produce a number immunosuppressive factors which alowed them to escape immune recognition. One of such potent factors is a group of gangliosides produced and sheded by different malignant cells. It has been demonstrated that gangliosides suppress the generation of cytotoxic T lymphocytes (Merritt et al. 1984), activation and proliferation of T helper cells (Offner et al. 1987), and alter differentiation and maturation of dendritic cells (DCs) (Shurin et al. 2001; Caldwell et al. 2003; Bennaceur et al. 2006).

5.1. T Cells

Finke et al. (2001) reported impaired activation of NF- κB in T cells from patients with RCCs. RCC-derived GM1 and GD1a gangliosides, as well as GM1 and GD1a gangliosides purified from bovine brain, suppressed NF-κB-binding activity in T cells reducing production of IL-2 and IFN-γ. Tumor-derived gangliosides triggered apoptosis in T cells through the mitochondrial pathways, in which cytochrome c release and caspase-9 activation led to initiation of caspase cascade. Activated caspases in turn stimulated non-caspase proteases to degrade the NF-kB transactivating complex, RelA/p50. The loss of NF-kB complex further promoted apoptosis through decreased expression of antiapoptotic genes, including Bcl-x_L and Bcl-2, which are controlled by NF-kB (Thornton et al. 2004). Gangliosides have been found to change cytokine profile in affected cells shifting it from Th1 toward the Th2 phenotype (Tourkova et al. 2005; Crespo et al. 2006). Negative modulation of CD4 molecule on T lymphocytes has been described for both the N-acetylated (Sorice et al. 1995) and the N-glycolylated variants of GM3 ganglioside (de Leon et al. 2006).

5.2. Antigen-Presenting Cells

Gangliosides have been reported to block the nuclear translocation of NF-κB in human monocytes and DCs (Caldwell et al. 2003). Transcriptional factor NF-κB is important for induction and maintenance of antigen-presenting function and responsible for the expression of cell surface molecules and production of cytokines. It has been demonstrated that GD1a ganglioside suppresses the nuclear translocation of NF-κB proteins (p50, p65, RelB, and C-Rel) in DCs. This results in inhibition of expression of different gene products critical for DC maturation and function. Thus, GD1a ganglioside inhibits DC differentiation and production of IL-12 (Bronnum et al. 2005).

Monocytes stimulated with LPS in the presence of GD1a ganglioside exhibited significant inhibition of CD80 and CD40 expression, as well as an impaired release of IL-12 and TNF-α (Caldwell et al. 2003). Furthermore, gangliosides derived from squamous cell carcinoma cell lines down-regulates antigen-presenting machinery components during ex vivo DC generation (Tourkova et al. 2005). GM3 and GD3 gangliosides purified from human melanomas down-regulates Langerhans cell maturation suppressing the expression of co-stimulatatory molecules and maturation markers such as chemokine receptor CCR7 (Bennaceur et al. 2006).

6. Role of Ganglioside Structure in Immunomodulation

It is known that the immunosuppressive properties of gangliosides are mediated by both carbohydrate and ceramide portions (Lengle et al. 1979; Ladisch et al. 1994). The structural features of gangliosides have a particular contribution to the immunosuppressive activity of ganglioside molecules.

Studies on the relationship between features of carbohydrate structure and immunosuppressive activity of gangliosides demonstrated the importance of sialic acid and its position in ganglioside molecules. The most immunosuppressive gangliosides are GM4, GM3, and GQ1b (compared to GM2, GM1, GD2, and GD1b) containing terminal sialic acid at the non-reducing end of oligosaccharide. In addition, immunosuppressive activity of di- and polysialogangliosides is lower in molecules with sialic acid molecules bound to each other (GD1b) than in molecules in which they are bound to separate monosaccharides (GD1a) (Ladisch et al. 1992). Furthermore, a fully intact sialic acid structure is not required for the immunosuppressive activity since both α (2,3)-KDN and α(2,6)-KDN derivatives of GM3 and GM4, which contain a 3-deoxy-D-glycero-D-galacto-2-nonulopyranosic acid instead of *N*-acetylneuraminic acid, are as active as the native forms (Ladisch et al. 1995). The size of oligosaccharide is another important structural component of gangliosides that determine their immunosuppressive activity. Among the four monosialogangliosides—GM4, GM3, GM2, and GM1—the activity diminishes with increased size of the carbohydrate portion (Ladisch et al. 1992). In fact, the most common change in the composition of gangliosides shed by tumors is the simplification of carbohydrate structure (Hakomori and Kannagi 1983).

The studies of the role of ceramide structure in determining ganglioside immunosuppressive activity demonstrated that shorter acyl chain lengths of naturally occurring gangliosides and their synthetic analogs are associated with greater immunosuppressive activity of corresponding gangliosides (Ladisch et al. 1995). Gangliosides that contain shorter-chain fatty acids are known to be preferentially shed into the tumor microenvironment by tumor cells (Li and Ladisch 1991).

7. Concluding Remarks

Strategies for the rational design of antitumor therapy do not consider only differential expression of a specific molecule in malignant cells versus normal counterpart tissue but also the relevance of a selected target for the tumor biology. Alterations in gangliosides content on malignant cells seem to be a response of tumors to a selective pressure exerted by host immunosurveillance mechanisms. In vivo advantages of these modifications in cancer progression need further evaluations. However, in vitro assays clearly demonstrated that tumor-derived gangliosides posses such structural characteristics that promote their shedding from tumor cells and also that increase their immunosuppressive properties. Passive or active immunotherapies for cancer-considering gangliosides as selected target molecules are currently under evaluation with the goal to eliminate malignant cells and/or restore immune response in tumor-bearing patients.

References

Angata, T. and Varki, A. (2002) Chemical diversity in the sialic acids and related alpha-keto acids: an evolutionary perspective. Chem. Rev. 102, 439–469.

Bennaceur, K., Popa, I., Portoukalian, J., Berthier-Vergnes, O. and Peguet-Navarro, J. (2006) Melanoma-derived gangliosides impair migratory and antigen-presenting function of human epidermal Langerhans cells and induce their apoptosis. Int. Immunol. 18, 879–886.

Bergelson, L.D. (1995) Serum gangliosides as endogenous immunomodulators. Immunol. Today 16, 483–486.

Boon, T., Coulie, P.G. and Van den Eynde, B. (1997) Tumor antigens recognized by T cells. Immunol. Today 18, 267–268.

Bronnum, H., Seested, T., Hellgren, L.I., Brix, S. and Frokiaer, H. (2005) Milk-derived GM(3) and GD(3) differentially inhibit dendritic cell maturation and effector functionalities. Scand. J. Immunol. 61, 551–557.

Caldwell, S., Heitger, A., Shen, W., Liu, Y., Taylor, B. and Ladisch, S. (2003) Mechanisms of ganglioside inhibition of APC function. J. Immunol. 171, 1676–1683.

Cheresh, D.A., Pierschbacher, M.D., Herzig, M.A. and Mujoo, K. (1986) Disialogangliosides GD2 and GD3 are involved in the attachment of human melanoma and neuroblastoma cells to extracellular matrix proteins. J. Cell Biol. 102, 688–696.

Colell, A., Garcia-Ruiz, C., Roman, J., Ballesta, A. and Fernandez-Checa, J.C. (2001) Ganglioside GD3 enhances apoptosis by suppressing the nuclear factor-kappa B-dependent survival pathway. FASEB J. 15, 1068–1070.

Crespo, F.A., Sun, X., Cripps, J.G. and Fernandez-Botran, R. (2006) The immunoregulatory effects of gangliosides involve immune deviation favoring type-2 T cell responses. J. Leukoc. Biol. 79, 586–595.

de Leon, J., Fernandez, A., Mesa, C., Clavel, M. and Fernandez, L.E. (2006) Role of tumour-associated N-glycolylated variant of GM3 ganglioside in cancer progression: effect over CD4 expression on T cells. Cancer Immunol. Immunother. 55, 443–450.

De Maria, R., Lenti, L., Malisan, F., d'Agostino, F., Tomassini, B., Zeuner, A., Rippo, M.R. and Testi, R. (1997) Requirement for GD3 ganglioside in CD95- and ceramide-induced apoptosis. Science 277, 1652–1655.

Deng, W., Li, R. and Ladisch, S. (2000) Influence of cellular ganglioside depletion on tumor formation. J. Natl. Cancer Inst. 92, 912–917.

Diatlovitskaia, E.V., Kliuchareva, E.V., Matveeva, V.A., Sinitsyna, E.V. and Akhmed-Zade, A.S. (1985) Effect of gangliosides on the cytotoxic activity of natural killers from Syrian hamsters. Biochemistry (Russia) 50, 1514–1516.

Feizi, T. (1985) Demonstration by monoclonal antibodies that carbohydrate structures of glycoproteins and glycolipids are onco-developmental antigens. Nature 314, 53–57.

Finke, J.H., Rayman, P., George, R., Tannenbaum, C.S., Kolenko, V., Uzzo, R., Novick, A.C. and Bukowski, R.M. (2001) Tumor-induced sensitivity to apoptosis in T cells from patients with renal cell carcinoma: role of nuclear factor-kappaB suppression. Clin. Cancer Res. 7, 940s–946s.

Fujitani, M., Kawai, H., Proia, R.L., Kashiwagi, A., Yasuda, H. and Yamashita, T. (2005) Binding of soluble myelin-associated glycoprotein to specific gangliosides induces the association of p75NTR to lipid rafts and signal transduction. J. Neurochem. 94, 15–21.

Gonwa, T.A., Westrick, M.A. and Macher, B.A. (1984) Inhibition of mitogen- and antigen-induced lymphocyte activation by human leukemia cell gangliosides. Cancer Res. 44, 3467–3470.

Gottschalk, A. (1955) Structural relation between sialic acid, neuraminic acid, and 2-carboxypyrrole. Nature 176, 881–882.

Hakomori, S. (1981) Glycosphingolipids in cellular interaction, differentiation, and oncogenesis. Ann. Rev. Biochem. 50, 733–764.

Hakomori, S. and Kannagi, R. (1983) Glycosphingolipids as tumor-associated and differentiation markers. J. Natl. Cancer Inst. 71, 231–251.

Hettmer, S., Ladisch, S. and Kaucic, K. (2005) Low complex ganglioside expression characterizes human neuroblastoma cell lines. Cancer Lett. 225, 141–149.

Hoon, D.S., Okun, E., Neuwirth, H., Morton, D.L. and Irie, R.F. (1993) Aberrant expression of gangliosides in human renal cell carcinomas. J. Urol. 150, 2013–2018.

Iber, H., Zacharias, C. and Sandhoff, K. (1992) The c-series gangliosides GT3, GT2 and GP1c are formed in rat liver Golgi by the same set of glycosyltransferases that catalyse the biosynthesis of asialo-, a- and b-series gangliosides. Glycobiology 2, 137–142.

Keenan, T.W., Morre, D.J. and Basu, S. (1974) Ganglioside biosynthesis. Concentration of glycosphingolipid glycosyltransferases in Golgi apparatus from rat liver. J. Biol. Chem. 249, 310–315.

Klenk (1941) Uber die ganglioside, eine neue gruppe von zukerhaltigen gehirnlipoiden. Z. Physiol. Chem. 273, 76.

Klenk (1942) Neuraminsaure, das spaltprodukt eines neuen gehirnlipoids. Z. Physiol. Chem. 268, 50.
Ladisch, S., Becker, H. and Ulsh, L. (1992) Immunosuppression by human gangliosides: I. Relationship of carbohydrate structure to the inhibition of T cell responses. Biochimica et Biophysica Acta 1125, 180–188.
Ladisch, S., Gillard, B., Wong, C. and Ulsh, L. (1983) Shedding and immunoregulatory activity of YAC-1 lymphoma cell gangliosides. Cancer Res. 43, 3808–3813.
Ladisch, S., Hasegawa, A., Li, R. and Kiso, M. (1995) Immunosuppressive activity of chemically synthesized gangliosides. Biochemistry. 34, 1197–1202.
Ladisch, S., Kitada, S. and Hays, E.F. (1987) Gangliosides shed by tumor cells enhance tumor formation in mice. J. Clin. Invest. 79, 1879–1882.
Ladisch, S., Li, R. and Olson, E. (1994) Ceramide structure predicts tumor ganglioside immunosuppressive activity. Proc. Natl. Acad. Sci. U.S.A. 91, 1974–1978.
Ladisch, S., Ulsh, L., Gillard, B. and Wong, C. (1984) Modulation of the immune response by gangliosides. Inhibition of adherent monocyte accessory function *in vitro*. J. Clin. Invest. 74, 2074–2081.
Ladisch, S., Wu, Z.L., Feig, S., Ulsh, L., Schwartz, E., Floutsis, G., Wiley, F., Lenarsky, C. and Seeger, R. (1987) Shedding of GD2 ganglioside by human neuroblastoma. Int. J. Cancer 39, 73–76.
Ledeen, R.W. and Yu, R.K. (1982) Gangliosides: structure, isolation, and analysis. Meth. Enzymol. 83, 139–191.
Lengle, E.E., Krishnaraj, R. and Kemp, R.G. (1979) Inhibition of the lectin-induced mitogenic response of thymocytes by glycolipids. Cancer Res. 39, 817–822.
Li, R.X. and Ladisch, S. (1991) Shedding of human neuroblastoma gangliosides. Biochim. Biophys. Acta 1083, 57–64.
Mandon, E.C., Ehses, I., Rother, J., van Echten, G. and Sandhoff, K. (1992) Subcellular localization and membrane topology of serine palmitoyltransferase, 3-dehydrosphinganine reductase, and sphinganine N-acyltransferase in mouse liver. J. Biol. Chem. 267, 11144–11148.
Merritt, W.D., Bailey, J.M. and Pluznik, D.H. (1984) Inhibition of interleukin-2-dependent cytotoxic T lymphocyte growth by gangliosides. Cell. Immunol. 89, 1–10.
Offner, H., Thieme, T. and Vandenbark, A.A. (1987) Gangliosides induce selective modulation of CD4 from helper T lymphocytes. J. Immunol. 139, 3295–3305.
Partington, C.R. and Daly, J.W. (1979) Effect of gangliosides on adenylate cyclase activity in rat cerebral cortical membranes. Mol. Pharmacol. 15, 484–491.
Pizzo, P. and Viola, A. (2003) Lymphocyte lipid rafts: structure and function. Curr. Opin. Immunol. 15, 255–260.
Portoukalian, J., David, M.J., Shen, X., Richard, M. and Dubreuil, C. (1989) Tumor size-dependent elevations of serum gangliosides in patients with head and neck carcinomas. Biochem. Int. 18, 759–765.
Ritter, G. and Livingston, P.O. (1991) Ganglioside antigens expressed by human cancer cells. Semin. Cancer Biol. 2, 401–409.
Senn, H.J., Orth, M., Fitzke, E., Wieland, H. and Gerok, W. (1989) Gangliosides in normal human serum. Concentration, pattern and transport by lipoproteins. Eur. J. Biochem .181, 657–662.
Shurin, G.V., Shurin, M.R., Bykovskaia, S., Shogan, J., Lotze, M.T. and Barksdale, E.M., Jr. (2001) Neuroblastoma-derived gangliosides inhibit dendritic cell generation and function. Cancer Res. 61, 363–369.
Somova, O.G., Tekieva, E.A., Diatlovitskaia, E.V., Bassalyk, L.S. and Bergel'son, L.D. (1991) Ganglioside (GD3) in serum of cancer patients. Vopr. Med. Khim. (Russia) 37, 21–23.

Sorice, M., Parolini, I., Sansolini, T., Garofalo, T., Dolo, V., Sargiacomo, M., Tai, T., Peschle, C., Torrisi, M.R. and Pavan, A. (1997) Evidence for the existence of ganglioside-enriched plasma membrane domains in human peripheral lymphocytes. J. Lipid Res. 38, 969–980.
Sorice, M., Pavan, A., Misasi, R., Sansolini, T., Garofalo, T., Lenti, L., Pontieri, G.M., Frati, L. and Torrisi, M.R. (1995) Monosialoganglioside GM3 induces CD4 internalization in human peripheral blood T lymphocytes. Scand. J. Immunol. 41, 148–156.
Steck, T.L. and Dawson, G. (1974) Topographical distribution of complex carbohydrates in the erythrocyte membrane. J. Biol. Chem. 249, 2135–2142.
Svennerholm, L. (1980) Gangliosides and synaptic transmission. Adv. Exp. Med. Biol. 125, 533–544.
Thornton, M.V., Kudo, D., Rayman, P., Horton, C., Molto, L., Cathcart, M.K., Ng, C., Paszkiewicz-Kozik, E., Bukowski, R., Derweesh, I., Tannenbaum, C.S. and Finke, J.H. (2004) Degradation of NF-kappa B in T cells by gangliosides expressed on renal cell carcinomas. J. Immunol. 172, 3480–3490.
Tourkova, I.L., Shurin, G.V., Chatta, G.S., Perez, L., Finke, J., Whiteside, T.L., Ferrone, S. and Shurin, M.R. (2005) Restoration by IL-15 of MHC class I antigen-processing machinery in human dendritic cells inhibited by tumor-derived gangliosides. J. Immunol. 175, 3045–3052.
Wang, R.F., Zeng, G., Johnston, S.F., Voo, K. and Ying, H. (2002) T cell-mediated immune responses in melanoma: implications for immunotherapy. Crit. Rev. Oncol. Hematol. 43, 1–11.
Yogeeswaran, G. and Hakomori, S. (1975) Cell contact-dependent ganglioside changes in mouse 3T3 fibroblasts and a suppressed sialidase activity on cell contact. Biochemistry 14, 2151–2156.
Yu, R.K. and Ledeen, R. (1969) Configuration of the ketosidic bond of sialic acid. J. Biol. Chem. 244, 1306–1313.

21

Functional Changes of Macrophages Induced by Dimeric Glycosaminylmuramyl Pentapeptide

Anna Ilinskaya, Natalia Oliferuk, Valerii Livov, and Rakhim M. Khaitov

National Research Center Institute of Immunology, Russian Federal Medical Biological Agency, Moscow, Russia, nsoliferuk@inbox.ru

Abstract. Under the influence of dimeric glucosaminylmuramyl pentapeptide (diGMPP), a component of bacterial cell wall, macrophages undergo certain changes similar to those associated with dendritic cell (DC) maturation. The effect of diGMPP on DCs resulted in maturation and expression of CD83. Macrophages treated with diGMPP displayed reduced phagocytic activity and elevated ability to kill ingested bacteria. Reduced phagocytosis may be due to phenotypic changes that occur in macrophages during the maturation process, such as reduced expression of receptors that mediate ingesting of microorganisms (CD16, CD64, and CD11b). Down-regulated expression of pattern-recognizing receptors (TLR2, TLR4, and CD206) was accompanied by elevated expression of antigen-presenting (HLA-DR) and costimulating molecules (CD86 and CD40), similar to alterations observed in maturating DCs. In addition, diGMPP treatment of macrophages resulted in enhanced synthesis of IL-12, TNF-α, and IL-1β.

1. Introduction

Macrophages and dendritic cells (DCs) are professional antigen-presenting cells that serve as an effective link between the innate and adaptive immunity. In normal conditions, DCs are presented either in immature or mature state. Immature DCs are unable to present antigen and stimulate T cells though can ingest antigens by means of phagocytosis, pinocytosis, and receptor-mediated capturing. Mature DCs cease antigens capturing but acquire the ability to present previously ingested antigenic material and to induce cellular response that is due to markedly increased expression of HLA and costimulating molecules (Thery and Amigorena 2001; Rescigno et al. 1999). Bacterial products and proinflammatory cytokines (e.g., IL-1 and TNF-α) stimulate DC maturation (Henderson et al. 1997; Thurnher et al. 1997; Reis e Sousa et al. 1999). In vivo experiments revealed a similar process of macrophage maturation. Resident macrophage from mice peritoneal cavity isolated on the second day following bacillus of calmette and guérin (BCG) infection exhibited high phagocytic activity though poor ability to generate nitric oxide and to produce TNF-α. Murine peritoneal macrophages isolated on the 12th day exhibited low phagocytic activity though produced high levels of reactive nitrogen intermediates and class II major his to compatibility class (MHC) molecules (Hamerman and Aderem 2001). The goal of our study was to demonstrate that macrophages prepared from human peripheral blood monocytes undergo functional changes similar to DCs after their treatment with a bacterial cell wall component, dimeric glucosaminylmuramyl pentapeptide (diGMPP).

2. Experimental Procedures

For isolation of diGMPP, acctonc-dricd Ty2 strain of *Salmonella typhi* (*S. typhi*) cells was twice extracted with 45% aqueous phenol at 65°C–68°C for 30 min. After washing, the pellet was re-suspended in 2% acetic acid and boiled for 2 h to decompose potential traces of LPS. Lipid A contamination was removed by washing in methanol and then in methanol/chloroform (1:3). Obtained cell walls were treated by Lysozyme (Sigma) (50:1, 24 h, 37° C) in 0.1 M ammonium acetate buffer (pH 8.3). The major component, a cross-linked diGMPP, and two minor components, glucosaminylmuramyl tetrapeptide and glucosaminylmuramyl tripeptide, were isolated using G50 chromatography (Pharmacia) equilibrated with 0.05 M ammonium acetate buffer. Quantitative analysis of diGMPP was carried out using amino acid analyzer (Biotronic) after hydrolysis (4 M HCl, 6 h, 100°C). The ratio of GlcN:MurA: Ala:Glu:DAP was 1.0:0.9:2.3:0.9:0.85. The ^{13}C-NMR spectrum data (Bruker 250) confirmed the presence of specific amino acids and monosaccharide residues. Molecular weight of diGMPP (2004 m/z) was determined by MALDI-TOF-mass spectrometry. These data proved the presence of two muramic acid residues, *N*-acetylglucosamin, glutamic acid, diaminopimelic acid, and 6 alanine residues.

Buffy coats from 15 healthy donors 20–50 years old were obtained from the Department of Blood Transfusion, Russian Oncology Research Center, Moscow. Mononuclear cells were isolated by Ficoll–Paque density gradient centrifugation. To deplete platelets, cells were washed five times (130 × *g*, 5 min). Mononuclear cells were left for adhesion in Petri dishes for 1 h in RPMI 1640 medium with 1% AB human serum, 2 mM glutamine, sodium pyruvate, vitamins, and 80 μg/ml gentamycin at 37°C. Non-adherent cells were removed, and adherent monocytes were cultured for 7 days in complete medium with granulocyte-macrophage colony stimulating factor (GM-CSF) (80 ng/ml, Shering-Plough). For DC preparations, monocytes were cultured with GM-CSF and IL-4 (20 ng/ml, R&D). diGMPP (100 μg/ml) was added to cell cultures on day 7 and 24 h later cell-free supernatant were collected and frozen, while cells were analyzed by flow cytometry using the following antibodies: CD16-FITC, CD32-FITC, CD64-FITC, CD83-FITC, CD86-FITC, CD40-FITC, CD54-FITC, HLA-DR, CD11b (Caltag), CD206 (Immunotech), TLR2-PE, and TLR4-PE (eBioscience).

Phagocytic activity of macrophages was assessed as described (Mazurov and Pinegin 1999). Cells (180 × 10^3/well) and FITC-labeled *Staphylococus aureus* (*S. aureus*) (9 × 10^6/well) were incubated for 30 min at 37°C, washed in ice-cold

PBS, and analyzed by FACScan. Bactericidal activity of macrophages was assessed after 3 h of incubation with FITC-labeled bacteria by flow cytometry analysis of PI-stained bacteria released from lyzed cells.

Cytokines were measured by ELISA kits: IL-12 p70 (Pharmingen), IL-1β, and TNF-α (Proteinovy Kontur). Paired Student *t*-test was used to determine the statistical significance of the data. p value <0.05 was chosen for rejection of the null hypothesis.

3. Results

In a first series of experiments, we evaluated the effects of diGMPP on the maturation process of DCs using a CD83 marker known to reflect DCs maturation in vitro (Zhou and Tedder 1996; Lechmann et al. 2002). As a positive control, we used DCs matured with 20 ng/ml TNF-α and 250 ng/ml PgE_2 as described (Jonuleit et al. 1997). The results revealed no differences in CD83 expression on DCs matured with diGMPP or TNF-α + PGE2.

Functional analysis of control and DiGMPP-stimulated macrophages revealed that the percentage of FITC+ cells in cultures of non-stimulated macrophages co-incubated with FITC-*S. aureus* was 87.0 ± 12.2, whereas this value decreased to 73.0 ± 13.2 ($p < 0.05$) in DiGMPP-stimulated cultures. Furthermore, bactericidal ability of control macrophages was 24.0 ± 4.5% and increased to 40.0 ± 17.7 ($p < 0.05$) in diGMPP-stimulated cells. Thus, decreased phagocytic activity of diGMPP-stimulated macrophages was associated with simultaneous increase in their bactericidal activity.

Next, we focused on the effects of DiGMPP on phenotype of cultured macrophages and revealed that DiGMPP markedly down-regulated expression of the pattern-recognition receptor CD206, receptors interacting with peptidoglycane and lipopolysaccharide (TLR2 and TLR4), receptors mediating interactions with immunoglobulins (CD16 and CD64) and complement receptor CD11b. At the same time, expression of CD32 remained unchanged, while expression of costimulatory molecule CD86 and CD40 and CD54 molecules was significantly elevated on macrophages treated with diGMPP. These alterations of macrophage phenotype were seen in all tested donors ($n = 15$) although up-regulation of HLA-DR on macrophages was detected only in 11 donors.

Since cytokine production is an important function of macrophages, we next tested whether DiGMPP affects cytokine synthesis in these cells. Macrophage-derived IL-12 plays a key role in regulating Th1/Th2 balance (Trinchieri 2003). Our results demonstrated that although non-stimulated macrophages did not synthesize IL-12 p70, diGMPP induced IL-12 production by macrophages. Similarly, no release of IL-1β and TNF-α was detected in cultures with non-stimulated macrophages (Table 1). However, addition of DiGMPP resulted in significant up-regulation of their concentrations in macrophage-conditioned medium.

TABLE 1. diGMPP induces cytokine production by macrophages

	IL-12 (pg/ml)	IL-1β (pg/ml)	TNF-α (ng/ml)
Spontaneous	10.0 ± 10.5	8.0 ± 5.4	0.7 ± 1.21
diGMPP	169.0 ± 102.9[1]	556.0 ± 178.7[1]	3.2 ± 1.37[1]

($p < 0.05$)[1]

4. Discussion

We have demonstrated here that diGMPP-stimulated human macrophages experience similar alterations when compared to DCs stimulated by proinflammatory cytokines and PgE_2. Treatment with diGMPP reduced the ability of macrophages to phagocyte *S. aureus* but enhanced their capacity to kill ingested bacteria. Moreover, DiGMPP-stimulated macrophages displayed decreased expression of pattern-recognizing receptors TLR2, TLR4, and CD206, as well as of the receptors involved in phagocytosis, such as CD16, CD64, and CD11b. However, expression of costimulatory molecules and the molecules comprising immunological synapse (CD86, HLA-DR, CD40, CD54) was markedly up-regulated. Finally, diGMPP enhanced production of proinflammatory cytokines IL-12, IL-1β, and TNF-α by macrophages.

Based on these and other data, we speculate that DiGMPP-treated cultures of macrophages comprise both mature and immature macrophages. It seems possible that diGMPP first activates TLR2 and NOD2 causing NF-κB activation and synthesis of proinflammatory cytokines. Indeed, the ability of muramyl peptides to interact with TLR2 and NOD and activate NF-κB was reported (Girardin et al. 2003; Inohara and Nunez 2003). Later on, due to the production of cytokines, macrophages are activated and mature by an autocrine manner (Bastos et al. 2004; Hehlgans and Pfeffer 2005).

There are two ways to activate macrophages, classical and alternative, which result in IFN-γ or IL-4/IL-13 production, respectively (Bach et al. 1997; Nelms et al. 1999). The classical way is also characterized by enhanced expression of MHC class II molecules, reduced expression of mannose receptor, elevated bactericidal action of macrophages, and increased expression of TNF-α (Gordon 2003; Ma et al. 2003). Our data allowed us to suggest that diGMPP- stimulated maturation of macrophages through the classical pathway of activation. Interestingly, it has been shown in vitro that macrophages could digest bacteria and secret muramyl peptides (Majcherczyk et al. 1999; Moreillon and Majcherczyk 2003; Wehner and Gray 1991). Similar processes of bacteria digestion and generation of biologically active muramyl peptides was observed in in vivo setting as well (Kool et al. 1994; Johannsen 1993). Marked immunostimulatory potential of muramyl peptides allowed considering them to be the peculiar vitamins of the immune system (Ellouz et al. 1974).

Acknowledgments

We thank RFFI for financial support of these studies (grant 04-04-48927).

References

Bach, E.A., Aguet, M. and Schreiber, R.D. (1997) The IFN-γ receptor: a paradigm for cytokine receptor signaling. Annu. Rev. Immunol. 15, 563–591.

Bastos, K., Marinho, G., Barboza, R., Russo, M., Alvarez, J. and Lima, M. (2004) What kind of message does IL12/IL-23 bring to macrophages and dendritic cells? Microbes Infect. 6, 630–636.

Ellouz, F, Adam, A, Ciorbaru, R and Lederer, E. (1974) Minimal structure requirements for adjuvant activity of bacterial peptidoglycan derivatives. Biochem. Biophys. Res. Commun. 59, 1317–1325.

Girardin, S.E., Travassos, L.H., Herve, M., Blanot, D., Boneca, I.G., Philpott, D.J., Sansonetti, P.J. and Mengin-Lecreulx, D. (2003) Peptidoglycan molecular requirements allowing detection by NOD1 and NOD2. J. Biol. Chem. 278, 41702–41708.

Gordon, S. (2003) Alternative activation of macrophages. Nat. Rev. 3, 23–35.

Hamerman, J.A. and Aderem, A. (2001) Functional transitions in macrophages during *in vivo* infection with Mycobacterium bovis bacillus Calmette-Guerin. J. Immunol. 167, 2227–2233.

Hehlgans, T. and Pfeffer, K. (2005) The intriguing biology of tumor necrosis factor/tumor necrosis factor receptor superfamily: players, rules and the games. Immunology. 115, 1–20.

Henderson, R.A., Watkins, S.C. and Flynn, J.L. (1997) Activation of dendritic cells following infection with Mycobacterium tuberculosis. J. Immunology 159, 635–643.

Inohara, N. and Nunez, G. (2003) NOD: intracellular protein involved in inflammation and apoptosis. Nat. Rev. 3, 371–382.

Johannsen, L. (1993) Biological properties of bacterial peptidoglycan. APMIS 101, 337–33

Jonuleit, H., Kuhn, U., Muller, G., Steinbrink, G., Paragnik, L., Schmitt, E., Knop, L. and Enk, A.H. (1997) Pro-inflammatory cytokines and prostaglandins induce maturation of potent immunostimulatory dendritic cells under fetal calf serum-free conditions. Eur. J. Immunol. 27, 3135–3142.

Kool, J, De Visser, H. and Gerrits-Boeye, M.Y. (1994) Detection of intestinal flora-derived bacterial antigen combinedes in splenic macrophages of rats. J. Histochem. Cytochem. 42, 1435–1441.

Lechmann, M., Berchtold, S., Hauber, J. and Steinkasserer, A. (2002) CD83 on dendritic cells: more than just a marker for maturation. Trends Immunol. 23, 273–275.

Ma, J., Chen, T., Mandelin, J., Ceponis, A., Miller, N.E., Hukkanen, M., Ma, G.F. and Kottinen, Y.H. (2003) Regulation of macrophage activation. Cell Mol. Life sci. 60, 2334–2346.

Majcherczyk, P.A., Langen, H., Heumann, D., Fountoulakis M., Glauser M.P. and Moreillon P. (1999) Digestion of Streptococcus pneumoniae cell wall with its major peptidoglycan hydrolase releases branched stem peptides carrying proinflammatory activity. J. Biol. Chem. 274, 12537–12543.

Mazurov, D. and Pinegin, B. (1999) Flow cytometry application for assessment of engulf and bactericidial function of peripheral blood granulocytes and monocytes. Immunology (Rus) 9, 154–156.

Moreillon, P. and Majcherczyk, P.A. (2003) Proinflammatory activity of cell wall constituents from gram-positive bacteria. Scand. J. Infect. Dis. 35, 632–641.

Nelms, K., Keegan, A.D., Zamorano, J., Ryan, J.J. and Paul, W.E. (1999) The IL-4 receptor: signaling mechanisms and biological functions. Annu. Rev. Immunol. 17, 701–738.
Reis e Sousa, C., Sher, C. and Kaye, P. (1999) The role of dendritic cells in the induction and regulation of immunity to microbial infection. Curr. Opin. Immunol. 11, 392–399.
Rescigno, M., Granucci, F., Citterio, S., Foti, M. and Ricciardi-Castagnoli, P. (1999) Coordinated events during bacteria-induced DC maturation. Immunol. Today 20, 200–203.
Thery, C. and Amigorena, S. (2001) The cell biology of antigen presentation in dendritic cells. Curr. Opin. Immunol. 13, 45–51.
Thurnher, M., Ramoner, R., Gastl, G., Radmayr, C., Bock, G., Herald, M., Klocker, H. and Bartsch, G. (1997) Bacillus Calmette-Guerin mycobacteria stimulate human blood dendritic cells. Int. J. Cancer. 70, 128–134.
Trinchieri, G. (2003) Interleukin-12 and the regulation of innate resistance and adaptive immunity. Nat. Rev. 3, 133–146.
Wehner, N.G. and Gray, G.R. (1991) *In vitro* stimulation of immune functions by lipids derived from macrophages exposed to bacterial peptidoglycan. J. Immunol. 147, 3595–3600.
Zhou, L.J. and Tedder, T.F. (1996) CD14+ blood monocytes can differentiate into functionally mature CD83+ dendritic cells. Proc. Natl. Acad. Sci. U.S.A. 93, 2588–2592.

New Insights in Immune Regulation

22

Myeloid-Derived Suppressor Cells

Srinivas Nagaraj and Dmitry I. Gabrilovich

H. Lee Moffitt Cancer Center, University of South Florida, Tampa, FL, USA, dmitry.gabrilovich@moffitt.org

Abstract. The development of tumor-specific T cell tolerance is largely responsible for tumor escape. Accumulation of myeloid-derived suppressor cells (MDSCs) in animal tumor models as well as in cancer patients is involved in tumor-associated T cell tolerance. In recent years, it has become increasingly evident that MDSCs bring about antigen-specific T cell tolerance by various mechanisms, which is the focus of this chapter.

1. Introduction

Successful cancer immunotherapy relies on the effective function of antigen-presenting cells (APCs) and T cells. This strategy is based on the concept that the quantitative and qualitative characteristics of a T cell response to an antigen depend on the signals that the T cell receives from an APC. The ability of a T cell to mount an immune response against a foreign pathogen or a cancerous cell forms the basis of immunity. Failure of such a response is one of the major problems of tumor immunology. This failure in part can be explained by the fact that T cells cannot react to the antigen due to the prevailing tumor environment (Wells 2003).

One group of mechanisms affecting tumor escape involves the dysfunction of T cells (Finke et al. 1999). Tumor-induced T cell abnormalities include antigen-specific nonresponsiveness (anergy/tolerance), deletion of T cells by apoptosis, and nonspecific suppression of T cell function. In both animal models of cancer and clinical settings, unresponsiveness to the specific antigens has been shown to be an early event in tumor progression (Gabrilovich and Pisarev 2003). It is now evident that this inadequate function of host immune system is one of the major mechanisms of tumor escape from immune control as well as an important factor limiting the success of cancer immunotherapy. More recently, it has become clear that there is an accumulation of myeloid cells in cancer patients and in mouse tumor models.

2. Myeloid Suppressor Cells

In mice, these myeloid cells are characterized as Gr-1$^+$CD11b$^+$ cells. Myeloid lineage differentiation antigen Gr-1 (Ly6G and C) is expressed on myeloid precursor cells, on granulocytes, and transiently on monocytes (Hestdal et al. 1991). CD11b receptor (Mac-1) is α_M integrin that is expressed on the surface of monocytes/macrophages, dendritic cells (DCs), granulocytes, and activated B and T lymphocytes. Gr-1$^+$CD11b$^+$ cells represent about 30–40% of normal bone marrow cells and only 2–4% of all nucleated normal splenocytes. Morphological analysis demonstrated that these cells are composed of a mixture of myeloid cells such as granulocytes and monocyte–macrophages as well as myeloid precursor cells at various stages of differentiation. In the presence of appropriate growth factors and/or cytokines, Gr-1$^+$ cells from tumor-bearing host could be differentiated in vitro into DCs or macrophages (Bronte et al. 2000; Kusmartsev et al. 2003; Li et al. 2004). It is clear that these cells could differentiate both in lymphoid organs and inside the tumor bed. In the lymphoid organs, these cells differentiate predominantly into APCs including DCs and macrophages, whereas in the tumor microenvironment, they become tumor-associated macrophages and/or endothelial cells. Inoculation of transplantable tumor cells (Subiza et al. 1989; Bronte et al. 1999; Kusmartsev et al. 2000; Gabrilovich et al. 2001) or spontaneous development of tumors in transgenic mice with tissue-restricted expression of oncogenes (Melani et al. 2003) results in marked systemic expansion of these cells. The proportion of this myeloid cell population in spleen of tumor-bearing mice may reach up to 50% of all splenocytes (Kusmartsev et al. 2003). Less impressive but significant transient increase of the Gr-1$^+$CD11b$^+$ cells was also demonstrated in normal mice after immunization with different antigens (Bronte et al. 1998; Cauley et al. 2000; Kusmartsev et al. 2003) or in mice with bacterial and parasitic infections (Mencacci et al. 2002).

Recent data from a number of groups have demonstrated that myeloid cells accumulating in tumor-bearing hosts play an important role in tumor nonresponsiveness by suppressing antigen-specific T cell responses (Kusmartsev et al. 2000; Pandit et al. 2000; Almand et al. 2001; Bronte et al. 2001; Melani et al. 2003; Gabrilovich 2004; Kusmartsev and Gabrilovich 2005). These cells contribute to the failure of immune therapy in patients with advanced cancer and in tumor-bearing mice. Since Gr-1$^+$CD11b$^+$ cell population displays features of undifferentiated myeloid cells and contains precursors of different myeloid cell subsets and have the ability to suppress T cell function, these cells have been termed as myeloid-derived suppressor cells (MDSCs).

In cancer patients, MDSCs are defined as cells that express the common myeloid marker CD33 but lack expression of markers of mature myeloid and lymphoid cells and the MHC class II molecule HLA-DR (Almand et al. 2001). An accumulation of MDSCs was associated with the decreased number of DCs in the peripheral blood of patients with head and neck, lung, or breast cancer (Almand et al. 2000). Advanced-stage cancer was found to promote the accumulation of these cells in the peripheral blood, whereas surgical resection of the tumor decreased the number of MDSCs. A similar effect of tumor resection was observed in mouse tumor models (Salvadori et al. 2000).

Several recent publications pointed to MDSCs as a source of endothelial cells and their direct role in tumor vasculogenesis. Yang and coworkers (2004) have demonstrated that Gr-1$^+$CD11b$^+$ cells could be incorporated into the vascular endothelium, promoting tumor vascularization and tumor progression. Furthermore, Gr-1$^+$CD11b$^+$ cells derived from tumor-bearing mice produced high levels of metalloproteinase-9 (MMP-9), which is involved in the regulation of angiogenesis. Authors suggested that MMP-9 produced by those immature cells regulates the bioavailability of VEGF in tumors and promotes tumor angiogenesis and vascular stability. Selective deletion of MMP-9 in Gr-1$^+$CD11b$^+$ cells eliminated their ability to promote tumor growth and led to the inhibition of tumor formation. The authors observed that Gr-1$^+$CD11b$^+$ cells constituted about 5% of the total cells in tumor tissues and could represent a significant source of endothelial cells inside the tumor. A study by Young (2004) demonstrated that tumors could skew differentiation of CD34$^+$ progenitor cells into endothelial cells. CD34$^+$ cells cultured in the presence of LLC tumor-conditioned medium under conditions that support myeloid lineage cells skewed the differentiation of these precursor cells toward endothelial cells expressing CD31 and CD144. In vitro differentiation of CD34$^+$ cells into endothelial cells was dependent on angiopoietin-1 in the tumor-conditioned medium. Adoptive transfer of LacZ$^+$ CD34$^+$ cells into tumor-bearing mice resulted in the accumulation of LacZ$^+$ cells within tumor mass. Differentiation of CD34$^+$ cells into endothelial cells was confirmed by coexpression of CD31 and CD144 by donor's LacZ$^+$ cells (Young 2004). MDSCs may contribute to tumor growth directly by differentiating toward endothelial cells and by producing proangiogenic factors.

Functional activity of MDSCs involves the inhibition of IFN-γ production by CD8$^+$ T cells in response to peptide epitopes presented by MHC class I in vitro and in vivo (Gabrilovich et al. 2001). This antigen-specific T cell tolerance depends on MHC class I, is not mediated by soluble factors, requires direct cell–cell contact, and is mediated by reactive oxygen species (ROS) (Kusmartsev et al. 2004, 2005). Interestingly, freshly isolated MDSCs were not able to suppress CD4$^+$ T cells, whereas when MDSCs were incubated for several days in vitro, they acquired the ability to eliminate CD4-mediated T cell responses via induction of apoptosis (Bronte et al. 2003).

Accumulation of Gr-1$^+$ MDSCs in tumor-bearing mice is a gradual time-dependent process, which directly correlates with time and tumor mass. Tumor-derived factors play a direct role in dysregulation of myelopoiesis in tumor host, inhibition of DC differentiation, and expansion of MDSCs in peripheral organs. Indeed, several tumor-derived factors including VEGF, GM-CSF, IL-10, IL-6, TGF-β, prostaglandins, and gangliosides have been implicated in this phenomenon. Importantly, these different tumor-derived factors affect myeloid cells at various stages of differentiation. The list of tumor-derived factors that affect myeloid cell differentiation is not complete and is constantly growing.

Several other potential mechanisms for tumor-induced immune suppression mediated by MDSCs have been described. MDSCs have been linked to the induction of T cell dysfunction in cancer through the production of *TGF-β* (Young et al. 1996; Beck et al. 2001; Terabe et al. 2003), *ROS* (Otsuji et al. 1996; Schmielau and Finn 2001; Kusmartsev et al. 2004), *L-arginine metabolism* (Young et al. 1996; Kusmartsev et al. 2000; Pelaez et al. 2001; Bronte et al. 2003; Liu et al. 2003; Rodriguez et al. 2004; Bronte et al. 2005; Zea et al. 2005), and *peroxynitrites* (Kusmartsev et al. 2000; Bronte et al. 2003; De Santo et al. 2005).

3. TGF-β

Early studies of Young and colleagues (Young et al. 1996) have demonstrated that myeloid progenitor cells derived from tumor-bearing mice produced increased amounts of TGF-β. These cells were immune suppressive and inhibited in vitro T cell proliferation induced by anti-CD3 antibodies. TGF-β, but not nitric oxide (NO), mediated the suppression of T cell proliferation by MDSCs. Beck and colleagues (2001) suggested that MDSCs derived from a tumor-bearing host acquire immune suppressive features and prevent cytotoxic T lymphocyte (CTL) response after contact with TGF-β present in blood serum. Terabe and colleagues (2003) described a pathway that might negatively regulate tumor immunity through MDSC-produced TGF-β. These cells were found to be a major source of TGF-β in tumor-bearing mice. They proposed that tumor inoculation in mice induced IL-13 production by $CD4^{+}$ CD1d-restricted T cells. MDSCs express IL-13 receptor, which is required for MDSCs to produce TGF-β that inhibits CTL induction.

4. L-Arginine Metabolism

Arginine metabolism in myeloid cells is linked to tumor-associated T cell dysfunction. L-Arginine serves as a substrate for two enzymes: NO synthase, which generates NO and citrulline, and arginase, which converts L-arginine into urea and L-ornithine. Recent publications of Ochoa's group suggested a close correlation between the availability of arginine and the regulation of T cell proliferation (Rodriguez et al. 2002, 2003). They demonstrated that the increased activity of arginase I in myeloid cells led to enhanced L-arginine catabolism. The shortage of the non-essential amino acid L-arginine regulates T cell function through the modulation of CD3ζ expression (Rodriguez et al. 2002). Tumor growth is associated with upregulated expression and increased activity of arginase I in splenic myeloid cells (Bronte et al. 2003; Liu et al. 2003; Rodriguez et al. 2004), and especially in tumor-associated macrophages (TAMs) that are particularly effective in inhibition of T cell response including CTL- and antigen-induced T cell proliferation (Kusmartsev et al. 2005). Human prostate cancer (Bronte et al. 2005) and various murine tumors (Kusmartsev et al. 2005) have been shown to employ this mechanism to avoid T cell attack. As shown, in the murine tumor model, T cell deletion in tumor site could be mediated by TAMs (Saio et al. 2001; Kusmartsev and Gabrilovich 2005). This effect was dependent on STAT1 signaling, which controls iNOS and arginase I activity in TAMs.

5. Reactive Oxygen Species

Myeloid cells in tumor hosts produce high levels of reactive oxygen species. Oxidative stress, caused by MDSCs derived from tumor-bearing mice, inhibited ζ-chain expression in T cells and inhibited antigen-induced cell proliferation (Otsuji et al. 1996). Recent studies demonstrated that MDSCs freshly isolated from tumor-bearing mice but not their control counterparts were able to inhibit antigen-specific response of $CD8^+$ T cells (Kusmartsev et al. 2005). MDSCs obtained from tumor-bearing mice had significantly higher levels of ROS than MDSCs isolated from tumor-free animals. Since ROS production by MDSCs can be blocked by arginase inhibitors, it appears that arginase I activity played an important role in ROS accumulation in these cells. This suggests that arginase could be involved in mechanisms of T cell inhibition through the generation of ROS and may be linked with the role of arginase I in T cell deletion observed in the tumor site. What could be the potential mechanism of the link between arginase I activity and ROS production? Arginase catalyzes the hydrolysis of L-arginine to urea and L-ornithine. L-Arginine is used by NO synthase as a substrate for generation of NO (Wu and Morris 1998). However, low concentrations of L-arginine result in low NO formation and high generation of superoxide (O_2^-) (reviewed in Boucher et al. 1999). Thus, it is possible that high arginase activity in MDSCs may have lowered the level of L-arginine and resulted in increased production of O_2^- instead of NO. Superoxide itself is very unstable and is converted to H_2O_2 and oxygen. This is consistent with our data showing that in MDSCs ROS accumulates primarily in the form of H_2O_2 but not O_2^- (Kusmartsev et al. 2004).

Inhibition of ROS in MDSCs completely abrogated the negative effect of these cells on T cells. This suggested that MDSCs generated in tumor-bearing hosts could suppress $CD8^+$ T cell response via the release of ROS. Interaction of MDSCs with antigen-specific T cell in the presence of specific but not control antigens resulted in a significant increase of ROS production. That increase was independent of IFN-γ production by T cells but was mediated by integrins CD11b, CD18, and CD29. Blockage of these integrins abrogated ROS production and MDSCs-mediated suppression of $CD8^+$ T cell responses. Importantly, no T cell apoptosis or T cell deletion has been observed (Kusmartsev et al. 2004).

Schmielau and Finn (2001) observed that, in peripheral blood samples from cancer patients, an unusually large number of myeloid cells with a granulocyte phenotype co purified with low-density peripheral blood mononuclear cells (PBMCs). They found that reduced CD3ζ expression and decreased cytokine production by T cells correlated with the presence of activated myeloid cells in the PBMC population. Freshly isolated granulocytes from healthy donors, if activated, could also inhibit cytokine production by T cells. This action was abrogated by the addition of an H_2O_2 scavenger, catalase, implicating H_2O_2 as the effector molecule.

6. Peroxynitrites

Peroxynitrite ($ONOO^-$) is a product of NO and superoxide (O^-_2) reaction. Peroxynitrite is a powerful oxidant that can inhibit T cell activation and proliferation by impairment of tyrosine phosphorylation and apoptotic death. We demonstrated that $ONOO^-$ was involved in T cell inhibition by Gr-1$^+$ MDSCs derived from tumor-bearing mice (Gabrilovich 2004). It has been demonstrated that peroxynitrite production by myeloid cells could play a major role in preventing an antigen-stimulated T cell expansion in tumor-bearing hosts. Bronte and colleagues (2005) reported that human prostatic adenocarcinomas are infiltrated by terminally differentiated CTLs. These lymphocytes, however, are in an unresponsive status. Authors demonstrated the presence of high levels of nitrotyrosines in prostatic tumor infiltrating lymphocytes (TILs), suggesting a local production of peroxynitrites. Restoration of TIL responsiveness to tumor could be achieved by simultaneous inhibition of iNOS and arginase activity. Thus, local peroxynitrite production could represent one of the important mechanisms by which tumor escapes immune response.

Different signaling pathways may alter myeloid cell differentiation and maturation via different surface receptors. It is possible that signaling from various receptors converge on common signal transduction pathway, for instance Jak2/STAT3. Tumor-derived factors can activate the production of ROS and inhibit transcription factors NF-κB. In addition, these pathways may interact with each other. For instance, ROS may activate STAT3 and NF-κB, STAT3 can inhibit NF-κB, and activation of STAT3 may result in an increased synthesis of the members of NADPH complex, which in turn may result in an increased production of ROS. More studies are needed to clarify these complex interactions. In order to improve cancer vaccination strategies and enhance immune response against tumors, it is critically important to identify molecular targets and signaling pathways utilized by tumor-derived factors, which affect the process of differentiation and/or the function of APCs. This probably will be the focus of research in the near future.

7. Improvement of Antitumor Immunity Through MDSC Elimination or Stimulation of Their Differentiation

Since expansion of MDSCs in tumor-bearing mice is associated with profound inhibition of antitumor immune response, it seems logical that elimination of those immune suppressive cells may help enhance the immune defense mechanisms. Indeed, earlier experiments from Schreiber group (Seung et al. 1995) demonstrated that depletion of Gr-1$^+$ cells significantly improved CD8$^+$ T cell immune response and allowed for eradication of the tumor. Berzofsky's group demonstrated that depleting of Gr-1$^+$ myeloid cells or blocking TGF-β in vivo prevented the tumor recurrence, implying that TGF-β produced by MDSCs is necessary for downregulation of tumor immunosurveillance (Terabe et al. 2003).

Another promising approach to reduce the proportion of MDSCs in tumor-bearing hosts might be the use of agents that promote the differentiation of myeloid progenitors.

1-α 25-dihydroxyvitamin D3 (1α,25(OH)2D3), a biologically active metabolite of vitamin D3, is known as a stimulator of myeloid cell differentiation. In a series of publications, Young and colleagues demonstrated that in both clinical and experimental settings vitamin D3 is effective in diminishing the levels of immune suppressive myeloid cells and increases the effectiveness of tumor immunotherapy (Young and Lathers 1999; Wiers et al. 2000; Lathers et al. 2004). When DCs were generated from the $CD34^+$ cells in vitro in the presence of 1α,25(OH)2D3, their antigen-presenting ability was enhanced (Young and Lathers 1999). Treatment of lung cancer-bearing mice with vitamin D3 led to reduced tumor production of GM-CSF and lowered proportion of myeloid immune suppressive cells (Young et al. 1995). Administration of vitamin D3 in combination with adoptive immunotherapy significantly reduced metastases in mice with established tumors, and also reduced metastases and recurrence after surgical excision of the primary tumors (Wiers et al. 2000). Studies in cancer patients with head and neck cancer also demonstrated the ability of 1α,25(OH)2D3 to promote differentiation of myeloid cells and to reduce the proportion of immature myeloid immune suppressive cell population in peripheral blood (Lathers et al. 2004).

Another compound that is able to stimulate differentiation of the myeloid progenitors into myeloid DCs is vitamin A or retinoic acid (Gabrilovich et al. 2001; Hengesbach and Hoag 2004). Mice with vitamin A deficiency (Kuwata et al. 2000) and mice treated with a pan-RAR antagonist (Walkley et al. 2002) show accumulation of $Gr\text{-}1^+CD11b^+$ myeloid cells similar to the MDSCs that accumulate in cancer patients. Physiologic concentrations of all-*trans* retinoic acid (ATRA) induced in vitro differentiation of MDSCs in humans and mice (Gabrilovich et al. 2001; Mirza et al. 2006). In vivo administration of ATRA significantly reduced the presence of MDSCs in two different tested tumor models. This was not caused by direct antitumor effect of ATRA or decreased production of growth factors by tumor cells. Experiments with adoptive transfer demonstrated that ATRA differentiated MDSCs in vivo into mature DCs, macrophages, and granulocytes. Decreased presence of MDSCs in tumor-bearing mice noticeably improved CD4- and CD8-mediated tumor-specific immune response. Combination of ATRA with two different types of cancer vaccines in two different tumor models significantly prolonged the antitumor effect of the treatment (Kusmartsev and Gabrilovich 2003). These data suggest that elimination of MDSCs with ATRA may open an opportunity to improve the effect of cancer vaccines (Sinha et al. 2005).

Immune suppression mediated by MDSCs requires the presence of three factors: MDSCs, activated antigen-specific $CD8^+$ or $CD4^+$ T cells, and tumor-associated antigen. This hypothesis may also explain the difficulties in generating tumor-specific immune response to vaccination in cancer patients. MDSCs in tumor-bearing hosts have full access to tumor-associated antigens used for vaccination and are able to inhibit the very same tumor-specific immune response a vaccination is trying to induce. This underscores the necessity of combining cancer vaccines with strategies to eliminate MDSCs.

Understanding of precise mechanisms that tumors use to affect differentiation of APCs from myeloid precursors and inhibit T cell responses could help develop new approaches to cancer therapy and substantially improve the efficiency of existing cancer vaccination strategies.

References

Almand, B., Clark, J.I., Nikitina, E., English, N.R., Knight, S.C., Carbone, D.P. and Gabrilovich, D.I. (2001) Increased production of immature myeloid cells in cancer patients. A mechanism of immunosuppression in cancer. J. Immunol. 166, 678–689.

Almand, B., Resser, J., Lindman, B., Nadaf, S., Clark, J., Kwon, E., Carbone, D. and Gabrilovich, D. (2000) Clinical significance of defective dendritic cell differentiation in cancer. Clin. Cancer Res. 6, 1755–1766.

Beck, C., Schreiber, K., Schreiber, H. and Rowley, D. (2001) C-kit+ FcR+ myelocytes are increased in cancer and prevent the proliferation of fully cytolytic T cells in the presence of immune serum. Eur. J. Immunol. 33, 19–28.

Boucher, J.L., Moali, C. and Tenu, J.P. (1999) Nitric oxyde biosynthesis, nitric oxide synthase inhibitors and arginase competition for L-arginine utilization. Cell. Mol. Life Sci. 55, 1015–1028.

Bronte, V., Apolloni, E., Cabrelle, A., Ronca, R., Serafini, P., Zamboni, P., Restifo, N. and Zanovello, P. (2000) Identification of a CD11b(+)/Gr-1(+)/CD31(+) myeloid progenitor capable of activating or suppressing CD8(+) T cells. Blood 96, 3838.

Bronte, V., Casic, T., Gri, G., Gallana, K., Borsellino, G., Marrigo, I., Battistini, L., Iafrate, M., Prayer-Galletti, U., Pagano, F. and Viola, A. (2005) Boosting antitumor responses of T lymphocytes infiltrating human prostate cancers. J. Exp. Med. 201, 1257–1268.

Bronte, V., Chappell, D.B., Apolloni, E., Cabrelle, A., Wang, M., Hwu, P. and Restifo, N.P. (1999) Unopposed production of granulocyte-macrophage colony-stimulating factor by tumors inhibits CD8+ T cell responses by dysregulating antigen-presenting cell maturation. J. Immunol. 162, 5728–5737.

Bronte, V., Serafini, P., Appoloni, E. and Zanovello, P. (2001) Tumor-induced immune dysfunctions caused by myeloid suppressor cells. J. Immunother. 24, 431–446.

Bronte, V., Serafini, P., De Santo, C., Marigo, I., Tosello, V., Mazzoni, A., Segal, D.M., Staib, C., Lowel, M., Sutter, G., Colombo, M.P. and Zanovello, P. (2003) IL-4-induced arginase 1 suppresses alloreactive T cells in tumor-bearing mice. J. Immunol. 170, 270–278.

Bronte, V., Wang, M., Overwijk, W., Surman, D., Pericle, F., Rosenberg, S.A. and Restifo, N.P. (1998) Apoptotic death of CD8+ T lymphocytes after immunization: induction of a suppressive population of Mac-1+/Gr-1+ cells. J. Immunol. 161, 5313–5320.

Cauley, L., Miller, E., Yen, M. and Swain, S. (2000) Superantigen-induced CD4 T cell tolerance mediated by myeloid cells and IFN-gamma. J. Immunol. 165, 6056.

De Santo, C., Serafini, P., Marigo, I., Dolcetti, L., Bolla, M., Del Soldato, P., Melani, C., Guiducci, C., Colombo, M., Iezzi, M., Musiani, P., Zanovello, P. and Bronte, V. (2005) Nitroaspirin corrects immune dysfunction in tumor-bearing hosts and promotes tumor eradication by cancer vaccination. Proc. Natl. Acad. Sci. USA 102, 4185–4190.

Finke, J., Ferrone, S., Frey, A., Mufson, A. and Ochoa, A. (1999) Where have all the T cells gone? Mechanisms of immune evasion by tumors. Immunol. Today 20, 158–160.

Gabrilovich, D. (2004) Mechanisms and functional significance of tumour-induced dendritic-cell defects. Nat. Rev. Immunol. 4, 941–952.

Gabrilovich, D. and Pisarev, V. (2003) Tumor escape from immune response: mechanisms and targets of activity. Curr. Drug Targets 4, 525–536.

Gabrilovich, D.I., Velders, M., Sotomayor, E. and Kast, W.M. (2001) Mechanism of immune dysfunction in cancer mediated by immature Gr-1+ myeloid cells. J. Immunol. 166, 5398–5406.

Hengesbach, L. and Hoag, K. (2004) Physiological concentrations of retinoic acid favor myeloid dendritic cell development over granulocyte development in cultures of bone marrow cells from mice. J. Nutr. 134, 2653–2659.

Hestdal, K., Ruscetti, F., Ihle, J., Jacobsen, S., Dubois, C., Kopp, W., Longo, D. and Keller, J. (1991) Characterization and regulation of RB6-8C5 antigen expression on murine bone marrow cells. J. Immunol. 147, 22–28.

Kusmartsev, S., Cheng, F., Yu, B., Nefedova, Y., Sotomayor, E., Lush, R. and Gabrilovich, D.I. (2003) All-trans-retinoic acid eliminates immature myeloid cells from tumor-bearing mice and improves the effect of vaccination. Cancer Res. 63, 4441–4449.

Kusmartsev, S. and Gabrilovich, D.I. (2003) Inhibition of myeloid cell differentiation in cancer: the role of reactive oxygen species. J. Leukoc. Biol. 74, 186–196.

Kusmartsev, S. and Gabrilovich, D.I. (2005) STAT1 signaling regulates tumor-associated macrophage-mediated T cell deletion. J. Immunol. 174, 4880–4891.

Kusmartsev, S., Li, Y. and Chen, S.-H. (2000) Gr-1+ myeloid cells derived from tumor-bearing mice inhibit primary T cell activation induced through CD3/CD28 costimulation. J. Immunol. 165, 779–785.

Kusmartsev, S., Nagaraj, S. and Gabrilovich, D.I. (2005) Tumor-associated CD8+ T cell tolerance induced by bone marrow-derived immature myeloid cells. J. Immunol. 175, 4583–4592.

Kusmartsev, S., Nefedova, Y., Yoder, D. and Gabrilovich, D.I. (2004) Antigen-specific inhibition of CD8+ T cell response by immature myeloid cells in cancer is mediated by reactive oxygen species. J. Immunol. 172, 989–999.

Kuwata, T., Wang, I., Tamura, T., Ponnamperuma, R., Levine, R., Holmes, K., Morse, H., De Luca, L. and Ozato, K. (2000) Vitamin A deficiency in mice causes a systemic expansion of myeloid cells. Blood 95, 3349–3356.

Lathers, D., Clark, J., Achille, N. and Young, M. (2004) Phase 1B study to improve immune responses in head and neck cancer patients using escalating doses of 25-hydroxyvitamin D3. Cancer Immunol. Immunother. 53, 422–430.

Li, Q., Pan, P.Y., Gu, P., Xu, D. and Chen, S.H. (2004) Role of immature myeloid Gr-1+ cells in the development of antitumor immunity. Cancer Res. 64, 1130–1139.

Liu, Y., Van Ginderachter, J., Brys, L., De Baetselier, P., Raes, G. and Geldhof, A. (2003) Nitric oxide-independent CTL suppression during tumor progression: association with arginase-producing (M2) myeloid cells. J. Immunol. 170, 5064–5074.

Melani, C., Chiodoni, C., Forni, G. and Colombo, M.P. (2003) Myeloid cell expansion elicited by the progression of spontaneous mammary carcinomas in c-erbB-2 transgenic BALB/c mice suppresses immune reactivity. Blood 102, 2138–2145.

Mencacci, A., Montagnoli, C., Bacci, A., Cenci, E., Pitzurra, L., Spreca, A., Kopf, M., Sharpe, A. and Romani, L. (2002) CD80+Gr-1+ myeloid cells inhibit development of antifungal Th1 immunity in mice with candidiasis. J. Immunol. 169, 3180–3190.

Mirza, N., Fishman, M., Fricke, I., Dunn, M., Neuger, A.M., Frost, T.J., Lush, R.M., Antonia, S. and Gabrilovich, D.I. (2006) All-trans-retinoic acid improves differentiation of myeloid cells and immune response in cancer patients. Cancer Res. 66, 9299–9307.

Otsuji, M., Kimura, Y., Aoe, T., Okamoto, Y. and Saito, T. (1996) Oxidative stress by tumor-derived macrophages suppresses the expression of CD3 zeta chain of T cell receptor complex and antigen-specific T cell responses. Proc. Natl. Acad. Sci. USA 93, 13119–13124.

Pandit, R., Lathers, D., Beal, N., Garrity, T. and Young, M. (2000) CD34+ immune suppressive cells in the peripheral blood of patients with head and neck cancer. Ann. Otol. Rhinol. Laryngol. 109, 749–754.

Pelaez, B., Campillo, J., Lopez-Asenjo, J. and Subiza, J. (2001) Cyclophosphamide induces the development of early myeloid cells suppressing tumor growth by a nitric oxide-dependent mechanism. J. Immunol. 166, 6608.

Rodriguez, P.C., Quiceno, D.G., Zabaleta, J., Ortiz, B., Zea, A.H., Piazuelo, M.B., Delgado, A., Correa, P., Brayer, J., Sotomayor, E.M., Antonia, S., Ochoa, J.B. and Ochoa, A.C. (2004)

Arginase I production in the tumor microenvironment by mature myeloid cells inhibits T cell receptor expression and antigen-specific T cell responses. Cancer Res. 64, 5839–5849.

Rodriguez, P.C., Zea, A.H., Culotta, K.S., Zabaleta, J., Ochoa, J.B. and Ochoa, A.C. (2002) Regulation of T cell receptor CD3zeta chain expression by L-arginine. J. Biol. Chem. 277, 21123–21129.

Rodriguez, P.C., Zea, A.H., DeSalvo, J., Culotta, K.S., Zabaleta, J., Quiceno, D.G., Ochoa, J.B. and Ochoa, A.C. (2003) L-Arginine consumption by macrophages modulates the expression of CD3 zeta chain in T lymphocytes. J. Immunol. 171, 1232–1239.

Saio, M., Radoja, S., Marino, M. and Frey, A.B. (2001) Tumor-infiltrating macrophages induce apoptosis in activated CD8(+) T cells by a mechanism requiring cell contact and mediated by both the cell-associated form of TNF and nitric oxide. J. Immunol. 167, 5583–5593.

Salvadori, S., Martinelli, G. and Zier, K. (2000) Resection of solid tumors reverses T cell defects and restores protective immunity. J. Immunol. 164, 2214.

Schmielau, J. and Finn, O.J. (2001) Activated granulocytes and granulocyte-derived hydrogen peroxide are the underlying mechanism of suppression of T cell function in advanced cancer patients. Cancer Res. 61, 4756–4760.

Seung, L., Rowley, D., Dubeym, P. and Schreiber, H. (1995) Synergy between T cell immunity and inhibition of paracrine stimulation causes tumor rejection. Proc. Natl. Acad. Sci. USA 92, 6254–6258.

Sinha, P., Clements, V. and Ostrand-Rosenberg, S. (2005) Reduction of myeloid-derived suppressor cells and induction of M1 macrophages facilitate the rejection of established metastatic disease. J. Immunol. 174, 636–645.

Subiza, J., Vinuela, J., Rodriguez, R. and De la Concha, E. (1989) Development of splenic natural suppressor (NS) cells in Ehrlich tumor-bearing mice. Int. J. Cancer 44, 307–314.

Terabe, M., Matsui, S., Park, J.M., Mamura, M., Noben-Trauth, N., Donaldson, D.D., Chen, W., Wahl, S.M., Ledbetter, S., Pratt, B., Letterio, J.J., Paul, W.E. and Berzofsky, J.A. (2003) Transforming growth factor-beta production and myeloid cells are an effector mechanism through which CD1d-restricted T cells block cytotoxic T lymphocyte-mediated tumor immunosurveillance: abrogation prevents tumor recurrence. J. Exp. Med. 198, 1741–1752.

Walkley, C., Yuan, Y., Chandraratna, R. and McArthur, G. (2002) Retinoic acid receptor antagonism *in vivo* expands the numbers of precursor cells during granulopoiesis. Leukemia 16, 1763–1772.

Wells, A.D. (2003) Cell-cycle regulation of T cell responses—novel approaches to the control of alloimmunity. Immunol. Rev. 196, 25–36.

Wiers, K., Lathers, D., Wright, M. and Young, M. (2000) Vitamin D3 treatment to diminish the levels of immune suppressive CD34+ cells increases the effectiveness of adoptive immunotherapy. J. Immunother. 23, 115–124.

Wu, G. and Morris, S.M. (1998) Arginine metabolism: nitric oxide and beyond. Biochem. J. 336, 1–17.

Yang, L., DeBusk, L., Fukuda, K., Fingleton, B., Green-Jarvis, B., Shyr, Y., Matrisian, L., Carbone, D. and Lin, P. (2004) Expansion of myeloid immune suppressor Gr+CD11b+ cells in tumor-bearing host directly promotes tumor angiogenesis. Cancer Cell 6, 409–421.

Young, M. (2004) Tumor skewing of CD34+ progenitor cell differentiation into endothelial cells. Int. J. Cancer 109, 516–524.

Young, M., Ihm, J., Lozano, Y., Wright, M. and Prechel, M. (1995) Treating tumor-bearing mice with vitamin D3 diminishes tumor-induced myelopoiesis and associated immunosuppression, and reduces tumor metastasis and recurrence. Cancer Immunol. Immunother. 41, 37–45.

Young, M.R. and Lathers, D.M. (1999) Myeloid progenitor cells mediate immune suppression in patients with head and neck cancers. Int. J. Immunopharmacol. 21, 241–252.

Young, M.R.I., Wright, M.A., Matthews, J.P., Malik, I. and Pandit, R. (1996) Suppression of T cell proliferation by tumor-induced granulocyte-macrophage progenitor cells producing transforming growth factor-β and nitric oxide. J. Immunol. 156, 1916–1921.

Zea, A.H., Rodriguez, P.C., Atkins, M.B., Hernandez, C., Signoretti, S., Zabaleta, J., McDermott, D., Quiceno, D., Youmans, A., O'Neill, A., Mier, J. and Ochoa, A.C. (2005) Arginase-producing myeloid suppressor cells in renal cell carcinoma patients: a mechanism of tumor evasion. Cancer Res. 65, 3044–3048.

23

The Lytic NK Cell Immunological Synapse and Sequential Steps in Its Formation

Jordan S. Orange

University of Pennsylvania School of Medicine, Department of Pediatrics, The Children's Hospital of Philadelphia, Philadelphia, PA, USA, Orange@mail.med.upenn.edu

Abstract. Natural killer (NK) cells are lymphocytes of the innate immune system that are critical in host defense. They are best known for their ability to mediate cytotoxicity, which involves a coordinated series of events resulting in the directed secretion of lytic granules onto a target cell. This process requires the formation of an immunological synapse in NK cells. The NK cell immunological synapse involves the reorganization of the actin cytoskeleton and clustering of certain cell surface receptors in the NK cell at the interface with the target cell. The lytic NK cell immunological synapse, specialized for mediating cytotoxicity, is further distinguished by the polarization of lytic granules, which are then secreted through this region onto the target cell. These events unfold in a definitive sequence and lead to critical checkpoints that provide regulatory control at specific stages in the formation of the NK cell lytic synapse.

1. Introduction—Natural Killer Cells

Natural killer (NK) cells are lymphocytes that mediate cytotoxic activity, proliferate, provide costimulation to other cells, produce cytokines, and participate in inflammatory responses after being activated through receptors encoded in their germline DNA (Lanier 2005; Moretta et al. 2002). Thus, unlike T cells and B cells, they do not undergo genetic recombination events to attain specificity and are considered part of the innate immune system. NK cells are the most abundant innate immune lymphocytes representing 5–15% of the total human peripheral blood lymphocyte pool (Comans-Bitter et al. 1997). They do not express a T cell or B cell receptor complex and are characterized by the expression of a number of cell surface molecules including CD56 (NCAM1), the CD158 family (KIR or killer cell immunoglobulin-like receptors), the NKG2 family of C-type lectin-like receptors, and the natural cytotoxicity receptor family (NCR—CD335, CD336, and CD337) among others (reviewed in Orange and Ballas 2006). Importantly, the major subset of human NK cells expresses high levels of the pore-forming molecule perforin, which is contained within specialized organelles that serve as secretory lysosomes (also known as lytic granules). Lytic granules also contain granzymes, granulysins, and other molecules that can facilitate cell death. The directed secretion of lytic granules onto a neighboring cell enables NK cells to perform their best-known function: cell-mediated cytotoxicity.

NK cells function in the surveillance and destruction of tumor cells, as well as in protection against microbial pathogens (reviewed in Miller 2001; Tay et al. 1998). Induction of NK cell activity has been observed during infection in humans and other mammals (Biron et al. 1999; Tay et al. 1998) and the importance of NK cells in defense against viruses in particular is substantiated by the large number of specific strategies that viruses have evolved to evade NK cells (Orange et al. 2002a). Definitive roles for NK cells in protection against viral infection have been demonstrated experimentally using both the animal model depleted of NK cells and that genetically deficient in NK cells, as well as by the fact that enhancement of NK cell activities can result in favorable outcomes (Biron et al. 1999).

In humans, the importance of NK cells has been brought to light by a variety of NK cell deficiencies, which are generally associated with susceptibility to infection (reviewed in Orange 2002, 2006). Some of these human NK cell deficiencies occur in the context of a broader immunodeficiency syndrome that affects other components of immunity, but others appear to affect NK cells in isolation. This latter category of isolated NK cell deficiencies is very rare, but informative. In particular, there appears to be a definitive association with susceptibility to severe or recurrent infection with herpes viruses in these patients. Although a direct causal relation for NK cells in defense against viruses cannot be established in humans, in vivo enhancement of NK cell activities has been shown to improve cytotoxicity and antiviral defense (Fehniger et al. 2002; Smith 2001).

2. NK Cell Function

NK cell function results from the signals generated after the ligation of one or more NK cell receptors having various specificities. Since NK cell receptors will induce either activating or inhibitory signals, a critical balance between these is exploited to induce or restrain NK cell functions (Lanier 2005). The best known of the inhibitory receptors are the KIR family that recognize class I MHC and block activation signaling by inducing dephosphorylation (Moretta and Moretta 2004). To induce cytotoxic function and the secretion of lytic granules (also known as degranulation), NK cell activating receptors must be ligated in excess of NK cell inhibitory receptors. The exact ratios of ligated activation receptor to ligated inhibitory receptors required for degranulation have not been established and are likely to depend on the particular receptors ligated.

Experimentally, the ligation of some activating receptors in isolation can induce degranulation. These receptors include members of the NCR family, NKp30 (CD337) and NKp46 (CD335), which utilize signaling partners that contain immunotyrosine-based activation motifs (ITAMs). In the case of NKp30 and NKp46, the signaling partner is CD3ζ (Pende et al. 1999; Pessino et al. 1998), but other activation receptors can utilize different molecules, such as FcεRIγ or DAP12. Signals generated by these ITAM-containing partners then recruit kinases such as Syk or Zap70, which facilitate further downstream events that induce degranulation. Other receptors that do not partner with ITAM-containing molecules can also participate in inducing NK cell cytotoxicity or cytokine production and even degranulation under appropriate circumstances, but more commonly serve a costimulatory function. This group includes the C-type lectin-like NKG2D (Lanier 2005) and members of the CD2 subgroup of the immunoglobulin superfamily (Bryceson et al. 2005). These receptors utilize a variety of signaling intermediates including DAP10 (by NKG2D). DAP10 can induce phosphatidylinositol-3-kinase activity (Wu et al. 1999) but contributes to NK cell activation most likely by amplifying other concomitant signals. Thus, signals resulting from both ITAM-recruiting activating receptors and costimulatory receptors promote NK cell functions in most cases.

3. NK Cell Cytoskeleton and the Immunological Synapse

A major paradigm of NK cell function dependent on activating receptors is the immunological synapse (IS). The IS defines the dynamic arrangement of molecules at interface between an immune cell and the cell that it is engaging (Davis and Dustin 2004). To form an IS and ultimately generate functions such as cytotoxicity, NK cells rely upon rearrangements of the cell cytoskeleton. The cytoskeleton consists of structural and regulatory components that maintain or alter cell shape, approximate structures within cells, traffic specific organelles, and coordinate endocytosis and secretion. Key amongst the cytoskeletal proteins are actin, which is the chief component of the microfilaments (Pollard and Borisy 2003), and tubulin, which constitutes the microtubules (Desai and Mitchison 1997). These proteins have specific characteristics in immune cells (Vicente-Manzanares and Sanchez-Madrid 2004) and can be arranged to promote cell function as well as rearranged in response to appropriate signals to induce a different cell function. The dependence of NK cells on the cytoskeleton was originally demonstrated through the use of small-molecule inhibitors of cytoskeletal elements that interfere with NK cell activity. In particular, inhibition of actin filament reorganization by cytochalasins and depolymerization of microtubules by colchicine can completely abrogate NK cell cytotoxicity in a concentration-dependent manner (Ito et al. 1989; Katz et al. 1982; Lavie et al. 1985; Quan et al. 1982). Concentrations of these inhibitors that block cytotoxicity do not reduce NK cell conjugation with target cells (Ito et al. 1989; Katz et al. 1982). A key illustration of the importance of the cytoskeleton in NK cell cytotoxicity is that a major target of inhibitory KIR signaling is the central cytoskeletal activator Vav1 (Stebbins et al. 2003). Similarly, inhibitory signaling induced by a different NK cell inhibitory receptor, NKG2A, has also been shown to disrupt filamentous actin (F-actin) reorganization (Masilamani et al. 2006). Thus, NK cell inhibition can occur through the specific blockade of NK cell cytoskeletal reorganization that would be induced normally by an activation receptor.

When visualized directly, F-actin networks and the microtubule organizing center (MTOC) accumulate at the NK cell IS formed with a susceptible target cell (Blom et al. 2001; Carpen et al. 1983; Graham et al. 2006; Kupfer et al. 1983; Lou et al.

2001; McCann et al. 2003; Poggi et al. 1996; Radosevic et al. 1995; Sancho et al. 2000; Vyas et al. 2001). The collection of molecules at the IS is defined as the supramolecular activation cluster (SMAC) and is organized into a peripheral SMAC (pSMAC) and central SMAC (cSMAC). One purpose of the cytoskeletal rearrangements at the activating NK cell IS is to deliver lytic granules directionally via microtubules to the IS so that they can be secreted onto a target cell and mediate cytotoxicity (Davis and Dustin 2004; Trambas and Griffiths 2003). This particular type of IS defines a lytic NK cell IS and is where the directed secretion of lytic granules occurs (Bryceson et al. 2005; Davis et al. 1999; Orange et al. 2002b).

In order for focused actin and microtubule reorganization to occur in an NK cell after it binds to a susceptible target cell, a number of cytoskeletal events must unfold. In the case of actin, monomers are joined to existing actin filaments at 70° angles to create branch points via the Arp2/3 complex (Goley and Welch 2006). These branches are essential for actin reorganization and require the Wiskott–Aldrich syndrome (WAS) protein (WASp). WASp approximates Arp2/3 and an actin monomer and thus enables branch formation. WASp can be found in an inactive, "closed," or autoinhibited conformation and can be "opened" to mediate function. At least 22 proteins have been found to interact with WASp under different conditions (reviewed in Orange et al. 2004). This creates a complex protein unit required to generate functional alteration of the cytoskeleton, which can be referred to as the "actinosome".

4. Sequential Steps in Formation and Function of the Lytic NK Cell IS

Formation of the lytic NK cell IS requires that cytoskeletal events required for IS generation in NK cells occur in a sequential manner with specific stages and checkpoints (Bryceson et al. 2005; McCann et al. 2003; Orange et al. 2003; Vyas et al. 2004; Wulfing et al. 2003). Based on these data and others, the specific stages in formation and function of the lytic NK cell IS can be considered to include but are not limited to a number of steps. These are: **(1)** *adhesion* receptor ligation and resulting *signaling*; **(2)** activation receptor *ligation*; **(3)** activation receptor *signaling*; **(4)** *actinosome* formation and F-actin rearrangement; **(5)** receptor *clustering*; **(6)** possible additional activation receptor ligation and signaling (to result in an *amplification* loop by going back to step 4); **(7)** *MTOC movement*; **(8)** motor protein (such as kinesin)-dependent *granule movement* along microtubules to the IS; **(9)** granule *transit* through the IS, *fusion* with the cell membrane, and *release* of contents; and **(10)** IS *down modulation* (Figure 1). Study of the IS in T cells also reveals the existence of these specific steps in IS formation and function including (1) adhesion and signaling (Freiberg et al. 2002; Lee et al. 2002); (2) receptor ligation (Grakoui et al. 1999); (3) activation signaling (Barda-Saad et al. 2005; Krummel et al. 2000); (4) actinosome formation (Badour et al. 2003); (5) receptor clustering (Wulfing and Davis 1998); (6) amplification (Mossman et al. 2005; Purtic et al. 2005); (7) MTOC movement (Kuhn and Poenie 2002; Stinchcombe et al. 2006); (8) motor protein-dependent granule movement (Burkhardt et al. 1993; Clark et al. 2003);

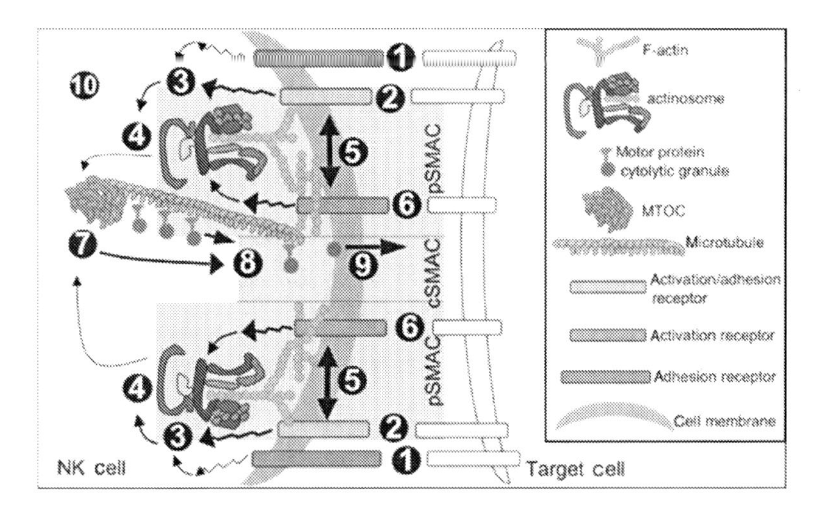

FIGURE 1. Proposed Sequential steps in the Formation and function of the secretory immunological synapse (IS). The specific steps are explained by numerical sequence in the text. The symbols are defined in the legend box on the right. The Natural killer (NK) cell is represented on the left and the target cell is on the right (in outline). The rectangles in the target cell represent ligands for the NK cell receptors. Jagged arrows represent intracellular signaling pathways and standard arrows depict influence or directionality.

(9) granule fusion/release (Feldmann et al. 2003); and (10) downmodulation (Lee et al. 2003). Although the results in T cells are not always the same as those in NK cells, they do offer parallels that underscore a series of distinct steps in IS formation. Importantly, the lytic NK cell IS is a sequential process with many potential checkpoints that can regulate cytolytic function.

5. Conclusions

Through their cytolytic activity NK cells participate in host defense and immunoregulation. Normal function of the lytic NK cell IS is required for both. Basic investigation of NK cell function and the cell biology of NK cell IS have provided insight into how these critical processes are regulated. This is instrumental in understanding diseases in which the process of NK cell IS formation is interrupted as well formulating strategies to improve formation and function of the lytic IS when more would be advantageous. As there are likely to be differences between NK cells and other types of cytolytic cells, specific studies of the lytic NK cell IS are important and may uncover regulatory mechanisms of general importance not found in T cells.

Acknowledgments

This work was supported by NIH grant AI055602 and a faculty development award from the Education Research Trust of the American Academy of Allergy and Immunology.

References

Badour, K., Zhang, J., Shi, F., McGavin, M.K., Rampersad, V., Hardy, L.A., Field, D. and Siminovitch, K.A. (2003) The Wiskott-Aldrich syndrome protein acts downstream of CD2 and the CD2AP and PSTPIP1 adaptors to promote formation of the immunological synapse. Immunity 18, 141–154.

Barda-Saad, M., Braiman, A., Titerence, R., Bunnell, S.C., Barr, V.A. and Samelson, L.E. (2005) Dynamic molecular interactions linking the T cell antigen receptor to the actin cytoskeleton. Nat. Immunol. 6, 80–89.

Biron, C.A., Nguyen, K.B., Pien, G.C., Cousens, L.P. and Salazar-Mather, T.P. (1999) Natural killer cells in antiviral defense: function and regulation by innate cytokines. Annu. Rev. Immunol. 17, 189–220.

Blom, W.M., de Bont, H.J., Meijerman, I., Kuppen, P.J., van Der Meulen, H., Mulder, G.J. and Nagelkerke, J.F. (2001) Remodeling of the actin cytoskeleton of target hepatocytes and NK cells during induction of apoptosis. Cell. Motil. Cytoskeleton 49, 78–92.

Bryceson, Y.T., March, M.E., Barber, D.F., Ljunggren, H.G. and Long, E.O. (2005) Cytolytic granule polarization and degranulation controlled by different receptors in resting NK cells. J. Exp. Med. 202, 1001–1012.

Burkhardt, J.K., McIlvain, J.M., Jr., Sheetz, M.P. and Argon, Y. (1993) Lytic granules from cytotoxic T cells exhibit kinesin-dependent motility on microtubules in vitro. J. Cell. Sci. 104, 151–162.

Carpen, O., Virtanen, I., Lehto, V.P. and Saksela, E. (1983) Polarization of NK cell cytoskeleton upon conjugation with sensitive target cells. J. Immunol. 131, 2695–2698.

Clark, R.H., Stinchcombe, J.C., Day, A., Blott, E., Booth, S., Bossi, G., Hamblin, T., Davies, E.G. and Griffiths, G.M. (2003) Adaptor protein 3-dependent microtubule-mediated movement of lytic granules to the immunological synapse. Nat. Immunol. 4, 1111–1120.

Comans-Bitter, W.M., de Groot, R., van den Beemd, R., Neijens, H.J., Hop, W.C., Groeneveld, K., Hooijkaas, H. and van Dongen, J.J. (1997) Immunophenotyping of blood lymphocytes in childhood. Reference values for lymphocyte subpopulations. J. Pediatr. 130, 388–393.

Davis, D.M., Chiu, I., Fassett, M., Cohen, G.B., Mandelboim, O. and Strominger, J.L. (1999) The human natural killer cell immune synapse. Proc. Natl. Acad. Sci. USA 96, 15062–15067.

Davis, D.M. and Dustin, M.L. (2004) What is the importance of the immunological synapse? Trends Immunol. 25, 323–327.

Desai, A. and Mitchison, T.J. (1997) Microtubule polymerization dynamics. Annu. Rev. Cell Dev. Biol. 13, 83–117.

Fehniger, T.A., Cooper, M.A. and Caligiuri, M.A. (2002) Interleukin-2 and interleukin-15: immunotherapy for cancer. Cytokine Growth Factor Rev. 13, 169–183.

Feldmann, J., Callebaut, I., Raposo, G., Certain, S., Bacq, D., Dumont, C., Lambert, N., Ouachee-Chardin, M., Chedeville, G. and Tamary, H. (2003) Munc13-4 is essential for cytolytic granules fusion and is mutated in a form of familial hemophagocytic lymphohistiocytosis (FHL3). Cell 115, 461–473.

Freiberg, B.A., Kupfer, H., Maslanik, W., Delli, J., Kappler, J., Zaller, D.M. and Kupfer, A. (2002) Staging and resetting T cell activation in SMACs. Nat. Immunol. 3, 911–917.

Goley, E.D. and Welch, M.D. (2006) The ARP2/3 complex: an actin nucleator comes of age. Nat. Rev. Mol. Cell Biol. 7, 713–726.

Graham, D.B., Cella, M., Giurisato, E., Fujikawa, K., Miletic, A.V., Kloeppel, T., Brim, K., Takai, T., Shaw, A.S., Colonna, M. and Swat, W. (2006) Vav1 controls DAP10-mediated natural cytotoxicity by regulating actin and microtubule dynamics. J. Immunol. 177, 2349–2355.

Grakoui, A., Bromley, S.K., Sumen, C., Davis, M.M., Shaw, A.S., Allen, P.M. and Dustin, M.L. (1999) The immunological synapse: a molecular machine controlling T cell activation. Science 285, 221–227.

Ito, M., Tanabe, F., Sato, A., Ishida, E., Takami, Y. and Shigeta, S. (1989) Inhibition of natural killer cell-mediated cytotoxicity by ML-9, a selective inhibitor of myosin light chain kinase. Int. J. Immunopharmacol. 11, 185–190.

Katz, P., Zaytoun, A.M. and Lee, J.H., Jr. (1982) Mechanisms of human cell-mediated cytotoxicity. III. Dependence of natural killing on microtubule and microfilament integrity. J. Immunol. 129, 2816–2825.

Krummel, M.F., Sjaastad, M.D., Wulfing, C. and Davis, M.M. (2000) Differential clustering of CD4 and CD3zeta during T cell recognition. Science 289, 1349–1352.

Kuhn, J.R. and Poenie, M. (2002). Dynamic polarization of the microtubule cytoskeleton during CTL-mediated killing. Immunity 16, 111–121.

Kupfer, A., Dennert, G. and Singer, S.J. (1983) Polarization of the Golgi apparatus and the microtubule-organizing center within cloned natural killer cells bound to their targets. Proc Natl. Acad. Sci. USA 80, 7224–7228.

Lanier, L.L. (2005) NK cell recognition. Annu. Rev. Immunol. 23, 225–274.

Lavie, G., Leib, Z. and Servadio, C. (1985) The mechanism of human NK cell-mediated cytotoxicity. Mode of action of surface-associated proteases in the early stages of the lytic reaction. J. Immunol. 135, 1470–1476.

Lee, K.H., Dinner, A.R., Tu, C., Campi, G., Raychaudhuri, S., Varma, R., Sims, T.N., Burack, W.R., Wu, H., Wang, J. and Shaw, A.S. (2003) The immunological synapse balances T cell receptor signaling and degradation. Science 302, 1218–1222.

Lee, K.H., Holdorf, A.D., Dustin, M.L., Chan, A.C., Allen, P.M. and Shaw, A.S. (2002) T cell receptor signaling precedes immunological synapse formation. Science 295, 1539–1542.

Lou, Z., Billadeau, D.D., Savoy, D.N., Schoon, R.A. and Leibson, P.J. (2001) A role for a RhoA/ROCK/LIM-kinase pathway in the regulation of cytotoxic lymphocytes. J. Immunol. 167, 5749–5757.

Masilamani, M., Nguyen, C., Kabat, J., Borrego, F. and Coligan, J.E. (2006) CD94/NKG2A inhibits NK cell activation by disrupting the actin network at the immunological synapse. J. Immunol. 177, 3590–3596.

McCann, F.E., Vanherberghen, B., Eleme, K., Carlin, L.M., Newsam, R.J., Goulding, D. and Davis, D.M. (2003) The size of the synaptic cleft and distinct distributions of filamentous actin, ezrin, CD43, and CD45 at activating and inhibitory human NK cell immune synapses. J. Immunol. 170, 2862–2870.

Miller, J.S. (2001) The biology of natural killer cells in cancer, infection, and pregnancy. Exp. Hematol. 29, 1157–1168.

Moretta, L., Bottino, C., Pende, D., Mingari, M.C., Biassoni, R. and Moretta, A. (2002) Human natural killer cells: their origin, receptors and function. Eur. J. Immunol. 32, 1205–1211.

Moretta, L. and Moretta, A. (2004) Killer immunoglobulin-like receptors. Curr. Opin. Immunol. 16, 626–633.

Mossman, K.D., Campi, G., Groves, J.T. and Dustin, M.L. (2005) Altered TCR signaling from geometrically repatterned immunological synapses. Science 310, 1191–1193.
Orange, J.S. (2002) Human natural killer cell deficiencies and susceptibility to infection. Microbes Infect. 4, 1545–1558.
Orange, J.S. (2006) Human natural killer cell deficiencies. Curr. Opin. Allergy Clin. Immunol. 6, 399–409.
Orange, J.S. and Ballas, Z.K. (2006) Natural killer cells in human health and disease. Clin. Immunol. 118, 1–10.
Orange, J.S., Fassett, M.S., Koopman, L.A., Boyson, J.E. and Strominger, J.L. (2002a) Viral evasion of natural killer cells. Nat. Immunol. 3, 1006–1012.
Orange, J.S., Harris, K.E., Andzelm, M.M., Valter, M.M., Geha, R.S. and Strominger, J.L. (2003) The mature activating natural killer cell immunologic synapse is formed in distinct stages. Proc. Natl. Acad. Sci. USA 100, 14151–14156.
Orange, J.S., Ramesh, N., Remold-O'Donnell, E., Sasahara, Y., Koopman, L., Byrne, M., Bonilla, F.A., Rosen, F.S., Geha, R.S. and Strominger, J.L. (2002b) Wiskott-Aldrich syndrome protein is required for NK cell cytotoxicity and colocalizes with actin to NK cell-activating immunologic synapses. Proc. Natl. Acad. Sci. USA 99, 11351–11356.
Orange, J.S., Stone, K.D., Turvey, S.E. and Krzewski, K. (2004) The Wiskott-Aldrich syndrome. Cell. Mol. Life Sci. 61, 2361–2385.
Pende, D., Parolini, S., Pessino, A., Sivori, S., Augugliaro, R., Morelli, L., Marcenaro, E., Accame, L., Malaspina, A., Biassoni, R., Bottino, C., Moretta, L. and Moretta, A. (1999) Identification and molecular characterization of NKp30, a novel triggering receptor involved in natural cytotoxicity mediated by human natural killer cells. J. Exp. Med. 190, 1505–1516.
Pessino, A., Sivori, S., Bottino, C., Malaspina, A., Morelli, L., Moretta, L., Biassoni, R. and Moretta, A. (1998) Molecular cloning of NKp46: a novel member of the immunoglobulin superfamily involved in triggering of natural cytotoxicity. J. Exp. Med. 188, 953–960.
Poggi, A., Panzeri, M.C., Moretta, L. and Zocchi, M.R. (1996) CD31-triggered rearrangement of the actin cytoskeleton in human natural killer cells. Eur. J. Immunol. 26, 817–824.
Pollard, T.D. and Borisy, G.G. (2003) Cellular motility driven by assembly and disassembly of actin filaments. Cell 112, 453–465.
Purtic, B., Pitcher, L.A., van Oers, N.S. and Wulfing, C. (2005) T cell receptor (TCR) clustering in the immunological synapse integrates TCR and costimulatory signaling in selected T cells. Proc. Natl. Acad. Sci. USA 102, 2904–2909.
Quan, P.C., Ishizaka, T. and Bloom, B.R. (1982) Studies on the mechanism of NK cell lysis. J. Immunol. 128, 1786–1791.
Radosevic, K., van Leeuwen, A.M., Segers-Nolten, I.M., Figdor, C.G., de Grooth, B. and Greve, J. (1995) Occurrence and a possible mechanism of penetration of natural killer cells into K562 target cells during the cytotoxic interaction. Cytometry 20, 273–280.
Sancho, D., Nieto, M., Llano, M., Rodriguez-Fernandez, J.L., Tejedor, R., Avraham, S., Cabanas, C., Lopez-Botet, M. and Sanchez-Madrid, F. (2000) The tyrosine kinase PYK-2/RAFTK regulates natural killer (NK) cell cytotoxic response, and is translocated and activated upon specific target cell recognition and killing. J. Cell. Biol. 149, 1249–1262.
Smith, K.A. (2001) Low-dose daily interleukin-2 immunotherapy: accelerating immune restoration and expanding HIV-specific T cell immunity without toxicity. Aids 15, S28–35.
Stebbins, C.C., Watzl, C., Billadeau, D.D., Leibson, P.J., Burshtyn, D.N. and Long, E.O. (2003) Vav1 dephosphorylation by the tyrosine phosphatase SHP-1 as a mechanism for inhibition of cellular cytotoxicity. Mol. Cell. Biol. 23, 6291–6299.
Stinchcombe, J.C., Majorovits, E., Bossi, G., Fuller, S. and Griffiths, G.M. (2006) Centrosome polarization delivers secretory granules to the immunological synapse. Nature 443, 462–465.

Tay, C.H., Szomolanyi-Tsuda, E. and Welsh, R.M. (1998) Control of infections by NK cells. Curr .Top. Microbiol. Immunol. 230, 193–220.
Trambas, C.M. and Griffiths, G. (2003) Delivering the kiss of death. Nat. Immunol 4, 399–403.
Vicente-Manzanares, M. and Sanchez-Madrid, F. (2004) Role of the cytoskeleton during leukocyte responses. Nat. Rev. Immunol. 4, 110–122.
Vyas, Y.M., Maniar, H., Lyddane, C.E., Sadelain, M. and Dupont, B. (2004) Ligand binding to inhibitory killer cell Ig-like receptors induce colocalization with Src homology domain 2-containing protein tyrosine phosphatase 1 and interruption of ongoing activation signals. J. Immunol. 173, 1571–1578.
Vyas, Y.M., Mehta, K.M., Morgan, M., Maniar, H., Butros, L., Jung, S., Burkhardt, J.K. and Dupont, B. (2001) Spatial organization of signal transduction molecules in the NK cell immune synapses during MHC class I-regulated noncytolytic and cytolytic interactions. J. Immunol. 167, 4358–4367.
Wu, J., Song, Y., Bakker, A.B., Bauer, S., Spies, T., Lanier, L.L. and Phillips, J.H. (1999) An activating immunoreceptor complex formed by NKG2D and DAP10. Science 285, 730–732.
Wulfing, C. and Davis, M.M. (1998) A receptor/cytoskeletal movement triggered by costimulation during T cell activation. Science 282, 2266–2269.
Wulfing, C., Purtic, B., Klem, J. and Schatzle, J.D. (2003) Stepwise cytoskeletal polarization as a series of checkpoints in innate but not adaptive cytolytic killing. Proc. Natl. Acad. Sci. USA 100, 7767–7772.

24

Infective, Neoplastic, and Homeostatic Sequelae of the Loss of Perforin Function in Humans

Joseph A. Trapani and Ilia Voskoboinik

Peter MacCallum Cancer Centre, St. Andrew's Place, East Melbourne, Australia, joe.trapani@petermac.org

Abstract. Perforin, a pore-forming protein toxin synthesized and stored in the cytoplasmic vesicles of cytotoxic T lymphocytes (CTLs) and natural killer (NK) cells, is secreted when these effector lymphocytes encounter virus-infected or neoplastic cells. Perforin is encoded by a single-copy gene and is critical for immune homeostasis and defense of the organism against intracellular sepsis. A complete deficiency of perforin expression in either mice or humans is associated with a syndrome of immune insufficiency and severely deregulated lymphoid homeostasis. Humans who inherit inactivating mutations of perforin or defects in various parts of the cellular machinery that delivers perforin to the target cell suffer from familial hemophagocytic lymphohistiocytosis (FHL), a fatal condition necessitating bone marrow transplantation, usually in infancy. In mice, a high incidence of spontaneous B cell lymphoma has also been noted as the animals age. Across human populations, a number of polymorphisms that result in measurable, but suboptimal CTL activity have been noted, and some of these predispose to attenuated FHL or susceptibility to infectious disease, but in many cases, to no discernible disease predisposition. This chapter discusses the significance of human perforin polymorphisms, particularly those associated with diseases other than FHL, and recent advances in our understanding of perforin biology and function.

1. Introduction

Although virtually all vaccine-induced immunity to viruses depends on the generation of neutralizing antibodies, once a virus enters and becomes established inside a cell, antibodies are ineffective and a cognate cytotoxic T lymphocyte (CTL) response must be raised in order to kill both the virus and the cell harboring it. CTLs share with natural killer (NK) cells the capacity to kill such cells by various contact-dependent cell death pathways that culminate in target cell apoptosis.

In the first pathway, ligands of the TNF family (FasL/CD95L and TRAIL) secreted or expressed on the CTL/NK cell surface induce the clustering of their receptors (Fas/CD95 and DR4/5) on the target cell, generating within it a caspase-dependent apoptotic cascade. Studies in mutant mice have shown that this pathway is most critical for maintaining immune (particularly lymphoid) homeostasis (Smyth et al. 2001). The second pathway, to be considered in some depth here, is known as the granule exocytosis mechanism, and defects of this pathway have been shown to cause marked susceptibility to various viruses and other intracellular pathogens, for example, *Listeria* (Trapani and Smyth 2002). Following T cell receptor-mediated conjugate formation, preformed toxins packaged within cytotoxic granules of the CTL migrate to the site of cell contact along the microtubular apparatus where a complex vesicle machinery facilitates its fusion with the plasma membrane to liberate toxins that include a pore-forming protein (perforin) and a battery of serine proteases (granzymes) into the synaptic cleft (Menager et al. 2007). Although its mechanism of action is poorly understood, perforin is responsible for admitting the granzymes into the target cell cytosol, where the cleavage of key substrates (including pro-caspases and molecules that result in mitochondrial destabilization) results in apoptosis (Bird et al. 2005; Trapani and Sutton 2003). The most potent proapoptotic granzyme is the Asp-ase, granzyme B; however, other granzymes such as granzyme A induce target cell death through parallel, non-caspase-dependent pathways (Lieberman and Fan 2003).

2. Molecular and Cellular Functions of Perforin: A Putative Domain Structure

The membrane pores formed by polymerized perforin are reminiscent of complement pores at an ultrastructural level, and perforin shares both antigenic cross-reactivity and limited amino acid similarity with the central portion of complement component C9. The structures of both perforin and C9 remain to be solved; however, it is predicted that the central region of both proteins forms an amphipathic alpha helix that spans the lipid bilayer (Lichtenheld et al. 1988; Liu et al. 1995). The action of perforin is entirely dependent on its prior binding of calcium ions, a function performed by a C2-domain motif at its carboxy terminus (see below) (Voskoboinik et al. 2005a). The final few residues of this region are not required for calcium binding but are crucial for the appropriate folding and acquisition of perforin's lytic function. It is believed that this "tailpiece" is cleaved, along with a bulky N-linked glycan when perforin reaches the secretory granules; however, the protease responsible for this processing is, as yet, unidentified (Uellner et al. 1997). The function of a cysteine-rich EGF-like motif is unknown, while synthetic peptides corresponding to the extreme amino terminus have been shown to possess some membranolytic properties (Rochel and Cowan 1996). While strongly conserved in perforins from species as diverse as fish and humans (Hwang et al. 2004), the remainder of the sequence, or about half of the 534 amino acid protein, is unique to perforin.

The development (by us and others) of robust expression methodologies for perforin over the past few years has for the first time enabled a meaningful mutagenic analysis of the perforin backbone (Shiver and Henkart 1991; Voskoboinik et al. 2004). This has enabled us both to rapidly assess the effects of inherited perforin mutations on its function, and permitted site-directed mutagenesis to determine the role of specific residues in functions such as calciumbinding (Voskoboinik et al.

2005a, b). Accordingly, we have confirmed that up to four calcium ions are coordinated by each perforin monomer through an ionic interaction with a "basket" of negatively charged aspartate residues in the C2 motif (Voskoboinik et al. 2005a). As aspartate residues are predicted to be uncharged at the acidic pH of secretory granules (5.5–6.0), perforin is effectively stored as an inactive (thus, harmless) protein until it is released into the synaptic cleft, an efficient and effective means of protecting the CTL/NK cell from inadvertent lysis prior to perforin release. It has also been elegantly recently shown that perforin release also requires the fusion of two distinct types of granule in the CTL: perforin- and granzyme-rich lysosome-like Rab11+ vesicles, and endosome-like Munc13-4 and Rab27a-containing vesicles required for final exocytosis of the granule contents (Menager et al. 2007).

3. Congenital perforin deficiency and other causes of FHL

Perforin is absolutely critical for the function of the granule exocytosis pathway. It is encoded by a single-copy gene PRF1, whose inactivation or deletion causes profound immunodeficiency in both humans and mice (Kagi et al. 1994; Stepp et al. 1999); this reflects perforin's pivotal and unique role in facilitating intracellular access for the granzymes. Gene-engineered perforin-deficient mice die of a variety of virus infections, are more susceptible to transplantable tumors, and develop spontaneous B cell lymphoma as they age, suggesting an additional role for perforin-dependent pathways in immune surveillance of transformed cells, particularly B lymphocytes (Street et al. 2004). In humans, congenital perforin deficiency is more generally associated with markedly disordered immune homeostasis arising in infancy, and less commonly with intractable virus infection (Katano and Cohen 2005; Voskoboinik and Trapani 2006). Familial Hemophagocytic Lymphohistiocytosis (FHL) is an autosomal recessive disorder characterized by early onset (<24 months) liver, spleen, and lymph node enlargement due to their engorgement with activated lymphocytes and macrophages secreting large quantities of inflammatory cytokines. Red cell phagocytosis by activated macrophages in the bone marrow is common and contributes to progressive anemia (Henter et al. 1998; Janka and Zur Stadt 2005; Janka 1989) (Figure 1). The condition can be treated with cytotoxic agents and immunosuppressants, but patients often relapse, requiring a bone marrow transplant (Henter et al. 2002). There are a number of FHL subtypes described, but the form associated with perforin gene mutation is known as Type 2 (Stepp et al. 1999).

4. Perforin Alleles Encoding Partial Loss of Function

Some perforin mutations can result in FHL-like syndromes with delayed onset and a variable degree of severity. Most commonly in full-blown disease, nonsense or frame-shift mutations in the gene result in protein truncation and invariably, a null phenotype. In such cases, disease manifestations are severe and the illness frequently commences prior to 6 months of age. Missense mutations (leading to single-amino acid substitutions) can also lead to disease (Voskoboinik et al. 2006). To date, the

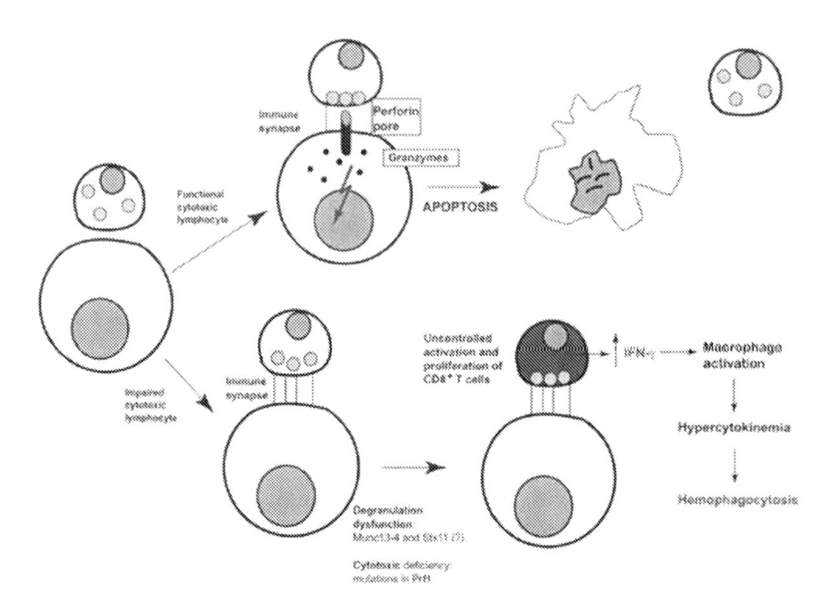

FIGURE 1. A mechanism for the pathogenesis of familial hemophagocytic lymphohistiocytosis (FHL). Immune competent cytotoxic lymphocytes release the content of cytotoxic granules into the immune synapse. The synergy between perforin and granzymes results in apoptotic target cell death. Effector cells with dysfunctional perforin or impaired degranulation pathway due to mutations in *PRF1* or *UNC13-4* genes, respectively, cannot clear the target causing uncontrolled activation and proliferation of T cells and macrophages and a marked release of inflammatory cytokines, resulting in hemophagocytosis.

function of more than 30 missense mutations has been analyzed in surrogate cytotoxicity assays. Typically, perforin function in a cellular context can be assessed by expressing it in rat basophil leukemia (RBL-2H4) cells (which are equipped to process, store, and secrete perforin upon conjugation with a target cell) (Shiver and Henkart 1991; Voskoboinik et al. 2004), or by complementing perforin function in lymphocytes isolated from perforin-deficient mice (our unpublished data). We have also pioneered the use of perforin purified from baculovirus-infected insect cells, providing a source of perforin whose lytic and proapoptotic capacities can be assessed by direct application to target cells in the presence or absence of granzyme B, respectively (Voskoboinik et al. 2004; Waterhouse et al. 2005). We have found that while the majority (about 80%) of missense mutations lead to protein instability and degradation, some lack function due to an inability to lyse the target cell membrane following secretion (Voskoboinik et al. 2005b). Unlike protein truncation, missense mutations frequently result in partial preservation of function, and by reviewing the clinical presentations of many such patients, we have found a close relationship between the residual degree of perforin function and the age of disease onset (Voskoboinik et al. 2006). As a "rule of thumb," if RBL cells expressing the mutated perforin have significant cytotoxic capacity and/or a patient's CTLs retain measurable lytic activity in vitro, disease onset may be delayed, typically to beyond 24 months of age, and as late as teenage or beyond (Voskoboinik et al. 2006).

5. Ala91Val, a Common Hypomorphic Perforin Allele

A number of independent population studies have found 9–17% of subjects to possess at least one perforin allele encoding the apparently conservative Ala91Val (A91V) substitution (Mehta et al. 2006; Zur Stadt et al. 2004). This high-frequency and the semiconservative nature of the substitution had led most to predict that the function of A91V would be equivalent to wild-type (Mehta et al. 2006). However, two recent studies have shown the A91V allele to impose a considerable loss of function. First, the NK cells of homozygous subjects have reduced cytotoxicity (Mancebo et al. 2006), and secondly, A91V-perforin expressed in RBL cells or primary mouse lymphocytes showed a similar reduction in function (Trambas et al. 2005; Voskoboinik et al. 2005b). We have recently found that while wild-type perforin expressed in baculovirus-infected insect cells is strongly lytic and remains potent with storage, A91V perforin has markedly reduced function (our unpublished observations). These findings collectively confirm the fact that A91V-perforin function is not normal and explain the observation that A91V-heterozygous individuals can develop an attenuated FHL-like syndrome when their second allele is either null or also encodes severely dysfunctional perforin (Busiello et al. 2004; Clementi et al. 2002). By contrast, individuals (such as the parents of FHL patients) who inherit one normal and a second null allele invariably remain healthy and immunocompetent. The possible association of A91V and cancer predisposition remains controversial (Mehta et al. 2006; Santoro et al. 2005). In the only large study to date, involving more than 1300 individuals with acute lymphoblastic leukemia, there was no increased frequency of A91V. However, A91V was significantly more common in a small subgroup of children with the Philadelphia chromosome translocation (Mehta et al. 2006).

Perforin mutations were also found in several patients with atypical forms of FHL and/or with various types of hematological cancer (Mehta et al. 2006; Santoro et al. 2005; Clementi et al. 2006; Clementi et al. 2002, 2005; Katano et al. 2004). To establish potential causative links between these mutant perforins and a disease, the properties of these mutants would have to be investigated on the molecular level. However, a partial loss of function would suggest a role for the cytotoxic granule-mediated pathway in tumor immune surveillance in humans, a mechanism that has been well documented in experimental animal models but has never been formally proven to exist in humans.

6. Conclusions

Over recent years, it has become increasingly possible to evaluate the effect of inherited point mutations on perforin function. In addition, it is now possible for a researcher to design specific mutations to test the role of individual perforin residues.

Overall, we have found that approximately 80% of missense mutations resulting in FHL cause a presynaptic defect in perforin function, usually manifested as reduced protein stability as a result of misfolding and/or mistrafficking. In the remainder of cases, the protein is synthesized, packaged, and released normally but possesses a postsynaptic defect that limits target cell lysis and/or the capacity to deliver pro-apoptotic granzymes to the target cell cytosol. Despite this recent progress, the molecular and cellular functions of most of perforin's putative domain remain unknown, and studies of this nature have been severely hampered by a lack of structural data (particularly X-ray crystallographic data) on monomeric and poly- perforin. Our group and a number of others are currently pursuing this endeavor with considerable enthusiasm.

References

Bird, C.H., Sun, J., Ung, K., Karambalis, D., Whisstock, J.C., Trapani, J.A. and Bird, P.I. (2005) Cationic sites on granzyme B contribute to cytotoxicity by promoting its uptake into target cells. Mol. Cell. Biol. 25, 7854–7867.

Busiello, R., Adriani, M., Locatelli, F., Galgani, M., Fimiani, G., Clementi, R., Ursini, M.V., Racioppi, L. and Pignata, C. (2004) Atypical features of familial hemophagocytic lymphohistiocytosis. Blood 103, 4610–4612.

Clementi, R., Emmi, L., Maccario, R., Liotta, F., Moretta, L., Danesino, C. and Arico, M. (2002) Adult onset and atypical presentation of hemophagocytic lymphohistiocytosis in siblings carrying PRF1 mutations. Blood 100, 2266–2267.

Clementi, R., Locatelli, F., Dupre, L., Garaventa, A., Emmi, L., Bregni, M., Cefalo, G., Moretta, A., Danesino, C., Comis, M., Pession, A., Ramenghi, U., Maccario, R., Arico, M. and Roncarolo, M.G. (2005) A proportion of patients with lymphoma may harbor mutations of the perforin gene. Blood 105, 4424–4428.

Clementi, R., Chiocchetti, A., Cappellano, G., Cerutti, E., Ferretti, M., Orilieri, E., Dianzani, I., Ferrarini, M., Bregni, M., Danesino, C., Bozzi, V., Putti, M.C., Cerutti, F., Cometa, A., Locatelli, F., Maccario, R., Ramenghi, U. and Dianzani, U. (2006) Variations of the perforin gene in patients with autoimmunity/lymphoproliferation and defective fas function. Blood 108, 3079–3084.

Henter, J.I., Arico, M., Elinder, G., Imashuku, S. and Janka, G. (1998) Familial hemophagocytic lymphohistiocytosis. Primary hemophagocytic lymphohistiocytosis. Hematol. Oncol. Clin. North. Am. 12, 417–433.

Henter, J.I., Samuelsson-Horne, A., Arico, M., Egeler, R.M., Elinder, G., Filipovich, A.H., Gadner, H., Imashuku, S., Komp, D., Ladisch, S., Webb, D. and Janka, G. (2002) Treatment of hemophagocytic lymphohistiocytosis with HLH-94 immunochemotherapy and bone marrow transplantation. Blood 100, 2367–2373.

Hwang, J.Y., Ohira, T., Hirono, I. and Aoki, T. (2004). A pore-forming protein, perforin, from a non-mammalian organism, Japanese flounder, Paralichthys olivaceus. Immunogenetics 56, 360–367.

Janka, G.E. (1989) Familial hemophagocytic lymphohistiocytosis: diagnostic problems and differential diagnosis. Pediatr. Hematol. Oncol. 6, 219–225.

Janka, G. and Zur Stadt, U. (2005) Familial and acquired hemophagocytic lymphohistiocytosis. Hematology (Am. Soc. Hematol. Educ. Program) 82–88.

Kagi, D., Ledermann, B., Burki, K., Seiler, P., Odermatt, B., Olsen, K.J., Podack, E.R., Zinkernagel, R.M. and Hengartner, H. (1994) Cytotoxicity mediated by T cells and natural killer cells is greatly impaired in perforin-deficient mice. Nature 369, 31–37.

Katano, H., Ali, M.A., Patera, A.C., Catalfamo, M., Jaffe, E.S., Kimura, H., Dale, J.K., Straus, S.E. and Cohen, J.I. (2004) Chronic active Epstein-Barr virus infection associated with mutations in perforin that impair its maturation. Blood 103, 1244–1252.

Katano, H. and Cohen, J.I. (2005) Perforin and lymphohistiocytic proliferative disorders. Br. J. Haematol. 128, 739–750.

Lichtenheld, M.G., Olsen, K.J., Lu, P., Lowrey, D.M., Hameed, A., Hengartner, H. and Podack, E.R. (1988) Structure and function of human perforin. Nature 335, 448-451.

Lieberman, J. and Fan, Z. (2003) Nuclear war: the granzyme A-bomb. Curr. Opin. Immunol. 15, 553–559.

Liu, C.C., Walsh, C.M. and Young, J.D. (1995) Perforin: structure and function. Immunol. Today 16, 194–201.

Mancebo, E., Allende, L.M., Guzman, M., Paz-Artal, E., Gil, J., Urrea-Moreno, R., Fernandez-Cruz, E., Gaya, A., Calvo, J., Arbos, A., Duran, M.A., Canet, R., Balanzat, J., Udina, M.A. and Vercher, F.J. (2006) Familial hemophagocytic lymphohistiocytosis in an adult patient homozygous for A91V in the perforin gene, with tuberculosis infection. Haematologica 91, 1257–1260.

Mehta, P.A., Davies, S.M., Kumar, A., Devidas, M., Lee, S., Zamzow, T., Elliott, J., Villanueva, J., Pullen, J., Zewge, Y. and Filipovich, A. (2006) Perforin polymorphism A91V and susceptibility to B-precursor childhood acute lymphoblastic leukemia: a report from the Children's Oncology Group. Leukemia 20, 1539–1541.

Menager, M.M., Menasche, G., Romao, M., Knapnougel, P., Ho, C.H., Garfa, M., Raposo, G., Feldmann, J., Fischer, A. and de Saint Basile, G. (2007) Secretory cytotoxic granule maturation and exocytosis require the effector protein hMunc13-4. Nat. Immunol. 8, 257–267.

Rochel, N. and Cowan, J. (1996) Negative cooperativity exhibited by the lytic amino-terminal domain of human perforin: implications for perforin-mediated cell lysis. Chem. Biol. 3, 31–36.

Santoro, A., Cannella, S., Trizzino, A., Lo Nigro, L., Corsello, G. and Arico, M. (2005) A single amino acid change A91V in perforin: a novel, frequent predisposing factor to childhood acute lymphoblastic leukemia? Haematologica 90, 697–698.

Shiver, J.W. and Henkart, P.A. (1991) A noncytotoxic mast cell tumor line exhibits potent IgE-dependent cytotoxicity after transfection with the cytolysin/perforin gene. Cell 64, 1175–1181.

Smyth, M.J., Kelly, J.M., Sutton, V.R., Davis, J.E., Browne, K.A., Sayers, T. and Trapani, J.A. (2001) Unlocking the secrets of cytotoxic granule proteins. J. Leukoc. Biol. 70, 18–29.

Stepp, S.E., Dufourcq-Lagelouse, R., Le Deist, F., Bhawan, S., Certain, S., Mathew, P.A., Henter, J.I., Bennett, M., Fischer, A., de Saint Basile, G. and Kumar, V. (1999) Perforin gene defects in familial hemophagocytic lymphohistiocytosis. Science 286, 1957–1959.

Street, S.E., Hayakawa, Y., Zhan, Y., Lew, A.M., MacGregor, D., Jamieson, A.M., Diefenbach, A., Yagita, H., Godfrey, D.I. and Smyth, M.J. (2004) Innate immune surveillance of spontaneous B cell lymphomas by natural killer cells and gammadelta T cells. J. Exp. Med. 199, 879–884.

Trambas, C., Gallo, F., Pende, D., Marcenaro, S., Moretta, L., De Fusco, C., Santoro, A., Notarangelo, L., Arico, M. and Griffiths, G.M. (2005) A single amino acid change, A91V, leads to conformational changes that can impair processing to the active form of perforin. Blood 106, 932–937.

Trapani, J.A. and Smyth, M.J. (2002) Functional significance of the perforin/granzyme cell death pathway. Nat. Rev. Immunol. 2, 735–747.

Trapani, J.A. and Sutton, V.R. (2003) Granzyme B: pro-apoptotic, antiviral and antitumor functions. Curr. Opin. Immunol. 15, 533–543.

Uellner, R., Zvelebil, M.J., Hopkins, J., Jones, J., MacDougall, L.K., Morgan, B.P., Podack, E., Waterfield, M.D. and Griffiths, G.M. (1997) Perforin is activated by a proteolytic cleavage during biosynthesis which reveals a phospholipid-binding C2 domain. EMBO J. 16, 7287–7296.

Voskoboinik, I., Thia, M.-C., De Bono, A., Browne, K., Cretney, E., Jackson, J.T., Darcy, P. K., Jane, S.M., Smyth, M.J. and Trapani, J.A. (2004) The functional basis for hemophagocytic lymphohistiocytosis in a patient with co-inherited missense mutations in the perforin (PFN1) gene. J. Exp. Med. 200, 811–816.

Voskoboinik, I., Thia, M.C., Fletcher, J., Ciccone, A., Browne, K., Smyth, M.J. and Trapani, J.A. (2005a) Calcium-dependent plasma membrane binding and cell lysis by perforin are mediated through its C2 domain: A critical role for aspartate residues 429, 435, 483, and 485 but not 491. J. Biol. Chem. 280, 8426–8234.

Voskoboinik, I., Thia, M.C. and Trapani, J.A. (2005b) A functional analysis of the putative polymorphisms A91V and N252S and 22 missense perforin mutations associated with familial hemophagocytic lymphohistiocytosis. Blood 105, 4700–4706.

Voskoboinik, I. and Trapani, J.A. (2006) Addressing the mysteries of perforin function. Immunol. Cell. Biol. 84, 66–71.

Waterhouse, N., Sedelies, K., Browne, K., Wowk, M., Newbold, A., Sutton, V., Clarke, C.J., Oliaro, J., Lindemann, R.K., Bird, P., Johnstone, R.W. and Trapani, J.A. (2005) A central role for Bid in granzyme B-induced apoptosis. J. Biol. Chem. 280, 4476–4482.

Zur Stadt, U., Beutel, K., Weber, B., Kabisch, H., Schneppenheim, R., Janka, G., Busiello, R., Galgani, M., De Fusco, C., Poggi, V., Adriani, M., Racioppi, L. and Pignata, C. (2004) A91V is a polymorphism in the perforin gene not causative of an FHLH phenotype. Blood 104, 1909–1910.

25

Natural Killer T (NKT) Cell Subsets in Chlamydial Infections

Xi Yang

Laboratory for Infection and Immunity, Departments of Medical Microbiology and Immunology, University of Manitoba, Winnipeg, Manitoba, Canada, yangxi@cc.umanitoba.ca

Abstract. Natural killer T (NKT) cells are a newly identified unique subset of cells that express both αβ T cell receptor and NK cell markers. We investigated the role of NKT cells in modulating adaptive T cell responses in chlamydial infections using human-disease-related chlamydial species. Our study provides in vivo evidence that even closely related pathogens may activate different functional NKT subsets, which can further polarize $CD4^+$ and $CD8^+$ cells in adaptive immune responses.

1. Introduction

Natural killer T (NKT) cells are a newly identified unique subset of T cells, which express both αβ T cell receptor and NK cell markers. NKT cells are capable of recognizing lipid antigens in the context of CD1, a molecule belonging to the non-MHC- encoded, MHC class I-like gene family. The CD1-dependent NKTcells are characterized by semi-invariant αβ TCR expression. Based on whether it responds to α-galactosylceramide (α-GalCer), NKT cells can be classified as type 1 or type 2 NKT cells. Type 1 NKT cells are the classical NKT cells that express αβ TCR with the use of Vα14 and Jα18, where as type 2 NKT are less common and often express Vα3.2-Jα9. α-GalCer in the context of CD1 can activate type 1 NKT cells, but not type 2 NKT cells.

2. NKT Cells

NKT cells have been found to be able to play an important role in immune regulation (Godfrey et al. 2005; Bendelac et al. 2007). The immune regulatory function is believed to be largely dependent on their capacity to quickly produce large amount of cytokines following activation, which can regulate both Th1 and Th2 responses. NKT cells are also found to play important roles in tumor rejection and the prevention of autoimmune diseases. The role of NKT in host defense against various infections including virus, parasites, and bacteria has also been reported. For example, studies on NKT with α-GalCer inhibited hepatitis B viral replication in HBV transgenic mice while CD1 gene-knockout mice (lack of NKT) showed substantial impairment in antiviral response to encephalomyocarditis virus infection and herpes simplex type 1 virus infections. Moreover, α-GalCer treatment significantly enhanced protective antimalarial immunity and reduced bacterial burden and tissue injury in *Mycobacterium tuberculosis* mouse models. Furthermore, CD1 gene-knockout mice have been found to be less resistant to *Borrelia burgdorferi* infection and fail to eradicate *Pseudomonas aeruginosa* infection.

Chlamydiae are obligate intracellular bacteria with a unique developmental cycle. Two chlamydial species, *C. pneumoniae* and *C. trachomatis*, commonly cause human diseases. *C. pneumoniae* is the causing agent of a wide spectrum of acute and chronic respiratory diseases such as bronchitis, sinusitis, and pneumonia, where as *C. trachomatis* causes ocular, respiratory, and sexually transmitted diseases. Chlamydial infections are very prevalent worldwide. In particular, up to 70% of healthy human individuals are positive for serum antibodies specific for *C. pneumoniae.* More recently, *C. pneumoniae* has been implicated in the pathogenesis of atherosclerosis, Alzheimer's disease, and multiple sclerosis. No vaccine is available for human chlamydial infections. A clear understanding of the adaptive and innate immune responses to chlamydial infection is critical in the rational development of an effective vaccine to this infection. The differences in T cell cytokine patterns have been correlated with the severity of disease progression in both human subjects and animal models.

3. NKT Cells in Chlamydial Lung Infection

Our laboratory has recently studied the role of NKT in chlamydial lung infection mouse models (Bilenki et al. 2005; Joyee et al. 2007). We investigated the potential role of NKT cells in modulating adaptive T cell responses in chlamydial infections using the two human-disease-related chlamydial species, *C. pneumoniae* and *C. trachomatis.* The *C. pneumoniae* strain used in the study is AR-39 while the *C. trachomatis* strain is mouse pneumonitis (MoPn), more recently called *C. muridarum.* The role of NKT cells was examined using the combinational approaches of NKT-knockout (CD1d-knockout and Jα18-knockout) mice and specific NKT activation by α-GalCer treatment. NKT-deficient mice showed exacerbated susceptibility to *C. pneumoniae* infection, but more resistance to *C. trachomatis* (MoPn) infection. Consistently, activation of NKT using α-GalCer reduced *C. pneumoniae* in vivo growth but enhanced MoPn infection. Since only classical type 1, but not type 2 CD1-dependent NKT cells are responsive to α-GalCer stimulation, the results suggest that the NKT cells that play immunomodulatory role during chlamydial infection are type 1 NKT cells.

More detailed cytokine analyses on NKT and T cells were performed to examine the linkage between NKT cytokine production and T cell cytokine responses. We performed a direct cellular level analysis on NKT kinetics and cytokine patterns using ligand-loaded CD1d tetramer. NKT cells showed an expansion of about fivefold after chlamydial infection. The number and percentage of NKT in the lungs peaked at day 3 postinfection (p.i.), after which decreased and reached basal levels about days 9–10 p.i. Intracellular cytokine staining of lung and spleen NKT cells at 3 days p.i. from the wild-type mice for IFN-γ and IL-4 showed significantly increased proportion of IFN-γ-producing cells compared with that in the uninfected mice, especially in NKT cells from infected organ (lung). Thus, these results suggest that *C. pneumoniae* infection can skew the cytokine response of NKT cells to IFN-γ polarization. In contrast, NKT cells from the lung of MoPn-infected mice showed predominant IL-4 production. To further evaluate whether NKT cells influence T cells ($CD8^+$ and $CD4^+$) in the context of cytokine response to chlamydial infection, we analyzed the cellular cytokine patterns of $CD4^+$ and $CD8^+$ T cells in the wild-type and NKT knockout mice in the adaptive immune phase following *C. pneumoniae* infection. Intracellular cytokine staining for IFN-γ and IL-4 showed distinct pattern in the IFN-γ production by $CD8^+$ T cells between the wild-type and NKT-knockout mice. The wild-type mice showed significantly higher proportion of IFN-γ-producing $CD8^+$ T cells than did NKT-knockout mice following *C. pneumoniae* infection. In addition, $CD4^+$ T cells of wild-type mice showed a Th1 cytokine pattern with significantly higher IFN-γ and lower IL-4 levels, while $CD4^+$ T cells from NKT-knockout mice showed a Th2-like pattern with more IL-4, but less IFN-γ following the infection. Overall, these results suggested that different NKT subsets are activated by different chlamydial species and that NKT cells could modulate the cytokine pattern of both $CD8^+$ and $CD4^+$ T cells.

The findings that different NKT subsets are activated by different chlamydial species and that the cytokine patterns of NKT are correlated with host susceptibility to the infections have significant implications in our understanding of the host defense mechanisms against infectious diseases. NKT cells represent the intermediate phase of immune responses between innate and adaptive immunity following infectious agents' exposure to the host. They are expected to play a protective role in the early stage of infection by quick response to infectious agents. Indeed, numerous studies have demonstrated the potent protective roles of NKT cells in various infection models. It is also true in the host resistance to *C. pneumoniae* infection. However, in the model of *C. trachomatis* infection, NKT cells appear to promote chlamydial infection. Therefore, NKT cells play various roles in host susceptibility to infections depending on the nature of the infectious agents. Notably, some previous studies have generated different conclusions by using different approaches in investigating the role of NKT in a particular infection. For example, a recent study showed that CD1-knockout mice exhibited minimal change in susceptibility in *M. tuberculosis* infection. In contrast, the treatment of mice with α-GalCer remarkably improved protection against *M. tuberculosis* infection. To our knowledge, our study is the first showing the suppressive role of NKT in an infection using both gene-knockout (CD1d-knockout and NKT-knockout) mice and α-GalCer treatment approaches. Our data provide in vivo evidence for functionally diverse role of NKT cells and the correlation of the cytokine profile of NKT cells with the T cell cytokine patterns following chlamydial infection. It is particularly interesting that distinct NKT subsets are activated by even relatively closely related pathogens and that NKT cells play opposite roles in immune responses to different chlamydial species. Since NKT cells clearly influence the direction of adaptive immune responses, including CD4 and CD8 T cells, to chlamydial infection in vivo, the results indicate that NKT cells function not only in the innate immune phase but also in bridging to the phase of acquired immune responses.

The results which showed that the type of immune response enhanced by the same NKT stimulator, α-GalCer, may vary depending on the nature of the exposed pathogens give alert to the experimental therapeutic approaches attempting to activate NKT cells. Encouragingly, a single injection of α-GalCer before intranasal infection substantially reduced *C. pneumoniae* loads in the lungs and enhanced type1 immunological responses, thus showing protective effect. However, in the model of MoPn infection, the administration of α-GalCer enhanced type2, instead of Th1, responses, leading to promotion of infection. Therefore, particular attention has to be given on how to preferentially induce preferable type of immunity when using α-GalCer as an adjuvant for immunotherapy.

3. Conclusions

Our study provides in vivo evidence that even biologically, closely related pathogens may activate different functional NKT subsets, which can further polarize the immune cells in adaptive immune responses, such as $CD4^+$ and $CD8^+$ cells. Further studies revealing the nature of pathogens/stimuli involving the development and/or promotion of differential NKT responses will greatly enhance our understanding on the linkage between innate and adaptive immune responses in infectious diseases.

Acknowledgments

The study was supported by grants from Canadian Institutes of Health Research, Canada Research Chair Program, and Manitoba Health Research Council. Several laboratory members are involved in this study, including Antony George Joyee, Laura Bilenki, Hongyu Qiu, Shuhe Wang, Yijun Fan, and Jie Yang. We are thankful to Dr. Y. Koezuka, Kirin Pharmaceuticals, Japan, for kindly providing α-galactosylceramide and NIH tetramer facility for providing CD1d tetramer.

References

Bendelac, A., Savage, P.B. and Teyton, L. (2007) The biology of NKT cells. Annu. Rev. Immunol. 25, 297–336.

Bilenki, L., Wang, S., Yang, J., Fan, Y., Joyee, A.G. and Yang, X. (2005) NK T cell activation promotes Chlamydia trachomatis infection in vivo. J. Immunol. 175, 3197–3206.

Godfrey, D.I., McCluskey, J. and Rossjohn, J. (2005) CD1d antigen presentation: treats for NKT cells. Nat. Immunol. 6, 754–756.

Joyee, A.G., Qiu, H., Wang, S., Fan, Y., Bilenki, L. and Yang, X. (2007) Distinct NKT cell subsets are induced by different Chlamydia species leading to differential adaptive immunity and host resistance to the infections. J. Immunol. 178, 1048–1058.

26

Memory T Cells in Allograft Rejection

Anna Valujskikh

The Cleveland Clinic Foundation, Department of Immunology, Cleveland, OH, USA, valujsa@ccf.org

Abstract. T cell repertoire of many humans contains high frequencies of memory T cells specific for alloantigens. The increasing evidence implicates these cells as a barrier to allograft survival and to the induction of transplantation tolerance. This review discusses several aspects of memory T cell immunobiology pertinent to their role in transplantation.

1. Memory T Cells: Generation, Phenotype, and Functions

In response to an invading pathogen, T cell receptors (TCRs) recognize pathogen-derived peptide determinants that are bound to MHC molecules on the surface of professional antigen-presenting cells (APCs). If appropriate costimulatory signals are provided through CD28 and CD154 molecules, such recognition results in T cell activation, multiple rounds of cell division, and differentiation into effector T cells (Bradley et al. 2000; Seder and Ahmed 2003). During these events, the T cells upregulate expression of activation markers CD25 and CD69 and the adhesion molecule CD44, and downregulate expression of the lymph node homing receptors, CD62L and CCR7 (Garcia et al. 1999; Perrin et al. 1999; Bradley et al. 2000). Such phenotypic changes permit the activated T cells to circulate widely to the peripheral organs where re-encounter with the same MHC–peptide complex expressed on an infected parenchymal cell triggers T cell effector functions (Masopust et al. 2001; Reinhardt et al. 2001). The tools employed by effector T cells include the production of proinflammatory cytokines and chemokines, cytotoxicity, activation of macrophages and subsequent mediation of delayed-type hypersensitivity (DTH) responses, and provision of B cell help for induction of antibody isotype switching. The result is destruction of infected parenchymal cells and, ultimately, control of the infection.

Upon resolution of the infection, the residual effector T cells rapidly undergo apoptosis leaving behind only a small proportion of memory T cells. There is an ongoing debate in current literature on whether memory T cells develop directly from effector T cells or whether the separate line of memory T cells is committed during the initial activation events (reviewed in Kaech et al. 2002; Sprent and Surh 2002). Regardless of origin, these memory T cells are smaller than typical effector cells and generally have a $CD25^{lo}CD69^{lo}CD44^{hi}$ phenotype (Dutton et al. 1998; Lanzavecchia and Sallusto 2001, 2002).

A dichotomy within memory T cell population has been observed based on the expression of surface markers, trafficking patterns, and functional properties. A model was proposed in which some memory T cells express high levels of CD62L and CCR7 (receptors for lymph node-specific adhesion molecule PNAd and chemokines CCL19 and CCL21, respectively) that allow these cells to migrate efficiently to peripheral lymph nodes, maintain the memory on a cell population level, and give rise to a second wave of effector T cells. Another population of memory T cells—$CD62L^{lo}CCR7^{lo}$—expresses receptors that direct migration to inflamed sites, including CCR1, CCR3, CCR5, and β1 and β2 integrins (Sallusto et al. 1999) and is typically found in nonlymphoid tissues where these cells can rapidly control invading pathogens. These subsets can be detected in both humans and mice and are generally referred to as "central" and "effector" memory T cells (Seder and Ahmed 2003; Lanzavecchia and Sallusto 2005). Consistent with the proposed model, memory CD4 T cells derived from nonlymphoid tissues predominantly secrete effector cytokines such as IFN-γ and IL-4 while central memory CD4 T cells primarily produce IL-2 (Reinhardt et al. 2001). Similarly, within the CD8 memory T cell population, effector memory cells have higher cytotoxic activity when compared with central memory cells (Masopust et al. 2001).

However, more recent studies of memory CD8 T cells demonstrated that both central and effector subsets had similar abilities to produce cytokines and exhibit cytotoxicity (Unsoeld et al. 2002; Ravkov et al. 2003; Wherry et al. 2003). These findings led to the conclusion that CCR7 or CD62L expression is not a reliable marker of memory CD8 T cell function but may be valuable for defining the anatomical location of these cells. Furthermore, factors defining the distribution and functions of memory CD4 T cells may be distinct from those described for memory CD8 T cells. Supporting this notion, recently published study (Kassiotis and Stockinger 2004) argues against a strong correlation between CD62L expression and preferential migration of memory CD4 T cells into the lymphoid or nonlymphoid compartments.

Despite massive apoptosis at the end of the immune response, the precursor frequency of antigen-specific memory T cells is significantly higher than that of naïve T cells, a fact that partially accounts for the stronger immune response after rechallenge with antigen (Dutton et al. 1998; Sprent and Surh 2002; Seder and Ahmed 2003). On a per-cell basis, memory T cells are more metabolically active than naïve T cells. This translates into their ability to rapidly enter cell cycle and to migrate into nonlymphoid tissues and express effector function faster than naïve T cells. Our laboratory, among many others, has previously shown that memory T cells produce effector cytokines, such as IFN-γ and IL-4, much faster than naïve cells (4–6 h vs 3–5 days), and we use this feature as one approach to differentiate memory from naïve T cells in our studies (Matesic et al. 1998; Cho et al. 1999; Zimmermann et al. 1999). Memory T cells can also kill antigen-expressing targets within hours after re-activation, whereas naïve T cells generally require several days of differentiation in vivo before they exhibit the same level of cytotoxicity (Watschinger et al. 1994; Kaech et al. 2002; Sprent and Surh 2002; Seder and Ahmed 2003). Memory CD4 T cells can efficiently provide help for antibody isotype switching, where naïve T cells have poor helper function (Swain 1994). Furthermore, upon reactivation, memory T cells are capable of reaching the inflammation site much faster than naïve cells—in a few hours versus several days (Kedl and Mescher 1998). All of these functions are essential for efficient host defense. However, it is becoming increasingly clear that the same features of memory T cells that are important for protection from pathogens represent significant hurdles for survival of transplanted organs and tissues.

2. Alloreactive T Cell Memory

T lymphocytes specific for allogeneic MHC molecules and polymorphic minor histocompatibility antigens are found at high frequency and are key mediators of allograft rejection (Lindahl and Wilson 1977; Matesic et al. 1998; Suchin et al. 2001). Theoretically, sensitization to alloantigens is not as common as immunizations and infections. Nonetheless, the immune repertoire of many humans contains memory T cells reactive to a variety of alloantigens (Heeger et al. 1999). These memory T cells were presumably primed through pregnancy, through blood transfusion, or by exposure to an infectious agent or immunization that cross-reacts with alloantigens (so-called heterologous immunity). In addition, recent studies in rodents demonstrated that T cells with memory phenotype can arise during homeostastic proliferation following lymphoablative therapy (Taylor et al. 2004).

The same features of memory T cells that permit rapid and efficient protection from pathogens are deleterious for transplanted organs. After reactivation, donor-specific memory CD8 and CD4 T cells rapidly proliferate and differentiate into secondary effectors that directly damage the graft tissue. In addition, memory CD4 cells can provide efficient help for the induction of naïve T cell responses reactive to the donor and for alloantibody production. Under selected circumstances, these downstream mechanisms can be responsible for memory T cell-initiated graft loss. Finally, interactions of memory T cells, especially memory CD8 T cells, with donor endothelium can lead to upregulation of adhesion molecules and chemokine expression, thus promoting infiltration of recipient's immune cells into the graft.

Because of the redundancy of effector mechanisms, controlling memory T cells in transplant setting represents a very challenging task. As a result, animals that have rejected an allograft (and have a pool of alloreactive memory T cells) reject a second graft from the same donor with accelerated kinetics ("second set rejection") (Nedelea et al. 1975; Gordon et al. 1976). Human recipients of multiple transplants following immune-mediated failure of the first graft also have poorer graft survival than recipients of first transplants (Cecka 2001). There is accumulating evidence that higher frequencies of alloreactive T cells prior to transplantation correlate with an increased risk of acute rejection episodes and inferior graft function (Heeger et al. 1999).

3. Memory T Cells and Costimulation

3.1. Classical Costimulatory Pathways

When compared with naïve T cells, polyclonal memory T cells exhibit lower activation thresholds and lower costimulatory requirements for activation (Flynn and Mullbacher 1996; Mullbacher and Flynn 1996; Pihlgren et al. 1996). However, careful examination of these issues became possible only with use of TCR transgenic mice which allowed direct comparison of naïve and memory T cells of the same specificity. Using this approach, it has been confirmed that TCR transgenic memory CD4 T cells are activated when exposed to lower antigen doses and with lower costimulatory requirements relative to naïve cells (Rogers et al. 2000). Memory CD4 T cells proliferate and produce cytokines when the antigen is presented by resting B cells, macrophages, and even endothelial cells that do not express high levels of costimulatory molecules (Croft et al. 1994; Khayyamian et al. 2002). In contrast, naïve CD4 cells can only be activated by professional APCs—dendritic cells and activated B cells (Croft et al. 1994). Studies using CD40-deficient or CD80/86-deficient APCs demonstrated that unlike naïve cells, CD4 memory T cells do not require signaling through these pathways for proliferation and cytokine production (Dutton et al. 1998; London et al. 2000).

In transplantation, blockade of T cell costimulation through CD28/CD80/86 and/or CD40/CD154 pathways seems to be one particularly effective tool for inducing long-term graft survival and, in some cases, donor-specific tolerance in a variety of model systems (Wekerle et al. 2002). Costimulatory blockade-based tolerance inducing protocols are generally effective in naïve animals with naïve T cell repertoires. However, as memory T cells have pre-committed cytokine profiles, lower thresholds for T cell activation, and lower costimulatory requirements than naïve cells (Dutton et al. 1998; Kaech et al. 2002; Sprent and Surh 2002), they are likely to be resistant to the effects of costimulatory blockade and subsequently to tolerance induction. Indeed, we and others have demonstrated that both CD4 and CD8 memory T cells induced either by previous exposure to donor antigens or by cross-reactive antimicrobial immunity prevented the beneficial effect of costimulatory blockade on prolonging allograft survival (Pantenburg et al. 2002; Valujskikh et al. 2002; Adams et al. 2003; Brehm et al. 2003). Our published data indicate that despite costimulatory blockade, alloreactive memory CD4 T cells contribute to allograft rejection through multiple pathways. Upon reactivation, memory CD4 T cells not only can become effector cells themselves but can also provide help for the priming of naïve T and B lymphocytes that in turn mediate graft destruction (Chen et al. 2004). In general, alloreactive memory T cells interfere with long-term allograft survival induced through conventional costimulatory blockade. This issue is highly relevant to clinical transplantation because memory T cells are present in human transplant recipients, costimulatory blockade-based trials in humans are ongoing, and initial results in humans and primates suggest that their efficacy is significantly lower than in naïve mice.

3.2. Alternative Costimulatory Pathways

Newly recognized alternative costimulatory pathways may be required for optimal restimulation of memory CD4 T cells. These receptors include members of the immunoglobulin superfamily represented by inducible T cell costimulator (ICOS) and the tumor necrosis factor receptor superfamily, including CD134 (OX40), CD27, CD137 (4-1BB), CD30, and herpes virus entry mediator (HVEM). Most of these molecules are expressed only upon T cell activation, suggesting a particularly important role in the effector/memory phases of T cell response rather than the initial T cell priming. Several lines of evidence indicate that memory T cells may rely on such alternative costimulatory molecules for activation, survival, or effector functions.

One of the recently characterized pathways is interaction of ICOS on T cells with the B7RP-1 molecule on APCs (Yoshinaga et al. 1999; Tafuri et al. 2001). Studies of autoimmune responses demonstrated that costimulation through ICOS is crucial for activating previously primed (but not naïve) T cells and that blocking ICOS costimulation after the onset of the disease ameliorates signs of experimental allergic encephalomyelitis (EAE) and collagen-induced arthritis (Sporici et al. 2001; Iwai et al. 2002). In transplantation, the expression of ICOS is significantly upregulated during acute heart allograft rejection. Blockade of the ICOS/B7RP-1 costimulatory pathway prolongs allograft graft survival in unsensitized rodents and synergizes with DST/anti-CD154 therapy (Ozkaynak et al. 2001). Recently, Sayegh and colleagues demonstrated that blocking the ICOS/B7RP-1 pathway prolongs allograft survival in rodents even if the blockade is administered at a delayed time point after transplantation, suggesting that it interferes with the recall of primed T cells (Harada et al. 2003). These studies further revealed that blockade of ICOS or B7RP-1 limited expansion of alloreactive CD4 T cells, inhibited generation of CD8 effector cells, and decreased production of alloantibodies.

We investigated how therapeutic interference with ICOS/B7RP 1 costimulatory pathway affects functions of allospecific memory CD4 T cells during allograft rejection. Treatment with blocking anti-ICOS monoclonal antibody synergized with donor-specific transfusion plus anti-CD154 antibody therapy (DST/MR1) and prolonged murine cardiac allograft survival despite the presence of donor-reactive memory CD4 T cells. However, we found that ICOS blockade acted downstream of memory CD4 T cell reactivation. While anti-ICOS antibody had little effect on the expansion of pre-existing memory CD4 T cells or the induction of allospecific effector T cells, it inhibited T cell recruitment into the graft. Furthermore, anti-ICOS antibody treatment in combination with DST/MR1 entirely prevented B cell help provided by memory CD4 T cells and production of donor-specific IgG antibody. These results could be utilized to rationally combine ICOS blockade with other graft-prolonging strategies in sensitized transplant patients.

4. Altering Memory T Cell Trafficking

In general, factors determining the migration patterns of memory T cells are poorly understood. Recent findings implicated sphingosine-1-phosphate as a potent lymphocyte chemoattractant for naïve T and B cells. The concentration of S1P in blood and lymph is higher than in the lymphoid organs, thus causing lymphocytes to leave the lymphoid tissues and enter the circulation (Goetzl and Graler 2004; Hla 2004; Lo et al. 2005). Although five different S1P receptors have been described in mammals, T lymphocytes predominantly express type 1 S1P (S1P1) receptor (Hla et al. 2001). The significance of S1P/S1P1 receptor interactions for lymphocyte recirculation was lately emphasized by the findings that S1P1 receptor-deficient T cells were permanently trapped in the thymus or secondary lymphoid organs (Matloubian et al. 2004). In addition, S1P1 receptor is transiently down regulated after T cell activation, thus ensuring proper T cell priming and differentiation within lymphoid compartments (Matloubian et al. 2004; Chi and Flavell 2005).

Extensive studies of S1P receptor-mediated migration were prompted by the discovery of an immunosuppressive agent FTY720. Unlike other immunosuppressants, FTY720 does not inhibit activation and proliferation of T cells or induction of humoral immune responses (Pinschewer et al. 2000; Xie et al. 2003). Instead, in vivo FTY720 is converted to its phosphate ester metabolite that binds to S1P1 receptor with higher affinity than the natural ligand S1P. This interaction results in internalization of S1P1 receptor and prevents lymphocytes from leaving the secondary lymphoid organs and from infiltrating the inflammation sites.

Despite being the subject of intensive studies, the potential effect of FTY720 on the trafficking of antigen-experienced T cells is controversial. In theory, the ability of FTY720 to sequester T cells in the lymphoid organs depends on the ability of these cells to migrate into lymphoid organs. While FTY720 effectively inhibits the recruitment of newly generated effector T cells to peripheral tissues and thus prolongs survival of solid organ allografts, it has been reported to have no significant effect on pre-existing memory T cells (Pinschewer et al. 2000; Xie et al. 2003).

We have lately explored the possibility of controlling deleterious donor-reactive memory CD4 T cells through lymphoid sequestration (Zhang et al. 2006). We showed that FTY720 induces relocation of circulating memory CD4 T cells into secondary lymphoid organs. Lymphoid sequestration of donor-reactive memory CD4 T cells prolonged the survival of murine heterotopic cardiac allografts and synergized with conventional costimulatory blockade to further increase graft survival. Despite limited trafficking, memory CD4 T cells remained capable of providing help for the induction of anti-donor CD8 T cell and alloantibody responses. Elimination of anti-donor humoral immunity resulted in indefinite allograft survival, proving the pathogenicity of alloantibody under these conditions. These findings emphasize that memory CD4 T cells can influence anti-donor immune responses regardless of lymphoid sequestration and that additional strategies may be required to control alloantibody-mediated graft injury under these conditions.

5. Conclusions

Initially developed for enhanced host protection against pathogens, memory T cells endanger life-saving organ and tissue transplants. Moreover, donor-reactive memory responses enhance multiple redundant effector mechanisms of allograft rejection.

Understanding immunobiology of memory T cells in general and allospecific memory T cells in particular is essential to rational and efficient control of these cells in allograft recipients.

References

Adams, A.B., Williams, M.A., Jones, T.R., Shirasugi, N., Durham, M.M., Kaech, S.M., Wherry, E.J., Onami, T., Lanier, J.G., Kokko, K.E., Pearson, T.C., Ahmed, R. and Larsen, C.P. (2003) Heterologous immunity provides a potent barrier to transplantation tolerance. J. Clin. Invest. 111, 1887–1895.

Bradley, L.M., Harbertson, J., Freschi, G.C., Kondrack, R. and Linton, P.J. (2000) Regulation of development and function of memory CD4 subsets. Immunol. Res. 21, 149–158.

Brehm, M.A., Markees, T.G., Daniels, K.A., Greiner, D.L., Rossini, A.A. and Welsh, R.M. (2003) Direct visualization of cross-reactive effector and memory allo-specific CD8 T cells generated in response to viral infections. J. Immunol. 170, 4077–4086.

Cecka, J.M. (2001) The UNOS renal transplant registry. Clin. Transpl., 1–18.

Chen, Y., Heeger, P.S. and Valujskikh, A. (2004) In vivo helper functions of alloreactive memory CD4+ T cells remain intact despite donor-specific transfusion and anti-CD40 ligand therapy. J. Immunol. 172, 5456–5466.

Chi, H. and Flavell, R.A. (2005) Cutting edge: regulation of T cell trafficking and primary immune responses by sphingosine 1-phosphate receptor 1. J. Immunol. 174, 2485–2488.

Cho, B.K., Wang, C., Sugawa, S., Eisen, H.N. and Chen, J. (1999) Functional differences between memory and naïve CD8 T cells. Proc. Natl. Acad. Sci. USA 96, 2976–2981.

Croft, M., Bradley, L.M. and Swain, S.L. (1994) Naïve versus memory CD4 T cell response to antigen. Memory cells are less dependent on accessory cell costimulation and can respond to many antigen-presenting cell types including resting B cells. J. Immunol. 152, 2675–2685.

Dutton, R.W., Bradley, L.M. and Swain, S.L. (1998) T cell memory. Annu. Rev. Immunol. 16, 201–223.

Flynn, K. and Mullbacher, A. (1996) Memory alloreactive cytotoxic T cells do not require costimulation for activation in vitro. Immunol. Cell. Biol. 74, 413–420.

Garcia, S., DiSanto, J. and Stockinger, B. (1999) Following the development of a CD4 T cell response in vivo: from activation to memory formation. Immunity 11, 163–171.

Goetzl, E.J. and Graler, M.H. (2004) Sphingosine 1-phosphate and its type 1 G protein-coupled receptor: trophic support and functional regulation of T lymphocytes. J. Leukoc. Biol. 76, 30–35.

Gordon, R.D., Mathieson, B.J., Samelson, L.E., Boyse, E.A. and Simpson, E. (1976) The effect of allogeneic presensitization on H-Y graft survival and in vitro cell-mediated responses to H-y antigen. J. Exp. Med. 144, 810–820.

Harada, H., Salama, A.D., Sho, M., Izawa, A., Sandner, S.E., Ito, T., Akiba, H., Yagita, H., Sharpe, A.H., Freeman, G.J. and Sayegh, M.H. (2003) The role of the ICOS-B7h T cell costimulatory pathway in transplantation immunity. J. Clin. Invest. 112, 234–243.

Heeger, P.S., Greenspan, N.S., Kuhlenschmidt, S., Dejelo, C., Hricik, D.E., Schulak, J.A. and Tary-Lehmann, M. (1999) Pretransplant frequency of donor-specific, IFN-gamma-producing lymphocytes is a manifestation of immunologic memory and correlates with the risk of posttransplant rejection episodes. J. Immunol. 163, 2267–2275.

Hla, T. (2004) Physiological and pathological actions of sphingosine 1-phosphate. Semin. Cell. Dev. Biol. 15, 513–520.

Hla, T., Lee, M.J., Ancellin, N., Paik, J.H. and Kluk, M.J. (2001) Lysophospholipids—receptor revelations. Science 294, 1875–1878.

Iwai, H., Kozono, Y., Hirose, S., Akiba, H., Yagita, H., Okumura, K., Kohsaka, H., Miyasaka, N. and Azuma, M. (2002) Amelioration of collagen-induced arthritis by blockade of inducible costimulator-B7 homologous protein costimulation. J. Immunol. 169, 4332–4339.

Kaech, S.M., Wherry, E.J. and Ahmed, R. (2002) Effector and memory T cell differentiation: implications for vaccine development. Nat. Rev. Immunol. 2, 251–262.

Kassiotis, G. and Stockinger, B. (2004) Anatomical heterogeneity of memory CD4+ T cells due to reversible adaptation to the microenvironment. J. Immunol. 173, 7292–7298.

Kedl, R.M. and Mescher, M.F. (1998) Qualitative differences between naïve and memory T cells make a major contribution to the more rapid and efficient memory CD8+ T cell response. J. Immunol. 161, 674–683.

Khayyamian, S., Hutloff, A., Buchner, K., Grafe, M., Henn, V., Kroczek, R.A. and Mages, H.W. (2002) ICOS-ligand, expressed on human endothelial cells, costimulates Th1 and Th2 cytokine secretion by memory CD4+ T cells. Proc. Natl. Acad. Sci. USA 99, 6198–6203.

Lanzavecchia, A. and Sallusto, F. (2001) Antigen decoding by T lymphocytes: from synapses to fate determination. Nat. Immunol. 2, 487–492.

Lanzavecchia, A. and Sallusto, F. (2002) Opinion-decision making in the immune system: progressive differentiation and selection of the fittest in the immune response. Nat. Rev. Immunol. 2, 982–987.

Lanzavecchia, A. and Sallusto, F. (2005) Understanding the generation and function of memory T cell subsets. Curr. Opin. Immunol. 17, 326–332.

Lindahl, K.F. and Wilson, D.B. (1977) Histocompatibility antigen-activated cytotoxic T lymphocytes. I. Estimates of the absolute frequency of killer cells generated in vitro. J. Exp. Med. 145, 500–507.

Lo, C.G., Xu, Y., Proia, R.L. and Cyster, J.G. (2005) Cyclical modulation of sphingosine-1-phosphate receptor 1 surface expression during lymphocyte recirculation and relationship to lymphoid organ transit. J. Exp. Med. 201, 291–301.

London, C.A., Lodge, M.P. and Abbas, A.K. (2000) Functional responses and costimulator dependence of memory CD4+ T cells. J. Immunol. 164, 265–272.

Masopust, D., Vezys, V., Marzo, A.L. and Lefrancois, L. (2001) Preferential localization of effector memory cells in nonlymphoid tissue. Science 291, 2413–2417.

Matesic, D., Lehmann, P.V. and Heeger, P.S. (1998) High-resolution characterization of cytokine-producing alloreactivity in naïve and allograft-primed mice. Transplantation 65, 906–914.

Matloubian, M., Lo, C.G., Cinamon, G., Lesneski, M.J., Xu, Y., Brinkmann, V., Allende, M.L., Proia, R.L. and Cyster, J.G. (2004) Lymphocyte egress from thymus and peripheral lymphoid organs is dependent on S1P receptor 1. Nature 427, 355–360.

Mullbacher, A. and Flynn, K. (1996) Aspects of cytotoxic T cell memory. Immunol. Rev. 150, 113–127.

Nedelea, M., Dima, S., Sandru, G. and Nicolaescu, V. (1975) A comparative study of allogenic "first set" and "second set" skin rejection in mice. Morphol. Embryol. 21, 53–57.

Ozkaynak, E., Gao, W., Shemmeri, N., Wang, C., Gutierrez-Ramos, J.C., Amaral, J., Qin, S., Rottman, J.B., Coyle, A.J. and Hancock, W.W. (2001) Importance of ICOS-B7RP-1 costimulation in acute and chronic allograft rejection. Nat. Immunol. 2, 591–596.

Pantenburg, B., Heinzel, F., Das, L., Heeger, P.S. and Valujskikh, A. (2002) T cells primed by Leishmania major infection cross-react with alloantigens and alter the course of allograft rejection. J. Immunol. 169, 3686–3693.

Perrin, P.J., Lovett-Racke, A., Phillips, S.M. and Racke, M.K. (1999) Differential requirements of naïve and memory T cells for CD28 costimulation in autoimmune pathogenesis. Histol. Histopathol. 14, 1269–1276.

Pihlgren, M., Dubois, P.M., Tomkowiak, M., Sjogren, T. and Marvel, J. (1996) Resting memory CD8+ T cells are hyperreactive to antigenic challenge in vitro. J. Exp. Med. 184, 2141–2151.

Pinschewer, D.D., Ochsenbein, A.F., Odermatt, B., Brinkmann, V., Hengartner, H. and Zinkernagel, R.M. (2000) FTY720 immunosuppression impairs effector T cell peripheral homing without affecting induction, expansion, and memory. J. Immunol. 164, 5761–5770.

Ravkov, E.V., Myrick, C.M. and Altman, J.D. (2003) Immediate early effector functions of virus-specific CD8+CCR7+ memory cells in humans defined by HLA and CC chemokine ligand 19 tetramers. J. Immunol. 170, 2461–2468.

Reinhardt, R.L., Khoruts, A., Merica, R., Zell, T. and Jenkins, M.K. (2001) Visualizing the generation of memory CD4 T cells in the whole body. Nature 410, 101–105.

Rogers, P.R., Dubey, C. and Swain, S.L. (2000) Qualitative changes accompany memory T cell generation: faster, more effective responses at lower doses of antigen. J. Immunol. 164, 2338–2346.

Sallusto, F., Lenig, D., Forster, R., Lipp, M. and Lanzavecchia, A. (1999) Two subsets of memory T lymphocytes with distinct homing potentials and effector functions. Nature 401, 708–712.

Seder, R.A. and Ahmed, R. (2003) Similarities and differences in CD4+ and CD8+ effector and memory T cell generation. Nat. Immunol. 4, 835–842.

Sporici, R.A., Beswick, R.L., von Allmen, C., Rumbley, C.A., Hayden-Ledbetter, M., Ledbetter, J.A. and Perrin, P.J. (2001) ICOS ligand costimulation is required for T cell encephalitogenicity. Clin. Immunol. 100, 277–288.

Sprent, J. and Surh, C.D. (2002) T cell memory. Annu. Rev. Immunol. 20, 551–579.

Suchin, E.J., Langmuir, P.B., Palmer, E., Sayegh, M.H., Wells, A.D. and Turka, L.A. (2001) Quantifying the frequency of alloreactive T cells in vivo: new answers to an old question. J. Immunol. 166, 973–981.

Swain, S.L. (1994) Generation and in vivo persistence of polarized Th1 and Th2 memory cells. Immunity 1, 543–552.

Tafuri, A., Shahinian, A., Bladt, F., Yoshinaga, S.K., Jordana, M., Wakeham, A., Boucher, L.M., Bouchard, D., Chan, V.S., Duncan, G., Odermatt, B., Ho, A., Itie, A., Horan, T., Whoriskey, J.S., Pawson, T., Penninger, J.M., Ohashi, P.S. and Mak, T.W. (2001) ICOS is essential for effective T-helper-cell responses. Nature 409, 105–109.

Taylor, D.K., Neujahr, D. and Turka, L.A. (2004) Heterologous immunity and homeostatic proliferation as barriers to tolerance. Curr. Opin. Immunol. 16, 558–564.

Unsoeld, H., Krautwald, S., Voehringer, D., Kunzendorf, U. and Pircher, H. (2002) Cutting edge: CCR7+ and CCR7 – memory T cells do not differ in immediate effector cell function. J. Immunol. 169, 638–641.

Valujskikh, A., Pantenburg, B. and Heeger, P.S. (2002) Primed allospecific T cells prevent the effects of costimulatory blockade on prolonged cardiac allograft survival in mice. Am. J. Transplant. 2, 501–509.

Watschinger, B., Gallon, L., Carpenter, C.B. and Sayegh, M.H. (1994) Mechanisms of allorecognition. Recognition by in vivo-primed T cells of specific major histocompatibility complex polymorphisms presented as peptides by responder antigen-presenting cells. Transplantation 57, 572–576.

Wekerle, T., Kurtz, J., Bigenzahn, S., Takeuchi, Y. and Sykes, M. (2002) Mechanisms of transplant tolerance induction using costimulatory blockade. Curr. Opin. Immunol. 14, 592–600.

Wherry, E.J., Teichgraber, V., Becker, T.C., Masopust, D., Kaech, S.M., Antia, R., von Andrian, U.H. and Ahmed, R. (2003) Lineage relationship and protective immunity of memory CD8 T cell subsets. Nat. Immunol. 4, 225–234.

Xie, J.H., Nomura, N., Koprak, S.L., Quackenbush, E.J., Forrest, M.J. and Rosen, H. (2003) Sphingosine-1-phosphate receptor agonism impairs the efficiency of the local immune response by altering trafficking of naïve and antigenActivated CD4(+) T cells. J. Immunol. 170, 3662–3670.

Yoshinaga, S.K., Whoriskey, J.S., Khare, S.D., Sarmiento, U., Guo, J., Horan, T., Shih, G., Zhang, M., Coccia, M.A., Kohno, T., Tafuri-Bladt, A., Brankow, D., Campbell, P., Chang, D., Chiu, L., Dai, T., Duncan, G., Elliott, G.S., Hui, A., McCabe, S.M., Scully, S., Shahinian, A., Shaklee, C.L., Van, G. and Mak, T.W. (1999) T cell co-stimulation through B7RP-1 and ICOS. Nature 402, 827–832.

Zhang, Q., Chen, Y., Fairchild, R.L., Heeger, P.S. and Valujskikh, A. (2006) Lymphoid sequestration of alloreactive memory CD4 T cells promotes cardiac allograft survival. J. Immunol. 176, 770–777.

Zimmermann, C., Prevost-Blondel, A., Blaser, C. and Pircher, H. (1999) Kinetics of the response of naïve and memory CD8 T cells to antigen: similarities and differences. Eur. J. Immunol. 29, 284–290.

27

Differences in Dendritic Cell Activation and Distribution After Intravenous, Intraperitoneal, and Subcutaneous Injection of Lymphoma Cells in Mice

Alexandra L. Sevko[1], Nadzeya Barysik[2], Lori Perez[3], Michael R. Shurin[3], and Valentin Gerein[4]

[1]R.E. Kavetsky Institute of Experimental Pathology, Oncology and Radiobiology, Ukrainian Academy of Sciences, Kyiv, Ukraine
[2]Department of Pediatric Pathology, University of Mainz, Mainz, Germany
[3]Department of Pathology, University of Pittsburgh Medical Center, Pittsburgh, PA, USA
[4]Department of Pediatric Pathology, Institute of Pathology, University of Mainz, Mainz, Germany, v.gerein@web.de

Abstract. Dendritic cells (DCs) are key antigen-presenting cells (APCs) for initiating immune responses. However, in recent years, several groups have shown the defective function of DCs in tumor-bearing mice and in cancer patients. Our aim was to study the effects of lymphoma on DC differentiation and maturation and to assess the input of the tumor microenvironment and intravasation of tumor cells on DC precursors. EL-4 lymphoma cells were administrated via different routes (intraperitoneal, subcutaneous, and intravenous) and DC phenotype was investigated. Bone marrow-derived DCs and APCs obtained from the spleen were examined by flow cytometry, and immunohistochemical analysis of lymphoma, lungs, livers, and spleens was also performed. Intravenous administration of lymphoma cells induced suppression of DC differentiation and maturation assessed as a significant decrease of the IAb, CD80, CD86, CD11b, and CD11c expression on DCs and IAb on splenic APCs. Up-regulation of APC differentiation was observed in animals after subcutaneous and intraperitoneal administration of lymphoma cells determined as increased expression of CD40 and CD86 in spleen APCs. These data suggest that the development of antitumor immune response might differ in the host receiving tumor vaccines via different injection routes.

1. Introduction

Dendritic cells (DCs) are key antigen-presenting cells (APC) regulating immune responses and play a central role in antitumor immunity by taking up tumor antigens and stimulating antigen-specific T cells (McKechnie et al. 2004). In recent years, the defective function of DCs in tumor-bearing mice and in cancer patients has been repeatedly reported (Della Bella et al. 2003; Gabrilovich et al. 1997; Ishida et al. 1998; Tsuge et al. 2005; Gabrilovich et al. 1996b; Pospisilova et al. 2005; Gerlini et al. 2004). The major finding of these studies was the lack of expression of costimulatory molecules and activation markers, such as CD40, CD80, CD86, and

CD83, in tumor-associated DCs, consistent with the phenotype of immature and nonactivated DCs (Nestle et al. 1997; Chaux et al. 1996, 1997; Neves et al. 2005; Shurin et al. 2002). In agreement with these reports, DC function was impaired in patients with breast, head and neck, and lung cancer. The defective APC function in cancer patients was associated with the dramatic decrease of immunocompetent DCs in the peripheral blood and accumulation of cells lacking maturation markers (Almand et al. 2000).

Our aims were to study the effects of lymphoma on DC differentiation and maturation and evaluate the input of such factors as tumor microenvironment and intravasation of tumor cells on APCs. The EL-4 lymphoma model was used and tumor cells were administrated intraperitoneally, subcutaneously, and intravenously.

2. Experimental Procedures

To detect the effects of route of tumor administration on function and phenotype of DCs, three groups of C57BL/6 mice (Taconic, USA; N = 6/group) were studied. Group A was injected intraperitoneally (IP), group B subcutaneously (SC), and group C intravenously (IV) in the tail vein with 10^5 syngeneic EL-4 lymphoma cells in 0.2 ml of PBS. Three animals from each group were killed on the 7th and 14th day after lymphoma cell administration, and immunohistochemistry of lungs, livers, and spleens and of tumor tissue (in case of SC transplantation) was performed. Tumor-free mice were used as a control. All experiments were repeated and the results are given as mean ± SD. Differences between experimental and control values were assessed by Student's *t*-test or ANOVA after evaluating for normal distribution.

Murine bone marrow cells were obtained from flushed marrow cavities and depleted of B and T cells by incubating with anti-B220, anti-CD4, and anti-CD8 monoclonal antibodies plus rabbit complement for 1 h at 37°C. After washing and removing adherent cells, enriched population of hematopoietic precursors were cultured in RPMI 1640 complete medium with 1000 U/ml GM-CSF and 1000 U/ml IL-4 (PeproTech, USA) for 6–7 days.

For splenocyte isolation, the spleens were homogenized in RPMI 1640 medium and filtered through 70-μm nylon mesh (BD, USA), and red blood cells were removed with lysing buffer (Sigma, USA). Next, DCs and splenocytes were washed in PBS, and 3×10^5 cells/tube were suspended in FACScan buffer double stained for 30 min with the following monoclonal antibodies (mAbs): IAb, CD86, CD80, CD40, CD11c, and CD11b (PharMingen, USA). Samples and the appropriate isotype-matched controls were fixed in 2% PFA and analyzed on the FACScan (BD, USA).

Immunohistochemical analysis of lymphoid and nonlymphoid tissues was performed using 4-μm cryostat sections, and after the application of avidin/biotin blocking kit (Vector Laboratories, USA) CD11c (PharMingen, USA), and NLDC-145 (Serotec, USA), mAbs were used followed by biotinylated secondary mouse anti-rat antibodies (Jackson ImmunoResearch Laboratories Inc., USA). The color reaction was developed using the peroxidase chromogen kit (BioMega Corp., USA) and all sections were counterstained with hematoxylin. The number of positive cells per tissue sections was determined semiquantatively as follows: 0, no positive staining; +, weak but positive staining of scattered cells; ++, positive staining; ++++, strongly positive staining; and +++, a staining intensity between ++ and ++++.

3. Results

We have observed disseminated lymphoma growth upon IV administration, local growth upon SC, and partially circumscribed growth upon IP injection. After SC injection, EL-4 lymphoma cells disseminated in the skin and regional lymph nodes. Mice died on 24th–28th day after SC injection and no distant metastases were observed; the tumor was encapsulated and located at the place of injection. After IP administration of lymphoma cells, ascites was developed and tumor growth was restricted within the abdominal cavity and mesenteric lymph nodes. However, IV administration of tumor cells led to the systemic cell distribution, and on the 24th day after injection, we observed lymphoma infiltration in mesenteric, inguinal, and axillary lymph nodes, 10–20% of lymphoma cells in spleen, and tumor formation in the liver and, in some cases, in the kidneys.

The results of the immunohistochemical analysis showed the increase in CD11c expression in the spleen 7 days after SC and IV lymphoma cell injections, in the lungs after IP and IV administration, and in the liver after IV administration (Table 1). On day 14, CD11c expression was decreased in the spleen and lymph nodes in all groups irrespectively of the route of administration but was increased in the liver upon IV administration. In the tumor tissues, $CD11c^{+}$ cells were revealed in mice receiving SC injections of tumor cells on day 7 but were undetectable by day 14 (Table 1). These data suggest marked redistribution of mature $CD11c^{+}$ DC in tumor-bearing host with significant attenuation of these cells with tumor progression. Furthermore, on day 14, expression of NLDC-145 (DEC205) increased in the spleen in IV and IP groups, while it decreased in SC group of mice. However, on day 14, expression of NLDC-145 decreased in the lymph nodes in all groups when compared with tumor-free control tissues (Table 1).

The flow cytometry analysis of splenocyte phenotype revealed considerable decrease of IAb levels on splenic CD11c+ DC on day 7 after IV injection of lymphoma cells, while expression of the IAb was similar to control values upon SC and IP administration of tumor cells (Table 2; Figure 1).

Interestingly, the phenotype of bone marrow-derived DCs differed substantially in all tumor-bearing groups when compared with tumor-free mice (Table 3). The dramatic decrease of IAb, CD80, CD86, CD11b, and CD11c expression on DCs was revealed in IV group, while 42% increase in CD86 and 73% increase in CD40 expression was observed in DCs obtained from mice with Subcutaneously injected lymphoma cells. In addition, 35% increase in CD40 expression was shown in IP group (Table 3).

Thus, expression of all tested markers on bone marrow-derived DCs (IAb, CD80, CD86, CD40, CD11b, and CD11c) obtained from mice after IV administration of tumor cells and associated with systemic tumor spreading and multiple metastases was

TABLE 1. Immunohistochemical analysis of dendritic cell (DC) markers in lymphoid and non-lymphoid tissues in lymphoma-bearing and tumor-free mice

Organs and route of tumor cell injection	Markers			
	CD11c		NLDC-145	
	7th day of lymphoma growth	14th day of lymphoma growth	7th day of lymphoma growth	14th day of lymphoma growth
Spleen				
Control	+++	+++	++	++
SC	++++	0	++	+
IP	+++	+	++	+++
IV	++++	++	++	+++
Lung				
Control	0	0	0	0
SC	0	0	0	0
IP	+	0	0	0
IV	++	0	0	0
Liver				
Control	0	0	0	0
SC	0	0	0	0
IP	0	0	0	0
IV	+	+	0	0
Lymph nodes				
Control	++	++	+++	+++
SC	–	+	–	+
IP	–	+	–	0
IV	–	+	–	0
Tumor:				
SC	+	0	0	0

significantly different from their expression on DCs obtained from mice after SC or IP administration of lymphoma cells ($p < 0.001$). These data suggest that the route of lymphoma cell administration determines the tumor effects on differentiation and maturation of DCs.

TABLE 2. FACScan analysis of splenocytes isolated 7 days after lymphoma cell administration through the different routes

Route of lymphoma cell administration	Expression of markers on spleen cells	
	IAb	CD86
Control	13.68 ± 1.07	0.18 ± 0.06
SC	15.43 ± 2.03	0.21 ± 0.04
IV	6.65 ± 0.80***	0.16 ± 0.03
IP	17.54 ± 2.62	0.27 ± 0.06

*** $p < 0.001$ vs. tumor-free controls.

TABLE 3. Phenotype of bone marrow-derived dendritic cells (DCs) generated from hematopoietic precursors harvested from mice 14 days after lymphoma cell administration

Route of tumor injection	Expression of markers on DC (in %)					
	IAb	CD86	CD80	CD40	CD11b	CD11c
Control	42.9 ± 1.5	28.4 ± 2.2	28.1 ± 3.9	15.4 ± 2.3	38.1 ± 1.9	51.8 ± 1.5
SC	48.6 ± 1.7*	40.2 ± 1.4*	31.3 ± 1.2	26.7 ± 2.9*	31.0 ± 1.5*	50.7 ± 1.5
IV	21.5 ± 1.1*	15.4 ± 1.8*	17.7 ± 2.0*	14.9 ± 1.5	17.3 ± 1.0*	31.0 ± 2.2*
IP	49.4 ± 2.1*	32.9 ± 3.0	31.1 ± 2.3	20.8 ± 1.3*	39.4 ± 2.9	56.4 ± 1.6*

* $p < 0.05$ vs. tumor-free controls.

4. Discussion

Although DCs are able to induce cytotoxic T lymphocyte reaction against tumors, there are a number of mechanisms by which tumor cells can avoid detection and destruction by the immune system. This includes downregulation of HLA class I expression on the surface of tumor cells, down-regulation of tumor antigen processing and expression, lack of costimulatory molecules on tumors cells, release of immunosuppressive factors, and other mechanisms (Platsoucas et al. 2003). Some of the tumor-derived factors, such as vascular endothelial growth factor (VEGF) (Gabrilovich et al. 1996a; Ohm and Carbone 2001), transforming growth factor-beta (TGF-β) (Weber et al. 2005), cyclooxygenase-2 (Sharma et al. 2003), IL-6 (Hegde et al. 2004), and IL-10 (Qin et al. 1997; Yang and Lattime 2003), are able to downregulate the production and maturation

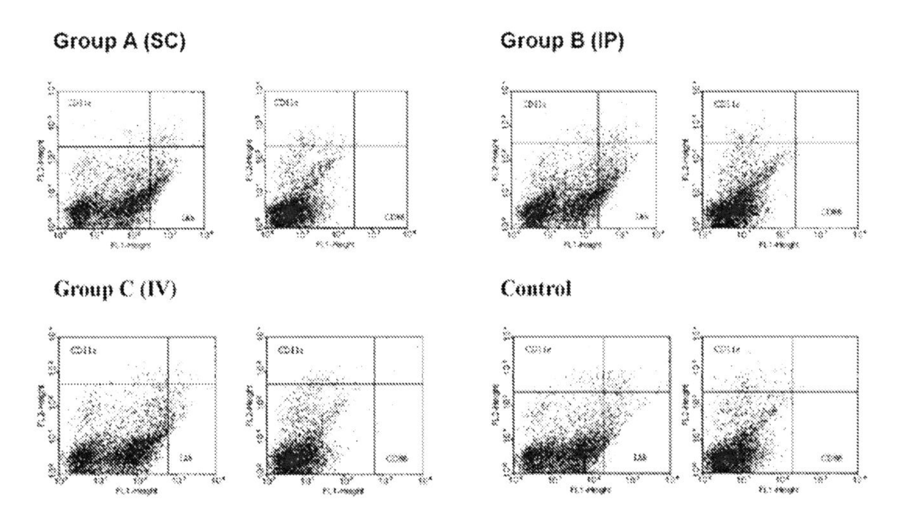

FIGURE 1. Flow cytometric analysis of murine splenocytes isolated and purified from control (tumor-free) and lymphoma-bearing mice 7 days after tumor cell administration. Cells were stained for the expression of IAb and CD86 markers (*X* axis, columns 1 and 3 and 2 and 4, respectively) and CD11c (*Y* axis).

of DCs or to induce DC apoptosis. More aggressive tumors induce higher levels of DC apoptosis and, therefore, more significant inhibition of antigen recognition, processing, and presentating by APCs, which are necessary for the initiation and maintenance of effective antitumor immune response (Esche et al. 1999).

Site of the origin has been described as a prognostic factor that determines clinical outcome of some tumors, such as T cell lymphomas, an anaplastic large cell lymphoma, cutaneous diffuse large B cell lymphoma, and others (ten Berge et al. 2000; Goodlad et al. 2003; Ko et al. 2004). We showed here that the route of lymphoma cell administration determined the patterns of tumor growth and distribution: lymphoma disseminated predominantly to the skin and regional lymph nodes upon SC administration, migrated to mesenteric lymph nodes and abdominal organs upon IP administration, and spread to different organs after IV injection. Furthermore, our results revealed that the effect of lymphoma cells on DC maturation and distribution also depended on the route of tumor cell administration. We have observed significant decrease in IAb, CD80, CD86, CD11b, and CD11c expression on bone marrow-derived DCs and lowering of IAb levels on splenic $CD11^{c}$+ DCs after IV administration of lymphoma cells. CD80 (B7-1) and CD86 (B7-2) are important costimulatory molecules in DC–T cell interaction (Caux et al. 1994). A low level of B7 molecules on DC in patients with cancer, however, may play a certain role in tumor evasion of host defencses (McKechnie et al. 2004).

CD11c is usually expressed at high density on DCs and is considered to be a marker of bone marrow-derived DCs (Hart 1997). The slight increase in CD11c expression on day 7 after lymphoma cell injection might be associated with DC activation, which was abrogated with tumor progression to day 14. Increase in NLDC-145^{+} cell numbers in the spleen on day 14 after IP and IV injection of EL-4 cells may be due to the increase in infiltration macrophages of that also express NLDC-145 molecules (Hart 1997).

Therefore, DC differentiation and maturation were downregulated upon IV injection of lymphoma cells, although some characteristics of DC activation were seen in mice that received SC or IP injections of EL-4 cells. These observations may reflect an initiation of the immune response in a more favorable microenvironment for DCs after SC or IP inoculations of tumor cells. Inhibition of DCs shown upon IV administration of tumor cells can be explained by the more aggressive and generalized disease process as lymphoma cells spread and disseminate to many organs. Thus, intravascular location of the lymphoma cells leads to the inhibition of DC maturation and activation and thus may result in the failure of immune surveillance and therefore worse clinical prognosis of the lymphoma process. Our findings are in agreement with Ferreri et al. (2004), who analyzing intravascular lymphoma, characterized by a predominant growth within the lamina of blood vessels, reported that patients with the disease limited to the skin ("cutaneous variant") exhibited a significantly better outcome and improved survival.

In summary, our results demonstrate that the effects of lymphoma cells on DC differentiation, activation, and distribution in vivo depend on the route of lymphoma cell inoculation and that IV administration of tumor cells induced the strongest downregulation of DC maturation. These data also suggest that antitumor immune response may significantly depend on the route of tumor cells administration, which will be a crucial notion for designing and testing novel tumor cell vaccines.

Acknowledgements

This work has been supported by the UICC International Cancer Technology Transfer Fellowship (contract NO2-CO-91012 to ALS) and NCI NIH (5RO1 CA84270 to MRS).

References

Almand, B., Resser, J.R., Lindman, B., Nadaf, S., Clark, J.I., Kwon, E.D., Carbone D.P. and Gabrilovich, D.I. (2000) Clinical significance of defective dendritic cell differentiation in cancer. Clin. Cancer Res. 6, 1755–1766.

Caux, C., Vanbervliet, B., Massacrier, C., Azuma, M., Okumura, K., Lanier, L.L. and Banchereau, J. (1994) B70/B7-2 is identical to CD86 and is the major functional ligand for CD28 expressed on human dendritic cells. J. Exp. Med. 180, 1841–1847.

Chaux, P., Favre, N. and Martin, M. (1997) Tumor-infiltrating dendritic cells are defective in their antigen-presenting function and inducible B7 expression in rats. Int. J. Cancer 72, 619–624.

Chaux, P., Moutet, M., Faivre, J. and Martin, F. (1996) Inflammatory cells infiltrating human colorectal carcinomas express HLA class II but not B7-1 and B7-2 costimulatory molecules of the T cell activation. Lab. Invest. 74, 975–983.

Della Bella, S., Gennaro, M., Vaccari, M., Ferraris, C., Nicola, S., Riva, A., Clerici, M., Greco, M. and Villa, M.L. (2003) Altered maturation of peripheral blood dendritic cells in patients with breast cancer. Br. J. Cancer 89, 1463–1472.

Esche, C., Lokshin, A., Shurin, G.V., Gastman, B.R., Rabinowich, H., Watkins, S.C., Lotze, M.T. and Shurin, M.R. (1999) Tumor's other immune targets, dendritic cells. J. Leukoc. Biol. 66, 336–344.

Ferreri, A.J., Campo, E., Seymour, J.F., Willemze, R., Ilariucci, F., Ambrosetti, A., Zucca, E., Rossi, G., Lopez-Guillermo, A., Pavlovsky, M.A., Geerts, M.L., Candoni, A., Lestani, M., Asioli, S., Milani, M., Piris, M.A., Pileri, S., Facchetti, F., Cavalli, F., Ponzoni, M. and International Extranodal Lymphoma Study Group (IELSG). (2004) Intravascular lymphoma, clinical presentation, natural history, management and prognostic factors in a series of 38 cases, with special emphasis on the 'cutaneous variant'. Br. J. Haematol. 127, 173–183.

Gabrilovich, D.I., Chen, H.L., Girgis, K.R. and Cunningham, H.T. (1996a) Production of vascular endothelial growth factor by human tumors inhibits the functional maturation of dendritic cells. Nat. Med. 2, 1096–1103.

Gabrilovich, D.I., Ciernik, I.F. and Carbone, D.P. (1996b) Dendritic cells in antitumor immune responses 1 defective antigen presentation in tumor-bearing hosts. Cell. Immunol. 170, 101–110.

Gabrilovich, D.I., Corak, J., Ciernik, I.F., Kavanaugh, D. and Carbone, D.P. (1997) Decreased antigen presentation by dendritic cells in patients with breast cancer. Clin. Cancer Res. 3, 483–490.

Gerlini, G., Tun-Kyi, A., Dudli, C., Burg, G., Pimpinelli, N. and Nestle, F.O. (2004) Metastatic melanoma secreted IL-10 down-regulates CD1 molecules on dendritic cells in metastatic tumor lesions. Am. J. Pathol. 165, 1853–1863.

Goodlad, J.R., Krajewski, A.S., Batstone, P.J., McKay, P., White, J.M., Benton, E.C., Kavanagh, G.M., Lucraft, H.H. and Scotland and Newcastle Lymphoma Group (2003) Primary cutaneous diffuse large B cell lymphoma, prognostic significance of clinicopathological subtypes. Am. J. Surg. Pathol. 27, 1538–1545.

Hart, D.N.J. (1997) Dendritic cells, unique leucocyte populations which control the primary immune response. Blood 90, 3245–3287.

Hegde, S., Pahne, J. and Smola-Hess, S. (2004) Novel immunosuppressive properties of interleukin-6 in dendritic cells, inhibition of NF-kappaB binding activity and CCR7 expression. FASEB J. 18, 1439–1441.

Ishida, T., Oyama, T., Carbone, D. and Gabrilovich, D.I. (1998) Defective function of Langerhans cells in tumor-bearing animals is the result of defective maturation from hematopoietic progenitors. J. Immunol. 161, 4842–4851.

Ko, Y.H., Cho, E.Y., Kim, J.E., Lee, S.S., Huh, J.R., Chang, H.K., Yang, W.I., Kim, C.W., Kim, S.W. and Ree, H.J. (2004) NK and NK-like T cell lymphoma in extranasal sites, a comparative clinicopathological study according to site and EBV status. Histopathology 44, 480–489.

McKechnie, A., Robins, R.A. and Eremin, O. (2004) Immunological aspects of head and neck cancer, biology, pathophysiology and therapeutic mechanisms. Surgeon 2, 187–207.

Nestle, F.O., Burg, G., Fah, J., Wrone-Smith, T. and Nickoloff, B.J. (1997) Human sunlight-induced basal-cell-carcinoma-associated dendritic cells are deficient in T cell co-stimulatory molecules and are impaired as antigen-presenting cells. Am. J. Pathol. 150, 641–651.

Neves, A.R., Ensina, L.F., Anselmo, L.B., Leite, K.R., Buzaid, A.C., Camara-Lopes, L.H. and Barbuto, J.A. (2005) Dendritic cells derived from metastatic cancer patients vaccinated with allogeneic dendritic cell-autologous tumor cell hybrids express more CD86 and induce higher levels of interferon-gamma in mixed lymphocyte reactions. Cancer Immunol. Immunother. 54, 61–66.

Ohm, J.E. and Carbone, D.P. (2001) VEGF as a mediator of tumor-associated immunodeficiency. Immunol. Res. 23, 263–272.

Platsoucas, C.D., Fincke, J.E., Pappas, J., Jung, W.J., Heckel, M., Schwarting, R., Magira, E., Monos, D. and Freedman, R.S. (2003) Immune responses to human tumors, development of tumor vaccines. Anticancer Res. 23, 1969–1996.

Pospisilova, D., Borovickova, J., Rozkova, D., Stary, J., Seifertova, D., Tobiasova, Z., Spisek, R. and Bartunkova, J. (2005) Methods of dendritic cell preparation for acute lymphoblastic leukaemia immunotherapy in children. Med. Oncol. 22, 79–88.

Qin, Z., Noffz, G., Mohaupt, M. and Blankenstein, T. (1997) Interleukin-10 prevents dendritic cell accumulation and vaccination with granulocyte-macrophage colonystimulating factor gene-modified tumor cells. J. Immunol. 159, 770–776.

Sharma, S., Stolina, M., Yang, S.C., Baratelli, F., Lin, J.F., Atianzar, K., Luo, J., Zhu, L., Lin, Y., Huang, M., Dohadwala, M., Batra, R.K. and Dubinett, S.M. (2003) Tumor cyclooxygenase 2-dependent suppression of dendritic cell function. Clin. Cancer Res. 9, 961–968.

Shurin, M.R., Yurkovetsky, Z.R., Tourkova, I.L., Balkir, L. and Shurin, G.V. (2002) Inhibition of CD40 expression and CD40-mediated dendritic cell function by tumor-derived IL-10. Int. J. Cancer 101, 61–68.

ten Berge, R.L., Oudejans, J.J., Ossenkoppele, G.J., Pulford, K., Willemze, R., Falini, B., Chott, A. and Meijer, C.J. (2000) ALK expression in extranodal anaplastic large cell lymphoma favours systemic disease with (primary) nodal involvement and a good prognosis and occurs before dissemination. J. Clin. Pathol. 53, 445–450.

Tsuge, K., Takeda, H., Kawada, S., Maeda, K. and Yamakawa, M. (2005) Characterization of dendritic cells in differentiated thyroid cancer. J. Pathol. 205, 565–576.

Weber, F., Byrne, S.N., Le, S., Brown, D.A., Breit, S.N., Scolyer, R.A. and Halliday, G.M. (2005) Transforming growth factor-beta1 immobilises dendritic cells within skin tumours and facilitates tumour escape from the immune system. Cancer Immunol. Immunother. 54, 898–906.

Yang, A.S. and Lattime, E.C. (2003) Tumor-induced interleukin 10 suppresses the ability of splenic dendritic cells to stimulate CD4 and CD8 T cell responses. Cancer Res. 63, 2150–2157.

28

Role of IL-1-Mediated Inflammation in Tumor Angiogenesis

Elena Voronov, Yaron Carmi, and Ron N. Apte

The Shraga Segal Department of Microbiology and Immunology, Faculty of Health Sciences and Cancer Research Center, Ben-Gurion University of the Negev, Beer-Sheva, Israel, elena@bgu.ac.il

Abstract. Angiogenesis, or generation of new blood vessels from pre-existing vessels, is an integral part of many physiological or pathological processes, including tumor growth. Physiological angiogenesis is a complex process controlled by different proangiogenic as well as antiangiogenic factors. For angiogenic induction, the balance between these pro- and anti-angiogenic factors in the microenvironment has to shift in favor of proangiogenic factors, either by upregulation of these pro-angiogenic factors or by downregulation of angiogenic inhibitors. Proinflammatory cytokines, such as IL-1 and TNFα, were found to be major pro-angiogenic stimuli of both physiological and pathological angiogenesis. The IL-1 family consists of pleiotropic proinflammatory and immunoregulatory cytokines, namely, IL-1α and IL-1β, and one antagonistic protein, the IL-1 receptor antagonist (IL-1Ra), which binds to IL-1 receptors without transmitting an activation signal and represents a physiological inhibitor of preformed IL-1. Previously, we described an important role for microenvironment IL-1, mainly IL-1β, in tumor angiogenesis. In this chapter, we analyze the role of microenvironment host- and tumor cell-derived IL-1 on angiogenesis and the role of inflammation in pathological angiogenesis.

1. Introduction

Angiogenesis comprises the formation of new capillaries from pre-existing vessels by endothelial cell migration and proliferation (Carmeliet and Jain 2000; Carmeliet 2003). This represents a complex series of events that include the local degradation of the basement membrane, directional migration of the underlying endothelial cells, invasion of the surrounding stroma, endothelial cell proliferation, capillary tube morphogenesis, coalescence of small capillaries into larger vessels, vascular pruning, and acquisition of a periendothelial coating. The angiogenic process has been considered to represent the outcome of a straightforward interaction of angiogenic factors with specific signaling receptors of the endothelial cell surface (Folkman and D'Amore 1996; Folkman 2003). Endothelial cells may be quiescent and inactive during months or years, and only upon angiogenic activation, they proliferate. The loss of quiescence is a common feature of different conditions, such as inflammation, atherosclerosis, restenosis, angiogenesis, and various types of vasculopathies, and might be a pathogenic mechanism, linking different diseases that are associated with endothelial cell activation. Blood vessel formation is downregulated in the healthy adult, compared to the embryo, and is manifested in steady-state homeostasis mainly in the female menstrual cycle and in wound healing or in pathological conditions. These include chronic inflammation, diabetic retinopathy, brain and myocardial infarction, other forms of severe ischemia and tissue damage, and tumor growth, when angiogenesis is induced by microenvironment factors in response to hypoxia (reviewed in Hanahan and Folkman1996; Carmeliet 2003; Fiedler and Augustin 2006).

Cross-talk between endothelial cells and microenvironment cellular elements (fibroblasts and other stromal cells and leukocytes) is a key mechanism of the angiogenic response, in both physiological and pathological conditions. In situ, in hypoxia, the net balance between pro- and antiangiogenic factors induces or inhibits the angiogenic switch. Proangiogenic factors include vascular endothelial growth factor (VEGF), platelet-derived growth factor, basic fibroblast growth factor, and proinflammatory molecules, while antiangiogenic factors include thrombospondin-1, interferon α/β, angiostatin, and endostatin. Inflammatory cells are major players in angiogenesis, as they secrete various cytokines and chemokines and induce the expression of adhesion molecules, thus affecting the physiology of endothelial cells. Thus, a close connection between inflammation and angiogenesis has been suggested in different experimental models and in patients (see below). Although previously it was thought that new blood vessel growth in the adult occurs only through angiogenesis, which is defined as the sprouting of new vessels from existing structures, it has now become widely recognized that circulating endothelial progenitor cells (EPCs) and other bone marrow-derived precursor cells can be recruited to injured ischemic tissue and form new blood vessels through a process of postnatal vasculogenesis (Moldovan and Asahara 2003; Ishikawa and Asahara 2004; Asahara and Kawamoto 2004). These hemopoietic precursor cells can either differentiate into endothelial cells that are incorporated into newly formed blood vessels or serve as accessory cells for the process of blood vessel formation (Murayama et al. 2002; Rafii et al. 2002). Studies on the mechanism of neovasculogenesis have only recently started to emerge.

2. IL-1 and Inflammation

IL-1 is a pleiotropic cytokine that is mainly involved in inflammation, immune regulation, and hemopoiesis. Three major gene products of the IL-1 family have been thoroughly studied, two agonistic proteins, namely, IL-1α and IL-1β, and one antagonistic protein, the IL-1 receptor antagonist (IL-1Ra). IL-1Ra, which binds to IL-1 receptors without transmitting an activation signal, represents a physiological inhibitor of preformed IL-1 (reviewed in Dinarello 1996; Apte and Voronov 2002; Apte et al. 2006 a, b).

Many cell types produce and secrete IL-1α, IL-1β, and IL-1Ra upon activation with microbes, microbial products, cytokines, and other environmental stimuli. Mononuclear cells secrete the highest levels of IL-1α and IL-1β, whereas non-phagocytic cells secrete IL-1 in only a very limited fashion, mainly IL-1β. As IL-1α is secreted to a much lesser extent than IL-1β, even in activated macrophages, it is not commonly detected in body fluids, except in severe inflammatory responses, in which case it is probably released by necrotizing cells.

Secreted IL-1 (mainly IL-1β), at low local doses, induces limited inflammatory responses followed by activation of specific immune mechanisms, while at high doses, broad inflammation accompanied by tissue damage and tumor invasiveness is evident (Song et al. 2003).

In distinct experimental systems, the association between neoangiogenesis and inflammation has been established and proinflammatory cytokines were shown to be involved in the control of angiogenesis, mainly by affecting endothelial cell proliferation and function. Of special relevance to angiogenesis are IL-1 and TNFα that are considered as "alarm cytokines"; they are generated by macrophages immediately after confronting inflammatory stimuli. IL-1 and TNFα cause inflammation, but more importantly, they induce the expression of proinflammatory genes in diverse stromal/inflammatory cells, which ultimately results in a local cascade of cytokines and small effector molecules that propagate and sustain inflammation (reviewed in Dinarello 1996; Balkwill 2002; Apte and Voronov 2002; Apte et al. 2006a,b). Of major importance are cyclooxygenase type 2 (COX-2), inducible nitric oxide synthase (iNOS), IL-6, and other cytokines/chemokines. In addition, IL-1 and TNFα increase the expression of adhesion molecules on endothelial cells and leukocytes, which promote leukocyte infiltration from the blood into tissues. IL-1 was shown to affect the physiology of endothelial cells, directly or through other cytokines and other proangiogenic molecules that it induces, for example, VEGF (Mantovani and Dejana 1989; Ben-Av et al. 1995; El Awad et al. 2000; Jung et al. 2001). The effects of IL-1 on neovasculogenesis are less characterized. However, IL-1 has pronounced and diverse effects on hemopoiesis, especially myelopoiesis, in the bone marrow and in extramedullary sites that are possibly involved in neovasculogenesis (Apte et al. 2006a,b).

3. IL-1 and Tumor Angiogenesis

The tumor microenvironment consists of malignant cells, immune, stromal, and inflammatory cells, all of which produce cytokines, growth factors, and metalloproteinases and induce expression of adhesion molecules that could promote tumor angiogenesis, progression, and metastasis. IL-1 is an abundant mediator at tumor sites, being produced by stromal/inflammatory cells, as well as by the malignant cells. In experimental tumors and in cancer patients, a positive correlation between local levels of IL-1, tumor invasiveness, and poor prognosis has been established (Apte et al. 2006a,b). We have shown that malignant cells overexpressing and secreting IL-1β are more invasive and metastatic than the parental cells (Song et al. 2003). IL-1β of tumor cell origin was shown to potently promote tumor angiogenesis, which largely account for the malignant phenotype of the cells. In addition, tumor-cell derived IL-1β was shown to promote extramedullary expansion of immature myeloid cells ($Gr1^+/CD11b^+$) that, on the one hand, induce tumor-mediated suppression and, on the other hand, were recently shown to promote angiogenesis by mechanisms not yet understood (Song et al. 2005). We have shown the critical role of microenvironment-derived IL-1β in the process of carcinogenesis and tumor invasiveness (Song et al. 2003, 2005; Krelin et al. 2007). Thus, IL-1β-deficient mice were much less susceptible to the development of fibrosarcomas than control mice, following treatment with the carcinogen 3-methylcholantrene (MCA). This was evidenced by relatively low frequency of tumors and the long lag period before tumor development. On the contrary, mice lacking the natural inhibitor of IL-1, the IL-1 Ra, developed tumors more quickly than did control mice. In the process of 3-MCA carcinogenesis, IL-1β-induced inflammatory responses consisted initially of neutrophils and later of macrophages and were directly correlated to tumor development. The role of IL-1β in tumor invasiveness was manifested by the inability of transplantable tumor cell lines to develop tumors in IL-1β-deficient mice, mainly due to the failure of inducing the angiogenic switch (Voronov et al. 2003). Most strikingly, 3-MCA-induced fibrosarcoma cells from IL-1β-deficient mice failed to form tumors in control mice, due to the inability to initiate an IL-1-dependent inflammatory response. Thus, IL-1β of both the malignant cell and the host are essential for tumor invasiveness, largely through inducing a local inflammatory response that promotes angiogenesis and other phenomena characteristic of the malignant process. IL-1α, which is largely cell-associated and almost not secreted, is less important for carcinogenesis and tumor invasiveness.

The IL-1Ra has been routinely used for alleviating clinical symptoms of rheumatoid arthritis patients, due to its anti-inflammatory features. We have hypothesized that it also has the potential to intervene in tumor growth. Indeed, we have shown that treatment of tumor-bearing mice with IL-1Ra, in controlled-release systems, reduced tumor growth and angiogenesis (Bar et al. 2004). These results were also confirmed by others (Elaraj et al. 2006).

In conclusion, we have demonstrated the role of secretable IL-1 in carcinogenesis and tumor invasiveness and have applied the basic findings to a cancer treatment protocol based on the application of the IL-1Ra. This is advantageous as IL-1 is an upstream inflammatory mediator that induces in the tumor's microenvironment many products with redundant proangiogenic function. Thus, neutralizing a single molecule, secretable IL-1, has potential to prevent the inflammatory and angiogenic cascade in the malignant process. Studies are being performed to optimize the use of the IL-1Ra as an anticancer agent, in aim to use it at the bedside in cancer patients.

Acknowledgments

The authors would like to thank their collaborators, the late Prof. Shraga Segal and Prof. Smadar Cohen, Ben-Gurion University, Israel; Prof. Charles A. Dinarello, University of Colorado, Denver, CO, USA; Prof. Yochiro Iwakura, University of Tokyo, Tokyo, Japan; Prof. Hynda K. Kleinman, NIH, Bethesda, MD, USA;

Prof. Margot Zoller, Deutsches Krebsforschungscentrum (DKFZ), Heidelberg, Germany, and Drs. Mina Fogel and Monika Huszar, Kaplan Hospital, Rehovot, Israel. Elena Voronov was supported by the Israel Cancer Association, the Israel Ministry of Health Chief Scientist's Office, and the Concern Foundation. Ron N. Apte was supported by the Israel Ministry of Science (MOS) jointly with the DKFZ, Heidelberg, Germany, the United States-Israel Bi-national Foundation (BSF), the Israel Science Foundation founded by the Israel Academy of Sciences and Humanities, the Israel Ministry of Health Chief Scientist's Office, Association for International Cancer Research (AICR), and the German-Israeli DIP collaborative program.

References

Apte, R.N., Dotan, S., Elkabets, M., White, M.R., Reich, E., Carmi, Y., Song, X., Dvorkin, T., Krelin, Y. and Voronov, E. (2006a) The involvement of IL-1 in tumorigenesis, tumor invasiveness, metastasis and tumor-host interactions. Cancer Metastasis Rev. 25, 387–408.

Apte, R.N., Krelin, Y., Song, X., Dotan, S., Recih, E., Elkabets, M., Carmi, Y., Dvorkin, T., White, R.M., Gayvoronsky, L., Segal, S. and Voronov, E. (2006b) Effects of microenvironment- and malignant cell-derived interleukin-1 in carcinogenesis, tumour invasiveness and tumour-host interactions. Eur. J. Cancer 42, 751–759.

Apte, R.N. and Voronov, E. (2002) Interleukin-1-a major pleiotropic cytokine in tumor-host interactions. Semin. Cancer Biol. 12, 277–290.

Asahara, T. and Kawamoto, A. (2004) Endothelial progenitor cells for postnatal vasculogenesis. Am. J. Physiol. Cell Physiol. 287, C572–579.

Balkwill, F. (2002) Tumor necrosis factor or tumor promoting factor? Cytokine Growth Factor Rev. 13, 135–141.

Bar, D., Apte, R.N., Voronov, E., Dinarello, C.A. and Cohen, S. (2004) A continuous delivery system of IL-1Ra reduces angiogenesis and inhibits tumor development. FASEB J. 18, 161–163.

Ben-Av, P., Crofford, L.J., Wilder, R.L. and Hla, T. (1995) Induction of vascular endothelial growth factor expression in synovial fibroblasts by prostaglandin E and interleukin-1: a potential mechanism for inflammatory angiogenesis. FEBS Lett. 372, 83–87.

Carmeliet, P. (2003) Angiogenesis in health and disease. Nat. Med. 9, 653–660.

Carmeliet, P. and Jain, R.K. (2000) Angiogenesis in cancer and other diseases. Nature 407, 249–257.

Dinarello, C.A. (1996) Biologic basis for interleukin-1 in disease. Blood 87, 2095–2147.

El Awad, B., Kreft, B., Wolber, E.M., Hellwig-Burgel, T., Metzen, E., Fandrey, J. and Jelkmann, W. (2000) Hypoxia and interleukin-1beta stimulate vascular endothelial growth factor production in human proximal tubular cells. Kidney Int. 58, 43–50.

Elaraj, D.M., Weinreich, D.M., Varghese, S., Puhlmann, M., Hewitt, S.M., Carroll, N.M., Feldman, E.D., Turner, E.M. and Alexander, H.R. (2006) The role of interleukin 1 in growth and metastasis of human cancer xenografts. Clin. Cancer Res. 12, 1088–1096.

Fiedler, U. and Augustin, H.G. (2006) Angiopoietins: a link between angiogenesis and inflammation. Trends Immunol. 27, 552–558.

Folkman, J. (2003) Fundamental concepts of the angiogenic process. Curr. Mol. Med. 3, 643–651.

Folkman, J. and D'Amore, P.A. (1996) Blood vessel formation: what is its molecular basis? Cell 87, 1153–1155.

Hanahan, D. and Folkman, J. (1996) Patterns and emerging mechanisms of the angiogenic switch during tumorigenesis. Cell 86, 353–364.

Ishikawa, M. and Asahara, T. (2004) Endothelial progenitor cell culture for vascular regeneration. Stem Cells Dev. 13, 344–349.

Jung, Y.D., Liu, W., Reinmuth, N., Ahmad, S.A., Fan, F., Gallick, G.E. and Ellis, L.M. (2001) Vascular endothelial growth factor is upregulated by interleukin-1 beta in human vascular smooth muscle cells via the P38 mitogen-activated protein kinase pathway. Angiogenesis 4, 155–162.

Krelin, Y., Voronov, E., Dotan, S., Elkabets, M., Reich, E., Fogel, M., Huszar, M., Iwakura, Y., Segal, S., Dinarello, C.A. and Apte, R.N. (2007) Interleukin-1beta-driven inflammation promotes the development and invasiveness of chemical carcinogen-induced tumors. Cancer Res. 67, 1062–1071.

Mantovani, A. and Dejana, E. (1989) Cytokines as communication signals between leukocytes and endothelial cells. Immunol. Today 10, 370–375.

Moldovan, N. I. and Asahara, T. (2003) Role of blood mononuclear cells in recanalization and vascularization of thrombi: past, present, and future. Trends Cardiovasc. Med. 13, 265–269.

Murayama, T., Tepper, O. M., Silver, M., Ma, H., Losordo, D.W., Isner, J.M., Asahara, T. and Kalka, C. (2002) Determination of bone marrow-derived endothelial progenitor cell significance in angiogenic growth factor-induced neovascularization in vivo. Exp. Hematol. 30, 967–972.

Rafii, S., Lyden, D., Benezra, R., Hattori, K. and Heissig, B. (2002) Vascular and haematopoietic stem cells: novel targets for anti-angiogenesis therapy? Nat. Rev. Cancer 2, 826–835.

Song, X., Krelin, Y., Dvorkin, T., Bjorkdahl, O., Segal, S., Dinarello, C.A., Voronov, E. and Apte, R.N. (2005) CD11b+/Gr-1+ immature myeloid cells mediate suppression of T cells in mice bearing tumors of IL-1β-secreting cells. J. Immunol. 175, 8200–8208.

Song, X., Voronov, E., Dvorkin, T., Fima, E., Cagnano, E., Benharroch, D., Shendler, Y., Bjorkdahl, O., Segal, S., Dinarello, C.A. and Apte, R.N. (2003) Differential effects of IL-1 alpha and IL-1 beta on tumorigenicity patterns and invasiveness. J. Immunol. 171, 6448–6456.

Voronov, E., Shouval, D.S., Krelin, Y., Cagnano, E., Benharroch, D., Iwakura, Y., Dinarello, C.A. and Apte, R.N. (2003) IL-1 is required for tumor invasiveness and angiogenesis. Proc. Natl. Acad. Sci. USA 100, 2645–2650.

News in Immunodiagnostics and Immunomonitoring

29

New Approaches for Monitoring CTL Activity in Clinical Trials

Anatoli Malyguine, Susan Strobl, Liubov Zaritskaya, Michael Baseler, and Kimberly Shafer-Weaver

Applied and Developmental Research Support Program, SAIC-Frederick, Inc., NCI-Frederick, Frederick, MD, USA, amalyguine@ncifcrf.gov

Abstract. We have developed a modification of the ELISPOT assay that measures Granzyme B (GrB) release from cytotoxic T lymphocytes (CTLs). The GrB ELISPOT assay is a superior alternative to the ^{51}Cr-release assay since it is significantly more sensitive and provides an estimation of cytotoxic effector cell frequency. Additionally, unlike the IFN-γ ELISPOT assay, the GrB ELISPOT directly measures the release of a cytolytic protein. We report that the GrB ELISPOT can be utilized to measure ex vivo antigen-specific cytotoxicity of peripheral blood mononuclear cells (PBMCs) from cancer patients vaccinated with a peptide-based cancer vaccine. We compare the reactivity of patients' PBMCs in the GrB ELISPOT, with reactivity in the tetramer, IFN-γ ELISPOT and chromium (^{51}Cr)-release assays. Differences in immune response over all assays tested were found between patients, and four response patterns were observed. Reactivity in the GrB ELISPOT was more closely associated with cytotoxicity in the ^{51}Cr-release assay than the tetramer or IFN-γ ELISPOT assays. We also optimized the GrB ELISPOT assay to directly measure immune responses against autologous primary tumor cells in vaccinated cancer patients. A perforin ELISPOT assay was also adapted to evaluate peptide-stimulated reactivity of PMBCs from vaccinated melanoma patients. Modifications of the ELISPOT assay described in this chapter allow a more comprehensive evaluation of low-frequency tumor-specific CTLs and their specific effector functions and can provide a valuable insight into immune responses in cancer vaccine trials.

1. Immunological Assays for Monitoring Cancer Vaccine Trials

Active specific immunotherapy is a promising but investigational modality in the management of cancer patients. Currently, several cancer vaccine formulations such as peptides, proteins, antigen-pulsed dendritic cells, and whole tumor cells, in combination with various adjuvants and carriers are being evaluated in clinical trials (Wang and Rosenberg 1999; Rosenberg 2001; Kwak 2003). Monitoring T cell responses in the course of clinical trials is widely used to assess the efficacy of cancer immunotherapy. Selection of an ex vivo monitoring method that provides the best measure of immune reactivity is important in determining correlations between clinical and immunological responses to specific immunotherapy. Standard immunological assays, such as cytokine induction and cell proliferation, can detect immune responses in vaccinated patients but are not suitable for evaluation of individual effector cell reactivity. The chromium-release assay (^{51}Cr-release) has long been the standard assay to measure natural killer (NK) and cytotoxic T lymphocyte (CTL) cytotoxicity; however, the traditional ^{51}Cr-release assay provides only semiquantitative data unless limiting dilution assays are performed and has a relatively low level of sensitivity. The tetramer assay identifies the number of epitope-specific CTLs (Altman et al. 1996), but to quantitate functional cells, it should be combined with intracellular cytokine staining. Assays that can monitor both CTL frequency and function, such as the IFN-γ ELISPOT assay, have gained increasing popularity for monitoring clinical trials (Scheibenbogen et al. 1997 a, b; Keilholz et al. 2002). The ELISPOT assay, a modification of the ELISA, utilizes antibody-coated membranes to detect locally secreted cytokines or other immune proteins by individual cells. The release of immune proteins from activated cells results in spot formation. At appropriate cellular concentrations, each spot formed represents a single reactive cell. Thus, the ELISPOT assay provides both qualitative (type of immune protein) and quantitative (number of responding cells) information. The assay is both reliable and highly sensitive as its detection limit is $1/10^5$ peripheral blood mononuclear cells (PBMCs). However, the IFN-γ ELISPOT assay is not an exclusive measure of CTL activity as non-cytotoxic cells can also secrete IFN-γ (Bachmann et al. 1999). Additionally, CTLs with lytic activity do not always secrete IFN-γ. A more relevant approach to assess functional activity of CTLs would be to measure the secretion of molecules associated with lytic activity.

2. Granzyme B ELISPOT Assay as an Alternative to ^{51}Cr Release

One of the major mechanisms of cell-mediated cytotoxicity involves exocytosis of preformed granules from the effector toward the target cell. These granules contain a number of cytotoxic proteins, including the pore-forming protein perforin and a family of serine proteases called granzymes, including Granzyme B (GrB). Gr B and perforin are present mainly in the granules of $CD8^+$ CTL and NK cells and the release of these factors in response to the appropriate target may be used as a surrogate to evaluate cell-mediated cytotoxicity.

Unlike the IFN-γ ELISPOT, which is widely utilized, the application of the GrB ELISPOT for monitoring clinical trials has been limited. The GrB ELISPOT assay was previously shown to measure GrB release by GrB-transfected CHO cells, T cell lines, and PBMCs from patients with AIDS (Rininsland et al. 2000; Kleen et al. 2004). Our laboratory assessed whether the GrB ELISPOT assay was a viable alternative to the ^{51}Cr-release assay for measuring antigen-specific CTLs (Shafer-Weaver et al. 2003). We generated αFMP-CTL as a clinically relevant model system to assess whether the GrB ELISPOT assay can reliably detect effector cell responses to specific peptides. We found excellent correlation between GrB release in the ELISPOT assay and cytotoxicity in the ^{51}Cr-release assay Figure 1.

To evaluate specificity of the assay, we tested CTL reactivity against FMP-pulsed C1R.A2 (specific targets) as well as nonpulsed and MART-1-pulsed C1R.A2 cells (nonspecific targets) in the GrB ELISPOT assay. K562 were utilized as a control for NK activity (data not shown). GrB secretion was antigen specific as only wells with CTLs and FMP-pulsed CIR.A2 contained a substantial number of spots Figure 2. To further confirm that we were measuring CTL activity, we removed CD8$^+$ cells from the cultures using anti-CD8 mAbs and magnetic beads. After depletion, the percentage of CD8$^+$ cells in the cultures was decreased from 24.8 ± 2.9% to 2.7 ± 0.4%. GrB secretion was assessed for both total and CD8$^+$-depleted CTL cultures. With CD8$^+$ cell depletion from the CTL cultures, there was almost complete abrogation of GrB measured in the ELISPOT. This demonstrates that CTLs are the main producers of GrB and confirm the specificity of the GrB ELISPOT assay.

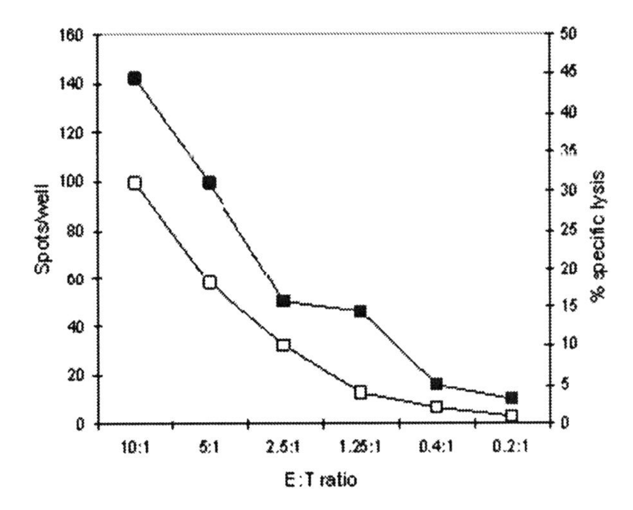

FIGURE 1. Granzyme B release in the ELISPOT assay correlates with cytotoxicity in the ^{51}Cr-release assay. Human anti-FMP-CTLs (7-day culture) were used as effector cells. Target cells were C1R.A2 pulsed with FMP. Closed squares, GrB ELISPOT; open squares, ^{51}Cr-release assay. Representative experiment is shown.

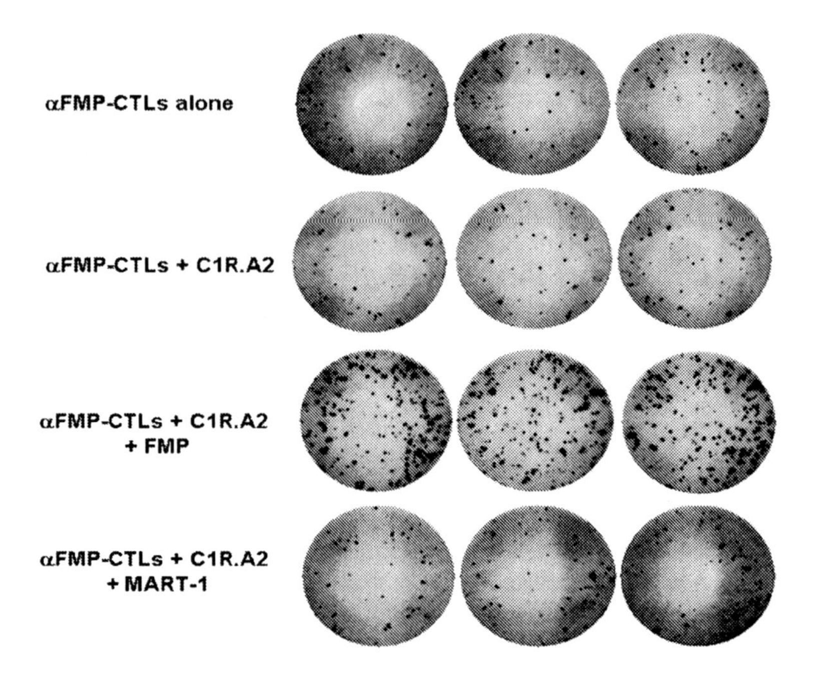

FIGURE 2. Specificity of Granzyme B (GrB) secretion by αFMP-CTL in the ELISPOT assay. Human αFMP-CTL (7-day culture, 5×10^3 cells/well) were run alone or against various target cells (5×10^4 cells/well): C1R.A2, C1R.A2 pulsed with 5 μg/ml FMP or C1R.A2 pulsed with 3 μM MART-1. Effector and target cells were incubated for 4 h at 37°C. Image from the plate scan generated by the CTL Analyzer is shown.

3. Advantages of the ELISPOT Method

Although the assays correlated, there are numerous advantages in utilizing the ELISPOT assays over the standard ^{51}Cr-release assay. The ELISPOT assays use a lower number of effector cells to accurately assess activity. The high sensitivity and specificity of the ELISPOT assays are beneficial for monitoring clinical trials where frequently there are limited numbers of patients' cells available. The ELISPOT assays also enumerate antigen - specific lymphocyte frequency by measuring secretion of a specific immune protein. Additionally, problems associated with the labeling efficiency of targets are not a concern with the ELISPOT assays. Therefore, the GrB assay is a superior alternative to the ^{51}Cr-release assay to assess a CTL response. Moreover, unlike the IFN-γ ELISPOT assay, the GrB ELISPOT directly measures the release of a cytolytic protein. Detection of low-frequency tumor-specific CTLs and their specific effector functions can provide a valuable insight in to immunological responses.

4. Application of the Gr B ELISPOT Assay for Monitoring Clinical Samples

4.1. CTL Reactivity to Cancer Vaccine Components

We further investigated whether the GrB ELISPOT assay can be applied to monitor the frequency and activity of CTL in PBMCs from patients with cancer (Shafer-Weaver et al. 2006). The clinical efficacy of cancer vaccines is likely to depend on several factors including the specificity, functional quality, and the magnitude of the induced antitumor T cell response. Therefore, we compared peptide-stimulated reactivity of PMBCs from vaccinated cancer patients in the GrB and IFN-γ ELISPOT assays as well as in the tetramer and ^{51}Cr-release assays. PBMCs from melanoma patients vaccinated with an HLA-A2*0201 binding peptide from the gp100 protein (gp100:209-2M) were utilized.

Four distinct response patterns were observed among the sixteen patients tested Figure 3. Five of the 16 patients were unresponsive in all four assays. Eleven responsive patients could be further categorized. Four were positive in all four assays, three were positive in the tetramer, IFN-γ, and GrB ELISPOT assays, and four were positive in only the tetramer and the IFN-γ assays. These data patterns demonstrate that vaccination can elicit differences in immune responses among patients. Correlations between the tetramer, IFN-γ GrB ELISPOT, and the ^{51}Cr-release assays differed. The IFN-γ and tetramer assays perfectly correlated (Phi Coefficient = 1.00, $p < 0.0001$), whereas the GrB ELISPOT assay was significantly associated with all three of the other assays (p values of 0.015, 0.015, and 0.0059 with tetramer, IFN-γ ELISPOT, and ^{51}Cr-release assays, respectively). Moreover, the ^{51}Cr-release assay significantly correlated only with the GrB ELISPOT. Although a limited number of patients were analyzed and direct comparisons were not made to clinical outcomes, the particular vaccination schema did not seem to correlate to discrete immune responses. Differences in the resulting vaccination-induced immune responses most likely are due to patient variation. Regardless, it may be speculated that vaccination is inducing the generation of different populations of effector cells or that the same effector cells are undergoing differentiation over time due to vaccination. Preliminary findings utilizing in vitro stimulated CTLs in a simultaneous GrB and IFN-γ dual-color ELISPOT approach show distinct populations of IFN-γ-secreting, GrB-secreting, or dual-secreting (both IFN-γ and GrB) populations in response to relevant peptide stimulation (unpublished data).

Previous research has demonstrated shifts in T cell phenotype from naïve to an activated/effector phenotype after continued vaccination. Tetramer-positive $CD8^+$ T cells from melanoma patients vaccinated with g209 peptide maintain an effector memory phenotype after 1 year postvaccination (Powell and Rosenberg 2004). However, only $CCR7^-CD45RA^-CD45RO^+$ tumor antigen-specific T cells produced IFN-γ and lysed tumor cells (Dunbar et al. 2000; Valmori et al. 2002). Although phenotypic analysis was not performed in our study, the functional data suggest that vaccination induced the generation of effector memory cells in some but not all patients. Significantly more antigen-specific T cells produced IFN-γ than GrB,

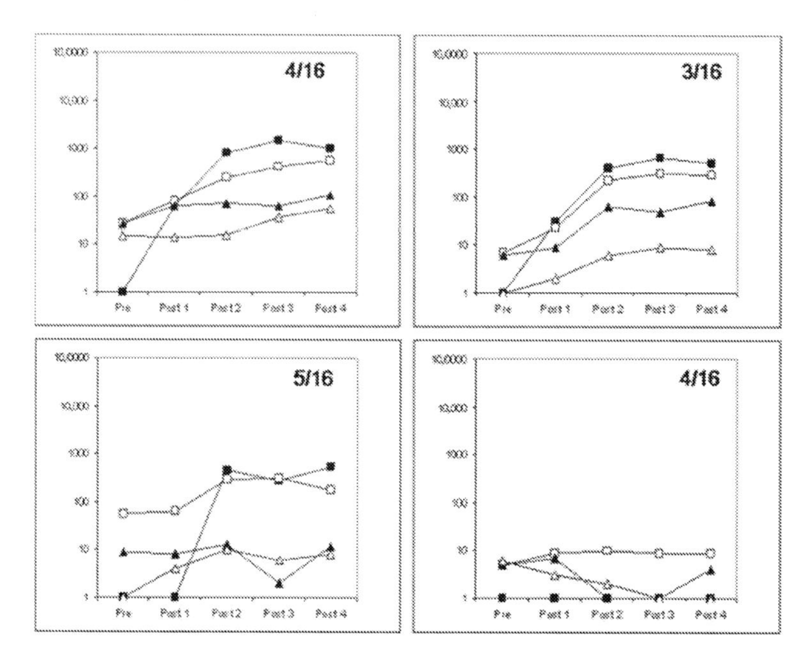

FIGURE 3. Granzyme B and IFN-γ release in the ELISPOT assays by peripheral blood mononuclear cells (PBMCs) from gp100:209-2M-vaccinated melanoma patients, cytotoxicity in the ^{51}Cr-release assay and tetramer data (Collaboration with Dr. Rosenberg). Patient PBMCs obtained before and 3 weeks after each vaccination (Post 1–4). Tetramer data were obtained by Surgery Branch (NCI) and shown as the number of tetramer-positive cells per 10^4 CD8^{bright} T cell. IFN-γ and GrB ELISPOT values are average number of IFN-γ - or GrB-secreting cells per 10^5 effector cells. The ^{51}Cr-release assay data are presented as percent specific lysis at E:T ratio of 50:1. The data presented are a representative patient for each pattern of response category. Closed squares, tetramer assay; open squares, IFN-γ ELISPOT assay; closed triangles, GrB ELISPOT assay; open triangles, ^{51}Cr-release assay.

suggesting that in addition to the magnitude of the immune response, the type of of immune response generated to peptide vaccination may also be important.

4.2. CTL Reactivity to Primary Tumors

The ELISPOT assay has been primarily used for the detection of T cell responses against vaccine components by using peptide- or protein-pulsed antigen-presenting cells as surrogate T cell targets. However, demonstration of reactivity to vaccine components does not necessarily equate to recognition and elimination of tumor cells. Accordingly, immunological assays that demonstrate recognition of native tumor cells (tumor-specific) may be more clinically relevant than assays that demonstrate recognition of tumor protein or peptide (antigen-specific). We tested antigen-specific CTL responses against autologous primary tumor cells in vaccinated cancer patients. We utilized PBMCs directly isolated from follicular lymphoma patients vaccinated with tumor-derived idiotype (Id) protein as a model system because Id vaccination has been

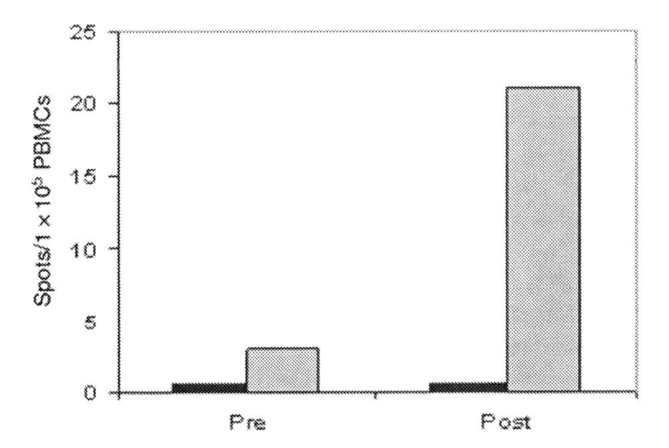

FIGURE 4. Granzyme B (GrB) secretion in ELISPOT assay by patients' peripheral blood mononuclear cells (PBMCs) vaccinated with Id and stimulated in vitro with CD40L-activated autologous follicular lymphoma cells. Pre- and postvaccine PBMC samples (1×10^5 cells/well) were cocultured with either autologous sCD40Lt-activated follicular lymphoma tumor cells or activated normal B cells (2×10^5 cells/well) for 24 h (GrB) or 48 h (IFN-γ) ELISPOT assays. Black bars, response to normal autologous B cells; gray bars, response to tumor cells. Data are presented as spots per 1×10^5 PBMCs and is representative of three separate experiments.

shown to induce tumor-specific T cell responses, including IFN-γ secretion in the ELISPOT assay (Malyguine et al. 2004) in these patients.

In this study, PBMCs isolated prior to vaccination and at subsequent time points throughout the vaccination schema were stimulated with patient tumor cells (relevant) or normal B cells (irrelevant) in the GrB assay. Our findings demonstrated that the GrB ELISPOT assay can be successfully applied to enumerate CTLs that can recognize and mount an immune response to tumor cells. PBMCs from several cancer patients vaccinated with Id secreted significant amounts of GrB when stimulated with autologous tumor cells but not normal B cells (Figure 4). Direct cyotoxicity was not assessed for these samples because primary tumor cells cannot be labeled efficiently with ^{51}Cr. Therefore, the GrB ELISPOT is a viable alternative to measure CTL cytotoxic ability against primary tumors.

4.3. Additional Applications of GrB ELISPOT

The role of GrB in immune surveillance and rejection of tumors is still controversial. Rejection of spontaneous or experimental tumors by CTLs in mice was recently shown to be independent of GrB secretion (Smyth et al. 2000, 2003). These findings are in contrast to numerous studies in which GrB contributes to controlling tumors in vivo (Sayers et al. 1992; Motyka et al. 2000; Davis et al. 2001; Medema et al. 2001; Smyth et al. 2001). Regardless of the role of GrB in cell-mediated killing, GrB expression is mainly restricted to cytolytic cells, and therefore, the release of GrB is a more specific measure of CTL than IFN-γ.

Currently, limited studies have utilized GrB ELISPOT assay to measure immune responses in cancer patients. In one study, $CD4^+$ T lymphocytes from one melanoma patient vaccinated with MHC II-restricted MART-1 peptide secreted GrB in the ELISPOT assay after ex vivo stimulation (Wong et al. 2004). In another study, $CD8^+$ T lymphocytes from two of five nonvaccinated hepatocellular carcinoma patients produced GrB against the NY-ESO-1b peptide (Shang, et al. 2004). Additionally, immunization of patients with acute myeloid leukemia led to increased GrB secretion in the ELISPOT assay but only after patients' PBMCs were prestimulated for 8 days with specific peptide or Id (Hasenkamp et al. 2006).

Our study utilized the GrB ELISPOT assay to evaluate the response of PBMCs (no preactivation) from cancer patients vaccinated with an MHC I-restricted peptide. We demonstrated that antigen-specific CTLs can be enumerated directly from peripheral blood samples without further in vitro manipulation including cell enrichment, expansion, or prolonged stimulation. This is an important finding because in vitro stimulation can hinder the accuracy of monitoring the frequency, phenotype and functions of T cells elicited to vaccination in vivo (Faure et al. 1998; thor Straten et al. 2000). Therefore, the GrB ELISPOT can directly assess vaccine-specific T cell frequency in vivo. Additionally, the ^{51}Cr-release assay significantly correlated only with the GrB ELISPOT, demonstrating that the GrB ELISPOT is a viable alternative for clinical monitoring of T cell cytolytic ability.

5. Development of a Perforin ELISPOT Assay

As mentioned earlier, one of the major mechanisms of cell-mediated cytotoxicity involves exocytosis of cytoplasmic granules from the effector toward the target cell. The granules contain, among others, the pore-forming protein perforin that is present mainly in the granules of $CD8^+$ CTL and NK cells (Trinchieri and Perussia 1984; Masopust et al. 2001), and therefore, similar to GrB, the release of this factor in response to the appropriate target may also be used to evaluate cell-mediated cytotoxicity by specific antitumor CTLs generated by vaccination. Since the release of perforin facilitates GrB entry into the target cell and its subsequent cytoxic effector functions (Froelich et al. 1996; Jans et al. 1996; Shi et al. 1997; Smyth et al. 2001; Trapani and Smyth 2002), perforin may be a better measure of overall cytolytic capability. Currently, the perforin ELISPOT assay has only been used to evaluate anti-viral immunity (Zuber et al. 2005; Burton et al. 2006). We compared peptide-stimulated reactivity of PMBCs from vaccinated cancer patients in the Perforin and GrB ELISPOT assays. PBMCs from melanoma patients vaccinated with an HLA-A2*0201 binding peptide from the gp100 protein (gp100:209-2M) were utilized as effectors. Our preliminary data show that perforin release by PBMCs from vaccinated melanoma patients correlated with GrB release Figure 5. Further comparison of these assays using more clinical samples may help us understand their relative values for immunological monitoring.

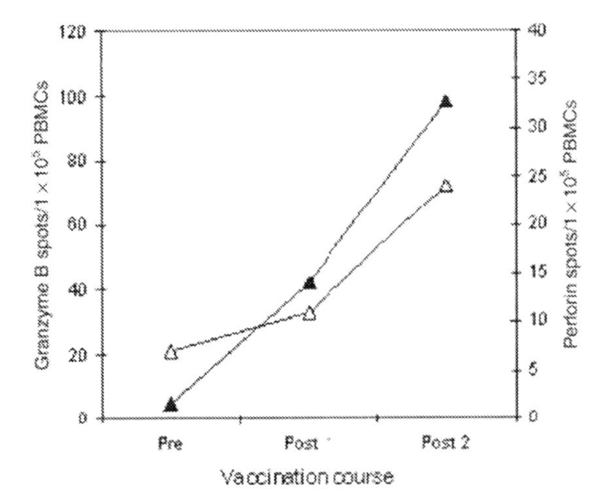

FIGURE 5. Secretion of perforin and Granzyme B in the ELISPOT assays by PBMCs from gp100:209-2M vaccinated melanoma patients. Patient PBMC obtained before (Pre) and 3 weeks after each vaccination (Post 1–2). ELISPOT values are average number of perforin- or GrB-secreting cells per 1×10^5 effector cells. Representative experiment shown.

6. Conclusions

It is important to emphasize that the ^{51}Cr-release assay and the ELISPOT assays measure different aspects of cell-mediated killing—target cell death and effector cell function, respectively. Therefore, these assays do not replace each other but could be used in concert. We believe even though the GrB and perforin ELISPOT assays indirectly measure cytolytic activity, they are still preferential for clinical monitoring because they provide both quantitative and qualitative information, are more sensitive than ^{51}Cr-release assay, utilize lower numbers of cells, and do not require target cell labeling that could be problematic in the clinical setting.

The particular immunological responses following vaccination needed for an effective response against cancer have not been fully elucidated. Detailed knowledge of specific immune responses that correlate with positive clinical outcomes will help identify better strategies to effectively activate the immune system against tumors. Taken together with published data, our results show that the simultaneous use of the ELISPOT assay with other immunological assays may provide additional immunological insight into patient responses to cancer vaccines.

Acknowledgments

The authors would like to thank Dr. Steven Rosenberg (NIH) and Dr. Larry Kwak (NIH) for providing clinical samples. This project has been funded in whole or in part with federal funds from the National Cancer Institute, National Institutes of

Health, under contract N01-CO-12400. The content of this publication does not necessarily reflect the views or policies of the Department of Health and Human Services, nor does it mention of trade names, commercial products, or organizations imply endorsement by the U.S. Government.

References

Altman, J., Moss, P., Goulder, P., Barouch, D., McHeyzer-Williams, M., Bell, J., McMichael, A. and Davis, M. (1996) Phenotypic analysis of antigen-specific T lymphocytes. Science 274, 94–96.

Bachmann, M., Barner, M., Viola, A. and Kopf, M. (1999) Distinct kinetics of cytokine production and cytolysis in effector and memory T cells after viral infection. Eur. J. Immunol. 29, 291–299.

Burkhardt, J., Hester, S., Lapham, C. and Argon, Y. (1990) The lytic granules of natural killer cells are dual-function organelles combining secretory and pre-lysosomal compartments. J Cell. Biol. 111, 2327–2340.

Burton, C., Gotch, F. and Imami, N. (2006) Rapid qualitative and quantitative analysis of T cell responses in HIV-1-infected individuals receiving successful HAART and HIV-1 sero-negative controls: concomitant assessment of perforin, IFN-gamma and IL-4 secretion. J. Immunol. Methods 308, 216–230.

Davis, J., Smyth, M. and Trapani, J. (2001) Granzyme A and B-deficient killer lymphocytes are defective in eliciting DNA fragmentation but retain potent *in vivo* anti-tumor capacity. Eur. J. Immunol. 31, 39–47.

Dunbar, P.R., Smith, C.L., Chao, D., Salio, M., Shepherd, D., Mirza, F., Lipp, M., Lanzavecchia, A., Sallusto, F., Evans, A., Russell-Jones, R., Harris, A.L. and Cerundolo, V. (2000) A shift in the phenotype of melan-A-specific CTL identifies melanoma patients with an active tumor-specific immune response. J. Immunol. 165, 6644–6652.

Faure, F., Even, J., and Kourilsky, P. (1998) Tumor-specific immune response: current *in vitro* analyses may not reflect the *in vivo* immune status. Crit. Rev. Immunol. 18, 77–86.

Froelich, C., Hanna, W., Poirier, G., Duriez, P., D'Amours, D., Salvosen, G.S., Alnemri, E., Earnshaw, W. and Shah, G. (1996) Granzyme B/perforin-mediated apoptosis of Jurkat cells results in cleavage of poly(ADP-ribose) polymerase to the 89-kDa apoptotic fragment and less abundant 64-kDa fragment. Biochem. Biophys. Res. Commun. 227, 658–665.

Hasenkamp, J., Borgerding, A., Wulf, G., Uhrberg, M., Jung, W., Dingeldein, S., Truemper, L. and Glass, B. (2006) Immunotherapy for patients with acute myeloid leukemia using autologous dendritic cells generated from leukemic blasts. Int. J. Oncol. 28, 855–861.

Jans, D., Jans, P., Briggs, L., Sutton, V. and Trapani, J. (1996) Nuclear transport of granzyme B (fragmentin-2). Dependence of perforin *in vivo* and cytosolic factors *in vitro*. J. Biol. Chem. 27, 30781–30789.

Keilholz, U., Weber, J., Finke, J., Gabrilovich, D., Kast, W., Disis, M., Kirkwood, J., Scheibenbogen, C., Schlom, J., Maino, V., Lyerly, H., Lee, P., Storkus, W., Marincola, F., Worobec, A. and Atkins, M. (2002) Immunologic monitoring of cancer vaccine therapy: results of a workshop sponsored by the Society for Biological Therapy. J. Immunother. 25, 97–138.

Kleen, T., Asaad, R., Landry, S., Boehm, B. and Tary-Lehmann, M. (2004) Tc1 effector diversity shows dissociated expression of granzyme B and interferon-gamma in HIV infection. AIDS 18, 383–392.
Kwak, L. (2003) Translational development of active immunotherapy for hematologic malignancies. Semin. Oncol. 30, 17–22.
Malyguine, A., Strobl, S., Shafer-Weaver, K., Ulderich, T., Troke, A., Baseler, M., Kwak, L. and Neelapu, S. (2004) A modified human ELISPOT assay to detect specific responses to primary tumor cell targets. J. Transl. Med. 2, 9.
Masopust, D., Vezys, V., Marzo, A.L. and Lefrancois, L. (2001) Preferential localization of effector memory cells in nonlymphoid tissue. Science 291, 2413–2417.
Medema, J., Jong, J.d., Peltenburg, L., Verdegaal, E., Gorter, A., Bres, S., Franken, K., Hahne, M., Albar, J., Melief, C. and Offringa, R. (2001) Blockade of the granzyme B/perforin pathway through overexpression of the serine protease inhibitor PI-9/SPI-6 constitutes a mechanism for immune escape by tumors. Proc. Natl. Acad. Sci. USA 98, 11515–11520.
Motyka, B., Korbutt, G., Pinkoski, M., Heibein, J., Caputo, A., Hobman, M., Barry, M., Shostak, I., Sawchuk, T., Holmes, C., Gauldie, J. and Bleackley, R. (2000) Mannose 6-phosphate/insulin-like growth factor II receptor is a death receptor for granzyme B during cytotoxic T cell-induced apoptosis. Cell 103, 491–500.
Powell, D.J., Jr. and Rosenberg, S.A. (2004) Phenotypic and functional maturation of tumor antigen-reactive CD8+ T lymphocytes in patients undergoing multiple course peptide vaccination. J. Immunother. 27, 36–47.
Rininsland, F., Helms, T., Asaad, R., Boehm, B. and Tary-Lehmann, M. (2000) Granzyme B ELISPOT assay for ex vivo measurements of T cell immunity. J. Immunol Methods 240, 143–155.
Rosenberg, S. (2001) Progress in human tumour immunology and immunotherapy. Nature 411, 380–384.
Sayers, T., Wiltrout, T., Sowder, R., Munger, W., Smyth, M. and Henderson, L. (1992) Purification of a factor from the granules of a rat natural killer cell line (RNK) that reduces tumor cell growth and changes tumor morphology. Molecular identity with a granule serine protease (RNKP-1). J. Immunol. 148, 292–300.
Scheibenbogen, C., Lee, K.H., Mayer, S., Stevanovic, S., Moebius, U., Herr, W., Rammensee, H. and Keilholz, U. (1997a) A sensitive ELISPOT assay for detection of CD8+ T lymphocytes specific for HLA class I-binding peptide epitopes derived from influenza proteins in the blood of healthy donors and melanoma patients. Clin. Cancer Res. 3, 221–226.
Scheibenbogen, C., Lee, K., Stevanovic, S., Witzens, M., Willhauck, M., Waldmann, V., Naeher, H., Rammensee, H. and Keilholz, U. (1997b) Analysis of the T cell response to tumor and viral peptide antigens by an IFNγ-ELISPOT assay. Int. J. Cancer 71, 932–936.
Shafer-Weaver, K., Rosenberg, S., Strobl, S., Alvord, G., Baseler, M. and Malyguine, A. (2006) Application of the granzyme B ELISPOT assay for monitoring cancer vaccine trials. J. Immunother. 29, 328–335.
Shafer-Weaver, K., Sayers, T., Strobl, S., Derby, E., Ulderich, T., Baseler, M. and Malyguine, A. (2003) The granzyme B ELISPOT assay: an alternative to the 51Cr-release assay for monitoring cell-mediated cytotoxicity. J. Transl. Med. 1, 14.
Shang, X., Chen, H., Zhang, H., Pang, X., Qiao, H., Peng, J., Qin, L., Fei, R., Mei, M., Leng, X., Gnjatic, S., Ritter, G., Simpson, A., Old, L. and Chen, W. (2004) The spontaneous CD8+ T cell response to HLA-A2-restricted NY-ESO-1b peptide in hepatocellular carcinoma patients. Clin. Cancer Res. 10, 6946–6955.
Shi, L., Mai, S., Israels, S., Browne, K., Trapani, J. and Greenberg, A. (1997) Granzyme B (GraB) autonomously crosses the cell membrane and perforin initiates apoptosis and GraB nuclear localization. J. Exp. Med. 185, 855–866.

Smyth, M., Kelly, J., Sutton, V., Davis, J., Browne, K., Sayers, T. and Trapani, J. (2001) Unlocking the secrets of cytotoxic granule proteins. J. Leukoc. Biol. 70, 18–29.

Smyth, M., Street, S. and Trapani, J. (2003) Cutting Edge: granzymes A and B are not essential for perforin-mediated tumor rejection. J. Immunol. 171, 515–518.

Smyth, M., Thia, K., Street, S., MacGregor, D., Godfrey, D. and Trapani, J. (2000) Perforin-mediated cytotoxicity is critical for surveillance of spontaneous lymphoma. J. Exp. Med. 192, 755–760.

Smyth, M. and Trapani, J. (1995) Granzymes: exogenous proteinases that induce target cell apoptosis. Immunol. Today 16, 202–206.

thor Straten, P., Kirkin, A.F., Siim, E., Dahlstrom, K., Drzewiecki, K.T., Seremet, T., Zeuthen, J., Becker, J.C. and Guldberg, P. (2000) Tumor infiltrating lymphocytes in melanoma comprise high numbers of T cell clonotypes that are lost during *in vitro* culture. Clin. Immunol. 96, 94–99.

Trapani, J. and Smyth, M. (2002) Functional significance of the perforin/granzyme cell death pathway. Nat. Rev. Immunol. 2, 735–747.

Trinchieri, G. and Perussia, B. (1984) Human natural killer cells: biologic and pathologic aspects. Lab. Invest. 50, 489–513.

Valmori, D., Scheibenbogen, C., Dutoit, V., Nagorsen, D., Asemissen, A.M., Rubio-Godoy, V., Rimoldi, D., Guillaume, P., Romero, P., Schadendorf, D., Lipp, M., Dietrich, P.Y., Thiel, E., Cerottini, J.C., Lienard, D. and Keilholz, U. (2002) Circulating tumor-reactive CD8(+) T cells in melanoma patients contain a CD45RA(+)CCR7(–) effector subset exerting ex vivo tumor-specific cytolytic activity. Cancer Res. 62, 1743–1750.

Wang, R. and Rosenberg, S. (1999) Human tumor antigens for cancer vaccine development. Immunol. Rev. 170, 85–100.

Wong, R., Lau, R., Chang, J., Kuus-Reichel, T., Brichard, V., Bruck, C. and Weber, J. (2004) Immune responses to a class II helper peptide epitope in patients with stage III/IV resected melanoma. Clin. Cancer Res. 10, 5004–5013.

Zuber, B., Levitsky, V., Jonsson, G., Paulie, S., Samarina, A., Grundstrom, S., Metkar, S., Norell, H., Callender, G., Froelich, C. and Ahlborg, N. (2005) Detection of human perforin by ELISpot and ELISA: ex vivo identification of virus-specific cells. J. Immunol. Methods 302, 13–25.

30

Serum Levels of Soluble HLA and IL-2R Molecules in Patients with Urogenital Chlamydia Infection

Victor V. Novikov[1], Natalya I. Egorova[1], Georgii Yu. Kurnikov[1], Irina V. Evsegneeva[2], Anatoly Yu. Baryshnikov[3], and Alexandr V. Karaulov[2]

[1] N.I. Lobachevsky State University of Nizhny Novgorod, Russia
[2] I.M. Sechenov Medical Academy of Moscow, Russia, karaulov@mtu-net.ru
[3] N.N. Blohin Russian Oncological Scientific Centre, Moscow, Russia

Abstract. Cellular immunity plays a central role in immune response to chlamydial infection, and soluble forms of immune cell membrane antigens take part in the regulation of immune response. Using an immunoenzymatic method, we determined serum levels of soluble HLA molecules (sHLA-I and sHLA-DR) and soluble CD25 molecules (sCD25) in patients with genital chlamydial infection. Specimens from patients with nonspecific inflammation of the urogenital tract were studied and healthy volunteers served as controls. We revealed that serum levels of sHLA-DR and sCD25 increased 3.5- and 2.3-fold, respectively, during chlamydial infection, while the levels of sHLA-I were not changed. Nonspecific inflammation of the urogenital tract was characterized by a 1.5-fold increase in sHLA-I, a 1.6-fold decrease in sCD25, and no changes of sHLA-DR levels in comparison with healthy volunteers. We concluded that Th1 immune responses might dominate during genital chlamydial infection contrary to the state of nonspecific inflammation of urogenital tract.

1. Introduction

The obligate intracellular bacterium *Chlamydia trachomatis* is the etiological agent of a common sexually transmitted disease—genital chlamydial infection. The frequent insidious infections and the increasing incidence of severe irreversible complications, which may be the first symptoms of an infection, support the rising concern that genital chlamydial disease poses a major threat to human reproduction and well-being. The central role in immune response to chlamydial infection is played by immune cells. Activation of T cells in the genital mucosa involves obligatory intimate interaction with accessory infected and noninfected cells via the cell surface molecules that include the gene products of the major histocompatibility complex (MHC), coreceptors, costimulatory, adhesion molecules, and receptors for cytokines. A detailed knowledge of the precise role of each of these cellular and molecular mechanisms in the overall host response against chlamydia is important for understanding the immune response to this disease.

MHC molecules on the cell surface present antigen peptides to T cell receptors. However, HLA class I and II molecules may be found in both membrane-bound and soluble forms. The mechanisms of appearance of soluble HLA class I and class II molecules (sHLA-I and sHLA-DR, respectively) involves the shedding of full-length molecules containing the hydrophobic transmembrane region, proteolytic cleavage from the cell surface, and alternative splicing of mRNA (Claus et al. 2000). Other membrane proteins may also be seen in a soluble form. Indeed, IL-2 receptor has a soluble form (sCD25), which binds with IL-2 and prevents its binding to the membrane IL-2 receptor (CD25). Increased expression of sHLA-I, sHLA-DR, and sCD25 may be associated with cell activation, and elevated serum levels of these molecules were found in different diseases (Symons et al. 1988; Brieva et al. 1990; Moore et al. 1997; Aultman et al. 1999). Increased production of soluble forms of MHC and IL-2R molecules might served as a sign of immune activation (Dummer et al. 1992; Carbone et al. 1996), and alterations of their concentrations in serum may be associated with bacterial infections, as was reported in patients with active pulmonary tuberculosis and syphilis (Inostroza et al. 1994; Vyasmina et al. 2001).

The aim of this study was to assess serum levels of sHLA-I, sHLA-DR, and sCD25 in patients with genital chlamydial infection and control patients with nonspecific inflammation of the urogenital tract.

2. Experimental Design

Thirty-five serum samples from patients with urogenital chlamydiosis and 33 serum samples from patients with nonspecific urogenital inflammation were studied. All patients were negative for HBV, HCV, syphilis, HIV, mycoplasma, ureaplasma, herpes simplex, cytomegalovirus, and gonorrhea. Serum samples from 151 healthy volunteers served as controls. Serum was stored at minus –60°C until the analysis.

The levels of sHLA-I, sHLA-DR, sCD25 molecules were determined using ELISA with the following monoclonal antibodies (mAbs): α-chain of HLA class I (ICO-53), HLA-DR (ICO-1), β_2-microglobulin (β_2m;ICO-216), CD25 (ICO-105), and goat polyclonal antibodies to human peripheral mononuclear cell surface antigens. Heterodimer β_2m-HLA class I was assessed using mAbs against α-chains of HLA class I and mAbs against β_2 conjugated with horseradish peroxidase (Lebedev et al. 2003). sHLA-DR and sCD25 antigens were determined using goat polyclonal antibodies against human peripheral blood mononuclear cell surface antigens and mAb ICO-1 against HLA-DR antigen conjugated with horseradish peroxidase and mAb ICO-105 against CD25 antigen conjugated with horseradish peroxidase. Samples from healthy volunteers with a high content of tested soluble antigen have been used for calibration curve. The results are expressed in Units per milliliter. Results are reported as mean ± SEM. $p < 0.05$ was considered significant.

3. Results

Healthy volunteers express 987.2 ± 96.0 U/ml of sHLA-I and 115.7 ± 21.9 U/ml sHLA-DR in serum. Patients with genital chlamydiosis demonstrated no alterations in serum levels of sHLA-I (Figure 1), while serum levels of sHLA-DR were significantly increased up to 3.5 times (475.9 ± 101.1 U/ml, $p < 0.05$). Levels of sCD25 in serum obtained from healthy donors were 113.4 ± 17.1 U/ml, while genital chlamydiosis was associated with their elevation in 2.3-fold ($p < 0.05$). In contrast to genital chlamydiosis, nonspecific inflammation of urogenital tract was characterized by increased levels of sHLA-I up to 150% ($p < 0.05$ vs. healthy controls), although the levels of sHLA-DR did not differ from control values (Figure 1). Serum sCD25 was decreased in 1.6-fold and was equal to 71.7 ± 42.0 U/ml. Concentrations of sCD25 in specimens obtained from patients with genital chlamydiosis and patients with nonspecific inflammation were significantly different.

Thus, serum concentrations of sHLA-I, sHLA-DR, and sCD25 molecules were differentially changed in patients with genital chlamydiosis or nonspecific inflammation of the urogenital tract.

4. Discussion

Protective immunity against *C. thrachomatis* can be associated with two pathways. In spite of the questionable role of $CD8^+$ cytotoxic T cells in genital chlamydial infection due to a vacuolar localization of chlamydiae, some authors considered that $CD8^+$ T cells play the major role. Indeed, Cap 1 protein of *C. thrachomatis* associated with vacuolar membrane could be presented by HLA-I–peptide complex. Furthermore, five clones of HLA class I-restricted $CD8^+$ cytotoxic T cells specific for MOMP, the major outer membrane protein of *C. thrachomatis*, were identified in patients with genital tract infection. In addition, protective cytotoxic T lymphocytes could be produced in mice infected with *Chlamidia thrachomatis* and $CD8^+$ T lymphocytes could mediate lysis of chlamydia-infected target cells in vitro (Igietseme et al. 1994; Starnbach et al. 1994; Bhushan et al. 1997; Morrison and Caldwell 2002).

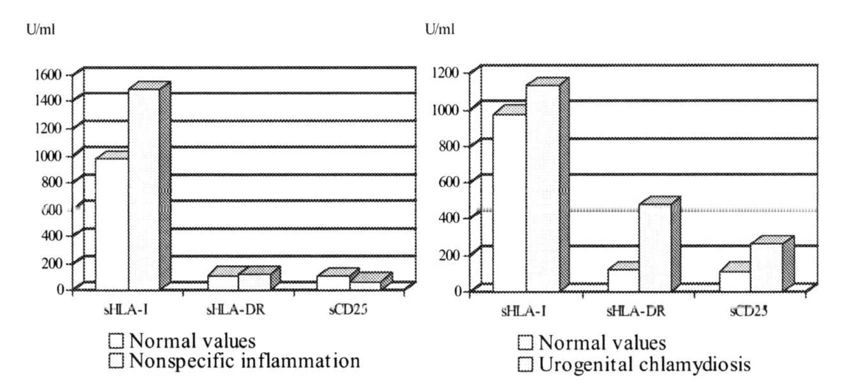

FIGURE 1. Serum level of sHLA-I, sHLA-DR, and sCD25 molecules in patients with nonspecific inflammation (left panel) and urogenital chlamydiosis (right panel).

Other authors speculate that antigen-specific CD8 T lymphocytes may be involved in the pathogenesis of *C. thrachomatis* infection more than they are involved in protective immunity. In accordance with this, $CD4^+$ Th1 cells activating macrophages and dendritic cells play a significant role in immunity to *C. thrachomatis* infection of the genital tract. Dendritic cells are the effective antigen-presenting cells and might internalize chlamydiae by macropinocytosis followed by a fusing of chlamydiae with dendritic lysosomes. After processing, chlamydia antigens may be presented by HLA class II molecules and recognized by specific $CD4^+$ T cells (Ojcius et al. 1998). The adoptive transfer of immune $CD4^+$ T cells, but not immune $CD8^+$ T cells, to naïve mice confers a level of protective immunity upon infectious challenge (Su and Caldwell 1995).

Since both serum concentrations of sHLA-DR and sCD25 reflect immune cell activation, simultaneous increase in sHLA-DR and sCD25 might corroborate Th1 activation processes in *C. thrachomatis* infection. Th1 cells secrete IFN-γ that induces ICAM-1 on epithelial cells, leading to enhanced epithelial/T cell interactions (Bevilacqua 1993). Thus, the antichlamydial effect of Th1 cytokines, such as IFN-γ, may be elaborated at multiple levels, including upregulation of co-stimulatory and adhesion molecules, and induction of antimicrobial processes, such as tryptophan catabolism, iron deprivation and nitric oxide secretion (Bhushan et al. 1997; Bilinska, Frydecka, and Podemski 2001). In fact, we have shown earlier significant elevation of soluble adhesion molecules in serum in patients with chlamydial infection (Egorova et al. 2003).

Thus, genital chlamydial infection in humans is characterized by increased concentrations of serum sHLA-DR and sCD25 and normal levels of sHLA-I molecules. These and other data allowed us to speculate that Th1 immune response predominates during genital chlamydial infection.

References

Aultman, D., Adamashvili, I., Yaturu, K., Langford, M., Gelder, F., Gautreaux, M., Ghali, G.E. and McDonald, J. (1999) Soluble HLA in human body fluids. Hum. Immunol. 60, 239–244.

Bevilacqua, M.P. (1993) Endothelial-leukocyte adhesion molecules. Annu. Rev. Immunol. 11, 767–804.

Bhushan, R., Kirkham, C., Sethi, S. and Murphy, T.F. (1997) Antigenic characterization and analysis of the human immune response to outer membrane protein E of Branhamella catarrhalis. Infect. Immun. 65, 2668–2675.

Bilinska, M., Frydecka, I. and Podemski, R. (2001) [Clinical course and changes of soluble interleukin-2 receptor and soluble forms of intercellular adhesion molecule-1 (ICAM-1) in serum of multiple sclerosis patients]. Neurol. Neurochir. Pol. 35, 47–56.

Brieva, J.A., Villar, L.M., Leoro, G., Alvarez-Cermeno, J.C., Roldan, E. and Gonzalez-Porque, P. (1990) Soluble HLA class I antigen secretion by normal lymphocytes: relationship with cell activation and effect of interferon-gamma. Clin. Exp. Immunol 82, 390–395.

Carbone, E., Terrazzano, G., Colonna, M., Tuosto, L., Piccolella, E., Franksson, L., Palazzolo, G., Perez-Villar, J.J., Fontana, S., Karre, K. and Zappacosta, S. (1996) Natural killer clones recognize specific soluble HLA class I molecules. Eur. J. Immunol. 26, 683–689.

Claus, R., Bittorf, T., Walzel, H., Brock, J., Uhde, R., Meiske, D., Schulz, U., Hobusch, D., Schumacher, K., Witt, M., Bartel, F. and Hausmann, S. (2000) High concentration of soluble HLA-DR in the synovial fluid: generation and significance in "rheumatoid-like" inflammatory joint diseases. Cell. Immunol. 206, 85–100.

Dummer, R., Posseckert, G., Nestle, F., Witzgall, R., Burger, M., Becker, J.C., Schafer, E., Wiede, J., Sebald, W. and Burg, G. (1992) Soluble interleukin-2 receptors inhibit interleukin 2-dependent proliferation and cytotoxicity: explanation for diminished natural killer cell activity in cutaneous T cell lymphomas *in vivo*? J. Invest. Dermatol. 98, 50–54.

Egorova, N.I., Kurnikov, G.Y., Babaev, A.A. and Novikov, V.V. (2003) Serum level of soluble molecules of intercellular adhesion in urogenital chlamidiosis. Cytokines Inflamm. 2, 32–36.

Igietseme, J.U., Magee, D.M., Williams, D.M. and Rank, R.G. (1994) Role for CD8+ T cells in antichlamydial immunity defined by Chlamydia-specific T lymphocyte clones. Infect. Immun. 62, 5195–5197.

Inostroza, J., Munoz, P., Espinoza, R., Millaqueo, L., Diaz, P., Leiva, L. and Sorensen, R. (1994) Quantitation of soluble HLA class I heterodimers and beta 2-microglobulin in patients with active pulmonary tuberculosis. Hum. Immunol. 40, 179–182.

Lebedev, M.J., Krizhanova, M.A., Vilkov, S.A., Sholkina, M.N., Vyasmina, E.S., Baryshnikov, A. and Novikov, V.V. (2003) Peripheral blood lymphocytes immunophenotype and serum concentration of soluble HLA class I in burn patients. Burns 29, 123–128.

Moore, C., Ehlayel, M., Inostroza, J., Leiva, L.E. and Sorensen, R.U. (1997) Elevated levels of soluble HLA class I (sHLA-I) in children with severe atopic dermatitis. Ann. Allergy Asthma Immunol. 79, 113–118.

Morrison, R.P. and Caldwell, H.D. (2002) Immunity to murine chlamydial genital infection. Infect. Immun. 70, 2741–2751.

Ojcius, D.M., Bravo de Alba, Y., Kanellopoulos, J.M., Hawkins, R.A., Kelly, K.A., Rank, R.G. and Dautry-Varsat, A. (1998) Internalization of Chlamydia by dendritic cells and stimulation of Chlamydia-specific T cells. J. Immunol. 160, 1297–1303.

Starnbach, M.N., Bevan, M.J. and Lampe, M.F. (1994) Protective cytotoxic T lymphocytes are induced during murine infection with Chlamydia trachomatis. J. Immunol. 153, 5183–5189.

Su, H. and Caldwell, H.D. (1995) CD4+ T cells play a significant role in adoptive immunity to Chlamydia trachomatis infection of the mouse genital tract. Infect. Immun. 63, 3302–3308.

Symons, J.A., Wood, N.C., Di Giovine, F.S. and Duff, G.W. (1988) Soluble IL-2 receptor in rheumatoid arthritis. Correlation with disease activity, IL-1 and IL-2 inhibition. J. Immunol. 141, 2612–2618.

Vyasmina, E.S., Ptytsina, Y.S., Matveeva, E.M., Komarova, V.D., Barysnikov, AY. and Novikov, V.V. (2001) Soluble forms of HLA class I in HIV-infected persons and patients with syphilis and/or hepatitis C. Russ. J. HIV/AIDS Relat. Probl. 5, 136–137.

31

Evaluation of Suspected Immunodeficiency

Thomas A. Fleisher

Department of Laboratory Medicine, NIH Clinical Center, National Institutes of Health, DHHS, Bethesda, MD, USA, tfleisher@mail.nih.gov

Abstract. The clinical utility and capacity to evaluate immunologic function has evolved significantly over the past few decades. This chapter summarizes screening methods and more sophisticated approaches to assess the immune system when there is a suspicion of an immune deficiency.

1. Introduction

The clinical utility and capacity to evaluate immunologic function has evolved significantly over the past few decades in parallel with the marked increase in understanding the human immune system (Fleisher and Oliveira 2004). In addition, the expanding range of characterized primary immune deficiency diseases and the secondary immunodeficiency pandemic resulting from HIV have added impetus to the development of new approaches for evaluating immunologic function.

Obtaining a thorough clinical history is the appropriate starting point to direct any laboratory evaluation for immunodeficiency. For example, a history of recurrent infections with encapsulated bacteria (e.g., *Haemophilus influenzae* and *Streptococcus pneumoniae*) usually affecting the sinuses and lungs suggests an antibody deficiency (Ballow 2002). In contrast, a clinical picture of recurrent infections with opportunistic organisms (e.g., *Pneumocystis jiroveci,* candida species, and cytomegalovirus) should focus the initial evaluation toward a T cell abnormality (Buckley 2002). The more recent identification that persistent nontuberculous mycobacterial (NTB) infections can be associated with defects in the IFN-γ–IL-12/23 circuit has opened a new appreciation of immune deficiencies that affect the interface between the adaptive and the innate immune systems (Filipe-Santos et al. 2006). In addition, the critical role of natural killer cells in host defense has been clarified more recently based on studies of patients demonstrating increased susceptibility to herpes family viruses (e.g., EBV and HSV) and in some cases accompanied by uncontrolled inflammation (Nichols et al. 2005; Filipovich 2006; Orange 2006).

Abnormalities in Toll-like receptor (TLR) function recently has been defined in patients with a specific pattern of bacterial infections (Ku et al. 2005). These findings suggest that additional innate immune defects impacting this early response pathway are likely. Innate immune defects affecting neutrophil function typically result in cutaneous and deep-seated abscesses, pneumonia, periodontitis, and osteomyelitis (Rosenzweig and Holland 2004). These infections are often caused by characteristic bacteria (e.g., *Staphylococcus aureus* and *Serratia marcesens*) and/or fungi (e.g., *Aspergillus* and *Nocardia* species). Congenital defects in specific complement components often are associated with autoimmunity as well as recurrent bacterial infections similar to the antibody deficiencies (Walport 2001). In addition, abnormalities of the late complement components are associated with a unique increase in susceptibility to *Neiserrial* infections (Figueroa and Densen 1991).

Thus, the clinical history and presentation should direct the immunologic evaluation and it is necessary to consider HIV infection in the differential diagnosis with appropriate questions during the history and laboratory testing. There are now more than 120 genetically linked immune disorders, so a careful family history is also extremely important (Notarangelo et al. 2006). Finally, the physical examination may provide clues regarding specific primary immunodeficiencies (e.g., typical facies in the hyper-IgE [Job] syndrome and scars from abscess drainage sites) and secondary immune disorders (e.g., oral hairy leukoplakia or Kaposi's sarcoma in HIV infection).

2. Screening Tests of Immune Function

The standard method for screening antibody-mediated immune function involves measuring the three major immunoglobulin classes: IgG, IgA, and IgM. These results must be compared with age-matched reference ranges and interpreted with the understanding that reference ranges are usually 95% confidence intervals, meaning that 2.5% of controls are above and below the stated range. Furthermore, serum immunoglobulin levels are the net of protein production, utilization, catabolism, and loss, so decreased levels can result from increased consumption or loss as well as from decreased production.

Measurement of a functional antibody response is particularly useful when the total immunoglobulin levels are modestly depressed or normal in the face of a strong history of recurrent bacterial infection. This can include evaluating "spontaneous" specific antibodies (e.g., antiblood group antibodies [isohemagglutinins] and antibodies to past immunizations). However, the definitive method is assessing pre- and postimmunization antibody levels using protein antigens (e.g., tetanus toxoid) or polysaccharide antigen (e.g., Pneumovac) vaccines (Go and Ballas 1996). Guidelines for interpreting the results are usually provided by the testing laboratory and typically consist of at least a fourfold increase in antibody and/or generation of protective antibody levels following immunization. Additional approaches to evaluating specific antibody production revolve around using a neoantigen such as a bacteriophage phi X 174 (Pyun et al. 1989).

An additional and readily available antibody screening test is quantitation of IgG subclass levels; however, in most settings detection of a laboratory finding of IgG subclass deficiency would still require demonstrating an abnormality in functional antibody production in order to establish a clinically meaningful diagnosis.

Screening of the T cell compartment has far fewer test options and generally includes obtaining an absolute lymphocyte count (i.e., white blood cell count with differential) and possibly testing the cutaneous delayed-type hypersensitivity (DTH) response to recall antigens. The significance of the former relates to the fact that T cells constitute approximately three-fourths of the circulating lymphocytes; thus, a substantial decrease in circulating T cells typically results in a decreased absolute lymphocyte count. This comparison must be made using age-matched reference intervals as the absolute lymphocyte count is substantially higher in infants and young children than in adults. The DTH response provides an in vivo window of T cell function in response to previously encountered antigens (recall antigens such as tetanus toxoid, candida antigen, and mumps antigen) (Blatt et al. 1993). However, failure to respond may reflect T cell dysfunction (T cell anergy), but it also could indicate, that the host has not been exposed (sensitized) to the antigen(s) being used. Consequently, it is prudent to use more than one recall antigen for this test and the reliability is dependent on the antigen preparations, test application, and interpretation (evaluation) of the response.

It is important to consider HIV infection as part of the screening process in an immune evaluation and this may require viral load testing in addition to serologic testing for antibodies to HIV.

Screening options of the innate immune system are rather limited consisting primarily of evaluating for the presence of neutrophils including reviewing their morphology and obtaining a total hemolytic complement assay (CH50).

Upon completion of medical history-directed screening tests, it may be necessary to utilize more sophisticated testing to develop or confirm the final diagnosis. This typically would involve quantitating and characterizing cells of the immune system, testing in vitro immune function, and potentially performing mutation analysis. These three general categories of testing will be considered in more detail.

3. Immune Cell Quantitation and Characterization

The capacity to study the various cells of the immune system depends on flow cytometry and is based on the surface expression of specific proteins. These can be detected using fluorochrome-labeled monoclonal antibodies with more than 300 specificities currently available. In the setting of primary immunodeficiency disorders, the application of this technology enables accurate determination of absent cell populations, cell subpopulations, or specific cell surface proteins as well as the biological effects of the primary immune deficiency. These studies are enhanced by the use of "multicolor" flow cytometry, and the future may see even greater possibilities with a recent report of 17-color analytical flow cytometry using a commercially available instrument (Perfetto et al. 2004). The utility of flow cytometry has been further extended by the introduction of intracellular flow cytometry to detect proteins found within the cell but not expressed on the cell surface. This approach can also be applied to cell function testing, a method that will be considered in the following section.

Evaluation of lymphocyte populations is accomplished by assessing the percentage and number of T, B, and NK cells. This is particularly useful in identifying and categorizing severe combined immunodeficiency (SCID) when evaluating an infant with failure to thrive and/or opportunistic infections usually in the setting of lymphopenia (Buckley 2002). From these studies, it is possible to generate four major groupings: T^- B^- NK^- SCID (typical of ADA deficiency), T^- B^- NK^+ SCID (characteristic of RAG 1/2 deficiency), T^- B^+ NK^- SCID (found in common γ-chain and JAK3 deficiency), and T^- B^+ NK^+SCID (seen with IL-7 receptor alpha-chain deficiency). Low T cell percentage and number is also found in severe DiGeorge syndrome. Low CD8 T cell number in the setting of recurrent infection is compatible with ZAP70-deficient combined immune deficiency. In the setting of recurrent sinopulmonary infections, the absence of B cells (<1%) is strongly suggestive of either X-linked agammaglobulinemia (XLA) or autosomal recessive agammaglobulinemia (Figure 1) (Ballow 2002). However, it is not diagnostic as some patients with common variable immunodeficiency have very low to absent B cells.

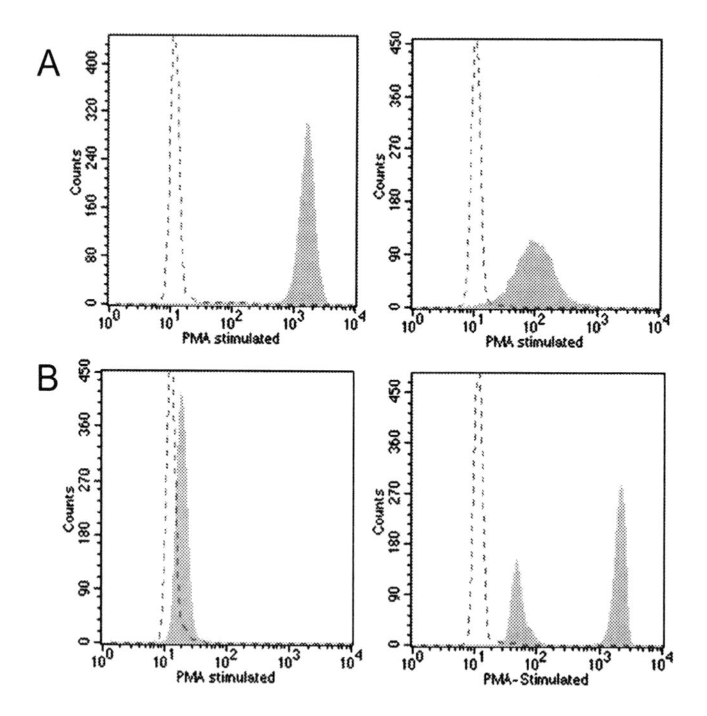

Figure 1. Panel A: Histograms from a dihydrorhodamine 123 (DHR) test showing normal oxidative burst (green histogram) on the left side and markedly decreased oxidative burst from a patient with autosomal recessive CGD (right side). Panel B: Histograms from a DHR test from an X-linked CGD patient (left side) and the maternal carrier (right side).

Focusing on the lack of expression of specific cell surface proteins can be an important approach to establishing the diagnosis of specific immunodeficiencies. As a group, these disorders usually involve genetic defects that do impact cell differentiation and release but also alter cell function. Examples include absence of CD40 on circulating B cells and monocytes or CD40 ligand (CD154) on activated CD4 T cells, findings characteristic of two specific forms of the hyper-IgM syndrome (Notarangelo et al. 2006).

A more recently described group of disorders characterized by recurrent nontuberculous mycobacterial (NTM) infection should direct the evaluation of such a patient to the function of IFN-γ and IL-12/23 circuit (Filipe-Santos et al. 2006). At the cell characterization level, this would involve assessment of specific cytokine receptor expression, but due to limited reagent availability, only specific receptor components can currently be evaluated by flow cytometry (e.g., IFNγR1).

Flow studies in a patient with recurrent bacterial infections involving the skin, oral cavity, and other internal organs in the face of neutrophilia and inability to form pus suggest a problem with β-2 integrin expression (Rosenzweig and Holland 2004). Decreased or absent CD18 expression is found in the leukocyte adhesion deficiency type 1 (LAD1) and results in diminished CD11a, b, and c on neutrophils and lymphocytes.

The application of intracellular flow cytometry has found its way into the evaluation of a number of primary immune deficiencies. This approach requires fixation and permeabilization of the cell membrane to allow intracellular entry of the antibody reagent. The primary application in this setting is linked to those disorders in which there is a defect in a specific cytoplasmic or nuclear protein. It is important to recognize that this approach has the same limitation as any protein detection method (e.g., immunoassay) in detecting disease—the absence of the protein is essentially diagnostic but the presence of the protein generally does not distinguish between the presence of a functional and that of a nonfunctional protein. An example of intracellular flow cytometry for the evaluation of an immunodeficiency is detecting the presence of Bruton's tyrosine kinase (BTK) expression in monocytes or platelets (Lopez-Granados et al. 2005). This is a useful screening test for patients suspected of having XLA or in potential carriers of XLA where one would expect a mixture of normal (BTK expressing) and abnormal (BTK nonexpressing) monocytes or platelets. This type of analysis has also been applied in the evaluation of patients suspected of having Wiskott–Aldrich syndrome-testing for WASp; X-linked lymphoproliferative syndrome (XLP)-evaluating for SH^2D^1A; hemophagocytic lymphohistiocytosis (HLH)-assaying for perforin; immune deregulation, polyendocrinopathy, and enteropathy X-linked (IPEX)-testing for FOXP3; and chronic granulomatous disease (CGD)-evaluating for gp91 phox and p47 phox.

Another approach to cell characterization focuses on evaluating the distribution of the T cell receptor (TCR) repertoire using reagents that distinguish between each of the Vβ families. This method is complementary to TcR spectratyping, a PCR-based method that evaluates the diversity of each Vβ family and has proven useful in characterizing patient T cells in disorders such as the DiGeorge syndrome (Davis et al.

1997) and Omenn syndrome as well as the response to hematopoietic stem cell reconstitution therapy. An additional means of evaluating T cells using flow cytometry has emerged to characterize the frequency of antigen-specific CD4 and CD8 T cells based on the binding of defined MHC–antigenic peptide tetramers (multimers) that are fluorescently labeled (Mallone and Nepom 2004).

Taken together, flow cytometry has become an effective laboratory adjunct in the evaluation and diagnosis of primary immune deficiencies. The addition of intracellular tests to evaluate for the presence of protein defects associated with a number of these disorders has expanded the utility of flow cytometry as an important tool in the diagnosis of immunologic disorders.

4. Testing of Immune Function

The standard method for ex vivo evaluation of immune function has historically consisted of evaluating lymphocyte proliferation in response to mitogens (e.g., PHA and ConA) and recall antigens (e.g., tetanus toxoid). This usually involves evaluating cell division poststimulation by measuring tritiated thymidine incorporation. In addition, one can assess early cell activation by evaluating specific receptor expression and later events such as cytokine production either intracellularly by flow cytometry or secreted by ELISPOT at the cell level or by immunoassay to evaluate cell free supernatants (Letsch and Scheibenbogen 2003). Testing of cytokine expression and secretion have applications beyond immunodeficiency evaluations and are being applied in the evaluation of immunomodulatory therapy and the characterization of inflammatory diseases. Assessment of cytokine production forms the mainstay of the approach to evaluating patients with recurrent NTM infections. This also resulted in a new application of flow cytometry, evaluation of protein phosphorylation associated with cytokine stimulation. In the setting of patients with persistent NTM infections, quantitating STAT1 phosphorylation following IFN-γ stimulation of monocytes has proven to be a very useful test in evaluating the IFN-γ–IL12/23 circuit (Fleisher et al. 1999). Evaluation of cell activation at the phosphorylation of intracellular messengers has been expanded to include other cytokines such as IL-2 and IL-12 and their specific signaling molecules (STAT5 and STAT3, respectively) as a method to assess for other possible immune defects. The list of intracellular targets using these methods continues to expand and has the distinct advantage of providing high-sensitivity results in real time.

Cytolytic function can be evaluated in vitro and is directed at cytotoxic T cells, NK cells, and myelomonocytic cells. T cell-mediated cytotoxicity is MHC and antigenic peptide restricted that requires prior exposure to develop the cytolytic function. In contrast, NK cell-mediated cytolysis and antibody-dependent cellular cytotoxicity do not require prior sensitization and are dependent either on naturally occurring receptors (NK cells) or on antibody bound to the target (NK cells, monocytes, and macrophages) in a process referred to as antibody-dependent cellular cytotoxicity (ADCC).

The function of the thymus in generating new T cells can be evaluated using a PCR-based method to assess the frequency of T cells displaying T cell receptor excision circles (TRECs) (McFarland et al. 2000). These small pieces of extra chromosomal DNA are generated in T cells during the thymic selection process. Because this DNA segment is episomal, it does not increase in frequency during post-thymic cell division. Hence, identifying the frequency of TREC-positive T cells is a reflection of the number of recent thymic immigrants and likewise a measure of thymic output. As one would expect, TREC analysis is useful in the setting of evaluating the DiGeorge syndrome as well as assessing thymic dependent immune reconstitution following hematopoietic stem cell and thymic transplantation.

The third arm of the lymphoid system consists of circulating cells distinct from B and T cells, the natural killer (NK) cells. Deficiency in NK cell function has been described in a limited number of patients with recurrent herpes infections (Orange 2006). In addition, experimental models point to a role for the NK cell in allograft and tumor rejection. Another category of NK cell defects found in disorders with an uncontrolled inflammatory response (initiated by an infection) that can lead to multiple organ damage (HLH and XLP) (Filipovich 2006; Nichols et al. 2005). Testing of NK cell function includes immunophenotyping NK cells by flow cytometry and assaying cytotoxic activity using standard in vitro assays.

An area of intense current investigation involves the identification of disorders potentially associated with defective signaling by TLRs (von Bernuth et al. 2006). This is a family of at least 10 receptors that represent a phylogenetically more primitive arm of the immune system dependent on pattern recognition of bacterial, fungal, and viral products. An example of such a process is the activation of monocytes and macrophages by bacterial lipopolysaacharide (LPS). This pathway of activating the immune system appears to be one of the first lines in host defense, as it does not require prior exposure to the pathogenic organism. Recently, two different clinical phenotypes have been identified with genetic defects specifically involving TLR signaling. In one, there is a genetic susceptibility to serious bacterial infection that presents in childhood and generally improves during adolescence and is associated with an autosomal recessive defect in the intracellular protein IRAK4 (Ku et al. 2005). The most recently described defect is associated with the development of herpes simplex encephalitis and is linked to an autosomal recessive defect in an intracellular protein (UNC 93B). This function is part of the TLR signaling pathway although its exact function has not yet been definitively elucidated (Casrouge et al. 2006).

Evaluation of neutrophil function after neutropenia (including cyclic neutropenia) has been ruled out is generally focused on NADPH oxidase activity in patients with recurrent bacterial and fungal infections that is consistent with CGD. The primary means of testing currently involves a flow cytometry test with leukocyte loading of a fluorochrome precursor (dihydrorhodamine 123, DHR) that fluoresces following normal activation of NADPH oxidase dependent electron transfer using an agonist like PMA (Vowells et al. 1995). The DHR assay is extremely reliable in diagnosing CGD and X-linked CGD carriers (Figure 1). Other methods have also been used to assess NADPH oxidase activity including the nitro blue tetrazolium (NBT) test and luminol-enhanced chemiluminescence. Although the clinical utility remains somewhat controversial, neutrophil-directed movement (chemotaxis) can be tested either in vivo using the Rebuck skin window or collection chamber technique or in vitro with a Boyden chamber or a soft agar system. Abnormalities of chemotaxis have been observed secondary to certain pharmacologic agents as well as the leukocyte adhesion deficiency, Chediak Higashi syndrome, Pelger–Huet anomaly, and juvenile periodontitis; however, the test is not specific for any diagnosis. A hallmark clinical feature of significantly abnormal chemotaxis is diminished neutrophil infiltration and decreased inflammation.

Evaluation of the complement pathway begins by screening the classical pathway using a CH50 assay and/or the alternate pathway using an AP50 test (Walport 2001). Interpretation of any result demonstrating decreased complement activity assumes correct handling of the serum sample (complement components are very heat labile) and should be repeated to confirm the abnormality. The next step would be to assess specific components using specific immunoassays recognizing the limitation of this approach relative to a component being produced that is dysfunctional. Ruling this possibility out requires component functional testing that is available only in very specialized complement laboratories.

5. Mutation Analysis

Identifying a specific gene mutation associated with a primary immunodeficiency provides the most definitive means of diagnosis. There is a growing list of specific genetic defects associated with primary immunodeficiencies (Notarangelo et al. 2006). Previously, screening methods (e.g., single-strand conformational polymorphism [SSCP] and dideoxy finger printing) were often applied in patient evaluations. However, with the ready availability of fluorescence-based sequencing and the advent of automated capillary sequencers, it is now practical to directly screen for mutations for virtually all genes associated with primary immunodeficiencies (Niemela et al. 2000). This is obviously applied only if there is strong evidence for a particular type of immunodeficiency based on a family history or a clinical history with definitive abnormalities on screening tests. The initial laboratory often serves to direct the mutation analysis by identifying specific characteristics associated with a diagnosis (e.g., absent or very low B cells in XLA and absent T and NK cells with B cells present in X-linked SCID). The information generated by mutation analysis has significant implications not only in establishing a definitive diagnosis but also for carrier assessment for prenatal diagnosis, and for offering genetic counseling.

6. Conclusions

The clinical pattern of recurrent infections provides the critical starting point in pursing a diagnosis of immunodeficiency and directing the best laboratory approaches for patient evaluation. Infections that are recurrent and difficult to treat or those that involve unusual organisms should raise suspicion of an underlying immunodeficiency. HIV infection has become the most likely cause of immunodeficiency, and early appropriate diagnostic testing for HIV is critical. The prudent use of laboratory tests requires that they be used in an orderly fashion, starting with the simpler screening tests chosen based on the patient history. The results generated are usually easy to interpret when they are either clearly normal or abnormal. The difficulty arises in determining the actual degree of immune dysfunction when the results fall in a gray zone, a circumstance that can be clarified by applying additional tests that can clarify the status of immune function or dysfunction. The increasing capacity to identify gene mutations linked to specific immunodeficiencies provides the most definitive evidence for a diagnosis and affords the opportunity to perform family studies and provide appropriate genetic counseling. The laboratory has emerged as the most critical source of diagnostic information in the characterization of primary immunodeficiency disorders.

References

Ballow, M. (2002) Primary immunodeficiency disorders: antibody deficiency. J. Allergy Clin. Immunol. 109, 581–591.

Blatt, S.P., Hendrix, C.W., Butzin, C.A., Freeman, T.M., Ward, W.W., Hensley, R.E., Melcher, G.P., Donovan, D.J. and Boswell, R.N. (1993) Delayed-type hypersensitivity skin testing predicts progression to AIDS in HIV-infected patients. Ann. Intern. Med. 119, 177–184.

Buckley, R.H. (2002) Primary cellular immunodeficiencies. J. Allergy Clin. Immunol. 109, 747–757.

Casrouge, A., Zhang, S.Y., Eidenschenk, C., Jouanguy, E., Puel, A., Yang, K., Alcais, A., Picard, C., Mahfoufi, N., Nicolas, N., Lorenzo, L., Plancoulaine, S., Senechal, B., Geissmann, F., Tabeta, K., Hoebe, K., Du, X., Miller, R.L., Heron, B., Mignot, C., de Villemeur, T.B., Lebon, P., Dulac, O., Rozenberg, F., Beutler, B., Tardieu, M., Abel, L. and Casanova, J.L. (2006) Herpes simplex virus encephalitis in human UNC-93B deficiency. Science 314, 308–312.

Davis, C.M., McLaughlin, T.M., Watson, T.J., Buckley, R.H., Schiff, S.E., Hale, L.P., Haynes, B.F. and Markert, M.L. (1997) Normalization of the peripheral blood T cell receptor V beta repertoire after cultured postnatal human thymic transplantation in DiGeorge syndrome. J. Clin. Immunol. 17, 167–175.

Figueroa, J.E. and Densen, P. (1991) Infectious diseases associated with complement deficiencies. Clin. Microbiol. Rev. 4, 359–395.

Filipe-Santos, O., Bustamante, J., Chapgier, A., Vogt, G., de Beaucoudrey, L., Feinberg, J., Jouanguy, E., Boisson-Dupuis, S., Fieschi, C., Picard, C. and Casanova, J.L. (2006) Inborn errors of IL-12/23- and IFN-gamma-mediated immunity: molecular, cellular, and clinical features. Semin. Immunol. 18, 347–361.

Filipovich, A.H. (2006) Hemophagocytic lymphohistiocytosis and related disorders. Curr. Opin. Allergy Clin. Immunol. 6, 410–415.

Fleisher, T.A., Dorman, S.E., Anderson, J.A., Vail, M., Brown, M.R. and Holland, S.M. (1999) Detection of intracellular phosphorylated STAT-1 by flow cytometry. Clin. Immunol. 90, 425–430.

Fleisher, T.A. and Oliveira, J.B. (2004) Functional and molecular evaluation of lymphocytes. J. Allergy Clin. Immunol. 114, 227–234; quiz 235.

Go, E.S. and Ballas, Z.K. (1996) Anti-pneumococcal antibody response in normal subjects: a meta-analysis. J. Allergy Clin. Immunol. 98, 205–215.

Ku, C.L., Yang, K., Bustamante, J., Puel, A., von Bernuth, H., Santos, O.F., Lawrence, T., Chang, H.H., Al-Mousa, H., Picard, C. and Casanova, J.L. (2005) Inherited disorders of human Toll-like receptor signaling: immunological implications. Immunol. Rev. 203, 10–20.

Letsch, A. and Scheibenbogen, C. (2003) Quantification and characterization of specific T cells by antigen-specific cytokine production using ELISPOT assay or intracellular cytokine staining. Methods 31, 143–149.

Lopez-Granados, E., Perez de Diego, R., Ferreira Cerdan, A., Fontan Casariego, G. and Garcia Rodriguez, M.C. (2005) A genotype-phenotype correlation study in a group of 54 patients with X-linked agammaglobulinemia. J. Allergy Clin. Immunol. 116, 690–697.

Mallone, R. and Nepom, G.T. (2004) MHC Class II tetramers and the pursuit of antigen-specific T cells: define, deviate, delete. Clin. Immunol. 110, 232–242.

McFarland, R.D., Douek, D.C., Koup, R.A. and Picker, L.J. (2000) Identification of a human recent thymic emigrant phenotype. Proc. Natl. Acad. Sci. USA 97, 4215–4220.

Nichols, K.E., Ma, C.S., Cannons, J.L., Schwartzberg, P.L. and Tangye, S.G. (2005) Molecular and cellular pathogenesis of X-linked lymphoproliferative disease. Immunol. Rev. 203, 180–199.

Niemela, J.E., Puck, J.M., Fischer, R.E., Fleisher, T.A. and Hsu, A.P. (2000) Efficient detection of thirty-seven new IL2RG mutations in human X-linked severe combined immunodeficiency. Clin. Immunol. 95, 33–38.

Notarangelo, L., Casanova, J.L., Conley, M.E., Chapel, H., Fischer, A., Puck, J., Roifman, C., Seger, R. and Geha, R.S. (2006) Primary immunodeficiency diseases: an update from the International Union of Immunological Societies Primary Immunodeficiency Diseases Classification Committee Meeting in Budapest, 2005. J. Allergy Clin. Immunol. 117, 883–896.

Orange, J.S. (2006) Human natural killer cell deficiencies. Curr. Opin. Allergy Clin. Immunol. 6, 399–409.

Perfetto, S.P., Chattopadhyay, P.K. and Roederer, M. (2004) Seventeen-colour flow cytometry: unravelling the immune system. Nat. Rev. Immunol. 4, 648–655.

Pyun, K.H., Ochs, H.D., Wedgwood, R.J., Yang, X.Q., Heller, S.R. and Reimer, C.B. (1989) Human antibody responses to bacteriophage phi X 174: sequential induction of IgM and IgG subclass antibody. Clin. Immunol. Immunopathol. 51, 252–263.

Rosenzweig, S.D. and Holland, S.M. (2004) Phagocyte immunodeficiencies and their infections. J. Allergy. Clin. Immunol. 113, 620–626.

von Bernuth, H., Ku, C.L., Rodriguez-Gallego, C., Zhang, S., Garty, B.Z., Marodi, L., Chapel, H., Chrabieh, M., Miller, R.L., Picard, C., Puel, A. and Casanova, J.L. (2006) A fast procedure for the detection of defects in Toll-like receptor signaling. Pediatrics 118, 2498–2503.

Vowells, S.J., Sekhsaria, S., Malech, H.L., Shalit, M. and Fleisher, T.A. (1995) Flow cytometric analysis of the granulocyte respiratory burst: a comparison study of fluorescent probes. J. Immunol. Methods 178, 89–97.

Walport, M.J. (2001) Complement. First of two parts. N. Engl. J. Med. 344, 1058–1066.

32

Frequently Ill Children

Tatiana Markova and Denis Chuvirov

Federal Educational Establishment Advanced Study Institute, FMBA, Moscow, Russia, markova@immune.umos.ru

Abstract. Frequently ill children (FIC) show persistence of infection in the nasopharynx, disbiosis of intestinal flora, and concomitant and allergic diseases. As per our results, FIC with acute respiratory diseases (ARD) frequency of 6–15 times a year plus chronic infection foci at the age of 2–15 y/o at the remission period have heterogeneous nature of immune system disorders. It depends on the age, frequency of ARD, and chronic infection foci. About 20–50% of children have low number of T cells and 70% of children have high number of activated T cells. About 5–23% of children have low level of serum IgG or IgA, while low level of saliva IgA has been determined in 94% of children and low synthesis of IFN-α in 80% and of IFN-γ in 30% of children. About 50% of kids have high level of common IgE (160–220 ME/ml) and diagnostic sensitization to various allergens. In contrast, only 25% of children with ARD frequency of 4–6 times a year without chronic infection foci had low synthesis of IFN-α, 30% had low IgA level in saliva, and 8.3% had low IgA level in serum. After vaccination against hepatitis B, antibody level to HBs-Ag and time of their circulation at FIC had been lower than in children with ARD frequency of 4–6 times a year. Examination of FIC at the remission period showed polymorphism of natural and adaptive immunity disorders associated with the immune system developmental delay and subsequent forming of chronic infection foci being an aggravating factor for these disorders.

1. Introduction

Frequently ill children (FIC) are kids liable to frequent respiratory infections, mainly due to transient deviations and age peculiarities of the child immune system, who go for regular medical check-up. FIC are characterized by a number of acute respiratory diseases (ARDs) per year (Albickiy and Baranov 1986). ARD prevalence in FIC is higher (once a month) than in occasionally ill children, thus being one cause for the frequent visits to pediatricians and hospitalization in the case of complications. Factors affecting the frequency of infections are developmental lag of the immune system, anatomic–physiological peculiarities of children's respiratory tract, and social living conditions. Localization of infection foci and nosologic forms of diseases in

FIC vary: (1) upper airways (nasopharyngitis, acute otitis, sinusitis, and tonsillitis), (2) "false" croup and laryngotracheobronchitis, and (3) inferior airway infections (bronchiolitis and pneumonia). Consecutive infections may be caused by (i) bacteriums, (ii) viruses, or (iii) pathogenic organisms: *Chlamydia pneumonia* and *Mycoplasma pneumonia.*

Up to 83% of all diseases in FIC refer to the orinasal and respiratory tract pathology. Viral infections are seen in 65–90% of cases, monoinfection in 52% of cases, and association of two and more infections in 36% of cases. A combination of carrier and sowing of *Streptococcus pneumonia* and *Haemophilus influenzae* is observed twice more often in FIC than in rarely sick kids (Markova et al. 2003; Makarova 2005).

Here, in order to further characterize children with ARD frequency from 6 to 15 times a year, we analyzed their immune status and evaluated the presence of immune-mediated diseases.

2. Experimental Procedures

Two hundred seventy FIC (160 girls and 110 boys, 2–15 y/o) have been selected during 1–3 years and examined. Most of them were diagnosed with a combined pathology at the orinasal side of the upper airways (Table 1). At the beginning of surveillance, 37 patients (13.7%) were diagnosed with concomitant recurrent herpetic infection (relapse number 5–9 times a year); 30 patients (11.1%) had manifestation of atopic dermatitis with localized foci. As a control, 60 children (35 girls and 25 boys, 2–15 y/o) with ARD frequency within 4–6 times a year and without chronic infections were also investigated. Furthermore, 60 FIC (35 girls and 25 boys, 6–15 y/o) without antibodies (Ab) to HBs-Ag have been vaccinated at the remission period against hepatitis B (Ingerix B vaccine). The control was represented by 30 children (17 girls and 13 boys, 6–15 y/o) with ARD frequency within 4–6 times a year and no foci of chronic infection who also received the same vaccine.

TABLE 1. Concomitant pathology in frequently ill children (FIC)

Concomitant pathology	Number of children
ARD more than 6 times a year	270
Chronic adenoiditis	67
Chronic tonsillitis	42
Chronic otitis	62
Chronic pharyngitis	67
Lingering laryngotrcheitis	50
Recurrent herpetic infection	37
Atopic dermatitis	30
Bowels dysbacteriosis	200

Immunodeficiency development has been estimated as described earlier (Markova et al. 1997). Immunoassay procedures included detection of CD3$^+$, CD4$^+$, CD8$^+$, CD16$^+$, CD19$^+$, HLA-DR$^+$, CD3$^+$HLA-DR$^+$, and CD4$^+$CD45$^+$RO$^+$ cells, phagocytosis, chemiluminescence indices, serum IgG, A, M and E, IgA in saliva, serum IFNs, and cellular IFN-α and IFN-γ (Khaitov et al. 1995; Ershov 1996). Cells were analyzed by FACScan (Calibur) or LKB-Wallac 1251 Luminometer. *Mycoplasma pneumonia* and *Chlamydia pneumonia* infections were determined by PTsR-diagnostic procedures (DNA-Technology, Russia). All immunoglobulins were assessed by ELISA (CHEMA or Vector, Russia).

3. Results

Bacterial infection persistence at the respiratory tract has been confirmed by a smear flora analysis taken from mucosa (Table 2). A single microbial agent was detected in 40% of children, two and more agents in 47.2% of patients, and combined bacterial and fungi flora in 10% of children. The persistence of infection was less in control group. Antibodies to *M. pneumonia* (IgM titer 1:8–1:32 and IgG titer 1:16–1:64) were seen in 50% of children. Antibodies to *C. pneumonia* (IgM titer 1:8–1:64 and IgG titer 1:16–1:128) were determined in 40% of kids. About 30% of children had combined infections. Verification of *Streptococcus haemolyticus-β*, *Chlamydia*, and *M. pneumonia* infections has been used as an indication for antibacterial therapy. Next, 74% of FIC had bowels dysbacteriosis, low index of *Bifido-* and *lactobacteriums,* and hemolyzed and/or lactose-negative *Escherichia coli* sowing. Our results thus show that FIC at the age of 2–15 y/o during the remission period have heterogeneous nature at immune disorders, which might depend on the age, ARD frequency, and type of chronic infections (Table 3).

Furthermore, 20% of kids at the age of 2–7 y/o had low number of T cells (CD3$^+$ CD4$^+$), while at the age of 7–15 y/o, 50% of children had these values out of normal

TABLE 2. Sowing microflora taken from fauces of examined children

Agent	% of children with sowing agent	
	ARD 6–12 times a year	ARD 4–6 times a year
Staphylococcus spp.	80	60
Staphylococcus aureus	66	50
Staphylococcus haemolyticus	40	20
Streptococcus spp.	60	40
Sreptococcus haemolyticus-β	30	6
Neisseria perflava	30	30
Corynebacterium pseudo-diphtheriae	14	6
Candida albicans	14	10
Flora has not been determined	10	20

TABLE 3. Immunological indices of immune system with chronic infection foci in frequently ill children (FIC)

Immunological parameters	% of children with immune disorder
IFN-α synthesis	Low level in 80% of children
IFN-γ synthesis	Low level in 30% of children
T cells	20–50% of kids have low number of $CD3^+$ or $CD4^+$ cells. 70% of kids have high number of activated T cells ($CD3^+HLA\text{-}DR^+$ cells)
Macrophages	Low index of spontaneous chemiluminescence in 20% of children
Immunoglobulins	IgG or IgA low blood level in 20% of children; IgA low level in saliva in 94% of children; IgE high level in 50% of children
EK cells	Low number of $CD16^+$ cells in 15% of children
B cells	Low number of $CD19^+$cells in 15% of children

range. In addition, 70% of children had high number of activated T cells ($CD3^+HLA\text{-}DR^+$), 15 kids (25%) with ARD frequency 4–6 times/year had down-regulated synthesis of IFN-α, 20 children (30%) had low IgA level in saliva, and 5 kids (8.3%) demonstrated decreased concentrations of IgA in the blood.

During a 2-year observation period, we detected an increase in total and specific IgE levels in dynamics, sensitization to allergens, and development of allergic diseases: 60% of children had rhinosinusopathy, 30% allergic rhinitis, 11.1% atopic dermatitis, 30% recurrent bronchitis, and 10% of children developed bronchial asthma; 14.8% of children had pharyngomycosis.

Next, FIC and age- and gender-matched control children were vaccinated during the remission period and levels of specific antibodies against HBs-Ag were measured 1, 6, and 12 months after vaccine administration (Table 4). A local reaction at the vaccination site was reported in five patients from control group and in three kids from FIC group; a system reaction — rise of temperature up to 37–37.2°C—was observed in two patients from control group and one child from FIC group. Antibody titers assessed 1, 6, and 12 months after vaccination were significantly lower in FIC when compared with control children.

TABLE 4. Antibody levels in frequently ill children after (FIC) vaccination against hepatitis B

Group	Antibodies level to HBs-Ag mME/ml			
	Before vaccination	In 1 month	In 6 months	In 1 year
FIC, mean ± SD	0	107 ± 1.9	113 ± 2.1	84 ± 2.5
Control, mean ± SD	0	672 ± 11.3	496 ± 10.1	440 ± 12.6
p-value		<0.001	<0.001	<0.001

TABLE 5. $CD45RO^+$ T cells (vaccination against hepatitis B)

Group	Indices		Before vaccination	In 1 month	In 6 months
FIC, mean ± SD	$CD4^+CD45^+RO^+$ cells	%	38 ± 1.9	39.6 ± 2.2	39 ± 2.8
		$\times 10^9$/l	0.84 ± 0.12	0.86 ± 0.1	0.81 ±0.08
Control, mean ± SD	$CD4^+CD45^+RO^+$ cells	%	36.5 ± 1.9	42.7 ± 2.3	41 ± 1.9
		$\times 10^9$/l	0.86 ± 0.07	0.99 ± 0.05	0.96 ± 0.05
p-value				<0.05	<0.05

Evaluation of T cells in vaccinated patients (Table 5) revealed increase of absolute number of $CD4^+CD45RO^+$ cells in control children 1 and 6 months after vaccination ($p < 0.05$).

4. Discussion

For the first time, we reported significant abnormalities in the immune status and immune responses in FIC in comparison with healthy controls. This confirms our earlier observations of decreased functional activity of T and B cells in these children (Markova and Kharianova 2001). High levels of IgA in the saliva and low levels of serum IgG and IgG_1 have been reported in children with recurrent infections (Mancini et al. 1996; Hewson-Bower and Drummond 1996). Results of multicenter examination of 180 FIC from various megapoleis (Moscow, Baku, and Krasnoyarsk) during acute period of ARD showed low levels of IFN-γ, IL-2, IL-4, IL-8, IgG, and IgM, as well as increase in $CD25^+$ T cells (Namazova et al. 2005). Interestingly, IFN-γ synthesis in rarely sick children has been shown to be much higher in fall and winter seasons, while it is lower in FIC. Systemic "early" synthesis of IFN seen in FIC indicates the functional immaturity of the system and may be due to genetic predisposition. Indeed, decreased synthesis of IFN has been determined in FIC's mothers and sibs. Low levels of secretory IgA were reported in 85% of FIC (Albickiy and Baranov 1986; Markova et al. 2003; Makarova 2005). It has been recently speculated that most chronic diseases seen in adults are the result of maturation abnormalities of the immune system in childhood (Holt and Sly 2002). Our own analysis of FIC suffering from frequent ARD and chronic infection foci during remissions showed a polymorphism of natural and adaptive immune abnormalities and suggests that delay of the immune system development and subsequent formation of chronic infection foci may be aggravating factors for these disorders.

References

Albickiy, V.Y. and Baranov, A.A. (1986) *Frequently Ill Children: Clinical and Social Aspects*. Saratov, Russia.

Ershov, F.I. (2005) *Interferons and their Inductors*. Goatar-Media Publishing, Moscow, Russia.

Hewson-Bower, B. and Drummond, P.D. (1996) Secretory immunoglobulin A increases during relaxation in children with and without recurrent upper respiratory infections. J. Dev. Behav. Pediatr. 17, 311–316.

Holt, P.G. and Sly, P.D. (2002) Interactions between RSV infection, asthma, and atopy: unraveling the complexities. J. Exp. Med. 196, 1271–1275.

Khaitov, R.M., Pinegin, B.V. and Istamov, H.I. (1995) *Ecological Immunology*. VNIRO Publishing, Moscow, Russia.

Makarova, Z.S. (2005) Frequently ill children and their rehabilitation at pediatric polyclinic. Polyclinic (Russ.) 14, 19.

Mancini, C., Iacovani, R. and Fiermonte, V. (1996) Evaluation of serum IgG subclasses at children with recurrent respiratory infections. Minerva Pediatr. 48, 79–83.

Markova, T.P., Chuvirov, D.G. and Chuvirov, G.N. (2003) Children getting ill frequently and for a long time. In: *Clinical Immunolology and Allergy Medicine*. Ed. Gianni Marone. JGC Publishing, Naples, Italy. 62, 425–429.

Markova, T.P., Khaitov, R.M. and Tchouvirov, G.N. (1997) Methodological approaches to immunological diagnosis. Immunol. Lett. 56, 332.

Markova, T.P. and Kharianova, M.E. (2001) Postvaccinal immunity forcing at frequently and lasting ill children. Allergy Asthma Clin. Immunol. 1, 75–77.

Namazova, A.S., Botvinjeva, V.V. and Torchoeva, R.M. (2005) Frequently ill children of megapolysis. Pediatr. Pharmacol. 2, 3–7.

33

Evaluation of Bactericidal Activity of Human Biological Fluids by Flow Cytofluorimetry

Anna Budikhina and Boris Pinegin

National Research Center Institute of Immunology, FMBA, Moscow, Russia, budihina@newmail.ru

Abstract. The ability of biological fluids to kill microbes is an important feature of the human immune system. Following incubation of fluorescein isothiocyanate-labeled *Staphylococcus aureus* with biological specimens and subsequent staining with propidium iodide, the proportions of killed bacteria were estimated by flow cytometry. FACScan is a simple, quick and reliable method to evaluate bactericidal activity of biological fluids. The results of the cytometric method correlated well with the results of the classic microbiological method. The proposed method is highly informative for evaluating bactericidal activity of sera in immunocompromised patients.

1. Introduction

The ability of sera and some other biological fluids to kill microbes is an important feature of the human immune system. The deficiency of bactericidal activity of sera may increase the susceptibility to infections (Li 1998).

The microbiological method was originally proposed to evaluate bactericidal activity of sera. This method involved incubation of serum with microbes and subsequent count of colony numbers (Taylor 1983). Another method is a photometric assay, which is based on the analysis of the optical density of bacterial cultures after incubation with serum. A drawback of the method is the common overestimation of the bactericidal activity due to sedimentation of bacteria in the cuvette (Muschel and Treffers 1956). Later, Fierer and colleagues designed a radiometric assay that detects the release of ^{51}Cr from ^{51}Cr-labeled bacteria. A shortcoming of this method is the use of γ radiation, which requires special safety procedures (Fierer et al. 1974). Laser flow cytometry offers considerable advantages over other methods, including low labor intensity, high speed (hours vs. days), and objective and accurate measurements. The aim of this study was to develop a method for the evaluation of biological fluid bactericidal activity by means of flow cytometry.

2. Experimental Procedures

Serum specimens were collected from 65 healthy donors (age 18–50), 35 patients with multiple myeloma (21, 10, and 4 patients with IgG, IgM, and IgA types, respectively, age 49–83), 2 patients with complement deficiency, and 5 children with Bruton's agammaglobulinemia lacking B cells (age 2–12). Saliva obtained from 50 healthy nonsmoking donors (age 18–50) with oral cavity sanitation was tested as well. Some of the patients were examined twice with a half-year interval.

Bactericidal activity of biological fluids was investigated using fluorescein isothiocyanate (FITC)-labeled *Staphylococcus aureus* (*S. aureus*, Cowan I strain, Tarasevich National Institute for Drugs Standardization, Russia, Moscow). *S. aureus* cultures were washed, resuspended in carbonate-bicarbonate buffer and the concentration of bacteria was brought to 1×10^8/ml according to the turbidity standard. FITC (1 mg/ml, Molecular Probes) was added to bacterial suspensions for 16 h at 4°C. *S. aureus*-FITC were washed, resuspended in PBS with 5% fetal calf serum (ICN Biomedicals) and 5% DMSO (Sigma), and diluted to 5×10^8 cells/ml. Aliquots of labeled *S. aureus* were stored at –70°C.

The assay was carried out in 96-well round-bottom plates with 90 µl of specimen fluid and 90 µl of *S. aureus*-FITC (1×10^6/ml) suspensions incubated at 37°C for various time intervals. Spontaneous death of *S. aureus*-FITC in PBS was used as a control. After incubation, *S. aureus*-FITC was precipitated by centrifugation at 1000 *g* for 10 min, washed, and resuspended in 200 µl PBS containing 2.5 µg/ml propidium iodide (PI, Sigma). Ten minutes later, samples were analyzed by FACS Calibur (BD) using CellQuest software package. The percentages of double-positive ($FITC^+PI^+$) bacteria among the all $FITC^+$ bacteria were calculated.

The reference assay utilized the analysis of colony formation (colony-forming units, CFUs) in Petri dishes with beef-extract agar (BD) after coincubation of *S. aureus* (90 µl, 1×10^4/ml) with serum (90 µl). Percentage of killing was calculated by the following formula: % killing = $(CFU_C - CFU_E)/\ CFU_E * 100$, where CFU_C is the number of colonies in control Petri dish and CFU_E the number of colonies in experimental Petri dish.

3. Results

3.1. Bactericidal Activity of Biological Fluids from Healthy Volunteers

Flow cytometry analysis of serum from 10 donors revealed that the percentage of bacteria killing were 1.14 ± 1.2%, 19.0 ± 6.6%, and 19.8 ± 10.5% for 1, 3, and 24 h of coincubation time, respectively. In the absence of serum, spontaneous death of bacteria was 0.5 ± 0.4%, 2.1 ± 0.2%, and 2.8 ± 0.1%, respectively. Thus, incubation for 3 h was the optimal time. The bactericidal activity of saliva markedly differed from serum activity: Saliva killed 23.8 ± 9.4% of bacteria after 3 h incubation and 43.3 ± 21.2% of bacteria after 24 h of incubation (n = 10). Considering the possibility of bactericidal substance release from neutrophils in the course of blood coagulation (Sorensen and Borregaard 1999), the bactericidal activity of serum and plasma was also compared.

TABLE 1. Bactericidal activity of serum in patients and healthy controls

Patients	*Staphylococcus aureus* killing by serum, % mean ± SD		
	Control	Native	Heated
Healthy donors ($N = 65$)	2.4 ± 1.2	19.0 ± 6.6	10.6 ± 5.3
Multiple myeloma, type IgG ($N = 21$)	2.5 ± 0.9	8.0 ± 2.3	6.1 ± 2.4
Multiple myeloma, type IgM ($N = 10$)	2.3 ± 1.0	11.4 ± 3.2	7.6 ± 3.6
Multiple myeloma, type IgA ($N = 4$)	2.5 ± 2.1	14.3 ± 3.7	10.3 ± 4.5
Complement deficiency ($N = 2$)	3.5 ± 0.7	16.3 ± 10.4	13.8 ± 5.7
Agammaglobulinemia ($N = 5$)	2.8 ± 1.0	10.4 ± 6.0	8.4 ± 6.7

In the group of 10 healthy donors, *S. aureus*-FITC death caused by serum and plasma was 22.7 ± 13.8% and 20.8 ± 14%, respectively, suggesting that the bacteria killing activity of sera and plasma are comparable.

Next, the cytometric and the bacteriological methods of evaluation of serum bactericidal activity were compared in a group of 11 healthy donors. Results of the cytometric method strongly correlated with the results of the classic microbiological method (Spearman $r = 0.84$, $p < 0.05$).

3.2. Bactericidal Activity of Biological Fluids from Patients with Immunological Disorders

Bactericidal activity of serum from patients with multiple myeloma was significantly reduced compared with that in healthy donors: by 57.89% in patients with IgG-secreting myeloma and by 40% in patients with IgM-secreting myeloma ($p < 0.001$). Reduction of the bactericidal activity in IgA-producing multiple myeloma was insignificant when compared with healthy donors (Table 1).

Heating of serum obtained from patients with IgG, IgM, and IgA multiple myeloma at 56°C for 30 min resulted in a marked decrease of bacteriolytic activity by 24%, 33%, and 28%, respectively.

Bactericidal activity of serum from agammaglobulinemia patients was 10.4% ($n = 5$). This value is significantly lower than the bactericidal activity observed in healthy donors ($p < 0.01$). Interestingly, bactericidal activity of serum from these patients did not change substantially after heating (10.4% vs. 8.4%). Finally, the bactericidal activity of sera from patients with complement deficiency did not differ from that of healthy donors and did not change considerably upon heating (Table 1).

4. Discussion

Heating of donors' serum at 56°C for 30 min reduced their bactericidal activity by 44% on average. Heating of donors' saliva at 56°C for 30 min did not change its bactericidal activity. The bactericidal activity of serum probably arises from thermolabile bacteriolytic activity of complement and natural antibodies and thermostable activity of proteins and peptides, such as defensins, catelicidins, lactoferrin, lysozyme, phospholipase A_2, and others (Caccavo et al. 2002; Ginsburg 2004; Gronroos et al. 2002; Jankowski 1995). The bactericidal activity of saliva depends on thermostable proteins and peptides by almost 100% (Ahmad et al. 2004; Nieuw Amerongen and Veerman 2002). It is possible that unlike healthy donors, the bactericidal activity of serum in patients with IgG multiple myeloma type, agammaglobulinemia, and complement deficiency rely on bactericidal proteins and peptides rather than complement and antibodies.

We believe that the cytometric method for evaluation of the bactericidal activity of biological fluids will be widely applied as one of the standard methods of the immune status assessment. This technique shows considerable promise for assessing the local immunity, particularly that of oral cavity. Tanida and coworkers (2003) recently demonstrated that salivary gland dysfunction resulting in a decrease of saliva antibacterial agents provokes candidosis.

References

Ahmad, M., Piludu, M., Oppenheim, F.G., Helmerhorst, E.J. and Hand, A.R. (2004) Immunocytochemical localization of histatins in human salivary glands. J. Histochem. Cytochem. Pharmacol. 52, 361–370.

Caccavo, D., Pellegrino, N.M., Altamura, M., Rigon, A., Amati, L., Amoroso, A. and Jirillo, E. (2002) Antimicrobial and immunoregulatory functions of lactoferrin and its potential therapeutic application. J. Endotoxin Res. 8, 403–417.

Fierer, J., Finley, F. and Braude, A.I. (1974) Release of C^{51}-endotoxin from bacteria as assay of serum bactericidal activity. J. Immunol. 112, 2184–2192.

Ginsburg, I. (2004) Bactericidal cationic peptides can also function as bacteriolysis-inducting agents mimicking beta-lactam antibiotics? It is enigmatic why this concept is consistently disregarded. Med. Hypotheses, 62, 367–374.

Gronroos, J.O., Laine Veli, J.O. and Nevalainen, T.J. (2002) Bactericidal group IIA phospholipase A_2 in serum of patients with bactericidal infections. J. Infect. Dis. 185, 1767–1772.

Jankowski, S. (1995) The role of complement and antibodies in impaired bactericidal activity of neonatal sera against gram-negative bacteria. Acta Microbiol. Pol. 44, 5–14.

Li, Y.M. (1998) Glycation ligand binding motif in lactoferrin. Implications in diabetic infection. Adv. Exp. Med. Biol. 443, 57–63.

Muschel, L.H. and Treffers, H.P. (1956) Quantitative studies on the bactericidal actions of serum and complement. A rapid photometric growth assay for bactericidal activity. J. Immunol. 76, 11–20.

Nieuw Amerongen, A.V. and Veerman, E.C.I. (2002) Saliva the defender or oral cavity. Oral Dis. 8, 12–22.

Sorensen, O. and Borregaard, N. (1999) Methods for quantitation of human neutrophil proteins, a survey. J. Immunol. Methods 232,179–190.

Tanida, T., Okamoto, T., Okamoto, A., Wang, H., Hamata, T., Ueta, E. and Osaki T. (2003) Decreased excretion of antimicrobial proteins and peptides in saliva of patients with oral candidiasis. Oral Pathol. Med. 32, 586–594.

Taylor, P.W. (1983) Bactericidal and bacteriolytic activity of serum against Gram-negative bacteria. Microbiol. Rev. 47, 46–83.

Immunotherapy and Vaccines

34

Rational Development of Antigen-Specific Therapies for Type 1 Diabetes

Georgia Fousteri, Damien Bresson, and Matthias von Herrath

La Jolla Institute for Allergy and Immunology, Department of Developmental Immunology, La Jolla, CA, USA, Matthias@liai.org

Abstract. Administration of autoantigens, especially via the mucosal route, can induce tolerance under certain circumstances. In autoimmune diabetes, mucosal vaccination with autoantigens was frequently effective in restoring tolerance in mice but has not yet succeeded in humans. Furthermore, in some instances, autoimmunity can be precipitated upon autoantigen administration. We will here briefly discuss the underlying reasons and delineate which efforts should be made in the future to rationally translate antigen-specific immunotherapy, for example, by establishing better assays to reduce the risk for possible adverse events in humans.

1. The Importance to Consider Pre-existing Autoimmunity When Administering Self-Antigens

Antigens administered via mucosal surfaces and administration of autoantigens overall can have strong therapeutic potential. The key issue is to gather sufficient knowledge to establish improved safeguards (i.e., improved predictive assays) that will help in directing the immune responses toward tolerogenic and away from pathogenic pathways. This information should reduce the risk for possible adverse events in humans. The main reasons, why this has proven to be difficult are because immunity per se comprises not one single but a multitude of potential classes of responses and the outcome of immunization depends on pre-existing immunity, dose and route of administration, and the autoantigen itself (O'Shea et al. 2002).

The precise class or effector molecules (i.e., cytokines) expressed by activated immune cells determines the outcome of a given response in context with the disease or pathogen that is being targeted (O'Shea et al. 2002). Thus, autoantigen-specific vaccination is not so much a question of immunity versus tolerance but rather a question of the class of response to be induced. In the absence of inflammatory stimuli, more TGF-β1 and other immunomodulatory cytokines are being produced. There are multiple reports in the literature that immune modulation is possible after antigenic vaccination. For example, mucosal immunization with self-antigens can induce regulatory T cells (Tregs) that are capable of mediating tolerance in a variety of animal models for autoimmune diabetes (Faria and Weiner 2005). In the contrary, if there is a pre-existing pool of activated effector T cells at the time of treatment, the autoaggressive T-helper 1 (Th1) response might not be altered by mucosal vaccination, and consequently, autoimmunity might become aggravated (Harrison and Hafler 2000). One of the paramount questions in autoimmunity remains: How can we induce safe and long-term tolerance upon vaccination with a self-antigen? Herein, we address some of the main issues to be overcome in order to reach this goal.

2. Immunity Versus Tolerance Viewed in an Operational Context

First, one should take a closer look at what "tolerance" means. In earlier days, tolerance to self- and foreign antigens was attributed to two major mechanisms: (1) deletion through activation-induced cell death, involving ubiquitously expressed self-antigens (Miller 1995; Mondino et al. 1996) and (2) tissue-specific functional inactivation of autoreactive lymphocytes also called anergy (Schwartz 1996). Moreover, there is also a state referred to as immunological ignorance, in which the lymphocytes "ignore" the existence of their specific self-antigen because of limited access (Ohashi and DeFranco 2002). Most of current research focuses on a third and probably more important mechanism known as "immune regulation," which results in functional tolerance upon antigenic encounter. In this last scenario, regulation is mediated by Tregs. Tregs are classified in to two major categories: (1) naturally arising "professional" $CD4^+CD25^+FoxP3^+$ Tregs ("nTregs") (Sakaguchi et al. 1995) and (2) adaptive (inducible) Tregs (also termed iTregs). The iTregs are characterized by cytokine effector mechanisms and comprise several important subgroups such as (a) $CD4^+$ Th2-like cells that produce IL-4/IL-10 cytokines and antagonize the Th1 effector cells, (b) $CD4^+$ Treg1 cells that function through IL-10 production and likely modulate antigen-presenting cells (APCs) (Roncarolo et al. 2006), (c) $CD4^+$ Th3 cells suppressing by a TGF-β1-dependent mechanism (Weiner 2001), and (d) $CD8^+$ Tregs that might act and in a contact-dependent manner by production of TGF-β1 (Chen et al. 1994; Bisikirska et al. 2005). Tregs can be generated in vivo following low levels of antigenic stimulation by encountering special "tolerogenic" APCs in the local microenvironment where they are induced (i.e., mucosal surfaces). Importantly, they have the capacity to exert their suppressive activity in a nonantigen-specific fashion by a mechanism termed "bystander suppression." Moreover, they have been shown to play a role in expanding a second "wave" of Tregs with different specificities, thus resulting in global amplification of suppression in vivo, a phenomenon termed "infectious tolerance" (Cobbold and Waldmann 1998). Other cells, especially Th2-like Tregs, promote deviation from a Th1 to a Th2 repsonse ("immune deviation"). Interestingly, mucosal antigenic administration can exhibit the capacity to induce all of these Treg subtypes in various autoimmune disease models.

However, it is still unclear under which circumstances immunization becomes detrimental in autoimmunity, forasmuch as we do not yet fully comprehend how vaccination will later impact the classes of immune responses. For example, antigenic vaccination may lead to different outcomes depending on the following important factors:

- The dose injected (higher dosages might favor cellular deletion and lower active Treg induction)
- The physical form (certain adjuvants such as IFA might favor tolerogenic responses and CTB [the cholera toxin B subunit] as a carrier molecule can enhance oral induction of Tregs)
- The route of immunization (mucosal antigens might favor tolerance)
- The affinity for the major histocompatibility complex (MHC) (lower avidity might favor nonpathogenic responses)
- The timing and duration of the treatment (more frequent antigen administration appears to favor tolerance)
- The frequency of antigen-specific precursor cells and T cell-receptor (TCR) repertoire and stage of the immune system (i.e., maturity or lymphopenia) (a pre-existing Th1 autoimmune response might be difficult to convert into a regulatory response).

Due to these multiple variables, transition "from bench to bedside" has proven to be complicated and suitable assays that confirm the desired outcome of immunization have to be refined and further developed.

3. Antigenic "Drivers" Versus "Followers" in Autoimmunity—How to Make the Correct Antigenic Choice for Immunotherapy?

During development of type 1 diabetes (T1D) for example, we know that the autoimmune response spreads intra- as well as intermolecularly to target several autoantigens, but there still is great debate as to whether only a single autoantigen ("driver" antigen) will be sufficient to trigger the disease (Jasinski and Eisenbarth 2005). This phenomenon can also be observed in antigenically induced autoimmune disease models such as experimental autoimmune encephalitis (EAE) (a model for multiple sclerosis) (Nakayama et al. 2005). The spreading hierarchy, not yet clear, seems to result from the activation and expansion of high-avidity β cell-reactive T cells, followed by T cells with progressively lower avidity (Lernmark and Agardh 2005). Autoreactive B cells that are able to present autoantigen to autoaggressive T cells might participate in this process by presenting cryptic epitopes (Melanitou et al. 2004). In T1D, autoaggressive responses develop toward insulin, glutamic acid decarboxylase 65 (GAD65), IA-2, heat shock protein 60 (HSP 60), and several other target autoantigens (Liblau et al. 1995; Tian et al. 2001; Dai et al. 2005). Current evidence indicates that CD8 responses to insulin develop first in nonobese diabetic (NOD) mice, followed by islet-specific glucose-6-phosphatase catalytic subunit-related protein (IGRP)-specific CD8 cells and other specificities.

Interestingly, by using proliferation assay and/or MHC class II tetramer analysis, T cell responses to GAD65 and proinsulin were also found in the peripheral blood of diabetic patients and at-risk individuals (Mannering et al. 2004; Oling et al. 2005). In a different study, naturally processed islet epitopes (NPPEs) were eluted from APCs of patients bearing HLA-DR4 haplotypes. NPPE-induced cytokines, measured by enzyme-linked immunosorbent spot assay (ELISPOT), revealed an extreme polarization toward a proinflammatory Th1 responses in T1D patients while HLA-matched control subjects responded by secreting a regulatory cytokine IL-10 (Arif et al. 2004). Thus, we can be rather certain that the class of effector molecules produced by the autoreactive T cells is more crucial than their absolute numbers in diabetes development. Therapeutically, this is encouraging in the sense that one could identify a protective response following peptide immunization. However, there are still significant concerns, since "wrong" autoantigens (i.e., those that are already being targeted by aggressive T cells or those that are not suitable for induction of Tregs or T cell deletion) could inadvertently be selected due to the lack of sufficient information. Still, no reliable universally standardized T cell assay is available to identify detrimental outcomes following mucosal immunization with autoantigens in humans early on, so that clinical immunizations with islet antigens would ideally be performed in those individuals first, where disease acceleration is unlikely (i.e., recent-onset T1D). However, prevention trials using oral or nasal insulin have produced no adverse outcomes at this point.

4. Human Studies of Antigen-Specific Interventions in T1D

Based on the success of mucosal islet-antigenic vaccines in inhibiting disease progression in animal models, a number of antigen-based immunotherapies have been conducted in human phase I/II clinical trials. Some of them were tested within the Diabetes Prevention Trial-1 (DPT-1), which focused on (i) assessing the effect of oral insulin in prediabetes and (ii) understanding of the natural history of β-cell responses (Tian et al. 1996). It is worth noting that glucose tolerance gradually deteriorates in a substantial number of type 1 diabetics over a period of at least 2 years, thus providing a quite large therapeutic window for interventions in newly diagnosed patients (Bresson et al. 2006). In the DPT-1, where 7.5 mg of human insulin was administered orally to prediabetic patients, no significant protective effect was observed. Although diabetes developed similarly in both oral insulin and placebo subjects, the annualized diabetes rate was slightly higher in the placebo group when a subgroup analysis of patients at higher risk was performed after opening the blinded trial (Ostroukhova et al. 2004). This finding suggests that there might be potential for oral insulin to prevent T1D in humans, when given at the right time and the correct dose. In the upcoming years, a repeat of the DPT-1 is planned by TrialNet and will include those individuals at risk and possibly at a higher insulin dose based on murine studies.

In a recent-onset diabetes trial, insulin-treated patients had similar residual β-cell function compared with placebo-treated controls. Moreover, the youngest patients (less than 15 years old) showed a tendency for a more pronounced decline of basal

C-peptide levels with an acceleration in the decline of β-cell function (Sherry et al. 2005). In a second trial, higher insulin dosages did not improve efficacy (Skyler et al. 2005; Sosenko et al. 2006). As seen with the murine models, oral insulin (as opposed to anti-CD3 therapy) has never been able to revert recent-onset T1D.

More recently, intranasal insulin or carrier solution was given to individuals with autoantigens to one or more pancreatic autoantigens (insulin, GAD65, or IA-2). Immunity to islet autoantigens as well as β-cell function in response to insulin treatments was studied. No local or systemic adverse effects were observed and diabetes developed in 12 of 38 participants. Generally intranasal insulin was associated with an increase in antibody and a decrease in T cell responses to insulin (Skyler et al. 2005).

Recently, phase II clinical data have been released by the Diamyd™ company at http://www.diamyd.com. In a double-blind randomized placebo-controlled clinical trial, 35 patients recently diagnosed with T1D received two subcutaneous injections (on days 1 and 30) with 20 μg of either GAD65 or placebo. The GAD65 treatment group showed higher C-peptide levels at 15 months compared with the placebo group. Although these promising results need to be confirmed in a multi-center study, they encourage us to believe that therapy using islet-autoantigen-derived vaccines might be effective in clinical settings.

5. Safety Issues

Short-term use of intranasal insulin without absorption enhancers was well tolerated, the risk of hypoglycemia was minimal, and no objective adverse effects were detected (Harrison et al. 2004). Progress in peptide immunotherapy for the treatment of diabetes has recently been hampered by reports of anaphylactic reactions in both mice and humans. For example, the insulin peptides InsB$_{9-23}$ and B$_{13-23}$, even when administered subcutaneously in the absence of adjuvant, can induce a dramatic humoral response, leading to fatal anaphylaxis in NOD mice (Liu et al. 2002). Therefore, peptide alterations are being introduced in order to decrease side effects. In addition, administration of autoantigens can, under certain circumstances, aggravate disease as reported in multiple sclerosis. However, such aggravation was never seen when the whole antigen was administered mucosally in humans and never reported in T1D trials.

In NOD mice, nasorespiratory insulin can induce CD8$^+$ α/αTCR γ/δ Tregs, whereas InsB$_{9-23}$ and proinsulin B24-C36 peptides that bind to the MHC class II molecule (IAg7) induced CD4$^+$ Tregs. It is noteworthy that most of these studies were conducted during early stages of the disease, between 4 and 5 weeks of age. On the contrary, preliminary data from our group showed an enhancement in disease progression when intranasal treatment with InsB$_{9-23}$, its altered peptide ligand APL,[16,19] and human proinsulin B24-C36 was started later (in 10-week-old NOD mice) (unpublished data Fousteri, Bresson, and von Herrath). Therefore, depending on the stage of the disease, similar treatments can lead to contradictory results.

6. Concluding Remarks

We are still struggling with obstacles to develop standardized reliable T cell assays in order to detect responses toward insulin and other islet autoantigens with sufficient sensitivity and precision to monitor mucosally vaccinated patients. This is urgently needed to predict the outcome of a given antigenic dose on a per-patient basis and in this way early detect potential trial failures or even adverse outcomes.

References

Arif, S., Tree, T.I., Astill, T.P., Tremble, J.M., Bishop, A.J., Dayan, C.M., Roep, B.O. and Peakman, M. (2004) Autoreactive T cell responses show proinflammatory polarization in diabetes but a regulatory phenotype in health. J. Clin. Invest. 113, 451–463.

Bisikirska, B., Colgan, J., Luban, J., Bluestone, J.A. and Herold, K.C. (2005) TCR stimulation with modified anti-CD3 mAb expands CD8+ T cell population and induces CD8+CD25+ Tregs. J. Clin. Invest. 115, 2904–2913.

Bresson, D., Togher, L., Rodrigo, E., Chen, Y., Bluestone, J.A., Herold, K.C. and von Herrath, M. (2006) Anti-CD3 and nasal proinsulin combination therapy enhances remission from recent-onset autoimmune diabetes by inducing Tregs. J. Clin. Invest. 116, 1371–1381.

Chen, Y., Kuchroo, V.K., Inobe, J., Hafler, D.A. and Weiner, H.L. (1994) Regulatory T cell clones induced by oral tolerance: suppression of autoimmune encephalomyelitis. Science 265, 1237–1240.

Cobbold, S. and Waldmann, H. (1998) Infectious tolerance. Curr. Opin. Immunol. 10, 518–524.

Dai, Y.D., Carayanniotis, G. and Sercarz, E. (2005) Antigen processing by autoreactive B cells promotes determinant spreading. Cell. Mol. Immunol. 2, 169–175.

Faria, A.M. and Weiner, H.L. (2005) Oral tolerance. Immunol. Rev. 206, 232–259.

Harrison, L. and Hafler, D.A. (2000) Antigen-specific therapy for autoimmune disease. Curr. Opin. Immunol. 12, 704–711.

Harrison, L.C., Honeyman, M.C., Steele, C.E., Stone, N.L., Sarugeri, E., Bonifacio, E., Couper, J.J. and Colman, P.G. (2004) Pancreatic beta-cell function and immune responses to insulin after administration of intranasal insulin to humans at risk for type 1 diabetes. Diabetes Care 27, 2348–2355.

Jasinski, J.M. and Eisenbarth, G.S. (2005) Hypothesis for the pathogenesis of type 1A diabetes. Drugs Today (Barc) 41, 141–149.

Lernmark, A. and Agardh, C.D. (2005) Immunomodulation with human recombinant autoantigens. Trends Immunol. 26, 608–612.

Liblau, R.S., Singer, S.M. and McDevitt, H.O. (1995) Th1 and Th2 CD4+ T cells in the pathogenesis of organ-specific autoimmune diseases. Immunol. Today 16, 34–38.

Liu, E., Moriyama, H., Abiru, N., Miao, D., Yu, L., Taylor, R.M., Finkelman, F.D. and Eisenbarth, G.S. (2002) Anti-peptide autoantibodies and fatal anaphylaxis in NOD mice in response to insulin self-peptides B:9-23 and B:13-23. J. Clin. Invest. 110, 1021–1027.

Mannering, S.I., Morris, J.S., Stone, N.L., Jensen, K.P., Van Endart, P.M. and Harrison, L.C. (2004) CD4+ T cell proliferation in response to GAD and proinsulin in healthy, prediabetic, and diabetic donors. Ann. NY Acad. Sci. 1037, 16–21.

Melanitou, E., Devendra, D., Liu, E., Miao, D. and Eisenbarth, G.S. (2004) Early and quantal (by litter) expression of insulin autoantibodies in the nonobese diabetic mice predict early diabetes onset. J. Immunol. 173, 6603–6610.

Miller, J.F. (1995) Autoantigen-induced deletion of peripheral self-reactive T cells. Int. Rev. Immunol. 13, 107–114.

Mondino, A., Khoruts, A. and Jenkins, M.K. (1996) The anatomy of T cell activation and tolerance. Proc. Natl. Acad. Sci. USA 93, 2245–2252.
Nakayama, M., Abiru, N., Moriyama, H., Babaya, N., Liu, E., Miao, D., Yu, L., Wegmann, D. R., Hutton, J.C., Elliott, J.F. and Eisenbarth, G.S. (2005) Prime role for an insulin epitope in the development of type 1 diabetes in NOD mice. Nature 435, 220–223.
Ohashi, P.S. and DeFranco, A.L. (2002) Making and breaking tolerance. Curr. Opin. Immunol. 14, 744–759.
Oling, V., Marttila, J., Ilonen, J., Kwok, W.W., Nepom, G., Knip, M., Simell, O. and Reijonen, H. (2005) GAD65- and proinsulin-specific CD4+ T cells detected by MHC class II tetramers in peripheral blood of type 1 diabetes patients and at-risk subjects. J. Autoimmun. 25, 235–243.
O'Shea, J.J., Ma, A. and Lipsky, P. (2002) Cytokines and autoimmunity. Nat. Rev. Immunol. 2, 37–45.
Ostroukhova, M., Seguin-Devaux, C., Oriss, T.B., Dixon-McCarthy, B., Yang, L., Ameredes, B.T., Corcoran, T.E. and Ray, A. (2004) Tolerance induced by inhaled antigen involves CD4+ T cells expressing membrane-bound TGF-β and FOXP3. J. Clin. Invest. 114, 28–38.
Roncarolo, M.G., Gregori, S., Battaglia, M., Bacchetta, R., Fleischhauer, K. and Levings, M.K. (2006) Interleukin-10-secreting type 1 regulatory T cells in rodents and humans. Immunol. Rev. 212, 28–50.
Sakaguchi, S., Sakaguchi, N., Asano, M., Itoh, M. and Toda, M. (1995) Immunologic self-tolerance maintained by activated T cells expressing IL-2 receptor alpha-chains (CD25). Breakdown of a single mechanism of self-tolerance causes various autoimmune diseases. J. Immunol. 155, 1151–1164.
Schwartz, R.H. (1996) Models of T cell anergy: is there a common molecular mechanism? J. Exp. Med. 184, 1–8.
Sherry, N.A., Tsai, E.B. and Herold, K.C. (2005) Natural history of beta-cell function in type 1 diabetes. Diabetes 54, S32–S39.
Skyler, J.S., Krischer, J.P., Wolfsdorf, J., Cowie, C., Palmer, J.P., Greenbaum, C., Cuthbertson, D., Rafkin-Mervis, L.E., Chase, H.P. and Leschek, E. (2005) Effects of oral insulin in relatives of patients with type 1 diabetes: The Diabetes Prevention Trial—Type 1. Diabetes Care 28, 1068–1076.
Sosenko, J.M., Palmer, J.P., Greenbaum, C.J., Mahon, J., Cowie, C., Krischer, J.P., Chase, H.P., White, N.H., Buckingham, B., Herold, K.C., Cuthbertson, D. and Skyler, J.S. (2006) Patterns of metabolic progression to type 1 diabetes in the Diabetes Prevention Trial-Type 1. Diabetes Care 29, 643–649.
Tian, J., Clare-Salzler, M., Herschenfeld, A., Middleton, B., Newman, D., Mueller, R., Arita, S., Evans, C., Atkinson, M., Mullen, Y., Sarvetnick, N., Tobin, A.J., Lehmann, P. and Kaufman, D. (1996) Modulating autoimmune responses to GAD inhibits disease progression and prolongs islet graft survival in diabetes-prone mice. Nat. Med. 2, 1348–1353.
Tian, J., Gregori, S., Adorini, L. and Kaufman, D.L. (2001) The frequency of high avidity T cells determines the hierarchy of determinant spreading. J. Immunol. 166, 7144–7150.
Weiner, H. L. (2001) Oral tolerance: immune mechanisms and the generation of Th3-type TGF-beta-secreting regulatory cells. Microbes Infect. 3, 947–954.

35

Interleukin-7 Immunotherapy

Claude Sportès and Ronald E. Gress

Experimental Transplantation and Immunology Branch, National Cancer Institute, National Institutes of Health, Bethesda, MD, USA, csportes@mail.nih.gov

Abstract. IL-7 is a member of the common γ-chain family of cytokines sharing a common γ-chain in their receptor. Beyond its long-established pivotal role in immune development, it has been more recently recognized as a critically important regulator of peripheral naïve and memory T cell homeostasis while its role in postdevelopment thymic function remains at best, poorly defined, and controversial. Its multiple immune-enhancing properties, most notably in the maintenance of T cell homeostasis, make it a very attractive candidate for immunotherapy in a wide variety of clinical situations. Following many years of rich preclinical data in murine and simian models, IL-7 is now emerging in human phase I trials as a very promising immunotherapeutic agent. Human in vivo data discussed here are derived from the phase I study initiated at the National Cancer Institute in collaboration with Cytheris, Inc., in a cohort of subjects with incurable malignancy.

1. Introduction

IL-7 is a multifunctional, hematopoietic cytokine first isolated as a 25-KDa glycoprotein produced by a murine bone marrow stromal cell line (Goodwin et al. 1989; Sakata et al. 1990). IL-7 is not produced by lymphoid cells but rather in bone marrow stroma (Sudo et al. 1989) as well as in other cell types including thymic stroma, keratinocytes, neurons, lymph node follicular dendritic cells, and endothelial cells. Several reviews on IL-7 in vivo and in vitro activity and mechanism of action are available (Appasamy 1999; Fry and Mackall 2002; Hofmeister et al. 1999; Komschlies et al. 1995; Rodewald and Fehling 1998). IL-7 signals through its specific receptor (IL-7R). IL-7R is composed of a unique α chain (CD127) and the common γ chain (γc). It shares its α chain with thymic stroma-derived lymphopoietin (TSLP) and shares the γc with other immunoregulatory cytokines of the γc family, specifically IL-2, IL-4, IL-9, IL-15, and IL-21 (Noguchi et al. 1993a). IL-7 has a broad range of biological activities in vivo (Komschlies et al. 1994b) that can be separated into broad categories of developmental and postdevelopmental.

1.1. IL-7 in Immune Development

IL-7 was first recognized as an important murine B and T cell maturation and differentiation factor in in vitro systems such as the Whitlock–Witte long-term marrow culture (Sudo et al. 1993) for B cells and in fetal thymic lobe culture for fetal and adult thymocytes (Murray et al. 1989). IL-7 is an essential factor in the ability of stromal cells to support murine B lymphopoiesis (Sudo et al. 1989). In vivo, daily injection into normal mice results in a reversible large increase of spleen, bone marrow, and lymph node cellularity mainly of $B220^{+}$/S Ig—precursor B cells (Komschlies et al. 1994a; Morrissey et al. 1991). IL-7 transgenic mice show expansion of immature B cells as well as lymphoproliferative disorders (Fisher et al. 1995). Conversely, the gene-deficient mouse models (IL-7 or IL-7R knockout) confirm the critical role of IL-7 by showing profound lymphopenia (Freeden-Jeffry et al. 1995; Peschon, et al. 1994). While IL-7 induces proliferation and differentiation from pro-B cell to pre-B cell on human bone marrow stroma (Dittel and LeBien 1995), it does not appear to be critical in human B cell development. Its much less prominent function in human B cell development was foreseen in in vitro experiments (Prieyl and LeBien 1996) but is best illustrated in the various phenotypes of severe combined immune deficiency (SCID) described below.

The critical role of IL-7 and IL-7R in early thymocyte development, first established in vitro (Murray et al. 1989), was subsequently demonstrated most directly in vivo in murine IL-7R-knockout experiments that show that IL-7 signaling is necessary for thymic development and normal thymopoiesis. IL-7R-deficient mice have thymic atrophy and arrest of T cell development at a double-positive stage (Peschon et al. 1994). In other IL-7R gene-deficient models, the animals are highly lymphopenic both in peripheral blood and in lymphoid organs and the thymic cellularity is reduced 20-fold (Freeden-Jeffry et al. 1995). Conversely, mice transgenic for IL-7 develop T cell lymphoproliferative/autoimmune diseases and T cell lymphomas (Fisher et al. 1995; Samaridis et al. 1991). The effects of IL-7 on T cell development are multifactorial, with at least some component due to protection from programmed cell death in developing lymphocytes since Bcl-2 overexpression rescues mice from the IL-7R-deficient phenotype (Akashi et al. 1997; Maraskovsky et al. 1997). In addition, in murine fetal thymocytes, IL-7 also promotes V(D)J rearrangement of the T cell receptor genes with sustained expression of the two genes critical in the control of gene rearrangements: RAG-1 and RAG-2 (Muegge et al. 1993). Furthermore, IL-7 also plays a role in the positive selection of $CD8^{+}$ T cells in the thymus. The effects on T cell development appear to be dose dependent as demonstrated in recent studies of three different murine transgenic founder lines with varying degrees of IL-7 overexpression (El Kassar et al. 2004); this phenomenon may have significant implications in the use of pharmacologic doses of rhIL-7 in the context of human immunotherapy.

The role of IL-7 in human T cell development was first studied in in vitro systems which suggested that defects in the IL-7 pathway may be responsible for some SCID phenotypes (Plum et al. 1996). The number of immature and mature T cells produced in neonatal human thymus culture is increased by the addition of exogenous recombinant IL-7 in combination with other growth factors, and anti-IL-7 antibodies inhibit T cell production (Yeoman et al. 1996). The critical role of IL-7 in human immune development is however best illustrated in children with SCID. Murine models of IL-7 deficiency can reproduce most of the abnormalities of T cell development found in SCID patients. At least three groups of children with SCID phenotypes have mutations involving the IL-7R or its signaling pathway. These phenotypes also do confirm, however, that IL-7 signaling is not essential for normal human B lymphocyte development. Children with X-linked SCID have a defect of the IL-7R γ chain. They have severe defects in development of both T and NK cells but have normal or even elevated numbers of peripheral blood B cells (Noguchi et al. 1993b). Autosomal recessive mutations of the Jak-3 tyrosine kinase (coupled to the receptor γc and required for γc-dependent signaling) lead to a similar SCID phenotype with normal numbers of B cells (Macchi et al. 1995; Russell et al. 1995). Finally, SCID patients with a defective expression of the IL-7R α chain have an abnormal T cell development but spared NK and B cells (Puel et al. 1998).

1.2. IL-7 as a Homeostatic Cytokine

The inverse correlation of IL-7 serum level with the circulating absolute lymphocyte count suggests a homeostatic function for IL-7. Although this correlation is well documented in experimental models and in many clinical settings in humans (Bolotin et al. 1999; Fry et al. 2001; Napolitano et al. 2001), it remains controversial whether it results from direct regulation of IL-7 stromal production through a lymphocyte mass responsive mechanism or from the variations of peripheral utilization of secreted IL-7 in IL-7R-expressing lymphocytes. As previously mentioned, IL-7 shares the receptor common γc with other cytokines of the γc family. Within this γc family, cytokines can be classified as activating versus homeostatic. In contrast to IL-2 that is the prototypic activating cytokine, IL-7 is a prototypic homeostatic cytokine. On the one hand, IL-2 selectively signals activated T cells, it is secreted by activated T cells, and IL-2 signaling upregulates its own receptor, thus amplifying the IL-2 response during immune activation. IL-2 also activates the $CD4^{+}CD25^{Hi}FoxP3^{+}$ regulatory T cell subset (T-regs). On the other hand, IL-7R is expressed on resting T cells but, in contrast to the relationship of other immunoregulatory cytokines with their receptor, IL-7R is downregulated following signaling by IL-7 itself, other prosurvival cytokines (IL-2, IL-4, IL-6, and IL-15) (Fry et al. 2003), or following T cell receptor (TCR) ligation. Decreased receptor surface expression is not simply due to increased receptor turnover and internalization since IL-7R α-chain mRNA transcription is reduced, in experimental models as well as in vivo in humans in our phase I study. This mechanism of receptor downregulation is congruent with the homeostatic role of IL-7 as it presumably prevents T cells that have already received a prosurvival signal from competing with unsignaled cells for its utilization (Park et al. 2004). IL-7 has been shown to be critical for in vivo homeostatic proliferation and prolonged survival of naïve T cells, the vast majority of which are circulating in a resting state and, in a model of IL-7-deficient host, adoptively transferred naïve T cells gradually disappeared over a month in the absence of IL-7 (Tan et al. 2001). Of note, IL-7Rα (CD127) expression on T-regs has been found recently to be constitutively downregulated and FoxP3 expression is inversely correlated with CD127 expression in these Tregs (Liu et al. 2006; Seddiki et al. 2006).

1.3. Role of IL-7 in Modulation of Immune Responses

Many immune-enhancing activities have been ascribed to IL-7 in in vitro murine and human experimental systems that could be relevant to clinical situations such as treatment of infectious diseases and cancer. IL-7 has a potent costimulatory effect on purified murine mature peripheral and spleen T cells from IL-7-treated mice that have enhanced Con A, allogeneic mixed lymphocyte reaction, and anti-CD3 proliferative responses in vitro as well as an enhanced allogeneic CTL response in vivo (Komschlies et al. 1994b). IL-7 has been shown to augment the effects of T cell - mediated antiviral adoptive immunotherapy in vivo in mice (Wiryana, Bui, Faltynek, and Ho 1997). In the murine model, IL-7 has been show to serve as a potent vaccine adjuvant, broadening the immune response by augmenting responses to subdominant antigens and by improving the survival of $CD8^+$ memory T cells (Melchionda et al. 2005).

IL-7's immunomodulatory effects may be particularly relevant in the treatment of cancer. In murine models, IL-7 has demonstrated an ability to enhance antitumor immune responses both in vitro and in mice previously injected with tumor cells (Komschlies et al. 1994a). Subsequently, murine models have demonstrated regression of established pulmonary metastases of Renca renal carcinoma with administration of IL-7 (Komschlies et al. 1994b). Mouse experiments modeled after the human clinical situation of autologous transplantation for metastatic breast cancer demonstrated that administration of rhIL-7 post-transplant resulted not only in improved T cell recovery but also in additive therapeutic activity resulting in prolonged survival of the animals beyond what was achieved with the chemotherapy and transplant regimen alone (Talmadge et al. 1993). In an immunodeficient mouse model of human colon carcinoma xenograft, the injection of the combination of recombinant human IL-7 and human lymphocytes resulted in a significant prolongation of survival (Murphy and Longo 1997). In in vitro human systems, IL-7 induces tumoricidal activity of peripheral monocytes (Alderson et al. 1991), and IL-7-activated killer cells can lyse allogeneic and autologous melanoma cells (Bohm et al. 1994) although a mechanism of direct IL-7 anti-tumor activity has not been demonstrated.

Specifically, administration of IL-7 to humans with cancer may be useful in generating and supporting T cell antitumor responses, including those generated by antitumor vaccination. Alternatively, IL-7's role in prolonging the survival of adoptively transferred T cells may be exploited in that clinical setting as well.

1.4. Patterns of Immune Reconstitution

Immune reconstitution following immune depletion has been shown to occur through two primary pathways. First, predominant in children, is the thymic pathway in which new T cell progenitors arise from pluripotent hematopoietic stem cells that will then home to the thymus and undergo a process of expansion, differentiation, and selection. The resulting T cells, which bear a naïve phenotype, display a diverse TCR repertoire and are poised to recognize an array of foreign antigens. The second pathway, which predominates in adults, is thymus independent (Mackall et al. 1993) consisting of the peripheral expansion of existing residual mature T cells which increase their number through mitotic division, driven in large part by exposure to their specific antigen and by homeostatic expansion. It results in a skewed T cell repertoire, poorly diversified and limited mostly to T cells that encounter their specific antigen during the period of immune reconstitution. Furthermore, this process is unable to restore CD4$^+$ T cell numbers to pretreatment levels (Mackall et al. 1996).

There is ample evidence that even short courses of standard dose chemotherapy have a significant impact on an individual's residual immune function (Mackall 2000). In particular, absolute peripheral CD4$^+$ T cells may remain below 200/mm^3 for several months following intensive chemotherapy (Mackall et al. 1994), but the recovery of other T cell populations may be delayed as well, for several months in some cases (Hakim et al. 1997; Mackall et al. 1997). The kinetics and the extent of the immune recovery correlate closely with the individual's amount of residual thymic function–the children reason for recovering a population of naïve T cells via a thymic expansion pathway much more readily than adults (Mackall et al. 1995). The return of a naïve CD4$^+$ cell population can be monitored by flow cytometry (CD45RO$^-$RA$^+$ phenotype) which correlates with other indicators of thymic function (Hakim et al. 2005), such as thymic size by CT scan, enumeration of TCR excision circles (TRECs), indicative of the thymic output of recent thymic emigrant (RTE) T cells having recently functionally rearranged their TCR (Douek et al. 2000) and TCR rearrangements spectratype patterns showing more widely distributed peaks indicative of the more diversity of TCR Vβ gene species, hence of the T cell repertoire. Following immune-depleting chemotherapy, the increase in naïve CD4$^+$ cells (CD45RO$^-$RA$^+$) correlates with increased thymic size, increased number of circulating TRECs, and increased normalization of TCR spectratyping patterns (Hakim et al. 2005).

Following more severe immune depletion such as post hematopoietic stem cell transplantation, the expansion of peripheral lymphoid progenitors is similarly contingent upon the functionality of the thymus (Mackall and Gress 1997). Furthermore, as indicated in clinical settings, the return to a normal CD4$^+$ lymphocyte count in adults is predicated upon the re-emergence of a pool of naïve T cells which is delayed 18–24 months and may occur only in individuals younger than 50 years (Hakim et al. 2005; Sportes et al. 2005).

The clinical implications of the kinetics, nature, and extent of immune reconstitution following standard or ablative chemotherapy are probably not fully appreciated, particularly with regard to the current efforts in immunotherapy development. One can speculate, however, that individuals who have suffered an immune injury, be it pathologic (HIV, graft vs. host disease), iatrogenic (radiochemotherapy), or physiologic (age), may be significantly impeded in their responses to active immunotherapy attempts, the therapeutic potential of which may be misjudged or overlooked (Hakim and Gress 2005).

2. Phase I Study of rhIL-7 in Humans

Based on the large body of preclinical evidence pointing to its multiple immune-enhancing properties, investigations of the possible roles of rhIL-7 in clinical immunotherapy were initiated through several phase I trials in varied subject populations as part of a long-standing collaboration for the clinical development of rhIL-7 between Cytheris, Inc. (Rockville, MD), and the National Cancer Institute (NIH,

HHS, Bethesda, MD). Three phase I clinical trials were initiated at the National Cancer Institute. The experience from a trial in adults with incurable malignancy is discussed below.

2.1. Subject Population and Study Design

Adults diagnosed with incurable nonhematologic malignancy (except primary carcinoma of the lung) were included and standard phase I eligibility criteria such as life expectancy greater than 3 months, adequate Karnofsky performance status, heart, lung, liver, kidney, and marrow functions were required. In addition, a mean value of 4 peripheral $CD3^+$ cell count determinations over a 2-week period was required to be above $300/mm^3$, with a coefficient of variation of less than 20% attesting to the stability of the peripheral lymphocyte count before therapy.

The study followed a classic phase I dose escalation design with four successive cohorts of three subjects to be expanded to six in the event of one occurrence of predefined dose-limiting toxicity (DLT). If two or more patients experienced DLT, the maximum tolerated dose (MTD) was deemed exceeded. The occurrence of neutralizing IL-7 antibodies in more than one subject would have been deemed unacceptable toxicity and prompted termination of the trial. Based on preclinical animal data, the four tested doses of "CYT 99 007" (rhIL-7; manufactured and provided by Cytheris, Inc., Rockville, MD) were 3, 10, 30, and 60 μg/kg of body weight per dose. It was given subcutaneously, every other day for 2 weeks (eight injections).

The primary study objective was the determination of safety and DLT. Secondary objectives aimed at the characterization of biological activity in humans and determination of a range of biologically active doses, based on known IL-7 properties from preclinical studies such as enlargement of lymphoid organs, increase in number of various peripheral blood lymphocyte subsets, and evidence of T cell effects by flow cytometry: (1) increase in T cell cycling (Ki67 expression), (2) downregulation of the IL-7R (CD127 expression), and (3) Bcl-2 upregulation. Possible antitumor effects and lymphoid organ enlargement were evaluated by CT scan at baseline, at the end of therapy (day 14), and at day 28, at a minimum. Subjects with stable disease at day 28 had the possibility of longer follow-up until disease progression. rhIL-7 effects on various bone marrow mononuclear subsets were studied by H&E and immunostaining on biopsy and flow cytometry on aspirate at baseline and at the end of treatment. Those with significant changes noted at day 14 underwent a follow-up biopsy between 4 and 8 weeks.

2.2. Results

Twelve men and 4 women, ages ranging from 20 to 71 years (median 49), were treated. "CYT 99 007" was overall well tolerated clinically and MTD was not reached. Serum from all subjects was tested for the presence of anti-IL-7 antibodies on day 28. Three subjects developed non-neutralizing antibodies, but none developed detectable anti-IL-7-neutralizing antibodies. One subject showed a possible antitumor effect, albeit minimal and questionable due to the somewhat unpredictable natural history of the rare disease. This subject with resected primary CNS hemangiopericytoma and recurrent abdominal metastatic disease had a substantial relief of his abdominal pain and 20% shrinkage of the mass was noted 3 months after treatment. The abdominal disease remained stably decreased until and beyond CNS recurrence, 9 months after rhIL-7. All other subjects showed disease progression at day 28 or at the 6-week follow-up.

Biological activity, defined per protocol as 50% increase over baseline of peripheral blood CD3$^+$ T cells, was seen with a clear dose–response effect starting with 10 μg/kg/dose, with already a trend suggesting activity at the 3μg/kg dose (Figure 1). A 50–80% transient drop of total circulating lymphocytes following the first injection, likely representing early trafficking changes and tissue redistribution as observed in most preclinical models, was observed in all individuals at all four dose levels. Other IL-7 clinical and biological ffects expected from murine and nonhuman primate preclinical models were seen in all subjects receiving 10 μg/kg/dose or more: spleen and lymph node enlargement (on CT scan), increase in circulating CD3$^+$, CD4$^+$, CD8$^+$ populations, IL-7R α-chain (CD127) downregulation, Bcl-2 up-regulation, and marked increase in lymphocyte proliferation and activation markers (Ki67, CD71) by flow cytometry. Consistent with animal data, the IL-7Rα down-regulation was observed after 1 week of treatment, not only by decreased CD127 expression on flow cytometry but also by a greater than 50% decline of IL-7Rα mRNA copy number in CD4$^+$ and CD8$^+$ cell-sorted populations.

In addition, more detailed evaluation of naïve, memory, and effector subsets in the CD4$^+$ and CD8$^+$ cell-sorted populations of T cells as well as of mature and transitional B cells was performed by multicolor flow cytometry at baseline, week 1, 2 and 3.

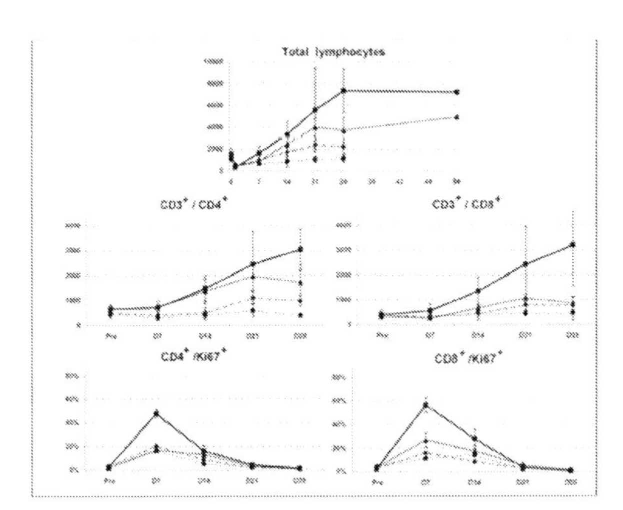

FIGURE 1. Kinetics of circulating lymphocytes during and after IL-7 treatment. *X* axis: days from start of treatment (treatment: days 0–14); *Y* axis: mean cohort value of absolute counts per mm^3 for top (total lymphocytes) and middle panels, left (CD4$^+$) and right (CD8$^+$). Lower panels: percentage of circulating cells in cycle (Ki67$^+$); CD4$^+$ (left) and CD8$^+$ (right) Cohort 1: 3 μg/kg ● ….. ; cohort 2: 10 μg/kg ♦ – – –; cohort 3: 30 μg/kg ▲——; cohort 4: 60 μg/kg ■——— .

Cell numbers, cycling (Ki67$^+$), and Bcl-2 expression were evaluated within each subset. RTEs, the most naïve circulating CD4$^+$ cells, were defined as CD45RA$^+$/CD31$^+$. Among CD4$^+$ and CD8$^+$ cells, the main naïve, memory, and effector populations were defined respectively as CD45RA$^+$/CD27$^+$, CD45RA$^-$/CD27$^+$, and CD45RA$^-$/CD27$^-$. rhIL-7 induced marked proliferation in all T cell subsets, but, although significant proliferation was induced in effector and memory CD4$^+$ and CD8$^+$ subsets, it resulted in a smaller net population expansion than for the CD4$^+$ RTE and other naïve subsets (Figure 2). The magnitude of the resulting biological effects was variable, but the kinetics were similar in all T cell subsets and at all effective doses. After 1 week of therapy, up to 70% of circulating CD4$^+$ RTE and naïve CD4$^+$ and CD8$^+$ T cells were in cycle and expressed elevated levels of Bcl-2. The combination of high proliferation rates, for the first week, and Bcl-2 upregulation, sustained throughout the 2 weeks of treatment, resulted in net T cell expansion (up to 20-fold in some individuals), persisting one to several weeks after treatment (Figure 1). Interestingly, in spite of continuing exposure to the pharmacologic doses of rhIL-7 during the second week of treatment and in the absence of IL-7 antibodies, the proliferation rates were halved by the end of treatment in all subsets, coinciding with IL-7Rα downregulation, then returned to baseline by week 3 (Figure 1), coinciding with the recovery of IL-7Rα surface expression and normalization of Bcl-2 expression after the cessation of treatment.

The T cell increase was primarily due to IL-7-induced peripheral expansion. There was no dilution effect of TREC numbers per 10^5 CD4$^+$ sorted cells, which, consistent with preclinical data, is more suggestive of cell recruitment or expansion but does not exclude a thymic contribution. Most importantly, this large expansion of naïve cells resulted in a broadening of T cell repertoire diversity demonstrated on spectratyping analysis of the TCR Vβ genes and, unlike what is observed with IL-2

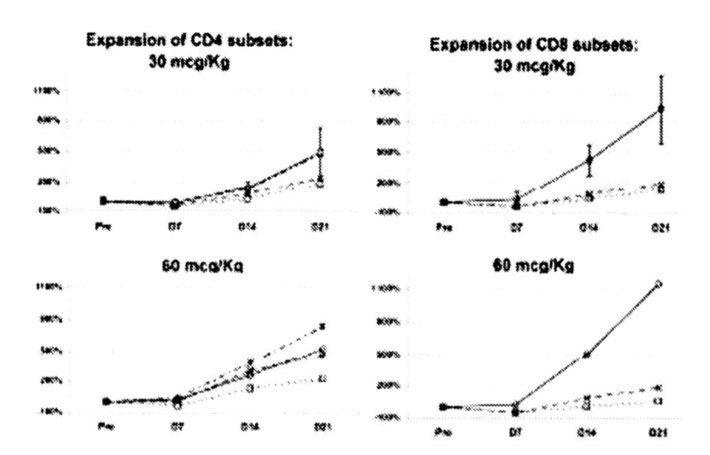

FIGURE 2. Expansion of CD4$^+$ and CD8$^+$ T cell subsets in cohorts 3 and 4 (30 and 60μg/kg). *Y* axis: mean cohort values of percentage increase over the baseline absolute number/mm^3; *X* axis: Days from start of treatment. CD4$^+$ subsets (left): RTE *_ .. _ .. ; Main naïve ○—— ; Memory x----- ; Effectors □...... ; CD8$^+$ subsets (right): Naïve ○—— ; Memory x----- ; Effectors: □......

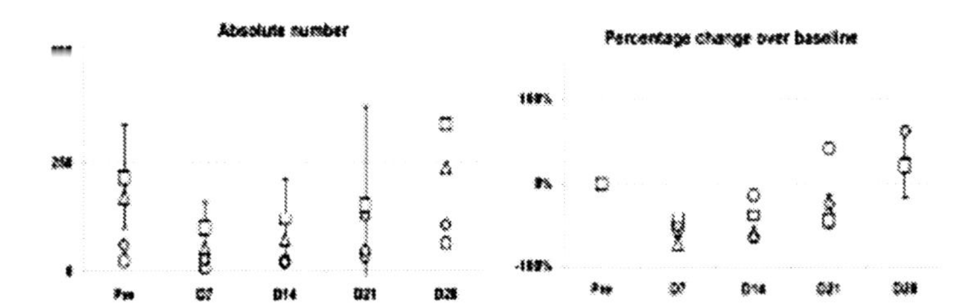

FIGURE 3. Kinetics of CD20^{+} B cell. X axis: Days from start of treatment; Y axis: mean cohort values of absolute number/mm^{3} (left panel) and percentage increase over baseline absolute number/mm^{3} (right panel); cohort 1: 3 μg/kg ○; cohort 2: 10 μg/kg ◊; cohort 3: 30 μg/kg Δ; cohort 4: 60 μg/kg □.

therapy, this IL-7-induced expansion resulted in a relative decrease in the proportion of CD4^{+}/CD25^{Hi}/FoxP3^{+} T-regs.

Mature CD20^{+} B cells showed a 40–70% decrease in absolute number of circulating cells during therapy, probably the result of trafficking alterations, with a gradual return to baseline numbers over the 2 weeks following treatment (Figure 3).

3. Conclusions

Central to the potential clinical use of IL-7 are its immunorestorative and immune-enhancing properties. As indicated above, spontaneous immune recovery following immune depletion, be it physiologic (age), pathologic (HIV), or iatrogenic (chemotherapy), requires evidence of some thymopoietic activity, in order to return to a normal number of lymphocytes. This has been shown to be inconsistent and often incomplete in adults younger than 50 years and, by and large, inexistent in individuals over 50. Consequently, various forms of immune therapeutic interventions under present or future investigations may be destined to failure if not preceded or accompanied by immunorestorative or immune-enhancing measures for the majority of concerned individuals left with an immune deficit from a variety of etiologies.

IL-7 may restore or even exceed the normal number of T cells in an immune depleted individual and, unlike IL-2, it appears to expand preferentially CD4^{+} RTE and other naïve subsets of both CD4^{+} and CD8^{+} T cells without increasing the proportion of CD4^{+}CD25^{Hi}FoxP3^{+} T-regs. This results in a broadening of the naïve T cell repertoire diversity. Therefore, IL-7 offers the prospect of returning to normal both the number and repertoire diversity of naïve T cells in adults with incomplete or inexistent potential for thymopoiesis. Furthermore, IL-7's ability to lower the threshold of immune responsiveness may allow, in combination with this broaden T cell repertoire, the development of new and more effective T cell anti tumor specificities in response to vaccines.

Subsequent clinical studies will need to focus on optimization of dose and delivery schedules based on the pharmacodynamics known from experimental models and those observed in the phase I studies as well as a better understanding of the likely heterogeneous relationship between dose and the various biological effects of IL-7.

IL-7 modalities of use in the clinical arena are likely to be highly dependent on a variety of clinical parameters such as the degree of initial immune depletion or its cause (physiologic, pathologic, or iatrogenic), the intended end results (e.g., global immune restoration vs. enhancement of specific response to a vaccine), and the timing of such interventions.

Acknowledgments

We wish to thank Rebecca Babb and Michael Krumlauf for their tireless dedication to the coordination of our clinical research and the staff of the Experimental Transplantation Immunology Branch of the National Cancer Institute for its contribution to our patients' care. We wish to thank Dr. Frances Hakim from the Experimental Transplantation Immunology Branch and Drs. Terry Fry and Crystal Mackall from the Pediatric Oncology Branch of the National Cancer Institute for their major contribution and invaluable collaboration in this endeavor. We also wish to thank Drs. Julie Engel, Renaud Buffet, and Michel Morre from Cytheris, Inc. (Rockville, MD, and Vanves, France), for allowing this long-standing and most fruitful scientific collaboration in the clinical development of rhIL-7.

References

Akashi, K., Kondo, M., Freeden-Jeffry, U., Murray, R. and Weissman, I.L. (1997) Bcl-2 rescues T lymphopoiesis in interleukin-7 receptor-deficient mice. Cell 89, 1033–1041.

Alderson, M.R., Tough, T.W., Ziegler, S.F. and Grabstein, K.H. (1991) Interleukin 7 induces cytokine secretion and tumoricidal activity by human peripheral blood monocytes. J. Exp. Med. 173, 923–930.

Appasamy, P.M. (1999) Biological and clinical implications of interleukin-7 and lymphopoiesis. Cytokines Cell Mol. Ther. 5, 25–39.

Bohm, M., Moller, P., Kalbfleisch, U., Worm, M., Czarnetzki, B.M. and Schadendorf, D. (1994) Lysis of allogeneic and autologous melanoma cells by IL-7-induced lymphokine-activated killer cells. Br. J. Cancer 70, 54–59.

Bolotin, E., Annett, G., Parkman, R. and Weinberg, K. (1999) Serum levels of IL-7 in bone marrow transplant recipients: relationship to clinical characteristics and lymphocyte count. Bone Marrow Transplant. 23, 783–788.

Dittel, B.N. and LeBien, T.W. (1995) The growth response to IL-7 during normal human B cell ontogeny is restricted to B-lineage cells expressing CD34. J. Immunol. 154, 58–67.

Douek, D.C., Vescio, R.A., Betts, M.R., Brenchley, J.M., Hill, B.J., Zhang, L., Berenson, J.R., Collins, R.H. and Koup, R.A. (2000) Assessment of thymic output in adults after haematopoietic stem-cell transplantation and prediction of T cell reconstitution. Lancet 355, 1875–1881.

El Kassar, N., Lucas, P.J., Klug, D.B., Zamisch, M., Merchant, M., Bare, C.V., Choudhury, B., Sharrow, S.O., Richie, E., Mackall, C.L. and Gress, R.E. (2004) A dose effect of IL-7 on thymocyte development. Blood 104, 1419–1427.

Fisher, A.G., Burdet, C., Bunce, C., Merkenschlager, M. and Ceredig, R. (1995) Lymphoproliferative disorders in IL-7 transgenic mice: expansion of immature B cells which retain macrophage potential. Int. Immunol. 7, 415–423.

Freeden-Jeffry, U., Vieira, P., Lucian, L.A., McNeil, T., Burdach, S.E. and Murray, R. (1995) Lymphopenia in interleukin (IL)-7 gene-deleted mice identifies IL-7 as a nonredundant cytokine. J. Exp. Med. 181, 1519–1526.

Fry, T.J., Connick, E., Falloon, J., Lederman, M.M., Liewehr, D.J., Spritzler, J., Steinberg, S.M., Wood, L.V., Yarchoan, R., Zuckerman, J., Landay, A. and Mackall, C.L. (2001) A potential role for interleukin-7 in T cell homeostasis. Blood 97, 2983–2990.

Fry, T.J. and Mackall, C.L. (2002) Interleukin-7: from bench to clinic. Blood 99, 3892–3904.

Fry, T.J., Moniuszko, M., Creekmore, S., Donohue, S.J., Douek, D.C., Giardina, S., Hecht, T.T., Hill, B.J., Komschlies, K., Tomaszewski, J., Franchini, G. and Mackall, C.L. (2003) IL-7 therapy dramatically alters peripheral T cell homeostasis in normal and SIV-infected nonhuman primates. Blood 101, 2294–2299.

Goodwin, R.G., Lupton, S., Schmierer, A., Hjerrild, K.J., Jerzy, R., Clevenger, W., Gillis, S., Cosman, D. and Namen, A.E. (1989) Human interleukin 7: molecular cloning and growth factor activity on human and murine B-lineage cells. Proc. Natl. Acad. Sci. USA 86, 302–306.

Hakim, F.T., Cepeda, R., Kaimei, S., Mackall, C.L., McAtee, N., Zujewski, J., Cowan, K. and Gress, R.E. (1997) Constraints on CD4 recovery postchemotherapy in adults: thymic insufficiency and apoptotic decline of expanded peripheral CD4 cells. Blood 90, 3789–3798.

Hakim, F.T. and Gress, R.E. (2005) Reconstitution of the lymphocyte compartment after lymphocyte depletion: a key issue in clinical immunology. Eur. J. Immunol. 35, 3099–3102.

Hakim, F.T., Memon, S.A., Cepeda, R., Jones, E.C., Chow, C.K., Kasten-Sportes, C., Odom, J., Vance, B.A., Christensen, B.L., Mackall, C.L. and Gress, R.E. (2005) Age-dependent incidence, time course, and consequences of thymic renewal in adults. J. Clin. Invest 115, 930–939.

Hofmeister, R., Khaled, A.R., Benbernou, N., Rajnavolgyi, E., Muegge, K. and Durum, S.K. (1999) Interleukin-7: physiological roles and mechanisms of action. Cytokine Growth Factor Rev. 10, 41–60.

Komschlies, K.L., Back, T.T., Gregorio, T.A., Gruys, M.E., Damia, G., Wiltrout, R.H. and Faltynek, C.R. (1994a) Effects of rhIL-7 on leukocyte subsets in mice: implications for antitumor activity. Immunol. Ser. 61, 95–104.

Komschlies, K.L., Gregorio, T.A., Gruys, M.E., Back, T.C., Faltynek, C.R. and Wiltrout, R.H. (1994b) Administration of recombinant human IL-7 to mice alters the composition of B-lineage cells and T cell subsets, enhances T cell function, and induces regression of established metastases. J. Immunol. 152, 5776–5784.

Komschlies, K.L., Grzegorzewski, K.J. and Wiltrout, R.H. (1995) Diverse immunological and hematological effects of interleukin 7: implications for clinical application. J. Leukoc. Biol. 58, 623–633.

Liu, W., Putnam, A.L., Xu-Yu, Z., Szot, G.L., Lee, M.R., Zhu, S., Gottlieb, P.A., Kapranov, P., Gingeras, T.R., Fazekas de St Groth, B., Clayberger, C., Soper, D.M., Ziegler, S.F. and Bluestone, J.A. (2006) CD127 expression inversely correlates with FoxP3 and suppressive function of human CD4+ T reg cells. J. Exp. Med. 203, 1701–1711.

Macchi, P., Villa, A., Giliani, S., Sacco, M.G., Frattini, A., Porta, F., Ugazio, A.G., Johnston, J.A., Candotti, F. and O'Shea, J.J. (1995) Mutations of Jak-3 gene in patients with autosomal severe combined immune deficiency (SCID). Nature 377, 65–68.

Mackall, C.L. (2000) T cell immunodeficiency following cytotoxic antineoplastic therapy: a review. Stem Cells 18, 10–18.

Mackall, C.L., Bare, C.V., Granger, L.A., Sharrow, S.O., Titus, J.A. and Gress, R.E. (1996) Thymic-independent T cell regeneration occurs via antigen-driven expansion of peripheral T cells resulting in a repertoire that is limited in diversity and prone to skewing. J. Immunol. 156, 4609–4616.

Mackall, C.L., Fleisher, T.A., Brown, M.R., Andrich, M.P., Chen, C.C., Feuerstein, I.M., Horowitz, M.E., Magrath, I.T., Shad, A.T., Steinberg, S.M., Wexler, L.H. and Gress, R.E. (1995) Age, thymopoiesis, and CD4+ T lymphocyte regeneration after intensive chemotherapy [see comments]. N. Engl. J. Med. 332, 143–149.

Mackall, C.L., Fleisher, T.A., Brown, M.R., Andrich, M.P., Chen, C.C., Feuerstein, I.M., Magrath, I.T., Wexler, L.H., Dimitrov, D.S. and Gress, R.E. (1997) Distinctions between CD8+ and CD4+ T cell regenerative pathways result in prolonged T cell subset imbalance after intensive chemotherapy. Blood 89, 3700–3707.

Mackall, C.L., Fleisher, T.A., Brown, M.R., Magrath, I.T., Shad, A.T., Horowitz, M.E., Wexler, L.H., Adde, M.A., McClure, L.L. and Gress, R.E. (1994) Lymphocyte depletion during treatment with intensive chemotherapy for cancer. Blood 84, 2221–2228.

Mackall, C.L., Granger, L., Sheard, M.A., Cepeda, R. and Gress, R.E. (1993) T cell regeneration after bone marrow transplantation—differential CD45 isoform expression on thymic-derived versus thymic-independent progeny. blood 82, 2585–2594.

Mackall, C.L. and Gress, R.E. (1997) Pathways of T cell regeneration in mice and humans: implications for bone marrow transplantation and immunotherapy. Immunol. Rev. 157, 61–72.

Maraskovsky, E., O'Reilly, L.A., Teepe, M., Corcoran, L.M., Peschon, J.J. and Strasser, A. (1997) Bcl-2 can rescue T lymphocyte development in interleukin-7 receptor- deficient mice but not in mutant rag-1–/– mice. Cell 89, 1011–1019.

Melchionda, F., Fry, T.J., Milliron, M.J., McKirdy, M.A., Tagaya, Y. and Mackall, C.L. (2005) Adjuvant IL-7 or IL-15 overcomes immunodominance and improves survival of the CD8+ memory cell pool. J. Clin. Invest. 115, 1177–1187.

Morrissey, P.J., Conlon, P., Charrier, K., Braddy, S., Alpert, A., Williams, D., Namen, A.E. and Mochizuki, D. (1991) Administration of IL-7 to normal mice stimulates B-lymphopoiesis and peripheral lymphadenopathy. J. Immunol. 147, 561–568.

Muegge, K., Vila, M.P. and Durum, S.K. (1993) Interleukin-7: a cofactor for V(DJ) rrearrangement of the T cell receptor β gene. Science 261, 93–95.

Murphy, W.J. and Longo, D.L. (1997) The potential role of NK cells in the separation of graft-versus- tumor effects from graft-versus-host disease after allogeneic bone marrow transplantation. Immunol. Rev. 157, 167–176.

Murray, R., Suda, T., Wrighton, N., Lee, F. and Zlotnik, A. (1989) IL-7 is a growth and maintenance factor for mature and immature thymocyte subsets. Int. Immunol. 1, 526–531.

Napolitano, L.A., Grant, R.M., Deeks, S.G., Schmidt, D., De Rosa, S.C., Herzenberg, L.A., Herndier, B.G., Andersson, J. and McCune, J.M. (2001) Increased production of IL-7 accompanies HIV-1-mediated T cell depletion: implications for T cell homeostasis. Nat. Med. 7, 73–79.

Noguchi, M., Nakamura, Y., Russell, S.M., Ziegler, S.F., Tsang, M., Cao, X. and Leonard, W.J. (1993a) Interleukin-2 receptor gamma chain: a functional component of the interleukin-7 receptor [see comments]. Science 262, 1877–1880.

Noguchi, M., Yi, H., Rosenblatt, H.M., Filipovich, A.H., Adelstein, S., Modi, W.S., McBride, O.W. and Leonard, W.J. (1993b) Interleukin-2 receptor gamma chain mutation results in X-linked severe combined immunodeficiency in humans. Cell 73, 147–157.

Park, J.H., Yu, Q., Erman, B., Appelbaum, J.S., Montoya-Durango, D., Grimes, H.L. and Singer, A. (2004) Suppression of IL7Ralpha transcription by IL-7 and other prosurvival cytokines: a novel mechanism for maximizing IL-7-dependent T cell survival. Immunity 21, 289–302.

Peschon, J.J., Morrissey, P.J., Grabstein, K.H., Ramsdell, F.J., Maraskovsky, E., Gliniak, B.C., Park, L.S., Ziegler, S.F., Williams, D.E. and Ware, C.B. (1994) Early lymphocyte expansion is severely impaired in interleukin 7 receptor-deficient mice. J. Exp. Med. 180, 1955–1960.

Plum, J., De Smedt, M., Leclercq, G., Verhasselt, B. and Vandekerckhove, B. (1996) Interleukin-7 is a critical growth factor in early human T cell development. Blood 88, 4239–4245.

Prieyl, J.A. and LeBien, T.W. (1996) Interleukin 7 independent development of human B cells. Proc. Natl. Acad. Sci. USA 93, 10348–10353.
Puel, A., Ziegler, S.F., Buckley, R.H. and Leonard, W.J. (1998) Defective IL7R expression in T(-)B(+)NK(+) severe combined immunodeficiency. Nat. Genet. 20, 394–397.
Rodewald, H.R. and Fehling, H.J. (1998) Molecular and cellular events in early thymocyte development. Adv. Immunol. 69, 1–112.
Russell, S.M., Tayebi, N., Nakajima, H., Riedy, M.C., Roberts, J.L., Aman, M.J., Migone, T.S., Noguchi, M., Markert, M.L. and Buckley, R.H. (1995) Mutation of Jak3 in a patient with SCID: essential role of Jak3 in lymphoid development. Science 270, 797–800.
Sakata, T., Iwagami, S., Tsuruta, Y., Teraoka, H., Tatsumi, Y., Kita, Y., Nishikawa, S., Takai, Y. and Fujiwara, H. (1990) Constitutive expression of interleukin-7 mRNA and production of IL-7 by a cloned murine thymic stromal cell line. J. Leukoc. Biol. 48, 205–212.
Samaridis, J., Casorati, G., Traunecker, A., Iglesias, A., Gutierrez, J.C., Muller, U. and Palacios, R. (1991) Development of lymphocytes in interleukin 7-transgenic mice. Eur. J. Immunol. 21, 453–460.
Seddiki, N., Santner-Nanan, B., Martinson, J., Zaunders, J., Sasson, S., Landay, A., Solomon, M., Selby, W., Alexander, S.I., Nanan, R., Kelleher, A. and Fazekas de St Groth, B. (2006) Expression of interleukin (IL)-2 and IL-7 receptors discriminates between human regulatory and activated T cells. J. Exp. Med. 203, 1693–1700.
Sportes, C., McCarthy, N.J., Hakim, F., Steinberg, S.M., Liewehr, D.J., Weng, D., Kummar, S., Gea-Banacloche, J., Chow, C.K., Dean, R.M., Castro, K.M., Marchigiani, D., Bishop, M.R., Fowler, D.H. and Gress, R.E. (2005) Establishing a platform for immunotherapy: clinical outcome and study of immune reconstitution after high-dose chemotherapy with progenitor cell support in breast cancer patients. Biol. Blood Marrow Transplant. 11, 472–483.
Sudo, T., Ito, M., Ogawa, Y., Iizuka, M., Kodama, H., Kunisada, T., Hayashi, S., Ogawa, M., Sakai, K. and Nishikawa, S. (1989) Interleukin 7 production and function in stromal cell-dependent B cell development. J. Exp. Med. 170, 333–338.
Sudo, T., Nishikawa, S., Ohno, N., Akiyama, N., Tamakoshi, M., Yoshida, H. and Nishikawa, S. (1993) Expression and function of the interleukin 7 receptor in murine lymphocytes. Proc. Natl. Acad. Sci. USA 90, 9125–9129.
Talmadge, J.E., Jackson, J.D., Kelsey, L., Borgeson, C.D., Faltynek, C. and Perry, G.A. (1993) T cell reconstitution by molecular, phenotypic, and functional analysis in the thymus, bone marrow, spleen, and blood following split-dose polychemotherapy and therapeutic activity for metastatic breast cancer in mice. J. Immunother. 14, 258–268.
Tan, J.T., Dudl, E., LeRoy, E., Murray, R., Sprent, J., Weinberg, K.I. and Surh, C.D. (2001) IL-7 is critical for homeostatic proliferation and survival of naïve T cells. Proc. Natl. Acad. Sci. USA 98, 8732–8737.
Wiryana, P., Bui, T., Faltynek, C.R. and Ho, R.J. (1997) Augmentation of cell-mediated immunotherapy against herpes simplex virus by interleukins: comparison of in vivo effects of IL-2 and IL-7 on adoptively transferred T cells. Vaccine 15, 561–563.
Yeoman, H., Clark, D.R. and DeLuca, D. (1996) Development of CD4 and CD8 single positive T cells in human thymus organ culture: IL-7 promotes human T cell production by supporting immature T cells. Dev. Comp. Immunol. 20, 241–263.

36

Transmembrane Interactions as Immunotherapeutic Targets: Lessons from Viral Pathogenesis

Alexander B. Sigalov

University of Massachusetts Medical School, Department of Pathology, Worcester, MA, USA, Alexander.Sigalov@umassmed.edu

Abstract. Multichain immune recognition receptors (MIRRs) represent a family of structurally related but functionally different surface receptors expressed on different cells of the immune system. A distinctive and common structural characteristic of MIRR family members is that the extracellular recognition domains and intracellular signaling domains are located on separate subunits. How extracellular ligand binding triggers MIRRs and initiates intracellular signal transduction processes is not clear. A novel model of immune signaling, the Signaling Chain HOmoOLigomerization (SCHOOL) model, suggests possible molecular mechanisms and reveals the MIRR transmembrane interactions as universal therapeutic targets for a variety of MIRR-mediated immune disorders. Intriguingly, these interactions have been recently shown to play an important role in human immunodeficiency virus and cytomegalovirus pathogenesis. In this chapter, I demonstrate how the SCHOOL model, together with the lessons learned from viral pathogenesis, can be used practically for rational drug design and the development of new therapeutic approaches to treat a variety of seemingly unrelated disorders, such as T cell-mediated skin diseases and platelet disorders.

1. Introduction

Immune cells respond to the presence of foreign antigens with a wide range of responses. Antigen recognition by immune cells is mediated by the interaction of cell surface receptors with soluble, particulate, and cellular antigens. Key among these receptors is the family of multichain immune recognition receptors (MIRRs) that are expressed on many different immune cells, including T and B cells, natural killer (NK) cells, mast cells, macrophages, basophils, neutrophils, eosinophils, dendritic cells, and platelets (Sigalov 2006).

The most intriguing common structural feature of MIRR family members is that the extracellular recognition domains and intracellular signaling domains containing immunoreceptor tyrosine-based activation motifs (ITAMs) are located on separate subunits (Figure 1A). The association of the subunits in resting cells is mostly driven

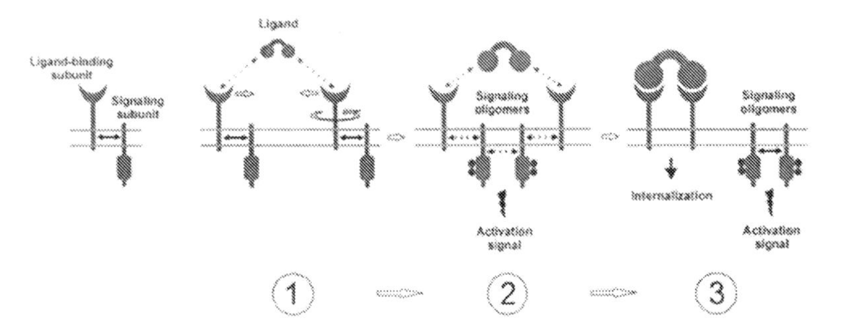

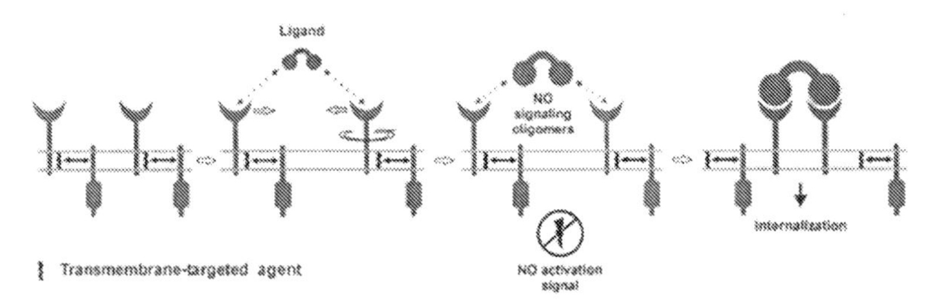

FIGURE 1. (A) Structural and functional organization of multichain immune recognition receptors (MIRRs). Transmembrane interactions between MIRR ligand-binding and signaling components (shown by solid arrow) play a key role in receptor assembly and integrity on resting cells. (B) The Signaling Chain HOmoOLigomerization (SCHOOL) model proposing that the homooligomerization of signaling subunits plays a central role in triggering MIRR-mediated signal transduction. Small solid black arrows indicate specific intersubunit hetero- and homointeractions between transmembrane and cytoplasmic domains, respectively. Circular arrow indicates ligand-induced receptor reorientation. All interchain interactions in a dimeric intermediate are shown by dotted black arrows reflecting their transition state. Phosphate groups are shown as dark circles. (C) Molecular mechanisms underlying proposed intervention by transmembrane-targeted agents. Specific blockade of transmembrane MIRR interactions between recognition and signaling subunits results in "predissociation" of the receptor complex, thus preventing the formation of signaling oligomers and inhibiting ligand-dependent immune cell activation. In contrast, stimulation of these "predissociated" MIRRs with crosslinking antibodies to signaling subunit should still lead to receptor triggering and cell activation (not shown). Similar mechanisms are proposed to be used by diverse viruses, such as human immunodeficiency virus and cytomegalovirus, in their pathogenesis to modulate the host immune response. See text for details.

by the noncovalent transmembrane (TM) interactions between recognition and signaling components (Figure 1A) and plays a key role in receptor assembly and integrity (reviewed in Sigalov 2006).

Crosslinking, or oligomerization, of the receptors after ligand binding results in phosphorylation of the ITAM tyrosines, which triggers the elaborate intracellular signaling cascade. However, the molecular mechanism by which clustering of the extracellular recognition domains of MIRRs leads to activation and phosphorylation of the intracellular domain ITAMs, thus triggering specific pathways and resulting in a immune cell functional outcome, remains to be identified.

MIRR-mediated signal transduction plays an important role in health and disease (Rudd 2006), making these receptors attractive targets for rational intervention in a variety of immune disorders. Thus, future therapeutic strategies depend on our detailed understanding of the molecular mechanisms underlying the MIRR triggering and subsequent TM signal transduction.

Despite numerous models of MIRR-mediated TM signal transduction suggested for particular MIRRs (e.g., T cell receptor, TCR; B cell receptor, BCR; Fc receptors, FcRs; and NK receptors), no current model fully explains at the molecular level how ligand-induced TM signal transduction commences. As a consequence, these models are mostly descriptive and do not reveal clinically important potential points of therapeutic intervention. In addition, no general model of MIRR-mediated immune cell activation has been suggested, thus preventing the potential transfer of therapeutic strategies between seemingly disparate immune disorders.

The central idea of this chapter is to show how the similar architecture of the MIRRs dictates similar mechanisms of MIRR-mediated signaling, thus building the basis for existing and future therapeutic strategies targeting MIRRs. In addition, this hypothesis significantly improves our understanding of the immune modulatory properties of human immunodeficiency virus (HIV) and human cytomegalovirus (CMV) pathogenesis and assumes that the lessons learned from a viral pathogenesis can be used for the development of new therapeutic approaches to treat a variety of seemingly unrelated disorders.

2. SCHOOL Model of MIRR Signaling

Recently, a novel biophysical phenomenon, the homointeractions of intrinsically disordered cytoplasmic domains of MIRR signaling subunits, has been discovered (Sigalov et al. 2004). Hypothesizing a crucial physiological role of these unique homointeractions, a novel mechanistic model of MIRR triggering and subsequent TM signal transduction, the Signaling Chain HOmoOLigomerization (SCHOOL) model (Sigalov 2004, 2005, 2006), suggests that MIRR engagement leads to receptor oligomerization coupled with a multistep structural reorganization driven by the homooligomerization of signaling subunits (Figure 1B).

The model also assumes that the diversity of the immune cell response is partly provided by the combinatorial nature of MIRR-mediated signaling. In other words, different patterns of MIRR signaling subunit oligomerization (Sigalov 2004, 2005) in combination with distinct functions of MIRR signaling modules and/or distinct activation signals provided by different ITAMs located on the same signaling module (TCR ζ chain) induce qualitatively and quantitatively distinct activation signals for the different functional outcomes in immune cell activation and apoptosis.

Within the model, MIRR triggering is considered to be the result of ligand-induced interplay between (1) intrareceptor TM interactions that stabilize and maintain receptor integrity and (2) interreceptor homointeractions between the cytoplasmic domains of MIRR signaling subunits. The plausible and easily testable SCHOOL model is fundamentally different from those previously suggested for particular MIRRs and has several important advantages.

First, this model is based on specific protein–protein interactions—biochemical processes that can be influenced and controlled, and specific inhibition and/or modulation of these interactions provides a promising novel approach for rational drug design (Loregian and Palu 2005).

Second, assuming that the general principles underlying MIRR-mediated TM signaling mechanisms are similar, the SCHOOL model can be applied to any particular receptor of the MIRR family.

Third, the SCHOOL model reveals new therapeutic targets for the treatment of a variety of disorders mediated by immune cells.

Finally, an important application of the SCHOOL model is that similar therapeutic strategies targeting key protein–protein interactions involved in MIRR triggering and TM signal transduction may be used to treat diverse immune-mediated diseases. This assumes that clinical knowledge, experience, and therapeutic strategies can be transferred between seemingly disparate immune disorders or used to develop novel pharmacological approaches.

3. Transmembrane Interactions as Immunotherapeutic Targets

Since it was first published in 2004 (Sigalov 2004), the SCHOOL model has revealed MIRR TM interactions as important therapeutic targets and provided a mechanistic explanation at the molecular level for specific processes behind "outside-in" MIRR signaling that were unclear (Sigalov 2004, 2005, 2006). Examples include molecular mechanisms of action of the therapeutically important TCR TM peptides (Enk and Knop 2000; Wang et al. 2002; Amon et al. 2006; Collier, et al. 2006) first introduced in 1997 (Manolios et al. 1997) and the mechanism underlying HIV-1 fusion peptide (FP)-induced inhibition of antigen-specific T cell activation (Quintana et al. 2005). The relevance of the latter mechanism has since been confirmed experimentally (Bloch et al. 2007).

As suggested by the SCHOOL model (Figure 1C), specific blockade or disruption of the TM interactions between MIRR recognition and signaling subunits causes a physical and functional disconnection of the subunits. Peptides and their derivatives, small-molecule disrupters of protein–protein interactions, site-specific mutations, and other similar agents/modifications can be used to affect the MIRR TM interactions. Antigen stimulation of these "predissociated" receptors leads to reorientation and clustering of the recognition but not signaling subunits. As a result, signaling oligomers are not formed, ITAM Tyr residues do not become phosphorylated, and the signaling cascade is not initiated. In contrast, this "predissociation" does not prevent the formation of signaling oligomers when signaling subunits are clustered by specific antibodies (e.g., anti-CD3 antibodies for TCR and anti-Igβ antibodies for BCR) that trigger cell activation (not illustrated).

Importantly, our current understanding of the MIRR structure and the nature and specificity of TM interactions between receptor recognition and signaling subunits allows us not only to block or disrupt these protein–protein interactions but also to modulate the interactions by sequence-based approach using corresponding peptides and/or their derivatives. Strengthening/weakening and/or selective disruption of the association between particular recognition and signaling subunits might allow us not to inhibit, but rather to modulate the ligand-induced cell response. In addition, selective "disconnection" of particular signaling subunits from their recognition partner would be invaluable in studies of MIRR-mediated cell activation.

3.1. Transmembrane Peptides and Immune Cell Activation

Selected agents suggested or predicted by the SCHOOL model to affect MIRR TM interactions, thus inhibiting or modulating MIRR-mediated immune cell activation, are listed in Table 1.

TM peptides capable of inhibiting MIRR-mediated cell activation were first reported in 1997 for antigen-stimulated TCR-mediated T cell activation (Manolios et al. 1997). Since that time, despite extensive basic and clinical studies of these peptides (Wang et al. 2002; Enk and Knop 2000), the molecular mechanisms of action of these clinically relevant peptides have not been elucidated until 2004 when the SCHOOL model was first introduced (Sigalov 2004).

The vast majority of basic and clinical findings were reported for the TCR TM core peptide (TCR-CP), which represents a synthetic peptide corresponding to the sequence of the TM region of the ligand-binding TCRα chain critical for TCR assembly and function. This region has been shown to interact with the TM domains of the signaling CD3δε and ζ subunits (Call et al. 2002), thus maintaining the integrity of the TCR in resting T cells.

Briefly, as suggested by the SCHOOL model (Sigalov 2004, 2005, 2006), the TCR-CP competes with the TCRα chain for binding to CD3δε and ζ subunits, thus resulting in disconnection and dissociation of the signaling subunits from the remaining receptor complex (Figure 1C). The proposed mechanism is the only mechanism consistent with all experimental and clinical data reported up to date for TCR TM peptides and their lipid and/or sugar conjugates.

The SCHOOL model predicts that the same mechanisms of inhibitory action can be applied to MIRR TM peptides corresponding to the TM regions of not only the MIRR recognition subunits but also the corresponding signaling subunits. This was recently confirmed experimentally (Collier et al. 2006; Vandebona et al. 2006) by showing that the synthetic peptides corresponding to the sequences of the TM regions of the signaling CD3 (δ, ε, or γ) and ζ subunits are able to inhibit the immune response in vivo (CD3 TM peptides) and NK cell cytolytic activity in vivo (ζ TM peptide) (Table 1).

TABLE 1. Selected agents reported to modulate the immune cell response and suggested or predicted by the SCHOOL model to affect MIRR transmembrane interactions

Agent	Receptor	Action	Mechanism
TCR–CP	TCR	Inhibits antigen-stimulated TM signal transduction (Manolios et al. 1997; Wang et al. 2002; Amon et al. 2006; Collier et al. 2006) Efficiently abrogates T cell-mediated immune responses in mice and humans in vitro and in vivo (Enk and Knop 2000)	Disrupts TCRα–CD3δε and TCRα–ζ TM interactions resulting in dissociation of these signaling subunits from the remaining complex, and thus preventing the formation of signaling oligomers upon antigen stimulation and, consequently, inhibiting T cell activation (Figure 1C; Sigalov 2004, 2005, 2006)
CD3δ–CP CD3ε–CP CD3γ–CP	TCR	CD3δ–CP and CD3γ–CP do not inhibit antigen-stimulated T cell proliferation and IL-2 secretion (Collier et al. 2006) CD3δ–CP, CD3ε–CP, and CD3γ–CP prevent disease development and progression in rats with adjuvant-induced arthritis (Collier et al. 2006)	Disrupt TCRα–CD3δ (CD3δ–CP), TCRα–CD3ε (CD3ε–CP), TCRβ–CD3ε (CD3ε–CP), and TCRβ–CD3γ (CD3γ–CP) TM interactions, resulting in selective dissociation of the particular signaling subunits from the remaining receptor complex
NK–CP ζ–CP	NKp44 NKp46 NKp30 NKG2D NKG2C KIR2DS	Inhibit NK cell cytolytic activity (Vandebona et al. 2006)	Disrupt the TM interactions between receptor ligand-binding subunits and associated homodimeric signaling subunits, such as ζ–ζ, γ–γ, or DAP-12 (Figure 1C; Sigalov 2006)
GPVI–CP	Platelet GPVI	Inhibits collagen-induced platelet activation and aggregation (Sigalov et al. unpublished data)	Disrupts the TM interactions between collagen-binding GPVI subunit and associated homodimeric γ chain (Figure 1C; Sigalov 2006)
HIV gp41–FP	TCR	Colocalizes with CD4 and TCR molecules, coprecipitates with the TCR, and inhibits antigen-specific T cell proliferation and proinflammatory cytokine secretion in vitro (Quintana et al. 2005) Blocks the TCR–CD3 TM interactions needed for anti-	Similarly to the TCR–CP, disrupts TCRα–CD3δε and TCRα–ζ TM interactions resulting in dissociation of these signaling subunits from the remaining complex, and thus preventing the formation of signaling oligomers upon antigen stimulation and, consequently, inhibiting

		gen-triggered T cell activation (Bloch et al. 2007)	T cell activation (Figure 1C; Sigalov 2006)
CMV pp65	NKp30	Interacts directly with NKp30, leading to dissociation of the linked ζ subunit and, consequently, to reduced killing (Arnon et al. 2005)	Affects the NKp30–ζ TM interactions resulting in dissociation of the ζ signaling subunit from the remaining complex, and thus preventing the formation of ζ signaling oligomers upon antigen stimulation and, consequently, inhibiting NK cell cytolytic activity (Figure 1C; Sigalov 2006)

CMV, human cytomegalovirus; CP, core peptide; DAP-12, DNAX activation protein 12; FP, fusion peptide; GPVI, glycoprotein VI; HIV, human immunodeficiency virus; NK cells, natural killer cells; TCR, T cell antigen receptor; TM, transmembrane.

Interestingly, the model suggests a molecular explanation for the apparent discrepancy in CD3 TM peptide activity between in vitro and in vivo T cell inhibition reported by Collier et al. (2006). It has been shown that the CD3δ and CD3γ TM peptides do not impact T cell function in vitro (the CD3ε TM peptide has not been used in the reported in vitro experiments because of solubility issues) but that all three CD3 TM peptides decrease signs of inflammation in the adjuvant-induced arthritis rat model in vivo and inhibit an immune response (Collier et al. 2006). Within the SCHOOL model, the CD3δ and CD3γ TM peptides disconnect the corresponding signaling subunits (CD3δ and CD3γ, respectively) from the remaining receptor complex (Table 1). Thus, these subunits do not participate in further processes upon antigen stimulation. On the contrary, the previously reported in vitro activation studies with T cells lacking CD3γ and/or CD3δ cytoplasmic domains indicate that antigen-stimulated induction of cytokine secretion and T cell proliferation are intact (Luton et al. 1997), thus explaining the absence of inhibitory effect of the CD3δ and CD3γ TM peptides in the in vitro activation assays used (Collier et al. 2006). However, in vivo deficiency of either CD3δ or CD3γ results in severe immunodeficiency disorders (Roifman 2004). This could explain the inhibitory effect observed in the in vivo studies for all three CD3 TM peptides (Collier et al. 2006). Thus, these experimental data confirm that our ability to selectively "disconnect" specific signaling subunits using the MIRR TM peptides in line with the SCHOOL model can provide a powerful tool to study MIRR functions and immune cell signaling (Sigalov 2006).

Similar molecular mechanisms of action of other MIRR TM peptides are suggested by the SCHOOL model to describe and/or predict the observed inhibitory/modulatory effect on MIRR-mediated cell activation (Table 1).

In summary, considering the high therapeutic potential of the MIRR TM peptides illustrated by the clinical results for the TCR–CP (Enk and Knop 2000), the SCHOOL model represents an invaluable tool in further development of this novel pharmacological approach targeting MIRR TM interactions.

3.2. Transmembrane Interactions in HIV Pathogenesis

The FP found in the N terminus of the HIV envelope glycoprotein gp41 functions together with other gp41 domains to fuse the virion with the host cell membrane. This peptide has been shown to inhibit antigen-specific T cell proliferation and proinflammatory cytokine secretion in vitro (Quintana et al. 2005). This effect is specific: T cell activation via PMA/ionomycin or mitogenic antibodies to CD3 is not affected by FP. As with TCR-CP, FP shows immunosuppressive activity, inhibiting the activation of arthritogenic T cells in the autoimmune disease model of adjuvant arthritis and reducing the disease-associated IFN-γ response (Quintana et al. 2005).

The SCHOOL model provides a molecular mechanism of action for FP (Sigalov 2006) and explains the observed difference in the response of T cells (Quintana et al. 2005). The peptide prevents the formation of signaling oligomers and thus inhibits antigen-dependent T cell activation, acting similarly in this respect to TCR CP. However, stimulation with anti-CD3 antibodies of these "predissociated" TCRs still results in receptor triggering and cell activation (Sigalov 2006). Thus, the model suggests that clinically relevant antibodies (OKT3) could be used to modulate the affected T cell response during HIV infection. Recently, OKT3 antibodies have been used successfully in HIV therapy to augment immune activation. More recent studies (Bloch et al. 2007) have confirmed the predicted mechanism of action of the HIV FP.

Thus, according to the SCHOOL model, the TCR TM interactions represent not only important therapeutic targets for immune-mediated diseases but also a point of HIV intervention. The molecular mechanisms revealed by the model can be used in rational antiviral drug design and the development of novel antiviral therapies.

3.3. Transmembrane Interactions in CMV Pathogenesis

To escape from NK cell-mediated surveillance, human CMV interferes with the expression of NKG2D ligands in infected cells. Despite considerable progress in the field, a number of issues regarding the involvement of NK receptors in the innate immune response to human CMV remain unresolved.

Recently, a specific and functional interaction between the human CMV tegument protein pp65 and the NK cell-activating receptor NKp30 has been reported (Arnon et al. 2005). Surprisingly, the recognition of pp65 by NKp30 does not lead to NK cell activation but instead results in a general inhibition mediated by the dissociation of the signaling ζ subunit from the NKp30–ζ complex (Arnon et al. 2005).

Within the context of SCHOOL model, the reported action of the human CMV pp65 protein may be due to its potential impact on the TM interactions between NKp30 and ζ, leading to the dissociation of the ζ subunit. This would prevent the formation of ζ signaling oligomers upon antigen stimulation and consequently inhibit NK cell cytolytic activity (Table 1; Figure 1). However, further experimental studies are needed to confirm the proposed mechanism.

4. Lessons from Viral Pathogenesis

There are several important lessons that we can learn from the SCHOOL model-revealed similarity of molecular mechanisms underlying viral pathogenesis and inhibitory effect of synthetic agents affecting MIRR TM interactions.

First, it is possible to design and produce highly effective TM agents that are able to affect specific TM interactions of a targeted MIRR and suppress and/or modulate the MIRR-mediated immune response.

Second, within the model, TCR–CP and HIV gp41–FP affect similar TCR TM interactions. Primary sequence analysis of these two peptides shows different primary sequences but a similarity in charged or polar residue distribution patterns, suggesting that a computational approach can and should be used in the rational design of effective inhibitory peptides. General well-known principles of designing TM peptides with an ability to insert into the membrane might be readily used at this stage.

Third, the SCHOOL model suggests that antibodies to MIRR signaling subunits can be used to modulate the affected immune cell response during viral infection.

Finally, two unrelated enveloped viruses, HIV and human CMV, use a similar mechanism to modulate the host immune response mediated by two functionally different MIRRs—TCR and NKp30. Thus, it is very likely that similar general mechanisms can be or are used by other viral and possibly nonviral pathogens.

5. Conclusions

Despite growing interest in targeting MIRR signaling as a potential treatment strategy for different immune-mediated diseases, the molecular mechanisms underlying MIRR triggering and subsequent transmembrane signal transduction are unknown, thus preventing the development of novel pharmacological approaches.

Suggesting MIRR triggering as a result of ligand-induced interplay between well-defined protein–protein interactions, the SCHOOL model reveals intrareceptor TM interactions as universal therapeutic targets for MIRR-mediated immune diseases. Importantly, the lessons learned from the SCHOOL model and viral pathogenesis indicate that a general drug design approach may be used to treat a variety of different and seemingly unrelated immune diseases. Application of this model to the platelet collagen receptor glycoprotein VI resulted in the invention of novel platelet inhibitors. This illustrates a novel view of potential possibility to transfer clinical knowledge, experience, and therapeutic strategies between seemingly disparate immune-mediated diseases.

Acknowledgments

I thank Walter M. Kim for critical reading of this manuscript. This work was partly supported by grant P30 AI42845-08 from the NIAID/UMass Center for AIDS Research.

References

Amon, M. A., Ali, M., Bender, V., Chan, Y. N., Toth, I. and Manolios, N. (2006) Lipidation and glycosylation of a T cell antigen receptor (TCR) transmembrane hydrophobic peptide dramatically enhances in vitro and in vivo function. Biochim. Biophys. Acta 1763, 879–888.

Arnon, T. I., Achdout, H., Levi, O., Markel, G., Saleh, N., Katz, G., Gazit, R., Gonen-Gross, T., Hanna, J., Nahari, E., Porgador, A., Honigman, A., Plachter, B., Mevorach, D., Wolf, D. G. and Mandelboim, O. (2005) Inhibition of the NKp30 activating receptor by pp65 of human cytomegalovirus. Nat. Immunol. 6, 515–523.

Bloch, I., Quintana, F. J., Gerber, D., Cohen, T., Cohen, I. R. and Shai. Y. (2007) T cell inactivation and immunosuppressive activity induced by HIV gp41 via novel interacting motif. FASEB J. 21, 393–401.

Call, M. E., Pyrdol, J., Wiedmann, M. and Wucherpfennig, K. W. (2002) The organizing principle in the formation of the T cell receptor-CD3 complex. Cell 111, 967–979.

Collier, S., Bolte, A. and Manolios, N. (2006) Discrepancy in CD3-transmembrane peptide activity between in vitro and in vivo T cell inhibition. Scand. J. Immunol. 64, 388–391.

Enk, A. H. and Knop, J. (2000) T cell receptor mimic peptides and their potential application in T cell-mediated disease. Int. Arch. Allergy Immunol. 123, 275–281.

Loregian, A. and Palu, G. (2005) Disruption of protein-protein interactions: towards new targets for chemotherapy. J. Cell. Physiol. 204, 750–762.

Luton, F., Buferne, M., Legendre, V., Chauvet, E., Boyer, C. and Schmitt-Verhulst, A. M. (1997) Role of CD3gamma and CD3delta cytoplasmic domains in cytolytic T lymphocyte functions and TCR/CD3 down-modulation. J. Immunol. 158, 4162–4170.

Manolios, N., Collier, S., Taylor, J., Pollard, J., Harrison, L. C. and Bender, V. (1997) T cell antigen receptor transmembrane peptides modulate T cell function and T cell-mediated disease. Nat. Med. 3, 84–88.

Quintana, F. J., Gerber, D., Kent, S. C., Cohen, I. R. and Shai, Y. (2005) HIV-1 fusion peptide targets the TCR and inhibits antigen-specific T cell activation. J. Clin. Invest. 115, 2149–2158.

Roifman, C. M. (2004) CD3 delta immunodeficiency. Curr. Opin. Allergy Clin. Immunol. 4, 479–484.

Rudd, C. E. (2006) Disabled receptor signaling and new primary immunodeficiency disorders. N. Engl. J. Med. 354, 1874–1877.

Sigalov, A. B. (2004) Multichain immune recognition receptor signaling: different players, same game? Trends Immunol. 25, 583–589.

Sigalov, A. (2005) Multi-chain immune recognition receptors: spatial organization and signal transduction. Semin. Immunol. 17, 51–64.

Sigalov, A. B. (2006) Immune cell signaling: a novel mechanistic model reveals new therapeutic targets. Trends Pharmacol. Sci. 27, 518–524.

Sigalov, A., Aivazian, D. and Stern, L. (2004) Homooligomerization of the cytoplasmic domain of the T cell receptor zeta chain and of other proteins containing the immunoreceptor tyrosine-based activation motif. Biochemistry 43, 2049–2061.

Vandebona, H., Ali, M., Amon, M., Bender, V. and Manolios, N. (2006) Immunoreceptor transmembrane peptides and their effect on natural killer (NK) cell cytotoxicity. Protein Pept. Lett. 13, 1017–1024.

Wang, X. M., Djordjevic, J. T., Kurosaka, N., Schibeci, S., Lee, L., Williamson, P. and Manolios, N. (2002) T cell antigen receptor peptides inhibit signal transduction within the membrane bilayer. Clin. Immunol. 105, 199–207.

37

Tumor Cell Vaccines

Patricia L. Thompson and Sophie Dessureault

University of South Florida, Department of Interdisciplinary Oncology, H. Lee Moffitt Cancer Center & Research Institute, Tampa, FL, USA, sophie.dessureault@moffitt.org

Abstract. This chapter reviews the history of tumor cell vaccines, both autologous and allogeneic, as well as adjuvants used with tumor cell vaccines. The chapter discusses various tumor cell modifications that have been tested over the years. The immune response to tumor vaccines is briefly described, as are some methods of immune monitoring after vaccine therapy. Finally, there is a description of various tumor cell-based vaccines that have been tested in clinical trials.

1. Introduction

In 1798 Edward Jenner described a vaccine against small pox (White and Shackelford 1983). Since then, vaccines have been used to prevent over 20 infectious diseases. Today, vaccines are being used in the setting of cancer to treat malignancies rather than prevent them. The goal of cancer vaccines is to activate the immune system so that it can destroy established cancer cells.

The concept that cancer cells are recognized by the immune system has been the subject of investigation for over 100 years. In the 19th century, Dr. William Coley attempted to treat tumor-bearing patients by injecting them with live cultures of *Serysipelas*. He reported a decrease of tumor burden in some patients (Wiemann and Starnes 1994). In the late 1950s, Gross et al. demonstrated that mice could be successfully immunized against subsequent challenges with methylcholanthrene-induced sarcomas (Klein et al. 1960; Prehn and Main 1957).

Over the past 15 years, advances in immunology and molecular biology have led to the identification of specific tumor antigens as well as improved methods of immunization. Whole-cell tumor vaccines have been investigated for decades. Some of these vaccines have already entered phase III clinical trials (Table 1). Melanoma vaccines have received the most attention to date, but vaccines for many solid and hematologic malignancies have also been studied and shown varied results (Mitchell 2002a,b).

TABLE 1. Whole-Cell cancer vaccines

Canvaxin, CancerVax Corporation, Carslbad, CA, USA
Melacine, Corixa Corp., Seatlle, WA, USA
M-Vax, AVAX, Overland Park, KS, USA
OncoVax, Intracel LLC, Frederick, MD, USA
ONYCR1-3, ONYvax, London, UK

2. Anticancer Vaccine Approaches Derived from Autologous and Allogeneic Tumor Cells

Many cancer vaccines currently being developed use autologous or allogeneic tumor cells as a source of antigen for immunizing patients. Live vaccines can be modified to enhance immunogenicity. Non–viable cell vaccines can be used as a source of antigenic peptides, RNA, or heat shock proteins.

An autologous tumor cell vaccine is derived from a patient's own tumor. It offers the potential to immunize against antigens generated by tumor-specific gene mutations. An allogeneic tumor vaccine is derived from a cancer cell line or another patient's tumor. Allogeneic tumor cells are a more reliable and uniform source of antigens for the preparation of vaccines. These vaccines target antigens that are shared between the cell lines in the vaccine and patients' tumors.

To avoid the possibility that live tumor cells may seed implants or metastasize, whole tumor cells are irradiated or otherwise killed before reinjection into patients. Thus, final preparations of autologous and allogeneic tumor cell vaccines in clinical use contain irradiated cells (e.g., CancerVax), tumor cell lysates (e.g., Melacine), hapten-treated cells (e.g., M-Vax), tumor cell extracts, or mixtures of these.

Manufacture of a tumor cell vaccine requires identity testing and assurance that the vaccine is replication incompetent. Whole-cell vaccines contain multiple antigens, and patient immune responses may occur to several different antigens, so it may be essential to test for multiple antigens in the final vaccine product. Flow cytometry, which can assess cell surface antigens on cells, or polymerase chain reaction (PCR) assays, which can identify unique cell line-associated DNA, can be used to verify that each of the cell lines is present in the final product. Proliferation assays (e.g., ^{3}H-thymidine uptake) can be used to ensure that cells have been rendered replication incompetent.

3. Adjuvants

Adjuvants are agents that are mixed with tumor cells or tumor cell extracts to ensure adequate presentation of the antigens to the immune system and enhance an immune response after immunization (Durrant and Spendlove 2003). Adjuvants are capable of stimulating antigen-presenting cells (APCs) and natural killer (NK) cells into producing cytokines and promoting the survival of antigen-specific T cells. Examples include BCG, *Corynebacterium parvum* (Halpern et al. 1966), DETOX, and alum (aluminum-based salts). Cytokines such as interleukin-2 (IL-2) and granulocyte-macrophage colony—stimulating factor (GM-CSF) have also been used as adjuvants (Salgaller and Lodge 1998).

4. Modified Tumor Cells

Attempts to increase the immunogenicity of tumor cell vaccines have included the infection of cancer cells with viruses (e.g., vaccinia virus [Arroyo et al. 1990; Bash 1993] and vesicular stomatitis virus [Livingston et al. 1985]) to create oncolysates that contain both viral and tumor antigens. The strong viral antigens act as immunologic adjuvants, which enhance immune responses to the tumor antigens.

Transfection of certain genes into tumor cells can also increase their immunogenicity (Colombo and Forni 1994; Colombo and Rodolfo 1995; Pardoll 1995). Patients have been treated with tumor cells genetically modified ex vivo or in vitro to express different classes of genes, including cytokines, T cell costimulatory molecules, as well as allogeneic or xenogeneic MHC class I molecules.

An alternative to the use of whole tumor cells in vaccines has been the use of relevant and immune-activating portions of tumor cells and removal of portions proven to be irrelevant and immune suppressing. In experimental murine models, partially purified fractions of allogeneic tumor cell lines that shed tumor antigen have been shown to yield preparations capable of stimulating immune responses (Johnston et al. 1994). This method of preparing a tumor vaccine has been used in clinical trials (Bystryn 1993; Bystryn et al. 1992a,b), in combination with adjuvants such as alum or DETOX (Schultz et al. 1995).

5. Immune Response to Tumor Cell Vaccines

Dendritic cells (DCs) are key mediators in vaccine function (Banchereau and Steinman 1998; Bronte et al. 1997). They endocytose tumor cells and tumor cell lysates and express antigenic peptide in the context of MHC molecules. This antigen–MHC complex is recognized by the T cell receptor (TCR) and provides the first signal for T cell activation. DCs also express the costimulatory molecules B7-1 (CD80) and B7-2 (CD86), which interact with CD28 and CTLA-4 receptors on T cells and provide the second signal for T cell activation (Bluestone 1998). Engagement of CD28 receptor is associated with T cell differentiation and proliferation; engagement of the CTLA-4 receptor is associated with the attenuation of T cell differentiation and proliferation. (Blocking CTLA-4 engagement has been reported to enhance immune responses to tumor cells (Leach et al. 1996). Activated T cells can then circulate and kill tumor cells systemically.

T cells that accumulate within tumors have been observed to recognize autologous tumor cells in vitro. They are able to proliferate and secrete cytokines—such as IL-2, GM-CSF, interferon gamma (IFN-γ), and tumor necrosis factor α (TNF-α)—in response to stimulation with autologous tumor cells (Hom et al. 1993; Itoh et al. 1988). Mouse studies suggest that tumor rejection seems largely dependent on CD8$^+$ cytolytic T cells (Prehn 1975). CD4$^+$ helper T cells regulate antigen-specific immune responses by regulating functions of other components of the immune system, including B lymphocytes and CD8$^+$ T lymphocytes.

6. Allogeneic Vaccines

Allogeneic tumor cell vaccines consist of either whole tumor cells or lysates of tumor cells bearing multiple antigens with immunogenic potential. Allogeneic tumor cell vaccines have several therapeutic and manufacturing advantages, such as the presence of multiple tumor-associated antigens (TAAs), the potential to minimize the tumor cell's immune escape, and the ability to be manufactured consistently and in lots large enough to treat multiple patients. These vaccines can be designed using multiple cell lines to increase different HLA antigen content, making them applicable to a broader group of patients (Marshall 2003).

6.1. Allogeneic Tumor Cell Vaccines for Melanoma

6.1.1. Canvaxin™ (CancerVax Corporation, Carlsbad, CA, USA)

This is an irradiated polyvalent allogeneic melanoma cell vaccine (PMCV) (Mitchell 1998) tested extensively in phase I and II clinical trials (Van Epps 2004). Results show a statistically significant increase in median and 5-year survival of stage III/IV surgically resected patients with melanoma as compared with matched historical controls. Multicenter phase III randomized double-blind trials are in progress (Morton et al. 2002).

A study by Hsueh et al. evaluated patients who received PMCV following complete regional lymph node dissection. Patients' serum was tested for in vitro delta complement-dependent cytotoxicity (CDC) against M-14 cells, a melanoma cell line not used in the vaccine. CDC was increased in 82% of patients. Median 5-year survival was over 54 months for patients with delta-CDC levels at or above 10% (44 patients) but only 7 months for those with delta-CDC below the 10% level (56 patients; $p = 0.0001$). Disease-free survival (DFS) showed a similar correlation (Hsueh et al. 1998).

6.1.2. Melacine (Corixa Corporation, Seattle, WA, USA)

Melacine is a vaccine composed of lysates from two allogeneic melanoma cells lines admixed with DETOX adjuvant. Phase I and II trials in patients with stage IV melanoma have shown a 10% to 20% response rate. Results of clinical trials with stage II/III patients show improvement in DFS for stage II patients only (Vermorken et al. 1999). Difference in outcome between trials has been attributed to improved quality control in vaccine preparation and in the administration of an additional late dose to boost the immune response. A multicenter phase II study comparing Melacine with a four-drug chemotherapy combination showed comparable response rates and overall survival (Mitchell 1998), but the Melacine was better tolerated. A combination of IFN-α2b and Melacine appears to enhance antitumor response in advanced melanoma (Sondak and Sosman 2003).

6.2. Allogeneic Tumor Cell Vaccines for Pancreatic Cancer

Jaffe et al. (2001) have created a GM-CSF-secreting pancreatic vaccine. They evaluated the safety and immune activation in 14 patients with stage I–III resected adenocarcinoma of the pancreas. The vaccine was well tolerated and it induced an autologous tumor-specific delayed-type hypersensitivity (DTH) response in three patients. These patients showed a prolonged DFS of at least 25 months post-diagnosis.

6.3. Allogeneic Tumor Cell Vaccines for Prostate Cancer

An example of a prostate tumor cell vaccine is the multivalent cancer vaccine from the GVAX family (Cell Genesys, Inc.). This vaccine is designed from two cell lines genetically modified to secrete GM-CSF and cloned based on the expression of prostate cancer markers. Each cell line is produced and inoculated separately.

7. Autologous Vaccines

Autologous tumor cell vaccines are individualized and contain relevant antigens for any particular patient's tumor. These vaccines carry antigens unique to the patient's tumor cells, even ones that may be unknown to investigators. These vaccines, however, present some limitations. They depend on the availability of an adequate number of tumor cells. They require individualized manufacturing, incurring significant labor, expense, and time. Each lot of vaccine cells must meet FDA manufacturing guidelines.

7.1. Autologous Tumor Cell Vaccines for Melanoma

7.1.1. M-Vax (Overland Park, KS, USA)

M-Vax is a hapten-modified autologous melanoma cell vaccine using BCG as the adjuvant (Mitchell 2002a,b). Clinical trials have been done in the metastatic setting (stage IV) as well as in stage III patients. In a study of 83 patients with stage IV melanoma, 11 were reported to show antitumor response (two complete responses [CRs], four partial responses [PRs], and five mixed responses). In 214 patients with stage III melanoma, the 5-year overall survival rate was reported to be approximately 44%, which compares favorably with survival rates after surgery of 20–25% (Berd 2004).

7.1.2. Melanoma transduced with Ad-GM-CSF

This vaccine is an autologous GM-CSF-secreting melanoma cell vaccine engineered with recombinant-incompetent adenovirus. The transduced melanoma cells are irradiated to prevent cell proliferation in vivo. A phase I trial by Kusumoto et al. (2001) showed that the vaccine was well tolerated by patients and led to an antitumor response in some. CTL activity was detected in five of nine patients. One patient showed a partial clinical response.

7.2. Autologous Tumor Cell Vaccines for Colon Cancer

Irradiated autologous colon cancer cells injected with BCG as adjuvant were evaluated for stage II and III colon cancer. Three multi-institutional, prospective, randomized, controlled studies assessed the therapeutic effects of postsurgical adjuvant vaccines. DTH responses were observed in most patients. No overall survival benefit was reported (Hanna et al. 2001).

7.3. Autologous Tumor Cell Vaccines for Renal Cell Carcinoma

Various autologous tumor cell vaccine approaches have been explored in patients with advanced renal cell carcinoma (RCC). In a study of 20 evaluable patients immunized with autologous tumor cells and a bacterial (*Corynebacterium parvum*) adjuvant, one CR and four partial responses PRs were observed, all in patients with lung metastases (Sahasrabudhe et al. 1986). Positive DTH responses to the autologous tumor cells were observed to correlate with response and survival (McCune et al. 1990).

At our institution, a B7-1-transduced autologous tumor cell vaccine in combination with systemic IL-2 has been tested in patients with metastatic renal cell carcinoma. In the phase I portion of the trial (n = 15), two patients had a partial response and two had stable disease. DTH skin tests showed perivascular T cell infiltrates in three of the four patients with partial response or stable disease. No significant toxicity was observed with the vaccine (Antonia et al. 2002). In the phase II portion of the trial (n = 37), the median survival was 23 months, an improvement over historical median survival (manuscript in preparation).

7.4. Autologous Tumor Cell Vaccines for Lung Cancer

A phase I trial in which 35 patients with non-small cell lung cancer were vaccinated with irradiated autologous tumor cells engineered to secrete GM-CSF showed post-vaccine infiltration of metastatic sites with macrophages, lymphocytes, and granulocytes. Most patients demonstrated a positive DTH response. The clinical correlation of this response is less clear, as only 5 of the 35 patients showed stable disease (Salgia et al. 2003).

8. Assessing Effect of Cancer Vaccines: Immune Monitoring

Cancer cell vaccines stimulate the host's immune system to affect a response against tumor cells. They may produce local inflammatory responses that can induce tumor regression. The most sensitive technique for assessing a tumor-specific T cell response after immunization relies on the detection of the effects of T cell activation in response to tumor antigen, such as the secretion of cytokines or T cell proliferation. If immunization with a tumor cell vaccine results in an increase in the number of T cells against the immunizing antigen, then post-treatment cultures of T cells exposed to the appropriate target should produce a larger number of cytokines than those seen in pretreatment cultures. This can be detected by ELISA or ELISPOT assays.

Methods to evaluate immune responses to vaccine therapy include DTH enzym and ELISPOT assays. DTH testing is commonly used to measure responses to autologous and allogeneic cell vaccines. A cellular immune response to the antigen(s) injected into the skin can occur within 24–72 h. T cells and monocytes may be found infiltrating the area.

In ELISPOT testing, T cells are plated with targets and specific antibodies to predetermined cytokines (e.g., IL-2 and IFN-γ). The plate is then developed so that a spot appears on the plate wherever a T cell recognizes a target and is stimulated to produce the predetermined cytokine. Although this study has some limitations in sensitivity, it is still a good test to monitor cancer vaccine responses.

9. Conclusion

Many vaccine approaches are undergoing testing at this time (Table 2). The increasing knowledge in tumor antigens provides new strategies in vaccine design. With this increasing knowledge and new strategies, we should be able to overcome the challenges of immunotherapy caused by the ability of cancer cells to evade immune recognition and destruction through genetic alterations. Possible solutions include treating patients at earlier stages, with the hope that the cancer cells will not have undergone significant mutations; combining vaccines with other modes of cancer treatment; and producing polyvalent vaccines to target several antigens. Another concern is that vaccines may, through manipulation of the immune system, trigger autoimmune responses. This has not been a significant clinical problem to date but may become more real as vaccines become more effective. Many promising approaches are currently undergoing testing and more approaches will surely to develop in the future. It is clear that vaccine therapy as part of the treatment approach against cancer is gaining momentum in the 21th century.

Table 2. Various vaccine approaches

Phase	Title	Principal Investigator	Institution
III	Phase III Randomized Study of Immunotherapy with Polyvalent Melanoma Vaccine (Canvaxin™) plus BCG versus BCG plus Placebo After Surgery in Patients with Stage IV Melanoma	Donald L. Morton, MD	John Wayne Cancer Institute at Saint John's Health Center
III	PANVAC TM-VF Vaccine for the Treatment of Metastatic Pancreatic Cancer After Failing a Gemcitabine-Containing Regimen	Therion Biologics Corp. (sponsor)	Multicenter
III	GVAX® Prostate Cancer Vaccine vs Docetaxel and Prednisone in Patients with Metastatic Hormone-Refractory Prostate Cancer	Cell Genesys, Incorporated (sponsor)	Multicenter
I/II	Vaccine Treatment for Advanced Breast Cancer	NewLink Genetics Corp/ Charles Joseph Link, Jr., MD	Multicenter
I/II	Phase I/II Study of Immunization With Autologous In Vitro-Treated Tumor Cells and Dendritic Cells in Combination with Sargramostim (GM-CSF) in Patients With Stage IV or Recurrent Melanoma	Robert Dillman, MD	Hoag Cancer Center at Hoag Memorial Hospital Presbyterian
I/II	Phase I/II Study of Immunization With In Vitro-Treated Autologous Tumor Cells and Dendritic Cells in Combination with Sargramostim (GM-CSF) in Patients with Stage III or IV or Recurrent Renal Cell Cancer	Robert Dillman, MD	Hoag Cancer Center at Hoag Memorial Hospital Presbyterian
I/II	Phase I/II Study of Vaccination Comprising a-1 3-Galactosyltransferase-Expressing Allogeneic Tumor Cells (HyperAcute™ Lung Cancer Vaccine) in patients With Advanced Refractory or Recurrent NSCLC	John Morris, MD, and Charles Joseph Link, MD	NCI – Center for Cancer Research
II	Phase II Study of GVAX Pancreatic Cancer Vaccine in Combination with Adjuvant Chemoradiotherapy in Patients with Resected Stage I or II Adenocarcinoma of the Pancreas	Daniel Laheru, MD	Sidney Kimmel Comprehensive Cancer Center at Johns Hopkins
II	Phase II Study of Autologous Dendritic Cells (DC) Loaded With Autologous Tumor Lysate (DC Vaccine) in Combination With Interleukin-2 and Interferon Alfa in Patients With Metastatic Renal Cell Carcinoma	Marc Stuart Ernstoff, MD	Norris Cotton Cancer Center at Dartmouth–Hitchcock Medical Center
II	Phase II Study of GVAX Lung Cancer Vaccine in Patients with Selected Stage IIIB or Stage IV Bronchoalveolar Carcinoma	Angela Davies, MD, and Raja Mudad, MD	Southwest Oncology Group
II	Study on the Feasibility to Derive Vaccine from Tumor Tissue in Patients with Non-Small-Cell Lung Cancer	Antigenics, Incorporated	London, United Kingdom
II	Phase II Study of Vaccine Therapy Comprising Autologous Tumor Cells and a GM-CSF-Producing and CD40L-Expressing Cell Line (GMCD40L) Combined With IL-2 in Patients w Malignant Melanoma	Sophie Dessureault, MD, PhD	H. Lee Moffitt Cancer Center and Research Institute at University of South Florida
I	Phase I Pilot Study of Vaccine Comprising a HER2/neu-Positive Allogeneic Tumor Cell Line Transfected With the Sargramostim (GM-CSF) Gene in Combination With Low-Dose Interferon alfa and Low-Dose Cyclophosphamide in Women with Stage IV Breast Cancer	Charles Wiseman, MD	St. Vincent Medical Center–Los Angeles
I	Phase I Study of Vaccination Comprising Allogeneic Sargramostim (GM-CSF)-Secreting Breast Cancer Cells with or without Cyclophosphamide and Doxorubicin in Women with Stage IV Breast Cancer	Leisha Emens, MD, PhD	Sidney Kimmel Comprehensive Cancer Center at Johns Hopkins

Reference www.nci.org

References

Antonia, S.J., Seigne, J., Diaz, J., Muro-Cacho, C., Extermann, M., Farmelo, M.J., Friberg, M., Alsarraj, M., Mahany, J.J., Pow-Sang, J., Cantor, A. and Janssen, W. (2002) Phase I trial of a B7-1 (CD80) gene modified autologous tumor cell vaccine in combination with systemic interleukin-2 in patients with metastatic renal cell carcinoma. J. Urol. 167, 1995–2000.

Arroyo, P.J., Bash, J.A., Wallack, M.K., Arroyo, P.J., Bash, J.A. and Wallack, M.K. (1990) Active specific immunotherapy with vaccinia colon oncolysate enhances the immunomodulatory and antitumor effects of interleukin-2 and interferon alpha in a murine hepatic metastasis model. Cancer Immunol. Immunother. 31, 305–311.

Banchereau, J. and Steinman, R.M. (1998) Dendritic cells and the control of immunity. Nature 392, 245–252.

Bash, J.A. (1993) Active specific immunotherapy of murine colon adenocarcinoma with recombinant vaccinia/interleukin-2-infected tumor cell vaccines. Ann. NY Acad. Sci. 690, 331–333.

Berd, D. (2004) M-Vax: an autologous, hapten-modified vaccine for human cancer. Expert Rev. Vaccines 3, 521–527.

Bluestone, J.A. (1998) Cell fate in the immune system: decisions, decisions, decisions. Immunol. Rev. 165, 5–12.

Bronte, V., Carroll, M.W., Goletz, T.J. Wang, M., Overwijk, W.W., Marincola, F., Rosenberg, S.A., Moss, B. and Restifo, N.P. (1997) Antigen expression by dendritic cells correlates with the therapeutic effectiveness of a model recombinant poxvirus tumor vaccine. Proc. Nat. Acad. Sci. USA 94, 3183–3188.

Bystryn, J.C. (1993) Immunogenicity and clinical activity of a polyvalent melanoma antigen vaccine prepared from shed antigens. Ann. of the NY Acad. Sci. 690, 190–203.

Bystryn, J.C., Henn, M., Li, J. and Shroba, S. (1992a) Identification of immunogenic human melanoma antigens in a polyvalent melanoma vaccine. Cancer Res. 52, 5948–5953.

Bystryn, J.C., Oratz, R., Roses, D., Harris, M., Henn, M. and Lew, R. (1992b) Relationship between immune response to melanoma vaccine immunization and clinical outcome in stage II malignant melanoma. Cancer 69, 1157–1164.

Colombo, M.P. and Forni, G. (1994) Cytokine gene transfer in tumor inhibition and tumor therapy: where are we now? Immunol. Today 15, 48–51.

Colombo, M.P. and Rodolfo, M. (1995) Tumor cells engineered to produce cytokines or cofactors as cellular vaccines: do animal studies really support clinical trials? Cancer Immunol. Immunother. 41, 265–270.

Durrant, L.G. and Spendlove, I. (2003) Cancer vaccines entering phase III clinical trials. Expert Opin. Emerg. Drugs 8, 489–500.

Halpern, B.N., Biozzi, G., Stiffel, C. and Mouton, D. (1966) Inhibition of tumour growth by administration of killed Corynebacterium parvum. Nature 212, 853–854.

Hanna, M.G., Jr., Hoover ,H.C., Jr., Vermorken, J.B., Harris, J.E. and Pinedo, H.M. (2001) Adjuvant active specific immunotherapy of stage II and stage III colon cancer with an autologous tumor cell vaccine: first randomized phase III trials show promise. Vaccine 19, 2576–2582.

Hom, S.S., Schwartzentruber, D.J., Rosenberg, S.A. and Topalian, S.L. (1993) Specific release of cytokines by lymphocytes infiltrating human melanomas in response to shared melanoma antigens. J. Immunother. 13, 18–30.

Hsueh, E.C., Famatiga, E., Gupta, R.K., Qi, K. and Morton, D.L. (1998) Enhancement of complement-dependent cytotoxicity by polyvalent melanoma cell vaccine (CancerVax): correlation with survival. Ann. Surg. Oncol. 5, 595–602.

Itoh, K., Platsoucas, C.D. and Balch, C.M. (1988) Autologous tumor-specific cytotoxic T lymphocytes in the infiltrate of human metastatic melanomas. Activation by interleukin 2 and autologous tumor cells, and involvement of the T cell receptor. J. Exp. Med. 168, 1419–1441.

Jaffee, E.M., Hruban, R.H., Biedrzycki, B., Laheru, D., Schepers, K., Sauter, P.R., Goemann, M., Coleman, J., Grochow, L., Donehower, R.C., Lillemoe, K.D., O'Reilly, S., Abrams, R.A., Pardoll, D.M., Cameron, J.L. and Yeo, C.J. (2001) Novel allogeneic granulocyte-macrophage colony-stimulating factor-secreting tumor vaccine for pancreatic cancer: a phase I trial of safety and immune activation. J. Clin. Oncol. 19, 145–156.

Johnston, D., Rouby, S. and Bystryn, J.C. (1994) Identification of melanoma cell surface antigens immunogenic in mice. Cancer Biother. 9, 29–38.

Klein, G., Sjogren, H.O., Klein, E. and Hellstrom, K.E. (1960) Demonstration of resistance against methylcholanthrene-induced sarcomas in the primary autochthonous host. Cancer Res. 20, 1561–1572.

Kusumoto, M., Umeda, S., Ikubo, A., Aoki, Y., Tawfik, O., Oben, R., Williamson, S., Jewell, W. and Suzuki, T. (2001) Phase 1 clinical trial of irradiated autologous melanoma cells adenovirally transduced with human GM-CSF gene. Cancer Immunol. Immunother. 50, 373–381.

Leach, D.R., Krummel, M.F. and Allison, J.P. (1996) Enhancement of antitumor immunity by CTLA-4 blockade. Science 271, 1734–1736.

Livingston, P.O., Albino, A.P., Chung, T.J. Real, F.X., Houghton, A.N., Oettgen, H.F. and Old, L.J. (1985) Serological response of melanoma patients to vaccines prepared from VSV lysates of autologous and allogeneic cultured melanoma cells. Cancer 55, 713–720.

Marshall, M. (2003) Development of therapeutic cancer vaccines conference, 27–29 April 2003. Cancer Immunol. Immunother. 52, 648–654.

McCune, C.S., O'Donnell, R.W., Marquis, D.M. and Sahasrabudhe, D.M. (1990) Renal cell carcinoma treated by vaccines for active specific immunotherapy: correlation of survival with skin testing by autologous tumor cells. Cancer Immunol. Immunother. 32, 62–66.

Mitchell, M.S. (1998) Perspective on allogeneic melanoma lysates in active specific immunotherapy. Semin. Oncol. 25, 623–635.

Mitchell, M.S. (2002a) Cancer vaccines, a critical review—Part I. Curr. Opin. Invest. Drugs 3, 140–149.

Mitchell, M.S. (2002b) Cancer vaccines, a critical review—Part II. Curr. Opin. Invest. Drugs 3, 150–158.

Morton, D.L., Hsueh, E.C., Essner, R., Foshag, L.J., O'Day, S.J., Bilchik, A., Gupta, R.K., Hoon, D.S., Ravindranath, M., Nizze, J.A., Gammon, G., Wanek, L.A., Wang, H.J. and Elashoff, R.M. (2002) Prolonged survival of patients receiving active immunotherapy with Canvaxin therapeutic polyvalent vaccine after complete resection of melanoma metastatic to regional lymph nodes. Ann. Surg, 236, 438–448.

Pardoll, D.M. (1995) Paracrine cytokine adjuvants in cancer immunotherapy. Annu. Rev. Immunol. 13, 399–415.

Prehn, R.T. (1975) Relationship of tumor immunogenicity to concentration of the oncogen. J. Nat. Cancer Inst. 55, 189–190.

Prehn, R.T. and Main, J.M. (1957) Immunity to methylcholanthrene-induced sarcomas. J. Nat. Cancer Inst. 18, 769–778.

Sahasrabudhe, D.M., deKernion, J.B., Pontes, J.E., Ryan, D.M., O'Donnell, R.W., Marquis, D.M., Mudholkar, G.S. and McCune, C.S. (1986) Specific immunotherapy with suppressor function inhibition for metastatic renal cell carcinoma. J. Biol. Response Mod. 5, 581–594.

Salgaller, M.L. and Lodge, P.A. (1998) Use of cellular and cytokine adjuvants in the immunotherapy of cancer. J. Surg. Oncol. 68, 122–138.

Salgia, R., Lynch, T., Skarin, A., Lucca, J., Lynch, C., Jung, K., Hodi, F.S., Jaklitsch, M., Mentzer, S., Swanson, S., Lukanich, J., Bueno, R., Wain, J., Mathisen, D., Wright, C., Fidias, P., Donahue, D., Clift, S., Hardy, S., Neuberg, D., Mulligan, R., Webb, I., Sugarbaker, D., Mihm, M. and Dranoff, G. (2003) Vaccination with irradiated autologous tumor cells engineered to secrete granulocyte-macrophage colony-stimulating factor augments antitumor immunity in some patients with metastatic non-small-cell lung carcinoma. J. Clin. Oncol. 21, 624–630.

Schultz, N., Oratz, R., Chen, D., Zeleniuch-Jacquotte, A., Abeles, G. and Bystryn, J.C. (1995) Effect of DETOX as an adjuvant for melanoma vaccine. Vaccine 13, 503–508.

Sondak,V.K. and Sosman, J.A. (2003) Results of clinical trials with an allogenic melanoma tumor cell lysate vaccine: Melacine. Semin. Cancer Biol. 13, 409–415.

Van Epps, D. (2004) Characterization of polyvalent allogeneic vaccines. Dev. Biol. 116, 79–90.

Vermorken, J.B., Claessen, A.M., van Tinteren, H., Gall, H.E., Ezinga, R., Meijer, S., Scheper, R.J., Meijer, C.J., Bloemena, E., Ransom, J.H., Hanna, M.G., Jr. and Pinedo, H.M. (1999) Active specific immunotherapy for stage II and stage III human colon cancer: a randomised trial. Lancet 353, 345–350.

White, P.J. and Shackelford, P.G. (1983) Edward Jenner, MD, and the scourge that was. Am. J. Dis. Child. 137, 864–869.

Wiemann, B. and Starnes, C.O. (1994) Coley's toxins, tumor necrosis factor and cancer research: a historical perspective. Pharmacol. Ther. 64, 529–564.

38

T Cell Tolerance to Tumors and Cancer Immunotherapy

Kimberly Shafer-Weaver[1], Michael Anderson[2], Anatoli Malyguine[1] and Arthur A. Hurwitz[2]

[1]Applied and Developmental Research Support Program, SAIC-Frederick, Inc., NCI-Frederick, Frederick, MD, USA
[2]Tumor Immunity and Tolerance Section, Laboratory of Molecular Immunoregulation, Cancer and Inflammation Program, CCR, NIH, Frederick, MD, USA, hurwitza@ncifcrf.gov

Abstract. It is widely recognized that the immune system plays a role in cancer progression and that some tumors are inherently immunogenic. The identification of tumor-associated antigens (TAAs) has stimulated research focused on immunotherapies to mediate the regression of established tumors. Cancer-specific immunity has traditionally been aimed at activating $CD8^+$ cytotoxic T lymphocytes (CTLs) directed against major histocompatibility complex (MHC) class I-binding peptide epitopes. Other approaches utilize T cell adoptive therapy where autologous, tumor-specific T cells propagated in vitro are transferred back into recipients. However, these strategies have met with limited success in part due to the regulatory mechanisms of T cell tolerance, which poses a considerable challenge to cancer immunotherapy. Our laboratory utilizes the TRansgenic Adenocarcinoma of the Mouse Prostate (TRAMP) model, a murine model of prostate cancer, to study mechanisms of T cell tolerization to tumor antigens. We previously demonstrated that upon encounter with their cognate antigen in the tumor microenvironment, naïve T cell become tolerized. Our ongoing studies are testing whether provision of $CD4^+$ T cells can enhance tumor immunity by preventing $CD8^+$ T cell tolerance. A greater understanding of the interaction between various tumor-specific T cell subsets will facilitate the design of novel approaches to stimulate a more potent antitumor immune response.

1. Introduction

Tumors may be caused by a variety of defects that occur in genes encoding proteins involved in the regulation of cell growth. Cellular and humoral immune responses against tumors can be detected in tumor-bearing hosts, and these responses can be manipulated for therapeutic purposes (Rosenberg 1999). Immunosuppression has been associated with cancer development; interestingly, cancer is 100 times more likely to occur in patients who take immunosuppressive medications (Berkow and

Beers 1997; Tenderich et al. 2001). Further corroboration is provided by the high incidence of lymphomas in patients with HIV-1 infection and AIDS (Boshoff and Weiss 2002). Conversely, heightened immunity has been correlated to spontaneous cancer regression (Challis and Stam 1990). Prolonged survival and reduced metastases have been associated with tumors that are highly infiltrated with T cells (Naito et al. 1998).

Further delineation of a role for the immune system in antitumor responses has shown that IFN-γ production and cytotoxic activity by endogenous lymphocytes is pivotal to protect the host from tumor growth or induction (Kaplan et al. 1998). IFN-γ^{-} mice and those with a mutation in the IFN-γRI ligand binding subunit of the IFN-γ receptor show a 10 to 20-fold increase to methylcholanthrene-induced tumor formation (Kaplan et al. 1998). The importance of lymphocyte lytic ability for proper immunosurveillance has been demonstrated in mice deficient in perforin (pfp^{+}). Pfp^{+} mice are 2–3 times more susceptible to MCA-induced tumors and 50% of pfp^{+} mice develop spontaneous lymphomas, compared to less than 10% of wt mice (Smyth et al. 2000). Additionally, mice treated with an αTRAIL blocking antibody (Takeda et al. 2002) or the use of TRAIL^{+} mice (Cretney et al. 2002) both demonstrate an increased susceptibility to MCA-induced tumors.

Taken together, the above data demonstrate that the immune system plays an important role in tumor surveillance. The fact that immune cells can recognize and eliminate tumor cells has led to the belief that immunotherapy may be a realistic option for the treatment of cancer. This idea is further boosted by the identification of tumor antigens recognized by T cells.

2. Tumor Antigens

Extensive studies of tumor formation and progression have led to an understanding of the unique antigenic composition of tumors. Tumor development is associated with the acquisition of genetic mutations, expression of neoantigens, and the overexpression of some cellular proteins. The identification of tumor-associated antigens (TAAs) recognized by the immune system has led to the classification of TAAs into at least five categories: (1) differentiation antigens (e.g., melanocyte differentiation antigens; MART and TRP), (2) tumor-specific antigens (TSAs), which represent the consequence of genetic changes in cancer cells and provide unique antigens for the immune system to target (e.g., p53), (3) overexpressed antigens (e.g., HER-2/neu), (4) cancer-testis antigens, which represent embryonic antigens reexpressed by anaplastic tumors (e.g., MAGE and NY-ESO-1), and (5) viral antigens (HPV/EBV) (Boon and van der Bruggen 1996; Rosenberg 1999). Although many approaches to immunotherapy of cancer have focused on eliciting immunity to specific TAAs, clinical trials of TAA-specific vaccines have not yielded durable immunity to these antigens (Blattman and Greenberg 2004). Because the majority of TAAs tested are self-antigens, it is now widely recognized that immunological tolerance is a major obstacle in the induction of effective and curative antitumor immunity.

3. T Cell Tolerance and Antitumor Responses

Although generalized immunosuppression has been linked to tumor progression, it cannot fully explain the limited success of inducing a protective or curative antitumor immune response. Immunological tolerance has been shown to be a major limitation in tumor immunity (Mapara and Sykes 2004). In one study, for example, the immune response to the TAA HER-2/neu in cancer patients was demonstrated to be significantly lower than immunity resulting from a vaccine for tetanus toxin (Ward et al. 1999). While the cancer patients responded quite actively to the tetanus antigen vaccine, they exhibited limited reactivity to an endogenous vaccination of HER-2/neu overexpression, demonstrating an insufficient response to the oncogenic protein. Therefore, to improve antitumor immunity, a better understanding of immune tolerance to TAAs is necessary.

Immunological tolerance ensures control of self-reactive lymphocytes and prevention of autoimmune disease, while generating a competent immune repertoire capable of responding to a myriad of foreign antigens. Central tolerance occurs in the thymus and eliminates T cells with strong self-reactivity (negative selection) and promotes maturation of T cells capable of responding to foreign antigens (positive selection). Most TAAs are self- or altered self-antigens and as such may be expressed in the thymus during T cell development. Therefore, the majority of high-affinity, potentially tumor-reactive T cells may be deleted from the T cell repertoire prior to leaving the thymus (Mathis and Benoist 2004). Potentially self-reactive T cells that exit the thymus tend to have low-affinity T cell receptors (TCRs) and are further subjected to peripheral tolerance mechanisms.

4. Peripheral Tolerance to Self/Tumor Antigens

Peripheral tolerance is a second tolerogenic process that keeps potentially autoreactive cells from responding in the peripheral immune system. Mechanisms of peripheral tolerance that play a role in hindering the immune response to tumors include anergy, deletion, suppression, and ignorance. To be fully activated, T cells require signals both through their antigen receptor (signal 1) and through costimulatory molecules (signal 2), such as CD28 or ICOS. In the absence of costimulation, anergy is induced and anergic T cells are refractory to subsequent stimulation. One way of inducing tolerance involves cross-presentation of tumor antigens by bone-marrow-derived APCs (Sotomayor et al. 2001). Dendritic cells (DCs) in tumor-draining lymph nodes tend to be incompletely activated favoring T cell tolerance over activation (Melief 2003). In addition, most tumor cells express limited or no costimulatory molecules (Abken et al. 2002) and therefore do not serve as effective APCs for T cell activation.

Tumor-specific T cells may also undergo apoptotic deletion. Increased tumor burden may lead to sustained stimulation of T cells through their antigen receptor. This chronic stimulation upregulates expression of death-inducing ligands on the T cells that can result in activation-induced cell death (AICD) (Molldrem et al. 2003).

AICD of T cells may be by interactions between Fas and Fas ligand (FasL) and this process is often referred to as abortive proliferation.

In addition to deletion from the repertoire, tumor-specific T cells can be actively suppressed. Immunosuppressive activity can be mediated by regulatory cells within the tumor microenvironment (Rohrer et al. 1999; Liyanage et al. 2002) or immunosuppressive factors secreted by the tumor, such as interleukin-10 (IL-10) and transforming growth factor-beta (TGF-β) (Chen et al. 1994; Gorelik and Flavell 2001). Two different types of regulatory cells have been associated with a variety of cancers: $CD4^{+}25^{+}$ $FoxP3^{+}$ regulatory T cells (Tregs) and Gr-1^{+} $CD11b^{+}$ myeloid suppressor cells (MSCs) (Khazaie and von Boehmer 2006; Serafini et al. 2006). Both populations of cells are able to inhibit $CD8^{+}$ and $CD4^{+}$ T cells from proliferating and secreting IFN-γ via contact-dependent and contact-independent mechanisms (Huang et al. 2006; Miyara and Sakaguchi 2007). Both IL-10 and TGF-β have been shown to inhibit the maturation and function of DCs (Gabrilovich et al. 2004). IL-10 has also been implicated in the down regulation of MHC expression (Salazar-Onfray 1999). A variety of tumors have also been shown to secrete vascular endothelial growth factor (VEGF), which has diverse immunosuppressive effects. VEGF can inhibit the maturation of DCs, promote angiogenesis (Gabrilovich et al. 1998; Conejo-Garcia et al. 2004), and induce expression of indoleamine 2,3-dioxygenase (IDO) (Moretti et al. 1997). IDO can inhibit T cell proliferation through the metabolism of tryptophan. Finally, T cells may be tolerant of tumor antigen due to their inability to encounter their cognate antigen, a process referred to as ignorance. Ignorance may be due to reduced vascularity of the tumor, lack of tumor antigen trafficking to the lymph nodes, or physical isolation of the tumor from the immune system.

The first direct evidence demonstrating the capacity of tumor cells to induce T cell tolerance came in the mid-1990s, following the development of TCR-transgenic mice, which bear a T cell repertoire skewed toward a single antigenic specificity. Using a model system with TCR-transgenic T cells specific for an immunoglobulin idiotype expressed by a murine myeloma, Bogen et al. demonstrated both peripheral deletion and functional inactivation of Ig-specific $CD4^{+}$ T cells (Bogen 1996). Levitsky and colleagues demonstrated that hemaggutinin (HA)-specific $CD4^{+}$ T cells are rapidly rendered anergic by both influenza HA-expressing lymphomas and renal carcinomas (Staveley-O'Carroll et al. 1998). The majority of adoptive transfer experiments revealed that T cells undergo an unsustainable period of proliferation and activation, followed by functional inactivation (Shrikant et al. 1999). Together, these studies demonstrate that tumor-specific T cells are modulated by many different mechanisms of peripheral tolerance. Therefore, identifying the mechanisms underlying T cell tolerance in a particular type of cancer may facilitate the development of new strategies aimed at inducing antigen-driven T cell expansion, activity to tumors, and maintenance of antigen responsiveness.

5. Overcoming Tolerance Leads to Tumor Immunity

Numerous studies have shown that overcoming T cell tolerance results in the induction of antitumor immune responses. T cells receive both positive and negative signals during initial activation, and the balance of these signals can affect the functional fate of the T cell. Initially, it was demonstrated by many groups that conferring costimulatory ligands to T cells makes tumors more immunogenic (Townsend and Allison 1993; Zheng et al. 2006). Subsequently, it was shown that blocking negative signals favors more efficient T cell activation and can result in a potent antitumor response. Cytotoxic T lymphocyte-associated antigen-4 (CTLA-4) binds to B7.1/2 on APC with higher affinity than CD28 (Linsley et al. 1994) and delivers an inhibitory signal to T cells that serves to block IL-2 expression and cell cycle progression (Krummel and Allison 1996). Transient in vivo blockade of CTLA-4 at the time of tumor vaccination can enhance vaccine potency and antitumor immunity against melanoma and prostate cancer (van Elsas et al. 1999; Hurwitz et al. 2000). We and others previously reported that treatment of a mouse model of prostate cancer (TRansgenic Adenocarcinoma of the Mouse Prostate [TRAMP] model) with a GM-CSF-expressing cell-based vaccine in combination with CTLA-4 blockade can reduce tumor incidence and tumor grade (Hurwitz et al. 2000).

Tregs have been shown to limit the efficacy of vaccine-induced tumor responses. Many tumors, both human and experimental, have been demonstrated to recruit Tregs to both the draining lymph nodes and tumor bed (Curiel et al. 2004). Murine studies have demonstrated that depletion of Tregs can enhance immunotherapy directed toward tumor antigens. In these tumor models, injection of an anti-CD25 monoclonal antibody to deplete Tregs was shown to significantly enhance vaccine efficacy to a variety of tumors, including mammary adenocarcinoma (Comes et al. 2006) and melanoma (Sutmuller et al. 2001). Even in the absence of vaccination, intratumoral depletion of accumulating $CD4^{+}$ Tregs in a murine fibrosarcoma model also led to tumor rejection (Yu et al. 2005). Interestingly, Sharma and colleagues (2005) demonstrated that Cox-2 inhibitors, which can block accumulation of tumor-induced Tregs, can enhance the immune response in non-small cell lung cancer. The ability to deplete Tregs in humans remains controversial as targeting CD25-expressing cells may not be as efficient as in mice (Attia et al. 2005a; Barnett et al. 2005).

Generating antitumor immune responses can be a double-edged sword. Antitumor therapy is aimed at eliciting an immune response to cells derived from self-tissue. As such, inducing tumor immunity that modulates tolerance to self-antigens may result in autoimmunity. Several studies have shown that enhancing immune responsiveness to tumors can also result in autoimmune reactions. In both cancer patients and murine models of melanoma, immunotherapy for melanoma often results in an autoimmune depigmentation, referred to as vitiligo (Naftzger et al. 1996; Overwijk et al. 1999; Hurwitz and Ji 2004), in which T cells with antigenic specificity for pigmentation antigens destroy normal melanocytes. Rosenberg et al. have reported that treating melanoma patients with an mAb against CTLA-4 can results in clinical regression of the tumor. However, autoimmunity including colitis, dermatitis, hepatitis, hypophysitis, and uveitis can also occur with this treatment (Attia et al. 2005b). Despite these treatable autoimmune sequelae, the durability of the clinical responses with CTLA-4 blockade has generated further interest in using this therapy in the treatment of cancer.

6. Modulating Tolerance of Tumor-Specific T cells in a Murine Model of Prostate Cancer

The development of genetically modified murine models of cancer has profoundly aided in understanding T cell responses to tumor antigens. To study T cell tolerance and tumor immunity, we utilize the TRAMP model, an autochthonous model of prostate cancer that highly resembles the pathogenesis and progression of prostate cancer in humans. TRAMP mice carry the –426/+28 fragment of the androgen-driven, prostate-specific rat Probasin (PB) regulatory element fused to the SV40 T/t antigen gene (TAg) (Greenberg et al. 1995). The TRAMP model system has been proven by us and others to be a valid model to study immunological tolerance and potential immunotherapeutics aimed at overcoming T cell tolerance to tumors (Granziero et al. 1999; Hurwitz et al. 2000; Tourkova et al. 2004).

Traditionally, generating an effective antitumor immune response has primarily focused on priming $CD8^+$ T cells to tumor antigens. $CD8^+$ T cells have been strategically targeted because the majority of tumors express MHC I, not MHC II, and upon recognition, these cytotoxic T cells may be able to kill these MHC I^+ tumor cells. We recently characterized the fate of naïve tumor-specific $CD8^+$ T cells in TRAMP mice (Table 1). We demonstrated that naïve, tumor-specific $CD8^+$ T cells (TcR-I) adoptively transferred into TRAMP mice were inefficiently primed and underwent abortive proliferation. Addition of an ex vivo matured, peptide-pulsed DC vaccine resulted in effective TcR-I priming and protection of these cells from initial tolerization (Anderson et al. 2007). A slowing of prostate tumors in mice treated with the DC vaccine and tumor-specific TcR-I cells was also noted. However, these effects are not durable and T cells were eventually tolerized due to the immunosuppressive environment of the transgene-driven tumor that appears to continually exert suppressive effects on the infiltrating T cells (Anderson et al. 2007).

Table 1. Relative Infiltration and Responsiveness of TcR cells in the Prostate of Tumor-bearing Mice.

TcR Transfer[1]	Infiltration into Prostate[2]	T cell Responsiveness (tumor-infiltrating cells)[3]		
		D5	D10	D20
TcR-I only	D5, +	+/-	-	-
TcR-I + DC Vaccine	D5, ++++	++++	++	-
TcR-II only	D10, ++	++	++	-
TcR-I + TcR-II	D5, +++ (TcR-I)	+++ (TcR-I)	++ (TcR-I)	- (TcR-I)

[1]Twelve-week-old male TRAMP mice were transferred with 3.0×10^6 TcR-I (CD8+) and/or TcR-II (CD4+) T cells or given a s.c peptide-pulsed DC vaccine.
[2]Using magnetic beads, TcR T cells were isolated from prostate tissue on day 5 (D5), day 10 (D10), and day 20 (D20) after transfer and the number of TcR T cells were enumerated. Data is presented as the day of peak TcR T cell infiltration into the tumor and the relative number with respect to type of transfer.
[3]Using magnetic beads, TcR T cells were isolated from prostate tissue on day 5 (D5), day 10 (D10), and day 20 (D20) after transfer. Isolated TcR-I cells were assayed for functional capacit using IFN-γ and Granzyme B ELISPOT assays. TcR-II cells were assayed for functional capacity using an interleukin-2 and IFN-γ ELISPOT assays. Data is presented relative activity in the functional assays with respect to type of and day after transfer.

The signals CD8$^+$ T cells receive during their initial differentiation and maintenance of their immune response are critical for shaping their antitumor functions (Gattinoni et al. 2005). CD4$^+$ T lymphocytes play a pivotal role in activating DCs and other APCs to efficiently prime CD8 T cells. Activation of APCs by CD4$^+$ can be mediated through CD40/CD40L interactions that in turn upregulate the expression of costimulatory molecules (e.g., CD80, CD86, and ICAM) and cytokines (e.g., IL-12) on the APCs needed to generate CD8$^+$ effector and memory cells (Bennett et al. 1998; Schoenberger et al. 1998). This has been referred to as APC licensing. Additionally, CD4$^+$ T cells are a source of cytokines that are important for CTL differentiation, expansion, and survival (Kalams and Walker 1998; Moroz et al. 2004). With respect to tumor immunity, CD4$^+$ T cells may be important to maintain tumor-specific CD8$^+$ T cell numbers, to enhance their infiltration into the tumor microenvironment, and to sustain their effector functions once in the tumor microenvironment (Marzo et al. 2000). Taken together, these data suggest that CD4$^+$ T cells provide important and critical provisions for the generation of CD8$^+$ T cells with antitumor functions.

Given these findings, we are currently using the TRAMP model to test whether tumor-specific CD4$^+$ (TcR-II) T cells can affect tolerization of TcR-I cells. We first characterized TcR-II cell trafficking, activation, and functional kinetics in tumor-bearing mice. Naïve TcR-II T adoptively transferred into TRAMP mice encounter their cognate antigen in lymph nodes, undergo several rounds of proliferation, express activation markers (CD25 and CD44), and traffic to the prostate. Over time, TcR-II cells become functionally tolerant in the prostate, as measured by their inability to secrete IL-2 and IFN-γ in response to their cognate antigen (Table 1). Therefore, like CD8$^+$ TcR-I cells, some naïve TcR-II T cells can resist deletion and persist in the prostate in a tolerant state.

We have also tested the effect of CD4$^+$ TcR-II cells on the fate and function of TcR-I cells. Co-transfer of TcR-II cells enhanced the frequency, activation, survival, and function (IFN-γ and GrB secretion) of TcR-I cells in TRAMP mice. However, this protection was not durable, as TcR-I cell tolerance was noted 3 weeks after transfer into TRAMP mice. These data suggest that TcR-II cells facilitate efficient priming and differentiation of TcR-I cells, but CD4-mediated help is not sufficient for the maintenance of TcR-I antitumor effector functions in the tumor microenvironment (Table 1). Our current studies are testing whether CD4$^+$ cells can reverse CD8$^+$ T cell tolerance. By understanding the role of CD4$^+$ T helper cells in enhancing or rescuing tumor-specific cytotoxic T cells' responsiveness, we hope to facilitate the design of novel approaches to stimulate an immune response that would be therapeutically effective at controlling and eradicating tumors.

7. Conclusions

One of the common goals of tumor immunotherapy is to elicit a potent and durable T cell response directed against TAAs. T cell tolerance to TAAs is one of the many obstacles that prohibit the efficacy of many vaccination approaches. Historically, many studies have used transplantable murine tumors to study T cell tolerance to tumor antigens, but these models may not accurately reflect what occurs in humans.

Recently, we have employed a transgenic mouse model to study T cell tolerance to prostate cancer antigens. TRAMP mice develop autochthonous primary and metastatic prostate carcinomas. We reported the transfer of naïve, tumor-specific T cells that undergo an abortive proliferation, with residual cells trafficking to the prostate and persisting as tolerant. Interestingly, the kinetics for $CD8^+$ T cell tolerization was more rapid than for $CD4^+$ T cells. In addition, we have reported that a DC vaccine may temporarily delay tolerance, but over time, tolerance ensues. Our on-going studies will elucidate the interactions between tumor-specific $CD4^+$ and $CD8^+$ T cells and the effects of those interactions on T cell tolerance. We also examined the mechanisms by which T cells become tolerized by the prostatic tumors.

In conclusion, many studies have examined the mechanisms of T cell tolerance induction and have identified approaches to overcome tolerance to TAAs. However, despite this information, T cell tolerance still remains a major obstacle for eliciting durable antitumor immune responses. The use of more physiologically relevant tumor models and the application of the information from these studies into clinical trials will certainly assist in generating successful approaches to treating cancer.

Acknowledgments

This project has been funded in whole or in part with federal funds from the National Cancer Institute, National Institutes of Health, under contract N01-CO-12400. The content of this publication does not necessarily reflect the views or policies of the Department of Health and Human Services, nor does it mention of trade names, commercial products, or organizations imply endorsement by the U.S. Government. This research was also supported in part by the DOD PCRP and Prostate Cancer Foundation (CaP CURE). The authors thank Dr. Scott Durum for his critical review of the manuscript.

References

Abken, H., Hombach, A., Heuser, C., Kronfeld, K. and Seliger, B. (2002) Tuning tumor-specific T cell activation: a matter of costimulation? Trends Immunol 23, 240–245.

Anderson, M., Shafer-Weaver, K., Greenberg, N. and Hurwitz, A. (2007) Tolerization of tumor-specific T cells despite efficient initial priming in a primary murine model of prostate cancer. J Immunol 178, 1268–1276.

Attia, P., Maker, A., Haworth, L., Rogers-Freezer, L. and Rosenberg, S. (2005a) Inability of a fusion protein of IL-2 and diphtheria toxin to eliminate regulatory T lymphocytes in patients with melanoma. J Immunother 28, 582–592.

Attia, P., Phan, G., Maker, A., Robinson, M., Quezado, M., Yang, J., Sherry, R., Topalian, S., Kammula, U., Royal, R., Restifo, N., Haworth, L., Levy, C., Mavroukakis, S., Nichol, G., Yellin, M. and Rosenberg, S. (2005b) Autoimmunity correlates with tumor regression in patients with metastatic melanoma treated with anti-cytotoxic T lymphocyte antigen-4. J Clin Oncol 23, 6043–6053.

Barnett, B., Kryczek, I., Cheng, P., Zou, W. and Curiel, T. (2005) Regulatory T cells in ovarian cancer: biology and therapeutic potential. Am J Reprod Immunol 54, 369–377.

Bennett, S., Carbone, F., Karamalis, F., Flavell, R., Miller, J. and Heath, W. (1998) Help for cytotoxic-T cell responses is mediated by CD40 signalling. Nature 393, 478–480.

Berkow, R. and Beers, M. (1997). Cancer and the immune system. The Merck Manual of Medical Information. Whitehouse Station, Merck Research Laboratories: 792–794.

Blattman, J. and Greenberg, P. (2004) Cancer immunotherapy: a treatment for the masses. Science 305, 200–205.

Bogen, B. (1996) Peripheral T cell tolerance as a tumor escape mechanism: deletion of CD4+ T cells specific for a monoclonal immunoglobulin idiotype secreted by a plasmacytoma. Eur J Immunol 26, 2671–2679.

Boon, T. and van der Bruggen, P. (1996) Human tumor antigens recognized by T lymphocytes. J Exp Med 183, 725–729.

Boshoff, C. and Weiss, R. (2002) AIDS-related malignancies. Nat Rev Cancer 2, 373–382.

Challis, G. and Stam, H. (1990) The spontaneous regression of cancer. A review of cases from 1900 to 1987. Acta Oncol 29, 545–550.

Chen, Q., Daniel, V. and Maher, D. (1994) Production of IL-10 by melanoma cells: examination of its role in immunosuppression mediated by melanoma. Int J Cancer 56, 755–760.

Comes, A., Rosso, O., Orengo, A., Carlo, E.D., Sorrentino, C., Meazza, R., Piazza, T., Valzasina, B., B., Nanni, P., Colombo, M. and Ferrini, S. (2006) $CD25^{+}$ regulatory T cell depletion augments immunotherapy of micrometastases by an IL-21-secreting cellular vaccine. J Immunol 176, 1750–1758.

Conejo-Garcia, J., Benencia, F., Courreges, M, Kang, E., Mohamed-Hadley, A., Buckanovich, R., Holtz, D., Jenkins, A., Na, H., Wagner, D., Katsaros, D., Caroll, R. and Coukos, G. (2004) Tumor-infiltrating dendritic cell precursors recruited by a beta-defensin contribute to vasculogenesis under the influence of Vegf-A. Nat Med 10, 950–958.

Cretney, E., Takeda, K., Yagita, H., Glaccum, M., Peschon, J. and Smyth, M. (2002) Increased susceptibility to tumor initiation and metastasis in TNF-related apoptosis-inducing ligand-deficient mice. J Immunol 168, 1356–1361.

Curiel, T., Coukos, G., Zou, L., Alvarez, X., Cheng, P., Mottram, P., Evdemon-Hogan, M., Conejo-Garcia, J., Zhang, L., Burow, M., Zhu, Y., Wei, S., Kryczek, I., Daniel, B., Gordon, A., Myers, L., Lackner, A., Disis, M., Knutson, K., Chen, L. and Zou, W. (2004) Specific recruitment of regulatory T cells in ovarian carcinoma fosters immune privilege and predicts reduced survival. Nat Med 10, 942–949.

Gabrilovich, D. (2004) Mechanisms and functional significance of tumour-induced dendritic-cell defects. Nat Rev Immunol 4, 941–952.

Gabrilovich, D., Ishida, T., Oyama, T., Ran, S., Kravtsov, V., Nadaf, S. and Carbone, D. (1998) Vascular endothelial growth factor inhibits the development of dendritic cells and dramatically affects the differentiation of multiple hematopoietic lineages in vivo. Blood 92, 4150–4166.

Gattinoni, L., Klebanoff, C., Palmer, D., Wrzesinski, C., Kerstann, K., Yu, Z., Finkelstein, S., Theoret, M., Rosenberg, S. and Restifo, N. (2005) Acquisition of full effector function in vitro paradoxically impairs the in vivo antitumor efficacy of adoptively transferred CD8+ T cells. J Clin Invest 115, 1616–1626.

Gorelik, L. and Flavell, R. (2001) Immune-mediated eradication of tumors through the blockade of transforming growth factor-beta signaling in T cells. Nat Med 7, 1118–1122.

Granziero, L., Krajewski, S., Farness, P., Yuan, L., Courtney, M., Jackson, M., Peterson, P. and Vitiello, A. (1999) Adoptive immunotherapy prevents prostate cancer in a transgenic animal model. Eur J Immunol 29, 1127–1138.

Greenberg, N., DeMayo, F., Finegold, M., Medina, D., Tilley, W., Aspinall, J., Cunha, G., Donjacour, A., Matusik, R. and Rosen, J. (1995) Prostate cancer in a transgenic mouse. Proc Natl Acad Sci USA 92, 3439–3443.

Huang, B., Pan, P., Li, Q., Sato, A., Levy, D., Bromberg, J., Divino, C. and Chen, S. (2006) Gr-1+CD115+ immature myeloid suppressor cells mediate the development of tumor-induced Treg cells and T cell anergy in tumor-bearing host. Cancer Res 66, 1123–1131.

Hurwitz, A., Foster, B., Kwon, E., Truong, T., Choi, E., Greenberg, N., Burg, M. and Allison, J. (2000) Combination immunotherapy of primary prostate cancer in a transgenic mouse model using CTLA-4 blockade. Cancer Res 60, 2444–2448.

Hurwitz, A. and Ji, Q. (2004) Autoimmune depigmentation following sensitization to melanoma antigens. Methods Mol Med 102, 421–427.

Kalams, S. and Walker, B. (1998) The critical need for CD4 help in maintaining effective cytotoxic T lymphocyte responses. J Exp Med 188, 2199–2204.

Kaplan, D., Shankaran, V., Dighe, A., Stockert, E., Aguet, M., Old, L. and Schreiber, R. (1998) Demonstration of an interferon gamma-dependent tumor surveillance system in immunocompetent mice. Proc Natl Acad Sci USA 95, 7556–7561.

Khazaie, K. and von Boehmer, H. (2006) The impact of CD4+CD25+ Treg on tumor specific CD8+ T cell cytotoxicity and cancer. Semin Cancer Biol 16, 124–136.

Krummel, M. and Allison, J.P. (1996) CTLA-4 engagement inhibits IL-2 accumulation and cell cycle progression upon activation of resting T cells. J Exp Med 183, 2533–2540.

Linsley, P., Greene, J., Brady, W., Bajorath, J., Ledbetter, J. and Peach, R. (1994) Human B7-1 (CD80) and B7-2 (CD86) bind with similar avidities but distinct kinetics to CD28 and CTLA-4 receptors. Immunity 1, 793–801.

Liyanage, U., Moore, T. and Joo, H. (2002) Prevalence of regulatory T cells is increased in peripheral blood and tumor microenvironment of patients with pancreas or breast adenocarcinoma. J Immunol 169, 2756–2761.

Mapara, M. and Sykes, M. (2004) Tolerance and cancer: mechanisms of tumor evasion and strategies for breaking tolerance. J Clin Oncol 22, 1136–1151.

Marzo, A., Kinnear, B., Lake, R., Frelinger, J., Collins, E., Robinson, B. and Scott, B. (2000) Tumor-specific CD4+ T cells have a major "post-licensing" role in CTL mediated anti-tumor immunity. J Immunol 165, 6047–6055.

Mathis, D. and Benoist, C. (2004) Back to central tolerance. Immunity 20, 509–516.

Melief, C. (2003) Regulation of cytotoxic T lymphocyte responses by dendritic cells: peaceful coexistence of cross-priming and direct priming? Eur J Immunol 33, 2645–2654.

Miyara, M. and Sakaguchi, S. (2007) Natural regulatory T cells: mechanisms of suppression. Trends Mol Med. 13, 108–116.

Molldrem, J., Lee, P., Kant, S., Wieder, E., Jiang, W., Lu, S., Wang, C. and Davis, M. (2003) Chronic myelogenous leukemia shapes host immunity by selective deletion of high-avidity leukemia-specific T cells. J Clin Invest 111, 639–647.

Moretti, S., Pinzi, C., Berti, E., Spallanzani, A., Chiarugi, A., Boddi, V., Reali, U. and Giannotti, B. (1997) In situ expression of transforming growth factor beta is associated with melanoma progression and correlates with Ki67, HLA-DR and beta 3 integrin expression. Melanoma Res 7, 313–321.

Moroz, A., Eppolito, C., Li, Q., Tao, J., Clegg, C. and Shrikant, P. (2004) IL-21 enhances and sustains CD8+ T cell responses to achieve durable tumor immunity: comparative evaluation of IL-2, IL-15, and IL-21. J Immunol 173, 900–909.

Naftzger, C., Takechi, Y., Kohda, H., Hara, I., Vijayasaradhi, S. and Houghton, A. (1996) Immune response to a differentiation antigen induced by altered antigen: a study of tumor rejection and autoimmunity. Proc Natl Acad Sci USA 93, 14809–14814.

Naito, Y., Saito, K., Shiiba, K., Ohuchi, A., Saigenji, K., Nagura, H. and Ohtani, H. (1998) CD8+ T cells infiltrated within cancer cell nests as a prognostic factor in human colorectal cancer. Cancer Res 58, 3491–3494.
Overwijk, W., Lee, D., Surman, D., Irvine, K., Touloukian, C., Chan, C., Carroll, M., Moss, B., Rosenberg, S. and Restifo, N. (1999) Vaccination with a recombinant vaccinia virus encoding a "self" antigen induces autoimmune vitiligo and tumor cell destruction in mice: requirement for $CD4^{(+)}$ T lymphocytes. Proc Natl Acad Sci USA 96, 2982–2987.
Rohrer, J., Barsoum, A. and Dyess, D. (1999) Human breast carcinoma patients develop clonable oncofetal antigen-specific effector and regulatory T lymphocytes. J Immunol , 162, 6880–6892.
Rosenberg, S. (1999) A new era for cancer immunotherapy based on the genes that encode cancer antigens. Immunity 10, 281–287.
Salazar-Onfray, F. (1999) Interleukin-10: a cytokine used by tumors to escape immunosurveillance. Med Oncol 16, 86–94.
Schoenberger, S., Toes, R., Voort, E.v.d., Offringa, R. and Melief, C. (1998) T cell help for cytotoxic T lymphocytes is mediated by CD40-CD40L interactions. Nature 393, 480–483.
Serafini, P., Borrello, I. and Bronte, V. (2006) Myeloid suppressor cells in cancer: recruitment, phenotype, properties, and mechanisms of immune suppression. Semin Cancer Biol 16, 53–65.
Sharma, S., Zhu, L., Yang, S., Zhang, L., Lin, J., Hillinger, S., Gardner, B., Reckamp, K., Strieter, R., Huang, M., Batra, R. and Dubinett, S. (2005) Cyclooxygenase 2 inhibition promotes IFN-gamma-dependent enhancement of antitumor responses. J Immunol 175, 813–819.
Shrikant, P., Khoruts, A. and Mescher, M. (1999) CTLA-4 blockade reverses CD8+ T cell tolerance to tumor by a CD4+ T cell- and IL-2-dependent mechanism. Immunity 11, 483–493.
Smyth, M., Thia, K., Street, S., MacGregor, D., Godfrey, D. and Trapani, J. (2000) Perforin-mediated cytotoxicity is critical for surveillance of spontaneous lymphoma. J Exp Med 192, 755–760.
Sotomayor, E., Borrello, I., Rattis, F., Cuenca, A., Abrams, J., Staveley-O'Carroll, K. and Levitsky, H. (2001) Cross-presentation of tumor antigens by bone marrow derived antigen-presenting cells is the dominant mechanism in the induction of T cell tolerance during B cell lymphoma progression. Blood 98, 1070–1077.
Staveley-O'Carroll, K., Sotomayor, E., Montgomery, J., Borrello, I., Hwang, L., Fein, S., Pardoll, D. and Levitsky, H. (1998) Induction of antigen-specific T cell anergy: An early event in the course of tumor progression. Proc Natl Acad Sci USA 95, 1178–1183.
Sutmuller, R., Duivenvoorde, L.v., Elsas, A.v., Schumacher, T., Wildenberg, M., Allison, J., Toes, R., Offringa, R. and Melief, C. (2001) Synergism of CTLA-4 blockade and depletion of CD25(+) regulatory T cells in antitumor therapy reveals alternative pathways for suppression of autoreactive cytotoxic T lymphocyte responses. J Exp Med 194, 823–832.
Takeda, K., Smyth, M., Cretney, E., Hayakawa, Y., Kayagaki, N., Yagita, H. and Okumura, K. (2002) Critical role for tumor necrosis factor-related apoptosis-inducing ligand in immune surveillance against tumor development. J Exp Med 195, 161–169.
Tenderich, G., Deyerling, W. and Schulz, U. (2001) Malignant neoplastic disorders following long-term immunosuppression after orthotopic heart transplantation. Transplant. Proc 33, 3653–3655.
Tourkova, I., Yamabe, K., Foster, B., Chatta, G., Perez, L., Shurin, G. and Shurin, M. (2004) Murine prostate cancer inhibits both in vivo and in vitro generation of dendritic cells from bone marrow precursors. Prostate 59, 203–213.
Townsend, S. and Allison, J. (1993) Tumor rejection after direct costimulation of CD8+ T cells by B7-transfected melanoma cells. Science 259, 368–370.

van Elsas, A., Hurwitz, A. and Allison, J. (1999) Combination immunotherapy of B16 melanoma using anti-CTLA-4 and GM-CSF-producing vaccines induces rejection of subcutaneous and metastatic tumors accompanied by autoimmune depigmentation. J Exp Med 190, 355–366.

Ward, R., Hawkins, N., Coomber, D. and Disis, M. (1999) Antibody immunity to the HER-2/neu oncogenic protein in patients with colorectal cancer. Hum Immunol 60, 510–515.

Yu, P., Lee, Y., Liu, W., Krausz, T., Chong, A., Schreiber, H. and Fu, Y. (2005) Intratumor depletion of CD4+ cells unmasks tumor immunogenicity leading to the rejection of late-stage tumors. J Exp Med 201, 779–791.

Zheng, G., Liu, S., Wang, P., Xu, Y. and Chen, A. (2006) Arming tumor-reactive T cells with costimulator B7-1 enhances therapeutic efficacy of the T cells. Cancer Res 66, 6793–6799.

39

Herpes Simplex Virus: Treatment with Antimicrobial Peptides

Leonid V. Kovalchuk[1], Ludmila V. Gankovskaya[1], Oksana A. Gankovskaya[2], and Vyacheslav F. Lavrov[2]

[1] Russian State Medical University, Department of Immunology, Moscow, Russia
[2] Mechnikov Research Institute of Vaccines and Sera, Department of Viral Infection Diagnostics Moscow, Russia, oksgan@yandex.ru

Abstract. The herpes virus infection represents a significant challenge for public health. The innate immunity plays an important role in herpes simplex virus (HSV) elimination. The innate antiviral immunity has not been comprehensively studied. The recent investigations demonstrate that Toll-like receptors are actively involved in the virus recognition. The complement and natural antibodies, as well as cytokines and antimicrobial peptides, are the first molecules to bind to virions. In this chapter, some mechanisms of the innate antiviral immunity are discussed and treatment regimens are proposed. The complex of native cytokines and antimicrobial peptides (CCAP or Superlymph) proved to inhibit the virus reproduction in vitro. Protegrines, as a CCAP component, were active against the virus. Considering all the data, we conclude that the complex of native cytokines and antimicrobial peptides produces both immunomodulating and antiviral effects.

1. Herpes Simplex Virus Infection

Herpes simplex virus type 1 and 2 (HSV-1 and HSV-2) are common human pathogens infecting respectively more than 90% and 20% individuals in Russia, respectively (Barinskii et al. 2005). HSV-1 and HSV-2, producing primary and reactivation infections, are the causative agents of human diseases, including pharyngitis, herpes labialis, encephalitis, and eye and genital infection (Cunningham et al. 2006). HSV interacts with epithelial cells and replicates. Then HSV is transported within the axons of sensory nerve endings at the infection site to the peripheral ganglion, where the virus establishes latent infection (Garner 2003). Rarely, HSV leads to the development of encephalitis. After replicating in epithelial cells, HSV remains latent in sensory neurons. However, while HSV-1 usually affects oral mucosa, HSV-2 injures genital mucosa. Herpes virus infection in an expectant mother can result in prenatal transmission to a newborn (Hollier and Grissom 2005; Abrahams and Mor 2005).

HSV-1 and HSV-2 have double-stranded DNA, almost 150 kb each. The HSV genome is transported inside icosahedral capsid, surrounded by tegument peptides. The whole structure is followed by a host cell membrane (lipid bilayer) containing viral glycoproteins (Grunewald et al. 2003). In the first phase of infection, viral envelope glycoproteins (gB, gC, and gD) bind with specific cellular herpes virus entry mediator receptors (TNFαR [HveA]), polioviral receptors [HveB and HveC], and others) on plasma membrane. Tegument proteins are released into cytoplasm, after which the capsid is transported to the nuclear membrane, where the viral DNA penetrates the nucleus. Transcription of viral genes starts in the ordered cascade: α, β, γ1, and γ2 genes in turn. During expression of γ1 genes, the viral DNA replication also begins. Later, genome DNA is packaged into capsids within the nucleus and the virion envelopes by the proteins of tegument and plasma membrane with viral glycoproteins. The entire replication cycle takes place in 12–16 h (Spear 2004).

2. Immune Response Against Herpes Virus Infection

Mechanisms of innate immune are characterized by rapid induction and relatively undiversified response. The initial recognition is aimed to limit virus replication and spread to uninfected cells and involves additional cellular effectors in the infection control. Innate responses to the virus infection are composed of humoral components, including complement, natural antibodies, cytokines (IFN-α/β, TNF-α, IL-12, etc.), and antimicrobial peptides (defensins), and recruited cellular effectors (neutrophils, macrophages, resident dendritic cells, γ/δ T cells, natural killer (NK), and NKT cells). This mechanism restricts the number of infected cells. HSV replication in epithelial cells may stimulate the complement system and production of chemokines and IFN-α/β. These molecules may activate capillary endothelia, which begins expression of adhesion molecules, which alert dendritic cells (DCs) and resident macrophages to the presence of a pathogen (Duerst and Morrison 2003). Immature DCs transport HSV antigens to the regional lymph nodes, where the systemic adaptive immune system is alerted. DCs play an especially critical role in bridging innate to adaptive immunity through the presentation of microbial breakdown products to T lymphocytes in the context of the major histocompatibility complex (MHC), thereby initiating specific adaptive immune responses (Khanna et al. 2004). Neutrophils, monocytes, and NK cells destroy virus particles and infected cells and produce antiviral cytokines (TNF-α and IFN-α/β) and other substances such as defensins and nitric oxide (NO) that enhance their antiviral efforts (Hugseyin Baskin et al. 1997; Duerst and Morrison 2003).

Molecular mechanisms involved in the initial host responses to viral infection and recognition of viruses result in the innate immunity activation. The innate immune system utilizes Toll-like receptors (TLRs) to recognize and bind pathogen-associated molecular patterns (PAMPs). TLRs are the transmembrane proteins that detect redundant molecular patterns in broad classes of microbial pathogens. Currently, 11 members of the TLR family (TLR1–TLR11) are known. Recent studies revealed that human cells express TLRs, which are responsive to corresponding ligands including double-stranded RNA (dsRNA), TLR3, selected glycoproteins and lipopeptides (TLRs 1, 2, 4, and 6), and unmethylated cytosine–guanine (CpG) motifs (TLR9) (Bowie and Haga 2005; Herbst-Kralovetzl and Pyles 2006). TLRs alert the host to the presence of a pathogen by initiating intracellular signaling events by recruiting adaptor proteins, such as MyD88 (common to all TLRs), TIRAP (Toll–interleukin [IL]-1 receptor [TIR]-associated protein, TLRs 2 and 4), and TICAM (TIR-containing adaptormolecule-1; TLRs 3 and 4), to the intracellular TIR domain. Sequential phosphorylation and activation events result in nuclear translocation of nuclear factor NF-κB, which stimulates transcription of proinflammatory cytokines, antimicrobial peptides (defensins), and chemokine genes. Increased defensin synthesis is mediated by activation of receptors other than TLRs, such as NOD2, IL-17R, and PAR-2. These receptors are expressed by various cells: professional immune cells, epithelial cells, antigen-presenting cells, and others (Anumba 2005; Andersen et al. 2006).

At least six TLRs are known to be involved in viral detection. HSV glycoproteins interact with TLR2 and may interact with TLR1/2 and TLR2/6 heterodimers. CpG motifs of HSV-1 and HSV-2 stimulate TLR9, which is found in intracellular vesicles. HSV RNA molecules interact with TLR3 and TLR7/8. Our experiments showed that HSV led to the expression of high level of TLR9 by peritoneal mononuclear cells from patients.

The host reaction to infection is a highly regulated response aimed at eliminating the infectious agent from the organism. Cytokines are soluble secreted factors that play important roles in regulation of the immune response. In viral infections, the interferons (IFNs), a special subset of cytokines, have been ascribed particularly important roles (Malmgaard and Paludan 2003). In response to initial chemokine and cytokine production, cellular components of the innate immune system are activated and recruited to the site of infection. These include neutrophils, followed by monocytes and NK cells. Chemokines (IL-8 and others) and cytokines (TNF-α, IFN-γ, and GM-CSF), which are secreted by epithelial cells and resident macrophages in the infected mucosa, mediate neutrophil chemotaxis. In the place of infection, neutrophils secrete antiviral cytokines such as TNF-α and α-defensins. TNF-α, a major product of neutrophils, acts directly against HSV by causing lyses of infected cells (Milligan 1999) and indirectly by inhibiting viral replication. TNF-α also synergizes with IFN-β and IFN-γ in its antiviral activity, reducing HSV-2 yield by 1000-fold in pretreated epithelial cells. Treatment with TNF-α and IFN-β combination during lethal HSV-1 infection increases survival rates up to 70% (Malmgaard and Paludan 2003; Barinskii et al. 2005).

Macrophages become significant sources of inflammatory chemokines and cytokines, including TNF-α, IL-1, IL-6, IL-8, IL-12, IL-18, RANTES/CCL5, and IFNs. NK cells are important for controlling infection during the innate immune response to HSV-1 and HSV- 2. They are recruited to the sites of viral infection and activated within 2–3 days by chemokines and cytokines produced by activated resident cells, including IFNs, IL-12, IL-15, and IL-18. Activated NK cells produce IFN-γ and participate in direct cytolysis of virus-infected cells by perforin/granzyme-mediated processes (Ellermann-Eriksen 2005).

3. Antimicrobial Peptides

Cationic antimicrobial peptides (AMPs) have been gaining recognition as highly valuable therapeutic agents in fighting a wide range of microorganisms (Kokriakov et al. 2006). Some AMPs have been found to be antibacterial, antiviral, antifungal, or promoters of wound healing. They have emerged as central components of mammalian innate defenses and are of fundamental relevance to understanding host–microbe relationships. The importance of AMPs extends beyond their direct antimicrobial activity, as their broad biological activities indicate they are effector molecules providing communication between innate and adaptive immune systems (Zughaier et al. 2005).

AMPs—defensins are divided into three groups: α-, β-, and θ-defensins, based on their structure (Boman 2003; Lehrer and Ganz 2002). α-Defensins are particularly abundant in neutrophils and Paneth cells of the small intestine (Ouellette and Bevins 2001). They are secreted by neutrophils and demonstrate antiviral efficacy, potentially through the insertion into virion lipid envelopes or degradation of phagocytosed virions. In vitro evidence for α-defensin utility in the innate response to HSV-2 derives from experiments, in which inclusion of rabbit α-defensins during infection inactivated the virus by blocking fusion and entry steps of the replication cycle and by preventing cell-to-cell spread (Duerst and Morrison 2003). In vivo antagonism of HSV replication by α-defensins has not yet been demonstrated.

β-Defensins are expressed in several tissues, such as mucosa and skin (Zasloff 2002). hBD-1 is constitutively expressed, whereas hBD-2 and hBD-3 are inducible by bacterial and viral products and cytokines, such as IL-1β or TNF-α. hBD-4 has a more limited distribution than hBD-1, hBD-2, or hBD-3, and its expression can be upregulated by bacterial infection, but not by inflammatory factors, such as IL-1β or TNF-α (Fahlgren et al. 2004). Recently identified hBD-5 and hBD-6 are localized to the epidermis and airways (Boman 2003). Thus, β-defensins play a crucial role in

TABLE 1. Antiviral effect of antimicrobial peptides

Antimicrobial peptides	Viruses	Antiviral effects
HNP-1–HNP-3	Herpes simplex virus 1 and 2 types (Yasin et al. 2000)	Block early steps of viral replication
HNP-1	Cytomegalovirus	
θ-Defensin	Human immunodeficient virus (Garzino-Demo 2007)	Direct inactivation, block the entry (inactivate gp120 HIV)
Protegrins	Herpes simplex virus (Yasin et al. 2000; Koval'chuk et al. 2005)	Lyse viral envelope
Cathelicidin LL37	Vaccinia Virus (Howell et al. 2004)	

host defense against viral infections as constitutive or inducible components in the epithelial barrier (Yasin et al. 2001). θ Defensins have so far been identified only in rhesus monkeys (Garzino-Demo 2007).

There is another group of antimicrobial peptides so-called cathalicidins (Lehrer and Ganz 2002). In humans, this family is represented by cathalicidin LL37. It was shown that this peptide displays a direct antiviral effect by destroying the virus envelope (Table 1). In our study, we have demonstrated that HSV reproduction is inhibited by adding protegrins to cell cultures (Koval'chuk et al. 2005).

4. Treatment of HSV Infections

Nowadays, a wide range of drugs against herpes infection is manufactured, including acyclovir, which blocks viral DNA synthesis. However, since herpes virus is associated with lowering activity of NK and production of IFNs, it is conceivable to use cytokines, such as IFNs, for antiviral therapy. There are three TLR agonists (poly I:C, imiquimod, and CpG ODNs), which have proven efficacy against experimental HSV-2 infections (Pyles et al. 2002). Synthetic versions of these agonists have been tested in human clinical trials for other purposes and, generally, were well tolerated. Synthetic stabilized version of dsRNA triggers TLR3 pathways to elicit production of IFN and other antiherpetic cytokines. TLR3 signaling is mediated by IFN regulatory factor 3, a potent transcriptional regulator of the antiviral immune response, or by the MyD88 pathway. Poly I:C also significantly enhanced disease resolution in animals (Thompson et al. 1996).

Imiquimod is another TLR agonist, studied in experimental and clinical HSV-2 infections. Imidazoquinolines are recognized by TLR7 and TLR8, resulting in local antiviral cytokine elaboration and immune responses. In experimental HSV models, imiquimod reduced recurrent disease when applied therapeutically but failed in clinical trials for recurrent genital HSV disease (Gupta et al. 2004). Pathogen DNA containing unmethylated CpG motifs has been identified as a potent immunostimulator, which induces antiviral Th1 cytokine synthesis. CpG motifs are active in the context of synthetic oligodeoxynucleotides (CpG ODNs) stabilized by phosphorothionate and have been used as adjuvants. The CpG ODNs afford protection against HSV-involved T cells due to the lack of protection in T cell-deficient, but not B cell-deficient, mice. CpG ODNs activate TLR9$^+$ DCs that play key roles in vaginal responses to HSV-2 infection (Harandi et al. 2003).

We have recently demonstrated the effect produced by a complex of native cytokines and cationic antimicrobial peptides (protegrins) extracted from leukocytes from the pig peritoneal blood (CCAP). Our results have shown that CCAP activates both innate and adaptive immunity: stimulates NK cells, phagocytes, and Th1 cells and upregulates production of INF-γ and IL-12 (Koval'chuk et al. 2004). Consequently, the antiviral effect of native cytokines and antimicrobial peptide complex was proved in vitro. CCAP triggered molecular mechanisms suppressing viral reproduction in infected cells: lower synthesis of viral DNA and higher expression of viral thymidine kinase.

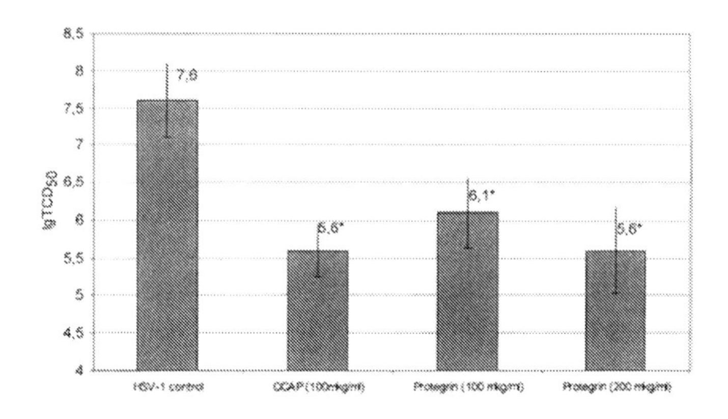

FIGURE 1. Antiviral effect of CCAP (100 μg/ml) and protegrine (100 and 200 μg/ml) (*$p \leq 0.05$).

The major results were obtained using preliminary incubation of CCAP with Vero cells for 24 h: HSV-1 titer was reduced up to 100-fold (Figure 1). This correlated with antiviral activity of protegrins and TNF-α involvement in CCAP, which blocked HVEM (TNFR) and impeded the virus penetration into cells. Cell culture treatment with high molecular weight fractions did not affect the viral titer, although low molecular weight peptides at 100 μg/ml reduced the viral titer by 1.5 lg TCD_{50}/0.1ml. The addition of 200 μg/ml peptides reduced the viral titer by 2.0 lg TCD_{50}/0.1ml versus the control (Figure 1). These experiments demonstrated that incubation of CCAP and HSV-1 decreased the viral titer by 10 times and led to the emergence of defective virions in the viral population. In control cultures, the ratio of virions to nucleocapsids was 1:2.2, although the estimation suggests 1:10 ratio. These results may be explained by the action of antimicrobial peptides on the viral capsid (Howell et al. 2004).

Synergistic effect of CCAP and acyclovir was noticed during clinical investigations of complex therapy of genital herpes infections. Two groups of patients were analyzed: (1) patients treated with CCAP (Superlymph) + acyclovir and (2) patients received acyclovir alone (standard regimen). The outcomes testified for the effectiveness of the combination treatment regimen. This treatment resulted in lengthy remission and less relapse cases. The viral thymidine kinase expression induced by CCAP stimulated the activity of acyclovir. Such an effect was due to the phosphorylation of acyclovir by thymidine kinase.

References

Abrahams, V.M. and Mor, G. (2005) Toll-like receptors and their role in the trophoblast. Placenta 26, 540–547.

Andersen, J.M., Al-Khairy, D. and Ingalls, R.R. (2006) Innate immunity at the mucosal surface: role of toll-like receptor 3 and toll-like receptor 9 in cervical epithelial cell responses to microbial pathogens. Biol. Reprod. 74, 824–831.

Anumba, D.O. (2005) Characterization of toll-like receptors in the female reproductive tract in humans. Hum. Reprod. 20, 1372–1378.

Barinskii, I.F., Posevaia, T.A., Shabalina, N.V. and Nikitina A.A. (2005) Herpes virus infection in patients with chronic glomerulonephritis. Vopr. Virusol. (Russ.) 50, 35–37.

Boman, H. (2003) Antibacterial peptides: basic facts and emerging concepts. J. Intern. Med. 254, 197–215.

Bowie, A.G. and Haga, I.R. (2005) The role of Toll-like receptors in the host response to viruses. Mol. Immunol. 42, 859–867.

Cunningham, A.L., Diefenbach, R.J., Miranda-Saksena, M., Bosnjak, L., Kim, M., Jones, C. and Douglas, M.W. (2006) The cycle of human herpes simplex virus infection: virus transport and immune control. J. Infect. Dis. 194, 11–18.

Duerst, R.J. and Morrison L.A. (2003) Innate immunity to herpes simplex virus type 2. Viral Immunol. 16, 475–490.

Ellermann-Eriksen, S. (2005) Macrophages and cytokines in the early defense against herpes simplex virus. Virology J. 2, 1–30.

Fahlgren, A., Hammarstrom, S., Danielsson, A. and Hammarstrom, M.L. (2004) Beta-Defensin-3 and -4 in intestinal epithelial cells display increased mRNA expression in ulcerative colitis. Clin. Exp. Immunol. 137, 379–385.

Garner, J.A. (2003) Herpes simplex virion entry into and intracellular transport within mammalian cells. Adv. Drug Deliv. Rev. 55, 1497–1513.

Garzino-Demo, A. (2007) Chemokines and defensins as HIV suppressive factors: an evolving story. Curr. Pharm. Des. 13, 163–172.

Grunewald, K., Desai, P., Winkler, D.C., Heymann, J.B., Belnap, D.M., Baumeister, W. and Steven, A.C. (2003) Three-dimensional structure of herpes simplex virus from Cryo–electron tomography. Science 302, 1396–1398.

Gupta, A.K., Cherman, A.M. and Tyring, S.K. (2004) Viral and nonviral uses of imiquimod: a review. J. Cutan. Med. Surg. 8, 338–352.

Harandi, A.M., Eriksson, K. and Holmgren, J. (2003) A protective role of locally administered immunostimulatory CpG oligodeoxynucleotide in a mouse model of genital herpes infection. J. Virol. 77, 953–962.

Herbst-Kralovetz1, M.M. and Pyles, R.B. (2006) Toll-like receptors, innate immunity and HSV pathogenesis. Herpes 13, 37–41.

Hollier, L.M. and Grissom, H. (2005) Human herpes viruses in pregnancy: cytomegalovirus, Epstein-Barr virus, and varicella zoster virus. Clin. Perinatol. 32, 671– 696.

Howell, M., Jones, J. and Kisich, K. (2004) Selective killing of vaccinia virus by LL-37: implications for eczema vaccinatum1. J. Immunol. 172, 1763–1767.

Hugseyin Baskin, H., Ellermann-Eriksen, S., Lovmand, J. and Mogensen, S.C. (1997) Herpes simplex virus type 2 synergizes with interferon-c in the induction of nitric oxide production in mouse macrophages through autocrine secretion of tumour necrosis factor-α. J. Gen. Virol. 78, 195–203.

Khanna, K.M., Lepisto, A.J., Decman, V. and Hendricks, R.L. (2004) Immune control of herpes simplex virus during latency. Curr. Opin. Immunol. 16, 463–469.

Kokriakov, V.N., Koval'chuk, L.V., Aleshina, G.M. and Shamova, O.V. (2006) Cationic antimicrobial peptides as molecular immunity factors: multi-functionality. Zh. Mikrobiol. Epidemiol. Immunobiol. (Russ.) 2, 98–105.

Koval'chuk, L.V., Gankovskaia, L.V., Moroz, A.F., Avedova, T.A. and Ukhina, T.V. (2004) Antistaphylococcal activity of the complex of natural cytokines. Zh. Mikrobiol. Epidemiol. Immunobiol. (Russ.) 1, 55–59.

Koval'chuk, L.V., Lavrov, V.F., Gankovskaia, L.V., Ebralidze, L.K. and Barkevich, O.A. (2005) In vitro inhibition of the cytopathic action of herpes simplex virus, type 1, with a natural cytokine complex. Zh. Mikrobiol. Epidemiol. Immunobiol. J 1, 57–60

Lehrer, R. and Ganz, T. (2002). Cathelicidins: a family of endogenous antimicrobial peptides. Curr. Opin. Hematol. 9, 18–22.

Malmgaard, L. and Paludan, S. (2003) Interferon (IFN)-a/b, interleukin (IL)-12 and IL-18 coordinately induce production of IFN-c during infection with herpes simplex virus type 2. J. Gen. Virol. 84, 2497–2500.

Milligan, G.N. (1999) Neutrophils aid in protection of the vaginal mucosae of immune mice against challenge with herpes simplex virus type 2. J. Virol. 73, 6380–6386.

Ouellette, A.J. and Bevins, C.L. (2001) Paneth cell defensins and innate immunity of the small bowel. Inflamm. Bowel Dis. 7, 43–50.

Pyles, R.B., Higgins, D., Chalk, C., Zalar, A., Eiden, J. and Brown, C. (2002) Use of immunostimulatory sequence containing oligonucleotides as topical therapy for genital herpes simplex virus type 2 infection. J. Virol. 76, 11387–11396.

Spear, P.G. (2004) Herpes simplex virus entry receptors and viral ligands. Cell. Microbiol. 6, 401–410.

Thompson, K.A., Strayer, D.R., Salvato, P.D., Thompson, C.E., Klimas, N. and Molavi, A. (1996) Results of a double-blind placebo-controlled study of the double-stranded RNA drug polyI:polyC12U in the treatment of HIV infection. Eur. J. Clin. Microbiol. Infect. 15, 580–587.

Yasin, B., Pang, M. and Turner, J. (2000) Evaluation of inactivation of infectious HSV by host-defense peptides. Eur. J. Clin. Microbiol. Infect. Dis. 19, 187–194.

Yasin, B., Wang, W., Pang, M., Cheshenko, N., Hong, T., Waring, A.J., Herold, B.C., Wagar, E.A. and Lehrer R.I. (2004) Theta defensins protect cells from infection by herpes simplex virus by inhibiting viral adhesion and entry. J. Virol. 78, 5147–5156.

Zasloff, M. (2002) Antimicrobial peptides of multicellular organism. Nature 415, 389–395.

Zughaier, S.M., Shafer, W.M. and Stephens, D.S. (2005) Antimicrobial peptides and endotoxin inhibit cytokine and nitric oxide release but amplify respiratory burst response in human and murine macrophages. Cell. Microbiol. 7, 1251–1261.

40

Vaccine Containing Natural TLR Ligands Protects from *Salmonella typhimurium* Infection in Mice and Acute Respiratory Infections in Children

Boris Semenov and Vitaly Zverev

Mechnikov Research Institute of Vaccines and Sera, Moscow, Russia, mech.inst@mail.ru

Abstract. It has been shown that a single parenteral administration of vaccine containing bacterial ligands for TLR1, TLR2, TLR4, TLR6, and TLR9 in mice induced rapid (24 h after administration) and effective (100%), but short-term (96 h) protection against lethal challenge with *Salmonella typhimurium*. Repeated mucosal applications of this vaccine stimulated long-term (up to 9 months) protection against acute respiratory infections in children of pre-school age.

1. Introduction

The threats of newly emerging infectious diseases as well as threats of bioterrorism have become one of the major challenges for the 21st century. From 1972 to 1999, 36 previously unknown infectious agents, that are pathogenic for humans, including highly pathogenic avian influenza viruses (H5N1) and human immunodeficiency virus, were isolated and identified (Sergiev et al. 2000).

The hypothesis about the use of innate immunity potentiators for both pre- and postexposure prophylaxis of infections caused by unknown microorganisms is widely discussed in the scientific literature (Hackett 2003; Alibek and Lobanova 2006; Semenov and Zverev 2007). Such nonspecific immunomodulators can activate innate immunity in an antigen-independent manner. A wide spectrum of recombinant, synthetic, and natural immunomodulators was investigated in preclinical and clinical trials. It was shown that the stimulation of innate immunity might provide pre- and post exposure protection against both bacterial and viral infections in laboratory animals (Hackett 2003).

We studied antibacterial protection in mice immunized with vaccine containing natural bacterial ligands for Toll-like receptors (TLRs). New results of immunization with this vaccine with the goal of prevention of acute respiratory infections (ARIs) in children are also discussed in this chapter.

2. Potentiators of Innate Immunity

Polycomponent bacterial vaccine (Immunovak VP-4®) licensed in Russia was used as a potentiator of innate immunity. The vaccine consists of antigen complexes of *Escherichia coli*, *Klebsiella pneumoniae*, *Proteus vulgaris*, and *Staphylococcus aureus*. VP-4 is a strong immunomodulator and was recommended for prophylaxis of infections caused by different microorganisms.

VP-4 contains diverse pathogen-associated molecular patterns, which are recognized by pattern-recognition receptors on cells from the innate immunity arm. This includes lipopeptides and lipoproteins (ligands for TLR1/TLR2 and TLR2/TLR6), lipoteichoic acid (TLR2 ligand), lipopolysaccharides (TLR2 and TLR4 ligands), unmethylated CpG ODN motifs (TLR9 ligand), and peptidoglycans (ligands for TLR2, NOD2).

Experiments on mice revealed that VP-4 stimulated innate immunity (Semenov and Zverev 2007). In fact, it induced maturation of murine dendritic cells (DCs) assessed by the expression of costimulatory molecules CD40, CD80, CD86, and MHC class I and II molecules and their ability to activate resting T cells. Furthermore, VP-4 stimulates the production of both proinflammatory cytokines TNF-α, IL-6, IL-12, and IFN-γ and anti-inflammatory cytokine IL-10.

3. VP-4 Induces Rapid but Short-Lasting Protection Against *Salmonella typhimurium* Infection

In these studies, CBA mice were immunized subcutaneously with VP-4 (400 mg per animal), and 24 h later, animals were infected with 40 LD_{50} of *S. typhimurium* and observed for 8 days. Animals were monitored daily, and lethality (%) was calculated.

The typical results of one of four experiments are presented in Table 1. As summarized in this table, VP-4 protected 100% of mice against *S. typhimurium* infection during a 96-h period, while the lethality in control group was 17% during the first 24 h after infection and 100% during a 96-h period. Lethality of immunized mice was registered from day 5 to day 7.

Thus, obtained results show that a single stimulation of innate immunity by the vaccine containing natural bacterial ligands for TLR induces rapid (24 h) and effective (100%), although short-lived protection against lethal *S. typhimurium* infection.

TABLE 1. VP-4 temporarily reduce lethality (%) due to *Salmonella typhimurium* infection

	Hours after infection						
	24	48	72	96	120	144	168
S. typhimurium	17%	33%	83%	100%			
VP-4 + *S. typhimurium*	0	0	0	0	60%	83%	100%

4. Repeated Mucosal Applications of VP-4 Protect Children from ARIs

ARIs represent a group of diseases with similar clinical features but caused by different pathogenic microorganisms. To date, more than 200 microorganisms are considered to cause ARIs, including ~150 viruses, various bacteria, and their combinations (Ison et al. 2002). Data about etiology of ARI are summarized in Table 2. Evidently, it is impossible to use traditional specific vaccines for preventing ARIs with multiple causes, with the only exception known for influenza infections.

We hypothesized that effective protection against ARIs can be induced by stimulation of innate immunity in the respiratory tract since this is the main entry point of all aerologic infections. Apparently, the stimulation of an innate immune response should be repeated because a single stimulation in the experimental conditions, as above, resulted in only a short-term protection.

In the placebo-controlled trial of VP-4 efficacy, 138 children were immunized and followed up for up to 14 months (Semenov et al. 2000). This study was approved by the Committee on Immunobiologic Preparations, Ministry of Health and Social Development, Russian Federation. ARI was diagnosed on the basis of clinical findings. The vaccine was administered intranasally (1–2 drops) on days 1, 4, and 7 and, then, orally on days 10 (0.5 ml), 13 (1 ml), 16 (2 ml), and on days 19, 22, 25, 28, and 31 (5 ml).

The results of this trial are presented in Table 3. It can be seen that repeated mucosal application of VP-4 vaccine induced a long-term immunity against ARIs.

Index reflecting the efficacy of VP-4 administration (i.e., ratio of ARIs incidence in control group to incidence in immunized group) was 9.2 when calculated 7 months after the completion of vaccinations. Protection against ARIs in immunized children lasted for at least 14 months (the duration of follow-up) but was less effective. Index of efficacy determined 14 months after therapy was only 3.

In another trial, VP-4 was administered to children with 8–10 registered cases of ARI per year (Semenov et al. 2000). Forty children were immunized, and placebo was administered to another 40 patients. The number of incidence of ARI in vaccinated group during 12 months of follow-up was 76% lower than in placebo group. Duration of ARI episode in vaccinated individuals decreased from ~16 to 6.8 days.

Others also reported a 6.3-fold decrease of ARI incidences in children with asthma after mucosal applications of VP-4 (Balabolkin et al. 1998). The periods of highly efficacious protection against ARI lasted for ~3 months followed by an efficacy drop to 2.6–2.9 and remained at this level for up to 9 months.

TABLE 2. Etiology of acute respiratory diseases in humans

Bacteria	Viruses		Combinations of bacteria and viruses
Haemophilus influenzae	Rhinoviruses	>100	
Mycoplasma pneumoniae	Adenoviruses	36	
Staphylococcus spp.	Parainfluenza viruses	4	Different, various
Streptococcus spp.	Coronaviruses	3	
and others	Reoviruses	3	
	Respiratory syncitial virus		

TABLE 3. Repeated mucosal applications of the VP-4 induce long-term protection against acute respiratory infections (ARIs) in children

Trial No.	Duration of follow-up (months)	VP-4	Children	ARI cases	ARI incidence	Index of efficacy[1]
1	7	Yes	138	2	1.4	9.2
		No	155	20	12.9	
2	14	Yes	89	12	13.5	3
		No	60	25	41.6	

[1] Index of efficacy: ratio of ARIs incidence in the control group to the same incidence in the immunized group

5. Conclusions

Presented data show that repeated mucosal applications of the vaccine containing ligands to TLRs stimulate a long-term protection against ARIs in children. It is not known however which mechanism underlies such long-term and, apparently, broad-spectrum preventive effect. It is possible that repeated (with short intervals) stimulation of TLR results in prolonged activation of an innate immunity. In addition, it is possible to suggest a formation of the adaptive immunity to potential causative agents of ARIs that dominated in certain specific populations (i.e., nursery schools).

References

Alibek, K. and Lobanova, K. (2006) Modulation of innate immunity to protect against biological weapon threat. In: B. Anderson, H. Fridman and M. Bendinelli (Eds.), *Microorganisms and Bioterrorism*. Springer Science Business Media, USA, N.Y, N.Y. pp. 39–61.

Balabolkin, I.I., Stepushina, M.A., Egorova, N.B., Krasnoproshina, L.I., Kurbatova, E.A., Skhodova, S.A. and Katosova, L.K. (1998) Use of polycomponent bacterial vaccine for treatment of asthma in children. Int. J. Immunorehabil. (Russ.) 10, 158–164.

Hackett, C.J. (2003) Innate immune activation as a broad-spectrum biodefence strategy: prospects and research challenges. J. Allergy Clin. Immunol. 112, 686–694.

Ison, M.G., Mills, J., Openshaw, P., Zambon, M., Osterhaus, A. and Hayden, F. (2002) Current research on respiratory viral infections: Fourth International Symposium. Antivir. Res. 55, 227–278.

Semenov, B.F., Egorova, N.B., Semenova, I.B. and Kurbatova, E.A. (2000) Therapeutic vaccines. Ros. Med. Vestn. (Russ.)3, 26–32.

Semenov, B.F. and Zverev, V.V. (2007) Concept of inducing of rapid immunological protection against pathogens. Zhurn. Microbiol. (Moscow), 4 (in press).

Sergiev, V.P., Malyshev, M.A. and Drynov, I.D. (2000) *Evolution of Infectious Diseases*. Nauka, Moscow (in Russian).

41

Lymphocyte Subpopulations in Melanoma Patients Treated with Dendritic Cell Vaccines

Zaira G. Kadagidze, Anna A. Borunova, and Tatiana N. Zabotina

N.N. Blokhin Russian Cancer Research Center, Russian Academy of Medical Sciences, Moscow, Russia, kad-zaira@yandex.ru

Abstract. The main goal of cancer immunotherapy is to induce or boost tumor-specific effector cells able to eliminate or reduce tumor progression. In this study, we characterized lymphocyte phenotypes in melanoma patients receiving dendritic cell (DC)-based vaccinotherapy. We found that several biological markers served as unfavorable prognostic factors for patients' response to therapy. This included decrease of CD4$^+$ and CD8$^+$ lymphocyte levels, 10% and higher increase of CD16$^+$CD3$^+$CD8$^+$ lymphocyte population, and increase of CD16$^+$CD8$^+$perforin$^+$ T lymphocytes, especially in combination with decreased levels of CD16$^+$CD8$^-$perforin$^+$ and CD8$^+$CD16$^-$perforin$^+$ cells. Increase in CD8$^+$CD16$^-$perforin$^+$ T lymphocytes with normal levels of CD16$^+$CD8$^-$perforin$^+$ cells and the absence of CD16$^+$CD8$^+$perforin$^+$ and regulatory lymphocytes were shown to be the positive prognostic markers for patients' response to DC vaccines.

1. Introduction

Along with traditional approaches to malignant neoplasm treatment, a great deal of attention is being currently paid to the development of novel biotherapeutic approaches, including vaccinotherapy. According to our recent experimental data, the number of regulatory lymphocytes might reflect the efficacy of vaccinotherapy (Kadagidze et al. 2006). The goal of this study was to characterize dynamic alterations of effector and regulatory lymphocyte subpopulations in patients with melanoma treated with autologous dendritic cells (DCs).

2. Experimental Design

The study was carried out at the Cancer Research Center according to the approved phase I protocol of a clinical trial focusing on autologous DC-based vaccine in patients with melanoma. Thirty-two patients were enrolled in this study: 11 patients were disease free and 21 patients have proven dissemination melanoma before the treatment. Initial immunological testing allowed allocating all patients in additional two groups: patients with reduced CD4/CD8 T cell ratio and patients with normal CD4/CD8 ratio. As a control group, 15 healthy donors were also evaluated.

Peripheral blood mononuclear cells (PBMCs) were washed in PBS containing 0.01% NaN_3 and 0.5% BSA, and 200,000 cells were incubated with fluorescent-labeled monoclonal antibodies for 30 min at 4°C. The following FITC-, PE-, or Cy5-labeled antibodies were used: anti-CD3, CD4, CD8, CD16, CD20, CD11b, and CD28. For the intracellular staining for perforin, cells were permeabilized according to the manufacturer's instructions. Cells were analyzed by FACScan and Lysis II software (BD Biosciences).

3. Results

3.1. Phenotypic Analysis of Lymphocytes in Melanoma Patients from Groups 1 and 2 (Disease-Free Patients Prior to Therapy)

Analysis of T cells in patients from group 1 revealed normal values of $CD3^+$ lymphocytes before and during the process of therapy. Subpopulational analysis of T lymphocytes ($CD3^+CD4^+$ and $CD3^+CD8^+$ cells) in this group of patients also revealed normal values. The ratio between the numbers of $CD3^+CD4^+$ and $CD3^+CD8^+$ lymphocytes, known as the immunoregulatory index, also remained within the normal range (1.29–1.99) before and during the treatment process. The number of $CD3^+$ cells in group 2 patients was significantly lower than the normal value before the therapy and was 54.9 ± 6.3%. At the same time, $CD3^+CD4^+$ subpopulation was 2.6 times lower than the normal value and was 15.4 ± 7.5% ($p < 0.002$). In contrast, the number of $CD3^+CD8^+$ T cells was above the normal value: 36.9 ± 5.9% ($p < 0.002$ vs. healthy donors).

During the process of vaccinotherapy, the level of $CD3^+$ T lymphocytes was slowly increased and reached 64.9 ± 4.5% and 76.3 ± 4.5% after 10 and 15 administrations of DC vaccines, respectively. The number of $CD3^+CD8^+$ cells was stabilized at the level of 34–36% during the vaccination process. Thus, the positive dynamics of $CD3^+$ T lymphocytes during the therapy was due to increased levels of $CD3^+CD4^+$ cells. After 15 vaccine injections, this population reached its normal value (48.7 ± 6.3%). Predominance of $CD3^+CD8^+$ cells over $CD3^+CD4^+$ cells before the vaccination accounted for the low immunoregulatory index (0.43 ± 0.2), which, however, reached its normal value (1.38 ± 0.3) after 15 DC administrations due to elevated levels of $CD4^+$ T lymphocytes.

Phenotypic analysis of NK cells demonstrated that the number of $CD16^+$ lymphocytes before the treatment was within the normal value in both patient groups (15.9 ± 1.6% and 17.1 ± 1.6%, respectively). During the vaccinotherapy, NK cell levels remained within the normal range (15.3–19.6%) in patients with disease stabilization. However, patients in group 2 demonstrated therapy-dependent increase in the number of $CD16^+$ NK cells reaching 23.3 ± 5.6% and 42.3 ± 3.3% ($p < 0.001$) after 3 and 16 vaccine administrations, respectively.

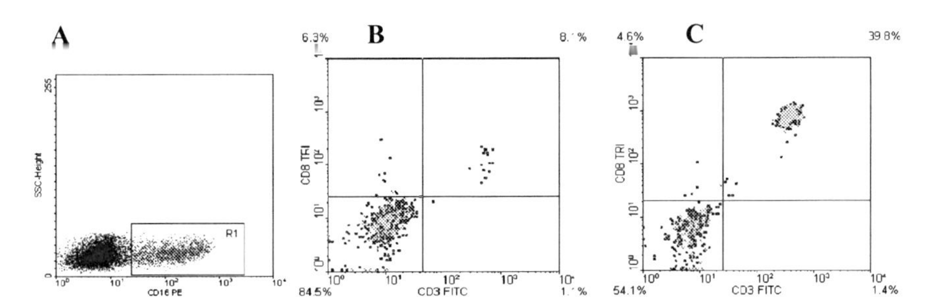

FIGURE 1. Coexpression of CD3 and CD8 Antigens on CD16$^+$ Cells in Melanoma Patients After 15 Administrations of Dendritic Cell (DC) Vaccines. (A) Gating of CD16$^+$ cells. (B) Less than 10% of CD16$^+$ cells coexpressed CD3 and CD8 molecules in melanoma patients from Group 1. (C) Approximately 50% of CD16$^+$ cells were also CD3$^+$CD8$^+$ in melanoma patient with disease progression.

Evaluation of coexpression of CD3, CD8, and CD16 antigens revealed 3.9 ± 1.2% and 2.3 ± 1.5% CD3$^+$CD8$^+$CD16$^+$ lymphocytes in melanoma patients from groups 1 and 2, respectively. During the therapy, this cell subpopulation remained within the normal range (5.2 ± 2.1%) in patients from group 1, while its statistically significantly ($p < 0.001$) increased in patients of group 2 reaching 10.1 ± 5.5% and 26.9 ± 5.2% after 6 and 16 vaccinations, respectively (Figure 1).

Expression of intracellular perforin in a subpopulation of effector lymphocytes has been also investigated. The number of lymphocytes containing perforin in patients from group 1 remained similar to the control values (17.6 ± 7.5) before and during therapy. Proportion of perforin$^+$ lymphocytes in group 2 patients was 1.6 times above the normal value at 28.3 ± 6.8% ($p < 0.002$) before the treatment and increased up to 38.6 ± 6.9% ($p < 0.05$) after six vaccine administrations.

The level of CD8$^+$CD16$^-$ perforin lymphocytes in group 1 patients was within the normal value (7–10%) before and during vaccinotherapy. The proportion of CD8$^+$CD16$^-$ perforin$^+$ cells in group 2 patients was 2.6 times higher than control values before the therapy (21.9 ± 6.4%, $p < 0{,}001$) but decreased to and remained at the level of 10.2 ± 3.3% ($p < 0.001$) after the third DC administration.

Furthermore, perforin population of NK cells in groups 1 and 2 was normal before the therapy (12.1 ± 3.2% and 7.3 ± 3.3%, respectively), but it significantly decreased to 3.1 ± 0.5% in patients from group 2 ($p < 0.001$). The number of perforin$^+$ CD16$^+$CD8$^+$ lymphocytes in group 1 patients did not exceed 1% before and during the vaccinotherapy. This cell subpopulation also accounted for 0.6 ± 0.2% of cells in group 2 patients before the therapy but increased to 19.9 ± 3.6% after 15 DC injections (P < 0.001).

3.2. Phenotypic Analysis of Lymphocytes in Patients with Disseminated Melanoma (Groups 3 and 4)

We demonstrated that the population of CD3$^+$ cells in patients from group 3 considerably exceeded the normal values before and during the treatment process, accounting for 80–87% of cells. CD3$^+$CD4$^+$ lymphocytes were within the normal range (38–45%), whereas the number of CD3$^+$CD8$^+$ cells exceeded the normal levels before the treatment (36.7 ± 4.7%, p<0.001). During the therapy, this subpopulation reached 41.6 ± 4.8% level (p < 0.01) after six DC injections and remained at the same level. The population of CD3$^+$ cells in group 4 patients was at the normal level before and during the treatment process. The number of CD3$^+$CD4$^+$ lymphocytes before the therapy decreased to 32.3 ± 4.3% (p < 0.01) and decreased further to 27.6 ± 5.1% (p < 0,01) after therapy. At the same time, the number of CD3$^+$CD8$^+$ cells exceeded the normal values before the treatment (35.9 ± 5.6%) and further increased to 43.9 ± 5.6% (p < 0.01) after therapy. This redistribution of T-lymphocyte subpopulations resulted in low immunoregulatory index before and during the treatment process.

Next, the level of CD16$^+$ NK cells in group 3 patients was normal (10.1 ± 2.2%) before the therapy and increased to 20.1 ± 4.9% (P < 0.01) after therapy. This increase was due to CD3$^+$CD8$^+$CD16$^+$ cells. In group 4, NK cells were 2.2-fold above the normal level before and during the therapy (30–36%).

The level of perforin cells in both groups was normal: 17.3 ± 3.7% and 17.6 ± 5.4%, respectively. During the therapy, patients from group 3 (stabilization of the disease) showed decrease in perforin$^+$ lymphocyte population to 22.8 ± 5.7% (p < 0.01). The number of perforin lymphocytes in patients with a progressive course of the disease (group 4) renamed within the initial level during the therapy. The number of CD8$^+$CD16$^-$ lymphocytes containing perforin was normal (9.2 ± 2.4%) in group 3 patients before the therapy and increased up to 20.1 ± 3.6% (p < 0.01) after

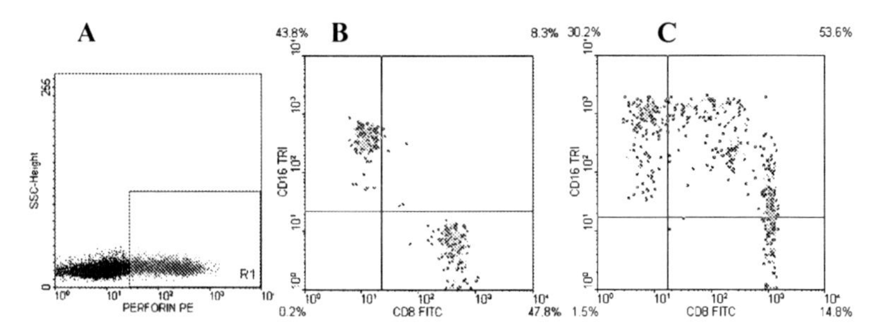

FIGURE 2. Coexpression of CD16 and CD8 antigens on the perforin$^+$ cells in patients with disseminated melanoma before vaccinotherapy. (A) Gating of perforin$^+$cells. (B) Less than 10% of perforin$^+$ cells coexpressed CD8 and CD16 molecules in patients from Group 3. (C) Approximately 50% of perforin$^+$ cells were CD8$^+$CD16$^+$ in patients from Group 3 with the disease progression.

the treatment with DCs. The proportion of $CD8^{+}CD16^{-}$perforin^{+} cells in patients with the progressive course of the disease (group 4) was normal (6.2 ± 3.6%) before the treatment and decreased to 2.5 ± 3.6% ($p < 0.01$) during the treatment process.

The population of perforin NK cells in patients from group 3 was normal (8.1 ± 3.2%) before the therapy and remained at the same level during the treatment. At the same time, the number $CD16^{+}CD8^{-}$perforin^{+} cells in patients with progressive stages of the disease was considerably lower than normal value before and during the treatment process ($p < 0.001$). In addition, the level of $CD16^{+}CD8^{+}$perforin^{+} cells in patients with progressive stages of the disease exceeded normal values (6.1 ± 1.1%, $p < 0{,}001$) before the treatment and progressively elevated during the therapy (Figure 2). The number of $CD16^{+}CD8^{+}$perforin^{+} lymphocytes in group 3 patients did not exceed 1% before and during vaccinotherapy. Thus, increase in perforin^{+} populations in group 3 patients was due to $CD8^{+}CD16^{-}$ and $CD16^{+}CD8^{-}$ lymphocytes, whereas in group 4 patients, it was due to $CD16^{+}CD8^{+}$ cells.

4. Discussion

Although malignant cells may express tumor antigens on their surface, they are often unable to stimulate tumor-specific immune response due to the defect in costimulatory molecules that are required for optimal T cell stimulation (Groscurth and Filgueira 1998). It was reported that immunotherapy with DCs loaded with tumor antigens was able to induce a complete immune response against the tumor as a result of NK cell activation as well as tumor-specific T lymphocyte activity (Kim et al. 1999). Our data revealed increased numbers of $CD8^{+}$ perforin T lymphocytes in the peripheral blood in melanoma patients during vaccinotherapy with DCs regardless of the clinical course of the disease. This, however, was not the case for melanoma patients with the disease progression. These patients showed 10-fold decrease in the number of $CD8^{+}$perforin^{+} cells. Smyth et al. (2000) have shown that tumor dissemination is mainly controlled by perforin-dependent cytotoxicity mediated by NK cells. There was an inverse correlation between the cytotoxic activity and tumor cell dissemination. Apparently, the observed interdependence between an increase in $CD16^{+}CD3^{+}CD8^{+}$ lymphocytes and progression of the main disease is caused not by the functional activity of these cells, but by the redistribution of effector lymphocytes. The population of perforin^{+} lymphocytes is heterogeneous and consists of $CD8^{+}CD16^{-}$, $CD16^{+}CD8^{-}$, and $CD16^{+}CD8^{+}$ subpopulations of cells. Increase in $CD16^{+}CD8^{+}$ lymphocytes in patients with the progressive stage of the disease took place with simultaneous decrease in $CD16^{+}CD8^{-}$ cell population. At the same time, $CD8^{+}CD16^{-}$perforin^{+} lymphocytes in patients with the clinical signs of cancer progression were three times higher than normal values before the vaccinotherapy but progressively decreased after repetitive administrations of DC vaccines.

Thus, increase in perforin^{+} cell populations in patients with the progressive disease was due to $CD16^{+}CD3^{+}CD8^{+}$ cells. We speculate that the simultaneous presence of two effector perforin subpopulations—$CD8^{+}CD16^{-}$ and $CD16^{+}CD8^{-}$ cells—is essential for an effective antitumor immune response.

5. Conclusions

Our data suggest that decrease in $CD4^+/CD8^+$ lymphocyte ratio, increase of $CD16^+CD3^+CD8^+$ lymphocytes, and increase of $CD16^+CD8^+perforin^+$ T lymphocytes, especially in combination with decreased levels of $CD16^+CD8^-perforin^+$ and $CD8^+CD16^-perforin^+$ cells, may serve as bad prognostic markers for melanoma patients and their response to immunotherapy. Increased levels of $CD8^+CD16^-perforin^+$ T lymphocytes with normal levels of $CD16^+CD8^-perforin^+$ cells and the absence of $CD16^+CD8^+perforin^+$ lymphocytes may be considered as favorable markers of the patients' response to immunotherapy.

Acknowledgments

We thank Drs. L.V. Demidov, G.Z. Chkadua, and I.N. Mikhailova (Russian Cancer Research Center) for providing vaccinotherapy.

References

Groscurth, P. and Filgueira, L. (1998) Killing mechanisms of cytotoxic T lymphocytes. News Physiol. Sci. 13, 17–21.

Kadagidze, Z.G., Chertkova, A.I. and Slavina, E.G. (2006) Immunoregulatory CD25+CD4+ cells. Russ. Biother. J., 7, 13–21.

Kim, K.D., Kim, J.K., Kim, S.J., Choe, I.S., Chung, T.H., Choe, Y.K. and Lim, J.S. (1999) Protective antitumor activity through dendritic cell immunization is mediated by NK cell as well as CTL activation. Arch. Pharm. Res. 22, 340–347.

Smyth, M.J., Thia, K.Y., Street, S.E., Cretney, E., Trapani, J.A., Taniguchi, M., Kawano, T., Pelikan, S.B., Crowe, N.Y. and Godfrey, D.I. (2000) Differential tumor surveillance by natural killer (NK) and NKT cells. J. Exp. Med. 191, 661–668.

42

Cell Technologies in Immunotherapy of Cancer

Vladimir Moiseyenko[1], Evgeny Imyanitov[2], Anna Danilova[1], Alexey Danilov[1], and Irina Baldueva[1]

[1] Prof. N.N. Petrov Research Institute of Oncology, Department of Biotherapy and Bone Marrow Transplantation, Saint-Petersburg, Russia, fedor@home.ru
[2] Prof. N.N. Petrov Research Institute of Oncology, Department of Molecular Genetics, Saint-Petersburg, Russia

Abstract. Tumor growth is accompanied by active immune reactions even on the early stages. Vaccine therapy implies the use of single antigen or combination of antigens, either with or without adjuvants, for the modulation of immune response. N.N. Petrov Institute of Oncology joined the field of antitumor vaccine therapy and related cellular technologies in 1998. The following activities are held: (1) Optimization of the preparation of autologous and allogeneic antitumor vaccines and development of tumor cell culture bank for the experiments on allogeneic vaccination. (2) Clinical evaluation of autologous vaccine therapy by (a) bone marrow precursors of dendritic cells (DCs), which are loaded with tumor lysates; (b) genetically modified tumor cells; (c) intact tumor cells used in combination with various adjuvants (BCG, IL-1β, and IL-1β combined with low doses of cyclophosphamide) in patients with disseminated melanoma, metastatic kidney cancer, and colorectal cancer. Total 117 patients have received non-modified vaccine (48 patients: 2–6 intracutaneous BCG injections; 54 patients: 4–6 intracutaneous IL-1β injections; 15 patients: up to 6 injections of IL-1β in combination with low doses of cyclophosphamide). Clinical trial of genetically modified vaccine included 59 patients (clinical results: 1 PR (partial response) / 8 SD (disease stabilization) – melanoma, 2 PR/ 2 MR (minimal response) / 3 SD – renal cancer). Vaccine prepared from tumor cell-activated DC bone marrow precursors was administered to 18 patients (clinical results: 2 MR and 6 SD).

1. Introduction

Despite impressive progress in surgical oncology and cancer therapy, recent decades did not bring dramatic changes in the outcome of patients suffering from locally advanced or metastatic neoplastic diseases (Sinkovics and Horvath 2000). Unfortunately, patients with potentially incurable tumor stages constitute a significant portion of cancer patients; therefore, the search for novel treatment strategies remains a priority for life science. Cell-based technologies hold a great promise for activities in this field.

So-called biotherapy of cancer holds particularly prominent position in cancer research. Biotherapy approaches include the ones that are based on either stimulating in-built human defense mechanisms (e.g., immunity) or external delivery of natural protective molecules (i.e., antibodies or cytokines) (Rosenberg 1997). It is indeed difficult to classify existing biotherapy approaches. Nevertheless, most of currently applied strategies fall within either "passive" therapy, that is, supplementation of the patient by apparently missing biologically active molecules, or "active" therapy aimed to stimulate natural defensive resources of the human body. In addition, there are some so-called indirect approaches, which are based on the inhibition of some natural tumor-promoting factors (Ribas et al. 2003).

2. Results and Discussion

N.N. Petrov Institute of Oncology joined the field of antitumor vaccine therapy and related cellular technologies in 1998. The following activities are held: (1) Optimization of autologous and allogeneic antitumor vaccine preparations and development of tumor cell bank for the experiments on allogeneic vaccination. (2) Clinical evaluation of autologous vaccine therapy by (a) bone marrow precursors of dendritic cells (DCs), which are activated by tumor lysate; (b) genetically modified tumor cells; and (c) intact tumor cells used in combination with various adjuvants (BCG, IL-1β, and IL-1β combined with low doses of cyclophosphamide) in patients with disseminated noma, metastatic kidney cancer, and colorectal cancer.

Due to genetic instability, tumor mass becomes highly heterogeneous during neoplastic progression. This heterogeneity is also applicable to the expression of specific antigens. Natural tumor cell selection leads to the changes in antigenic portrait and appears to be the main mechanism of evasion from the immune system surveillance. Tumor evolution is a serious obstacle for the efficacy of cancer vaccines (Moingeon 2001). If the treatment duration is long, or if the allogeneic vaccination is foreseen, long-living cultures are employed. In the latter case, cells cultivated ex vivo may significantly change the pattern of antigen expression. The divergence of antigenic properties of ex vivo cultivated cells and patients' metastases may seriously compromise the clinical efficiency of tumor vaccination. Our experiments have shown that melanoma cells may lose significant amount of antigens when cultivated in vitro for a long time. Indeed, while antigenic portrait was well preserved during early passages (1–5), further cultivation was accompanied by increasing both cellular heterogeneity and mitotic cell (Ki-67) count. By 25–30 passages, the following proportion of the cells expressed melanoma-specific antigens: CD63 25.7%, melan A/MART1 2.7%, tyrosinase 7.6%, MAGE1 10.0%, MITF 42.1%, S100 2.9%, and gp100 7.8%. It is known that melanoma antigen profile changes upon tumor progression (Barrow et al. 2006) and the expression of melanoma - associated antigens (MAAs) correlates with patients' survival (Murer et al. 2004). Therefore, accounting for MAA expression is essential for the tumor vaccine development.

Total 308 patients have been recruited for cancer vaccine trial in N.N. Petrov Institute of Oncology during 1998–2006. The following tumor types were included: melanoma 196, kidney cancer 108, colorectal cancer 3 and prostate cancer 1. Mean age of the patients was 53.7 years (time interval: 21–79 years). Total 117 patients received non-modified vaccine (48 patients: 2–6 intracutaneous BCG injections; 54 patients: 4–6 intracutaneous IL-1β injections; and 15 patients: up to 6 injections of IL-1β + low doses of cyclophosphamide. Clinical trial of genetically modified vaccine included 59 patients: 54 patients received 2–10 injections of the *tag7/PGRP-S* gene modified autologous tumor cells and 5 patients received 18–35 injections of this vaccine (Table 1).

Utilization of novel strategies, such as tumor cell modification by the cytokine-like *tag7/PGRP-S* gene (Kiselev et al. 1998) as well as the use of IL-1β (Veltri and Smith 1996) as immune-stimulating adjuvant, provides new opportunities to vaccine therapy for cancer.

Combination of IL-1β with vaccine administration has been proven to be nontoxic. Immune response, which was determined by peripheral blood tests, was observed in 90% of the patients. Disease stabilization was documented in 10 of 20 patients with metastases (6 for melanoma and 4 for kidney cancer); mean time to

TABLE 1. Results of clinical evaluation of genetically modified vaccines and vaccines combined with adjuvants

Autologous vaccine	Number of patients	Tumor type	Outcome
tag7/PGRP-S gene modified tumor cells, intracutaneous injection	45	Melanoma	1/31 PR / 8/31 SD
	14	Kidney cancer	2/10 PR/ 2/10 MR/ 3/10 SD
Tumor cells with BCG adjuvant, intracutaneous injection	10	Melanoma	3 SD
	25	Kidney cancer	1 PR / 5 SD
	8	Prostate cancer	1 SD
	3	Colon cancer	No effect
	1	Schwannoma	No effect
	1	Lung cancer	1 SD
Tumor cells with IL-1β adjuvant, intracutaneous injection	35	Melanoma	16 SD
	19	Kidney cancer	14 SD
Tumor cells with IL-1β adjuvant in combination with low-dose cyclophosphamide, intracutaneous injection	7	Melanoma	3 SD
	8	Kidney cancer	3 SD

MR, minimal response (less than 50% reduction of tumor mass, or increase of tumor lumps); PR, partial response (more than 50% reduction of tumor mass, absence of new metastases); SD, disease stabilization (less than 50% reduction of tumor mass, or increase of tumor lumps by no more than 25%; absence of new metastases).

progression approached to 6 months. Total 34 patients received this combination for the adjuvant treatment and 20 of vaccinated patients (10 with melanoma and 10 with kidney cancer) remained disease-free during the follow-up (mean: 16.4 months).

Use of the *tag7/PGRP-S* gene-modified vaccine was also proven to be safe. About 31 of 42 (74%) melanoma patients and 13 of 14 (93%) kidney cancer patients developed the delayed-type hypersensitivity (DTH) reaction. Three melanoma patients had the DTH reaction at the site of control injection, which included genetically nonmodified cells. This phenomenon was observed after 1 or 2 cycles of vaccine therapy and was interpreted as a "bystander effect." Correlation between intensity of DTH reaction (size and degree of hyperemia) and clinical effect was observed in most cases. In the majority of patients with disease stabilization, the diameter of DTH spot was 15–20 mm or larger. Patients with disease progression did not respond to the vaccination by DTH, or the latter was weak.

When peripheral blood was tested for the signs of immune response, appropriate changes were observed in 95% of patients. Almost all patients with clinical effect were characterized by the increase of absolute counts for $CD3^+$, $CD4^+$, $CD8^+$, and $CD16^+$ lymphocytes. Functional activity of CD8 lymphocytes was increased by the end of each vaccine therapy cycle.

When clinical evaluation was applied, no complete responses have been detected. Partial response (reduction of metastatic lumps by more than 50%) was observed in 2 of 10 patients with kidney cancer (duration of the effect was 23 months and 9 months, respectively) and 1 patient with skin melanoma (duration of the effect was 36 months). Disease stabilization was recorded for 11 (27%) patients (8 cases of skin melanoma and 3 cases of kidney cancer) (Table 1). Mean duration of the effect was 6 months (2–24 months). Two patients with kidney cancer had minimal regression (tumor reduction by less than 50%) of metastases in the lungs (duration of the effect: 12 months) and lymph nodes (duration of the effect: 4 months).

Theoretical assumptions suggest that bone marrow precursors of DCs may be particularly suitable for autologous vaccine therapy in terms of safety, efficacy, and feasibility. Bone marrow precursors of DCs have certain advantages in comparison with monocyte-derived DCs, as they preserve more options for further differentiation. We attempted to optimize the condition of accumulation of DC bone marrow precursors in cell cultures. It was revealed that DCs detach from the cell dish upon maturation and activation, while nonmature cells remain attached. Maximal detachment (75%) was observed when tumor cell lysates (0.4×10^6 cells/ml) was used together with TNF-α (25 ng/ml) and IL-1β (30 ng/ml). About 60–80% of detached cells expressed markers of activated DCs, such as HLA DR^{high}, $CD1a^{low}$, $CD80^{high}$, $CD86^{high}$, $CD40^{high}$, and $CD54^{high}$. Vaccine prepared from tumor cell-activated DC bone marrow precursors was administered (3×10^6/kg) to 18 patients (9 with kidney cancer, 7 with melanoma, and 2 with colorectal cancer) (Table 2). Several methods of vaccine delivery have been studied.

Six patients received intravenous injections of the vaccine. Complete or partial responses have not been observed. Disease stabilization for 3–4 months was documented in two patients, and tumor progression was noticed in remaining four

TABLE 2. Results of clinical evaluation of bone marrow precursors of DC vaccine

DC vaccine	DC dose $\times 10^6$/кг	Number of patients	Immune response (DHT)	Outcome
Intravenous injection	3	6	-	2 SD (3 and 4 mo.)
Intracutaneous or subcutaneous injection	3	5	2	2 SD (4 and 5 mo.)
Intracutaneous injection in combination with low doses of cytostatic drugs	3	2	0	No effect
Intracutaneous injection in combination with IL-1β	3	1	1	MR (12 mo.)
Intracutaneous injection in combination with IL-2	3	2	2	2 SD (7 and 16 mo.)
Intracutaneous injection in combination with IL-1β and IL-2	3	1	1	MR (36+ mo.)
Intralymphatic injection	1	1	1	No effect
Intralymphatic injection in combination with IL-2	1	1	1	No effect

MR, minimal response; SD, disease stabilization.

patients. The adverse effects included grade I fever during vaccine injections (one case), grade II fever during 4 days after vaccine administration (three patients), grade II hypotonia during vaccine injections (one case), grade II fatigue during treatment cycle (one case), grade I increase of platelets count, with the duration of 14 days (one case), grade I increase of blood transaminase concentration with the duration of 6 days (one case), and grade II increase of blood transaminase concentration with the duration of 14 days (one case). No grade III–IV adverse events have been detected.

Five patients received 2–12 subcutaneous and/or intracutaneous injections of DC vaccine. No adverse effects were observed. DHT reaction was documented in three of of five cases. Disease stabilization was achieved in two patients (4 and 5 months, respectively). In two patients, the administration of vaccine was combined with low doses of cytostatic drugs (5-FU 500 mg plus cisplatin 20 mg at days 1–5, followed by DC vaccine injection, or cyclophosphamide 300 mg/m^2 3 days prior to the vaccine administration). Four vaccine injections were performed. No toxic reactions were noticed. DTH reaction was not achieved, and tumor progression was observed in both cases. One patient received intracutaneous injection of DC vaccine in combination with IL-1β. Grade 1 fever and asthenia were observed within the first hour after injection but were easily compensated. Each vaccine administration was accompanied by DTH reaction of 2–14 mm in diameter. Minimal regress (<50%) of lung metastases was achieved for the period of 12 months. Two patients received intracutaneous vaccines in combination with IL-2. One patient experienced grade I headache, fever, arthralgia, and bone pains within several hours after injection. Both patients developed DTH reaction (2–20 mm). Disease stabilization was documented in both cases with the duration 7 and 16 months, respectively. Another patient has received the vaccine in combination with local administration of IL-1β and systemic administration of IL-2 (18 million units for 5 days). There was a DTH reaction at the sites of vaccine injection. Minimal regression was achieved with the duration 36+ months. Two patients have received intralymphatic injection of the vaccine. In one case, IL-2 was administered as well. There were no adverse effects; however, tumor progression occurred in both patients. The analysis of immune status revealed an activation of T cells ($CD3^+$), cytotoxic lymphocytes ($CD8^+$), and monocytes, whereas the status of humoral immunity was not changed significantly.

Thus, despite limited numbers of patients enrolled (18 cases), noticeable number of minimal responses (2 cases) and disease stabilizations (10 cases) was observed. The fact of persisting T cell stimulation suggests that higher immunogenicity of DC vaccine can be achieved, for example, with the use of adjuvants. The efficiency of the existing vaccine appears to be sufficient in order to recommend it to patients with preserved immune status.

3. Conclusions

In conclusion, all studied vaccines demonstrated certain degrees of immunological and clinical effects. It is important to emphasize that patients included in the trial suffered from chemo- and radio-resistant disseminated forms of neoplastic diseases. Use of immune therapy in less complicated cases, for example, patients with radically resected tumors or small residual mass left after cytoreductive surgery, appears to be promising.

Molecular and cell-based technologies hold a great promise for cancer therapy. These methods combine high efficacy and good safety profile. However, application of these methods requires careful preclinical studies as well as some support from clinical oncologists and general community.

References

Barrow, C., Browning, J., MacGregor, D., Davis, I.D., Sturrock, S., Jungbluth, A.A. and Cebon, J. (2006) Tumor antigen expression in melanoma varies according to antigen and stage. Clin. Cancer Res. 12, 764–771.

Kiselev, S.L., Kustikova, O.S. and Korobko, E.V. (1998) Molecular cloning and characterization of the mouse tag 7 gene encoding a novel cytokine. J. Biol. Chem. 273, 18633–18639.

Moingeon, P. (2001) Cancer vaccines. Vaccine 19, 1305–1326.

Murer, K., Urosevic, M., Willers, J., Selvam, P., Laine, E., Burg, G. and Dummer, R. (2004) Expression of Melan-A/MART-1 in primary melanoma cell cultures has prognostic implication in metastatic melanoma patients. Melanoma Res. 14, 257–262.

Ribas, A., Butterfield, L.H., Glaspy, J.A. and Economou, J.S. (2003) Current developments in cancer vaccines and cellular immunotherapy. J. Clin. Oncol. 21, 2415–2432.
Rosenberg, S.A. (1997) Principles of cancer management: biologic therapy. In: J.V.T. De Vita, S. Hellman and S.A. Rosenberg (Eds.), *Cancer: Principles and Practice of Oncology.* Lippincott-Raven Publishers, Philadelphia, pp. 349–373.
Sinkovics, J.G. and Horvath, J.C. (2000) Vaccination against human cancers. Int. J. Oncol. 16, 81–96.
Veltri, S. and Smith, J.W. (1996) Interleukin 1 trial in cancer patients: a review of the toxicity, antitumor and haematopoietic effects. Oncologist 1, 190–200.

43

Therapeutic Potential of Cannabinoid-Based Drugs

Thomas W. Klein and Catherine A. Newton

University of South Florida, Department of Molecular Medicine, Tampa, FL, USA, tklein@health.usf.edu

Abstract. Cannabinoid-based drugs modeled on cannabinoids originally isolated from marijuana are now known to significantly impact the functioning of the endocannabinoid system of mammals. This system operates not only in the brain but also in organs and tissues in the periphery including the immune system. Natural and synthetic cannabinoids are tricyclic terpenes, whereas the endogenous physiological ligands are eicosanoids. Several receptors for these compounds have been extensively described, CB_1 and CB_2, and are G protein-coupled receptors; however, cannabinoid-based drugs are also demonstrated to function independently of these receptors. Cannabinoids regulate many physiological functions and their impact on immunity is generally antiinflammatory as powerful modulators of the cytokine cascade. This anti-inflammatory potency has led to the testing of these drugs in chronic inflammatory laboratory paradigms and even in some human diseases. Psychoactive and nonpsychoactive cannabinoid-based drugs such as Δ^9-tetrahydrocannabinol, cannabidiol, HU-211, and ajulemic acid have been tested and found moderately effective in clinical trials of multiple sclerosis, traumatic brain injury, arthritis, and neuropathic pain. Furthermore, although clinical trials are not yet reported, preclinical data with cannabinoid-based drugs suggest efficacy in other inflammatory diseases such as inflammatory bowel disease, Alzheimer's disease, atherosclerosis, and osteoporosis.

1. Introduction

Historically, cannabis/marijuana has been recognized as an anti-inflammatory drug. For example, it was reported as a therapeutic for rheumatism in the third millennium BC by the Chinese emperor Shen-nung and was recommended by British physicians in the 19th century not only as an appetite stimulant and analgesic but also as a therapeutic in treating various infections, chronic cough, stomach pain, and convulsions (Tomida et al. 2004). The physiological basis for these therapeutic effects was not understood until the discovery of the endocannabinoid system of receptors and ligands in various organs and tissues in the body. It is now clear that this system is operating not only in the brain but also in the periphery in tissues such as joint, bone, spleen, gut, and vessels and is believed to help control the normal tone of these tissues. Therefore, cannabinoid-based drugs might be of value in the management of disease processes in these tissues. This chapter describes the endocannabinoid system and the general anti-inflammatory effects of cannabinoid-based drugs and then discusses current and future therapeutic applications of these agents.

2. Endocannabinoid System

2.1. Cannabinoid-Based Drugs

The main psychoactive substance in *Cannabis* extracts was isolated and synthesized in 1964 and shown to be Δ^9-tetrahydrocannabinol (THC) (Gaoni and Mechoulam 1964). THC and cannabinoids are tricyclic ring structures containing a phenol ring with attached 5-carbon alkyl chain, a central pyran ring, and a monounsaturated cyclohexyl ring; the structure and the function of THC and related compounds have been recently and extensively reviewed (Pertwee 2005; Pacher et al. 2006). Besides THC, other natural *Cannabis* products have been extracted and characterized such as Δ^8-THC (also psychoactive), cannabinol, and cannabidiol (both nonpsychoactive); in addition, synthetic cannabinoid derivatives have been produced and studied such as CP55,940, HU-210, HU-211, ajulemic acid, and abnormal cannabidiol (Figure 1). The defined physiological and pharmacological effects of marijuana cannabinoids

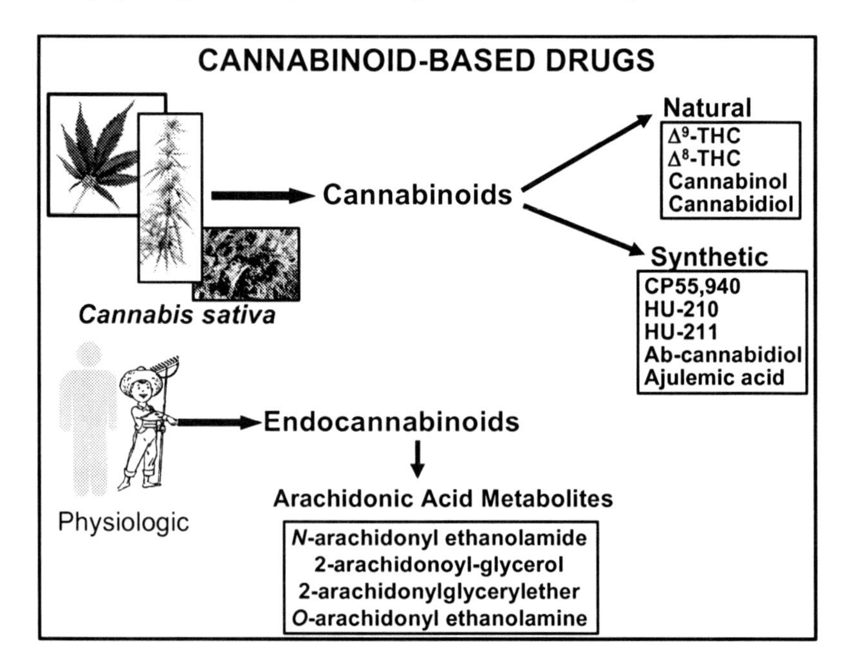

FIGURE 1. Cannabinoid-based drugs are either derived from the *Cannabis* plant or produced in the body from arachidonic acid. Many but not all have affinity for cannabinoid receptors. See text for explanation.

suggested the existence of an endogenous ligand with similar activity and the isolation of the first of these was reported in 1992 (Devane et al. 1992). This so-called endocannabinoid was demonstrated to be the arachidonic acid derivative, *N*-arachidonoyl ethanolamide (AEA), and since its discovery, several other similar compounds have been isolated and extensively studied including 2-arachidonoylglycerol (2-AG) (Mechoulam et al. 1995), 2-arachidonylglycerylether (noladin ether) (Hanus et al. 2001), and *O*-arachidonoyl ethanolamine (virohdamine) (Porter et al. 2002). Most of the cannabinoid-based drugs and endogenous ligands have immunomodulating activity and most, but not all, bind to one or both of the two known cannabinoid receptors, (see below). However, some, such as cannabidiol and HU211, have relatively low affinity for cannabinoid receptors, suggesting that other receptors or molecular mechanisms are involved in the action of these agents. A final class of cannabinoid-based drugs is the receptor antagonists/inverse agonist class. These are structurally quite different from cannabinoids and endocannabinoids and have played an important role in understanding the biology of the endocannabinoid system. A few examples of these are SR141716A (Rimonabant®), AM251, and SR144528 (Fowler et al. 2005).

2.2. Cannabinoid Receptors

Receptor pharmacology studies in the 1980s strongly suggested the existence of specific receptors in the brain mediating cannabinoid effects. In 1990, the first receptor (CB_1) was cloned from a rat brain cDNA library and the predicted amino acid sequence identified it as a seven-transmembrane G protein-coupled receptor (Matsuda et al. 1990). The human homolog of the receptor was cloned in 1991 (Gerard et al. 1991) and a second receptor, CB_2, was cloned in 1993 from the leukocyte cell line, HL60, and rat spleen (Munro et al. 1993). Both receptors are coupled through G_i and G_0 and inhibit adenylyl cyclase as well as a variety of other second messenger and signaling components found in neural and immune tissues (Klein et al. 2003; Howlett et al. 2004). CB_1 receptor orthologs have been demonstrated in many species from invertebrate sea squirt to humans and the structure is well conserved among these organisms (Anday and Mercier 2005). On the contrary, the structure of CB_2 is less conserved among species (Anday and Mercier 2005); however, critical similarities persist because many different cannabinoid ligands activate both receptors with similar affinities. The tissue distribution of CB_1 and CB_2 varies with CB_1 expressed extensively in the CNS with lower levels expressed in the heart, vessels, testis, immune system, and the peripheral nervous system. CB_2, on the contrary, shows little expression in nervous tissue and is primarily restricted to expression in the immune system. This type of distribution supports the prospect that anti-inflammatory and immunosuppressive CB_2-selective drugs with low psychoactivity can be developed for the management of chronic inflammatory diseases.

Besides CB_1 and CB_2, other receptors are stimulated by cannabinoid ligands. For example, the endocannabinoid, AEA, and cannabidiol have been shown to activate the transient receptor potential vanilloid 1 (TRPV1) in vascular tissues and neural circuits (Zygmunt et al. 1999). In addition, other poorly defined receptors have been suggested from studies with CB_1- and CB_2-knockout mice and in pharmacological studies wherein agonists for CB_1 and CB_2 displayed split potency. For example, in rat mesenteric arteries, AEA and abnormal cannabidiol (which has low affinity for CB_1 receptors) induced vasodilation that was inhibited by the CB_1 antagonist, SR141716A; however, other potent CB_1 agonists had no effect (Jarai et al. 1999; Offertaler et al. 2003). Studies with knockout mice also suggested a third receptor because signaling activity in brain membrane preparations occurred equally well in CB_1 knockout and wild-type mice (Breivogel et al. 2001). Furthermore, in an immune cell model, IL-12 production was suppressed equally by THC in CB_1-and CB_2-knockout mice and even the addition of receptor antagonists to knockout cells only partially attenuated the THC suppressive effect (Lu et al. 2006). A candidate for the third receptor was recently reported based on *in silico* patent searching (Baker, et al. 2006). The human gene for this receptor, termed GPR55, encodes a G protein-coupled receptor with very low identity (10–15%) to CB_1 and CB_2 and having binding affinity for only some but not all cannabinoid agonists and antagonists. Orthologs of the human gene have been identified in other mammals but not in lower forms such as avian and teleosts, suggesting later evolution of the gene. The tissue distribution of GPR55 transcripts is similar to that of cannabinoid receptors, and there appear to be some similarities in the vasodilation response between GPR55 and the unknown, abnormal-cannabidiol receptor; however, GPR55 does not appear to couple to G_i. At this point, it can be concluded that GPR55 may represent a class of receptors with overlapping affinity for endocannabinoids and other related lipid-based ligands and that additional studies are needed to define the function of these receptors and their relationship to CB_1 and CB_2.

Besides ligation and activity through G protein-coupled receptors, the action of cannabinoid-based drugs has been linked to interaction with other receptors such as peroxisome proliferator-activated receptors (PPARs). The members of the PPAR family of transcription factors were initially shown to be involved in regulating genes important in lipid metabolism; however, they are now known to also be involved in the regulation of inflammatory genes (Sun et al. 2006). The synthetic analog of THC-11-oic acid, ajulemic acid, was the first cannabinoid-based drug reported to interact with these proteins and was shown to bind to and activate PPARγ leading to changes in cell function including inhibition of the IL-8 promoter (Liu et al. 2003). The endocannabinoid, AEA, has also been reported to induce PPARγ transcriptional activation, and recently several cannabinoid-based drugs have been shown to interact with another PPAR, PPARα (Sun et al. 2006). What is important about these findings is that, as documented below, cannabinoid-based drugs with low affinity for CB_1 and CB_2 have quite marked anti-inflammatory activity and even some of the anti-inflammatory activity of drugs that bind cannabinoid receptors appears to be mediated by mechanisms other than CB_1 and CB_2 (Lu et al. 2006).

3. Anti-inflammatory Effects of Cannabinoids

The early experimental data involving marijuana cannabinoids and immunity suggested the drugs were immunosuppressive. For example, in 1974 it was reported that T cell responses to mitogens were suppressed in cultured peripheral blood mononu

TABLE 1. Cannabinoid-based drugs increase cytokines

Cannabinoid	Receptor	Model	Cytokine stimulus	Cytokines
Humans				
2-AG	CB2	HL-60	2-AG	CXCL8, CCL2
CP55,940	CB2	HL-60	CP55, 940	TNF, CXCL8, CCL2, CCL4
Marijuana smoking	ND	Peripheral blood monocytes	PHA, ConA	IL-10, TGF-β
Mice				
WIN55, 212-2 or HU-210	ND	Splenocytes	LPS, *Propionibacterium acnes*	IL-10
THC	ND	Splenocytes	*Legionella pneumophila*	TNF, IL-6
THC	ND	Peritoneal macrophages	LPS	IL-1α, IL-1β
THC	ND	Splenocytes	*Legionella pneumophila*	IL-4
THC	ND	Splenocytes	Tumor model	IL-10, TGF-β

References: (Klein et al. 1993; Zhu et al. 1994; Derocq et al. 2000; Klein et al. 2006; Smith et al. 2000; Zhu et al. 2000; Pacifici et al. 2003; Kishimoto et al. 2004).

clear cells (PBMCs) taken from chronic marijuana smokers (Nahas et al. 1974). Since then, cannabinoid-based drugs have been shown to suppress the function of T cells, B cells, macrophages, NK cells, and dendritic cells. The studies in this area have been numerous and in vivo and in vitro animal systems from humans to mice have been explored and extensively reviewed over the years (Cabral and Dove Pettit 1998; Klein et al. 1998, 2000a, 2003; Klein 2005; Klein and Cabral 2006). However, more recently, drug effects on the cytokine network have been focused on and in these reports the drugs have been shown to modulate cytokine production, suggesting their therapeutic potential for the management of chronic inflammatory diseases that are often due to a dysregulation of the cytokine network. Tables 1 and 2 are summaries of recent results from mouse and human studies showing that, depending on the system studied, cannabinoid-based drugs either increase or decrease various cytokines. From the tables, it is apparent that most studies show a decrease in proinflammatory cytokines such as IL-1, IL-2, IL-6, IL-12, TNF, GM-CSF, and IFN-γ production; however, increases in anti-inflammatory cytokines such as IL-4, IL-10, and TGF-β are also observed potentially contributing in some systems to an overall decrease in inflammatory tone. Based on these data, clinical trials have been initiated examining the safety and efficacy of these drugs in chronic inflammatory diseases.

4. Completed Clinical Trials

4.1. Multiple Sclerosis (MS)

MS is a neurodegenerative disease with a strong cell-mediated (CMI) autoimmune component. A number of animal models mimic the histopathological and immunological

TABLE 2. Cannabinoid-based drugs decrease cytokines

Cannabinoid	Receptor	Model	Cytokine stimulus	Cytokines
Humans				
Marijuana smoking	ND	Lung alveolar macrophages	LPS	TNF, GM-CSF, IL-6
Ajulemic acid	ND	Peripheral blood and synovial monocytes	LPS	IL-1β
Marijuana smoking	ND	Peripheral blood monocytes	PHA, ConA	IL-2
THC	CB2	Peripheral blood T cells	Allogeneic dendritic cells	IFN-γ
Mice				
WIN55, 212-2 or HU-210	ND	Splenocytes	LPS, Propionicacterium acnes	TNF, IL-12
THC	ND	Macrophage cell line RAW264.7	LPS	TNF
HU-211	NNDA	Brain	Closed head injury	TNF
WIN55, 212-2	CB2	Heart	Ischemia-reperfusion	IL-1β, CXCL8
THC, AEA, or 2-AG	ND	Macrophage cell line J774	LPS	IL-6
THC	ND	Splenocytes	Legionella pneumophila	IFN-γ
THC	CB1, CB2	Splenocytes	Legionella pneumophila	IL-12, IL-12Rβ2
THC	CB2	Splenocytes	Tumor model	IFN-γ
WIN55, 212-2	ND	Splenocytes	Theiler's murine encephalomyelitis virus	IFN-γ

References: Fischer-Stenger et al. 1993; Newton et al. 1994; Baldwin et al. 1997; Shohami et al. 1997; Klein et al. 2006; Smith et al. 2000; Zhu et al. 2000; Chang et al. 2001; Panikashvili et al. 2001; Yuan et al. 2002; Croxford and Miller 2003; Pacifici et al. 2003; Di Filippo et al. 2004)

features of MS, and in these models, cannabinoid-based drugs were shown to attenuate the disease. For example, natural cannabinoids were shown to suppress disease progression in rat models of EAE possibly by suppressing the HPA axis (Lyman et al. 1989; Wirguin et al. 1994); furthermore, in Theiler murine encephalomyelitis virus models of demyelinating disease, synthetic cannabinoids attenuated disease as well as suppressed infiltrating T cells and the brain production of proinflammatory cytokines (Arevalo-Martin et al. 2003; Croxford and Miller 2003). Besides the attenuation of neurodegeneration (Pryce et al. 2003) and immune reactivity, cannabinoid treatment has also been shown to suppress encephalitis-associated spasticity (Baker et al. 2000). This effect appears to involve CB_1 receptors more than CB_2 (Pryce and Baker 2007).

These animal studies suggested that cannabinoid treatment might be protective in neurodegenerative diseases by mechanisms involving suppression of immune and other neurodegenerative mechanisms. One of the first clinical trials in this area was reported in 2003 from a UK study with 630 stable MS patients with spasticity (Zajicek et al. 2003). The patients received either Δ^9 THC (Marinol), cannabis extract (Cannador) containing THC, cannabidiol, and 5% other cannabinoids, or placebo. There was no significant drug effect in the Ashworth score that is an objective measure of spasticity; however, the treated patients reported an improvement in status, suggesting cannabis might be clinically useful. In a 12-month follow-up to this study wherein patients elected to continue the medication, patients showed a small treatment effect on muscle spasticity as measured by a change in the Ashworth score (Zajicek et al. 2005). There were no major safety concerns in this study, and overall, the patients felt the drugs were helpful. Additional studies from Germany (57 subjects) (Vaney et al. 2004) and the UK (160 subjects) (Wade et al. 2004) using either cannabis extract capsules or oromucosal spray showed greatest beneficial effects on disease-related spasticity. Again, the drugs were well tolerated. These studies suggest that cannabinoid-based preparations of Δ^9-THC and cannabidiol can lessen certain symptoms in MS patients with relatively few side effects. It is not clear at this point if potency can be increased by different delivery methods or drug combinations and it is also not clear if the cannabinoids have a therapeutic effect and actually attenuate the underlying disease process.

4.2. Traumatic Brain Injury

Traumatic brain injury (TBI) is a major cause of death and disability and occurs following head injury from various causes. The resulting brain damage in TBI is due to many factors including disruption of the blood–brain barrier, hypotension, edema, and hypoxia; various treatment measures have been tried to lessen symptoms and damage including corticosteroids, mannitol, barbiturates, hyperventilation, and hypothermia (Mechoulam et al. 2002). The protective role of cannabinoid-based drugs in TBI was initially reported using the non-psychoactive, synthetic cannabinoid, HU-211 (dexanabinol), in a rat model of closed head injury (Shohami et al. 1995). Here, the drug was shown to reduce blood–brain barrier breakdown and brain edema and to improve measures of brain function. Some of these effects could be traced to drug action as an *N*-methyl-D-aspartate (NMDA) receptor antagonist and suppression of TNF-α in the brain (Shohami et al. 1997). In addition to these studies with HU-211, the endocannabinoid, 2-AG, was also shown to be involved in TBI. Following closed head injury in mice, 2-AG levels increased in the brain and the administration of exogenous 2-AG in this model reduced brain edema, cell death, and infarct volume while improving clinical recovery (Panikashvili et al. 2001).

These preclinical studies suggested that cannabinoid-based drugs might be of value in preventing brain inflammation and neuronal death due to excitotoxicity (Marsicano et al. 2003), and therefore, clinical trials have been undertaken to test the safety and efficacy of these agents in TBI. In a safety trial with 67 patients, a single dose of either 48 or 150 mg/kg of HU-211 was well tolerated showing no adverse effects and also produced significantly better intracranial pressure/cerebral perfusion pressure control in the treated group (Knoller et al. 2002). However, in a much larger study involving 861 patients with TBI, a single injection of HU-211 (150 mg/kg) although safe had no significant effect on disease outcome by 6 months after injury (Maas et al. 2006). Thus, in spite of the encouraging animals studies, a single injection of HU-211 in TBI appears to be safe, but without any significant clinical effect. However, higher or more frequent doses of the drug may have significant benefit in TBI, but this will have to await future trials with this well-tolerated drug.

4.3. Rheumatoid Arthritis

Rheumatoid arthritis (RA) is a chronic inflammatory disease of the synovial joints of unknown etiology that affects 1% of the US population (Nogid and Pham 2006). Currently, treatment of symptoms rests on the use of NSAIDs and corticosteroids in combination with non-biologicals such as methotrexate and biologicals such as etanercept. Many of these drugs have proven efficacy, but additional treatments for symptoms and the disease process are needed. Animal studies with cannabinoid-based drugs targeting joint disease were first reported in 1998 testing ajulemic acid (CT3) with low psychoactive potency in rodent models of acute and chronic inflammation (Zurier et al. 1998). The drug given orally significantly reduced the severity of joint inflammation and synovitis measured histologically in adjuvant-induced arthritis. In terms of anti-inflammatory mechanisms, ajulemic acid was shown to suppress IL-1β production by synovial fluid monocytes isolated from arthritis patients (Zurier et al. 2003) and to activate PPARγ in vitro (Liu et al. 2003). In addition to ajulemic acid, cannabidiol (a natural cannabinoid with low psychoactivity) and a synthetic derivative, HU-320, have been shown to attenuate the symptoms and disease in arthritis models. For example, cannabidiol, given orally or systemically, to collagen-induced arthritis rats, attenuated joint disease progression as well as suppressed lymphocyte immune sensitization and release of TNF-α by synovial cells (Malfait et al. 2000). A dimethyl-heptyl derivative of cannabidiol, HU-320, also with low psychoactivity, was shown following daily injections to attenuate symptoms and damage in collagen-induced arthritis and also to suppress immune reactivity (Sumariwalla et al. 2004).

These results in animal studies with cannabinoid-based drugs, particularly cannabidiol, have supported at least one trial in humans. Fifty-eight patients with RA were treated for 5 weeks with an oromucosal spray containing THC and cannabidiol (Sativex®) or placebo and efficacy assessed by measuring pain on movement, pain at rest, morning stiffness, and sleep quality by numerical rating scales (Blake et al. 2006). The results showed statistically significant improvements in pain and quality of sleep, with no effect on morning stiffness. Also, side effects were mild or moderate. From this short-term trial, it appears that natural cannabinoids have an attenuating effect on RA symptoms, which suggests further studies are warranted; however, the mechanisms of anti-inflammation are unclear and the ability of these drugs to suppress the pathological processes underlying RA has yet to be determined.

4.4. Neuropathic Pain

A variety of studies have demonstrated that endocannabinoids mediate antinociception in acute and chronic pain models involving tissue and nerve injury; furthermore, the role of CB_1, located in spinal and supraspinal areas, in regulating pain perception has been confirmed by a number of studies (Hohmann and Suplita 2006). However, there is also evidence from animal studies that CB_2 receptors and other mechanisms might also mediate analgesia especially in neuropathic and inflammatory pain. CB_2 involvement was demonstrated in 1998 in a formalin injection model in mice (Calignano et al. 1998). The two endocannabinoids, AEA and palmitoylethanolamide, when injected into treated animals reduced the pain responses of the animals; in addition, injection of either CB_1 or CB_2 antagonists prolonged and enhanced the pain behavior to formalin injection, suggesting that endocannabinoids working through both receptors played a role in the pain response in mice (Calignano et al. 1998). Similar involvement of CB_2 was reported in the carrageenan injection model in rats (Elmes et al. 2005) and the spinal nerve ligation model (Ibrahim et al. 2003). It is interesting to note that in the latter study, a role for CB_1 in attenuating the pain was excluded, suggesting an exclusive role of CB_2 in attenuating pain. Mechanisms not involving cannabinoid receptors have also been suggested by studies involving ajulemic acid, which has low affinity for these receptors. The analgesic action of the drug administered orally or systemically, was demonstrated in several mouse nociception tests (Burstein et al. 1998) and its potency was compared with morphine as well as its GI side effects studied in rats and mice (Dajani et al. 1999). Ajulemic acid was as potent as morphine in several measures of analgesia and was devoid of the GI ulceration effect seen with other anti-inflammatory drugs. More recently, in rat models of nerve-injury neuropathic pain and inflammatory pain, ajulemic acid showed good potency in relation to the high-affinity cannabinoid agonist, HU-210, but had none of the psychoactive effects associated with the HU compound (Mitchell et al. 2005).

The above animal models suggested that cannabinoid-based drugs might be of benefit in the management of pain, especially neuropathic pain. One of the first trials in this regard examined the efficacy of ajulemic acid given orally in a randomized, placebo-controlled, double-blind, crossover study of 21 patients with neuropathic pain of various etiologies (Karst et al. 2003). The treatment time was only a few weeks and the number of patients limited; however, the study reported significant differences in the drug-treated group in terms of visual analog scale (VAS) as a measure of pain, with no major drug side effects. This same group of patients was re-evaluated for other measures of pain and also for psychological and physical side effects of the drug (Salim et al. 2005) and the drug showed significant improvement relative to the placebo, suggesting that ajulemic acid has some effect in reducing chronic neuropathic pain without side effects.

Psychoactive cannabinoids have also been used in the treatment of central pain associated with MS. Twenty-four patients were given Δ^9-THC (dronabinol) orally for 3 weeks in a randomized, double-blind, placebo-controlled, crossover study, and median spontaneous pain intensity was assessed (Svendsen et al. 2004). The study concluded that the drug had a modest but clinically relevant analgesic effect in these patients with a few side effects such as dizziness. A similar study was done in 66 MS patients using the Sativex oromucosal spray (Rog et al. 2005). The results showed the drug was superior to placebo in reducing intensity of pain and in measures of sleep disturbance. The Sativex was generally well tolerated with few side effects.

Recently, a meta-analysis of cannabinoid-based drug trials for neuropathic and MS pain was published (Iskedjian et al. 2007). Seven studies were included in the analysis involving the use of Sativex or dronabinol and the report concluded that the cannabinoid-based drugs were effective in treating neuropathic pain in MS.

5. Future Clinical Trials

5.1. Intestinal Disorders

Anecdotal reports on the use of marijuana suggest usefulness in the treatment of intestinal disorders. This is in agreement with recent animal studies showing that CB_1 receptors are present on neurons of the enteric nervous system as well as on sensory nerves and that CB_2 receptors are located on immune cells. Furthermore, activation of CB_1 receptors has been reported to modulate intestinal functions such as reducing emesis, gastric secretion, and intestinal motility (Massa et al. 2005; Di Marzo and Izzo 2006). Endocannabinoids are detected in the digestive tract and experimentally they have been shown to inhibit colonic propulsion in an animal model (Pinto et al. 2002). Activation of CB_2 receptors have also been shown to regulate intestinal responses. For example, increased GI transit in rats due to the inflammatory action of LPS was attenuated by CB_2 activation rather than CB_1 (Mathison et al. 2004), and these receptors were significantly increased in colon biopsies from patients with ulcerative colitis (Wright et al. 2005). Interestingly, a recent report has shown that CB_2 receptors (along with opioid receptors) are induced in intestinal epithelial cells from rats fed the probiotic, *Lactobacillus acidophilus* (Rousseaux et al. 2007). This increase in receptor expression was shown to be related to maintaining an analgesic tone in experimentally induced gut inflammation in rats. All of these preclinical reports suggest that the gut endocannabinoid system may serve as a therapeutic target for many GI disorders including inflammatory bowel diseases such as Crohn's disease and functional bowel diseases such as irritable bowel syndrome. Although, no clinical trials have been reported in this area, it is interesting to note that in phase III clinical trials with the CB_1 antagonist, Rimonabant, one side effect noted was diarrhea (Van Gaal et al. 2005). For therapeutic purposes, CB_1 agonists are unlikely candidates because of their psychoactive potential. However, other approaches are possible including the use of agonists modified to prevent crossing the blood–brain barrier, the use of drugs designed to inhibit the degradation of endocannabinoids (e.g., fatty acid amide hydrolase inhibitors), and the use of CB_2 agonists that have no psychoactive potential.

5.2. Alzheimer's Disease

Alzheimer's disease (AD) is characterized by the deposition of β-amyloid (aβ) peptide in various areas of the brain, and senile plaques develop as a result of the local activation of microglia cells and neuron destruction (Mackenzie et al. 1995). It can be said, then, that AD is a chronic inflammatory disease much like AIDS dementia and MS. Because of this, it is not surprising that a few studies have been done in animals to examine the anti-inflammatory potential of cannabinoid-based drugs in AD models. One of the first studies utilized and in vitro model of aβ toxicity (Milton 2002). The NT2 human neuron cell line was treated with aβ and toxicity determined by the MTT assay. AEA and noladin ether co-treatment with aβ attenuated the toxicity effect of the peptide, and the attenuation was inhibited by treatment with a CB_1 antagonist. Thus, the endocannabinoids prevented aβ toxicity and the drug effect was mediated by CB_1. Studies in rats showed similar protective effects by CB_2 agonists (Ramirez et al. 2005), and in addition, the cannabinoids were shown to inhibit the activation of brain microglia cells stimulated by various agents including aβ (Ehrhart et al. 2005; Ramirez et al. 2005). Finally, a recent study showed that THC is a competitive inhibitor in vitro of acetylcholinesterase (AChE) (Eubanks et al. 2006). This has relevance to AD because this enzyme promotes the aggregation of aβ, thus promoting plaque formation, and in fact, several FDA-approved drugs target the inhibition of AChE. These preclinical studies support the possibility that cannabinoid-based drugs might have anti-inflammatory effects directed at mechanisms operating in the pathogenesis of AD and therefore be of ultimate use in the management of the disease.

5.3. Atherosclerosis

Atherosclerosis is a chronic inflammatory disease that involves disturbances in the biology of the cells lining the vessels as well as homing to and activation of the inflammatory cells that reside in the vascular plaque. The putative role of endocannabinoid system in vascular biology and disease stems from findings that endocannabinoids are produced by vascular endothelial cells and these and the natural cannabinoids have been shown to cause vasodilation, hypotension, and bradycardia (Wagner et al. 1998). This could be due to stimulation of presynaptic CB_1 receptors on peripheral sympathetic nerve terminals or as mentioned above to stimulation of TRPV1 on endothelium and nerves. Many of the elements of the endocannabinoid system are upregulated during systemic vascular inflammatory disease such as endotoxic shock (Wang et al. 2001) and experimentally induced endotoxin induced hypotension (Varga et al. 1998), suggesting that vascular inflammatory tone might be partly regulated by the endocannabinoid system. Besides changes in vascular biology, another major factor in the development of vascular inflammation and atherosclerosis is the influx of inflammatory cells. This may involve CB_2 receptors rather than CB_1. For example, leukocyte-dependent myocardial damage was suppressed by WIN55212-2 in a CB2-dependent way (Di Filippo et al. 2004). This was reported using a mouse model of myocardial ischemia/reperfusion, and when the drug was given 30 min before induction, infarct size was significant. The drug effect was inhibited by a CB_2 antagonist and was accompanied by reduction in the influx of leukocyte markers (Di Filippo et al. 2004). A recent study also reported a role for CB_2 in the suppression of atherosclerosis development in the apolipoprotein E (ApoE)-knockout model in mice (Steffens et al. 2005). Mice were fed THC along with the high-cholesterol diet and plaque development in cryosections of aortic roots was significantly depressed over time. In addition, CB_2 expression was observed in human and mouse atherosclerotic plaques and inflammatory cell function was suppressed in drug-treated ApoE-knockout mice. In total, these studies suggest that modulation of the endocannabinoid tone surrounding vascular plaque formation may have beneficial therapeutic effects in inflammatory diseases of blood vessels.

5.4. Osteoporosis

The overall bone mass is a product of the balanced activity of bone-building cells, the osteoblasts, and bone-resorbing cells, the osteoclasts. Many factors contribute to this steady state, and when out of balance, focal or generalized bone loss occurs, leading to diseases such as osteoporosis. An array of hormones and cytokines has been implicated in the maintenance of bone density as well as neuroendocrine pathways and neurotransmitters (Takeda et al. 2002). In this context, several studies have examined the role of endocannabinoid system in the regulation of bone mass and turnover. The first study reported that CB_1-knockout mice had increased bone mass and were protected from ovariectomy-induced bone loss (Idris et al. 2005). In addition, CB_1 and CB_2 antagonists prevented ovariectomy-induced bone loss *in vivo* and suppressed osteoclasts development in vitro. In another study, CB_2- knockout mice showed opposite effects with respect to bone density. These mice had an age-related bone loss and cortical changes similar to that seen in osteoporosis (Ofek et al. 2006). In addition, knockout mice showed a histological decrease in osteoblasts and CB_2-selective agonists in wild-type mice increased the number of endocortical osteoblasts, suppressed osteoclastogenesis, and attenuated ovariectomy-induced bone loss. The bone-remodeling cells all expressed CB_2 receptors. The pharmacology of these two studies is somewhat contradictory, with receptor antagonists and agonists both showing the same effect, that is, a decrease in osteoclast activation and attenuation of bone loss. However, many drugs have been targeted at inhibiting the action of osteoclasts (Rodan and Martin 2000), and it appears that cannabinoid receptors are involved in the development of these cells, possibly by different mechanisms, and cannabinoid-based drugs may be of value in inhibiting these cells in the management of disease. Finally, recently it was reported that a cohort of postmenopausal osteoporosis women displayed a significant association of single polymorphisms and haplotypes encompassing the CB_2 gene, whereas no association was found with the CB_1 gene (Karsak et al. 2005). The authors concluded that CB_2 plays a role in the etiology of osteoporosis, a fact that might provide novel drug targets for this disease.

6. Conclusions

The role of marijuana cannabinoids and the endocannabinoid system in health and disease has gone from anecdotal folk-medicine reports, through preclinical science testing in animals, to clinical trials in humans. A consistent theme is that these drugs affect mood and behavior through CNS effects but also significantly affect physiological systems outside of the CNS as well as the inflammatory response through effects on immune cells and the cytokine network. Clinical trials with MS patients using drugs ranging from psychoactive Δ^9-THC to nonpsychoactive cannabidiol showed some efficacy in reducing spasticity, sleep disturbances, and neuropathic pain with very few side effects. In addition to MS, other inflammatory diseases have been tested including traumatic brain injury and rheumatoid arthritis. In these studies, HU-211 and ajulemic acid (both nonpsychoactive) were moderately effective in reducing symptoms with few side effects. Because drugs with low affinity for CB_1 and CB_2 were effective in some of these studies, the role of known cannabinoid receptors in the clinical responses is uncertain, and therefore, other molecular targets may be involved such as other undefined cannabinoid receptors, vanilloid receptors, or PPAR proteins. In addition to these clinical trials, basic science and preclinical studies have suggested other therapeutic applications for these compounds. For example, animal models of inflammatory bowel disease have shown that the endocannabinoid system is activated during disease and regulates GI tone under pathological conditions. In addition, models of AD have also been probed for involvement of the endocannabinoid system. CB_1 and CB_2 agonists showed protective effects from damage due to aβ and the drugs were shown to inhibit the activity of microglia cells, important cellular mediators of the chronic inflammatory response in this disease. The regulation of inflammatory cells by CB_1 and CB_2 has also been shown in models of atherosclerosis and osteoporosis, pointing to applications in these widespread diseases. In conclusion, the endocannabinoid system appears to be involved in regulating the normal tone of many physiological processes, and it appears to be altered during disease development. This is suggestive that the manipulation of this system might be of benefit in either the management or the cure of these diseases; however, this will require further drug development in this area.

References

Anday, J.K. and Mercier, R.W. (2005) Gene ancestry of the cannabinoid receptor family. Pharmacol. Res. 20, 20.

Arevalo-Martin, A., Vela, J.M., Molina-Holgado, E., Borrell, J. and Guaza, C. (2003) Therapeutic action of cannabinoids in a murine model of multiple sclerosis. J. Neurosci. 23, 2511–2516.

Baker, D., Pryce, G., Croxford, J.L., Brown, P., Pertwee, R.G., Huffman, J.W. and Layward, L. (2000) Cannabinoids control spasticity and tremor in a multiple sclerosis model. Nature 404, 84–87.

Baker, D., Pryce, G., Davies, W.L. and Hiley, C.R. (2006) In silico patent searching reveals a new cannabinoid receptor. Trends Pharmacol. Sci. 27, 1–4.

Baldwin, G.C., Tashkin, D.P., Buckley, D.M., Park, A.N., Dubinett, S.M. and Roth, M.D. (1997) Marijuana and cocaine impair alveolar macrophage function and cytokine production. Am. J. Respir. Crit. Care Med. 156, 1606–1613.

Blake, D.R., Robson, P., Ho, M., Jubb, R.W. and McCabe, C.S. (2006) Preliminary assessment of the efficacy, tolerability and safety of a cannabis-based medicine (Sativex) in the treatment of pain caused by rheumatoid arthritis. Rheumatology (Oxford) 45, 50–52.

Breivogel, C.S., Griffin, G., Di Marzo, V. and Martin, B.R. (2001) Evidence for a new G protein-coupled cannabinoid receptor in mouse brain. Mol. Pharmacol. 60, 155–163.

Burstein, S.H., Friderichs, E., Kogel, B., Schneider, J. and Selve, N. (1998) Analgesic effects of 1',1' dimethylheptyl-delta8-THC-11-oic acid (CT3) in mice. Life Sci. 63, 161–168.

Cabral, G. and Dove Pettit, D. (1998) Drugs and immunity: cannabinoids and their role in decreased resistance to infectious diseases. J. Neuroimmunol. 83, 116–123.

Calignano, A., La Rana, G., Giuffrida, A. and Piomelli, D. (1998) Control of pain initiation by endogenous cannabinoids. Nature 394, 277–281.

Chang, Y.H., Lee, S.T. and Lin, W.W. (2001) Effects of cannabinoids on LPS-stimulated inflammatory mediator release from macrophages: involvement of eicosanoids. J. Cell. Biochem. 81, 715–723.

Croxford, J.L. and Miller, S.D. (2003) Immunoregulation of a viral model of multiple sclerosis using the synthetic cannabinoid R+WIN55,212. J. Clin. Invest. 111, 1231–1240.

Dajani, E.Z., Larsen, K.R., Taylor, J., Dajani, N.E., Shahwan, T.G., Neeleman, S.D., Taylor, M.S., Dayton, M.T. and Mir, G.N. (1999) 1',1'-Dimethylheptyl-delta-8-tetrahydrocannabinol-11-oic acid: a novel, orally effective cannabinoid with analgesic and anti-inflammatory properties. J. Pharmacol. Exp. Ther. 291, 31–38.

Derocq, J.M., Jbilo, O., Bouaboula, M., Segui, M., Clere, C. and Casellas, P. (2000) Genomic and functional changes induced by the activation of the peripheral cannabinoid receptor CB2 in the promyelocytic cells HL-60. Possible involvement of the CB2 receptor in cell differentiation. J. Biol. Chem. 275, 15621–15628.

Devane, W.A., Hanus, L., Breuer, A., Pertwee, R.G., Stevenson, L.A., Griffin, G., Gibson, D., Mandelbaum, A., Etinger, A. and Mechoulam, R. (1992) Isolation and structure of a brain constituent that binds to the cannabinoid receptor. Science 258, 1946–1949.

Di Filippo, C., Rossi, F., Rossi, S. and D'Amico, M. (2004) Cannabinoid CB2 receptor activation reduces mouse myocardial ischemia-reperfusion injury: involvement of cytokine/chemokines and PMN. J. Leukoc. Biol. 75, 453–459. Epub 2003 Dec 2004.

Di Marzo, V. and Izzo, A.A. (2006) Endocannabinoid overactivity and intestinal inflammation. Gut 55, 1373–1376.

Ehrhart, J., Obregon, D., Mori, T., Hou, H., Sun, N., Bai, Y., Klein, T., Fernandez, F., Tan, J. and Shytle, R.D. (2005) Stimulation of cannabinoid receptor 2 (CB2) suppresses microglial activation. J. Neuroinflammation 2, 29.

Elmes, S.J., Winyard, L.A., Medhurst, S.J., Clayton, N.M., Wilson, A.W., Kendall, D.A. and Chapman, V. (2005) Activation of CB1 and CB2 receptors attenuates the induction and maintenance of inflammatory pain in the rat. Pain 118, 327–335.

Eubanks, L.M., Rogers, C.J., Beuscher, A.E., Koob, G.F., Olson, A.J., Dickerson, T.J. and Janda, K.D. (2006) A molecular link between the active component of marijuana and Alzheimer's disease pathology. Mol. Pharm. 3, 773–777.

Fischer-Stenger, K., Dove Pettit, D.A. and Cabral, G.A. (1993) Δ^9-Tetrahydrocannabinol inhibition of tumor necrosis factor-α: suppression of post-translational events. J. Pharm. Exp. Ther. 267, 1558–1565.

Fowler, C.J., Holt, S., Nilsson, O., Jonsson, K.O., Tiger, G. and Jacobsson, S.O. (2005) The endocannabinoid signaling system: pharmacological and therapeutic aspects. Pharmacol. Biochem. Behav. 81, 248–262.

Gaoni, Y. and Mechoulam, R. (1964) Isolation, structure, and partial synthesis of an active constituent of hashish. J. Am. Chem. Soc. 86, 1646–1647.

Gerard, C.M., Mollereau, C., Vassart, G. and Parmentier, M. (1991) Molecular cloning of a human cannabinoid receptor which is also expressed in testis. Biochem. J. 279, 129–134.

Hanus, L., Abu-Lafi, S., Fride, E., Breuer, A., Vogel, Z., Shalev, D.E., Kustanovich, I. and Mechoulam, R. (2001) 2-Arachidonyl glyceryl ether, an endogenous agonist of the cannabinoid CB1 receptor. Proc. Natl. Acad. Sci. USA 98, 3662–3665.

Hohmann, A. and Suplita, R. (2006) Endocannabinoid mechanisms of pain modulation. AAPS J. 8, E693–E708.

Howlett, A.C., Breivogel, C.S., Childers, S.R., Deadwyler, S.A., Hampson, R.E. and Porrino, L.J. (2004) Cannabinoid physiology and pharmacology: 30 years of progress. Neuropharmacology 47, 345–358.

Ibrahim, M.M., Deng, H., Zvonok, A., Cockayne, D.A., Kwan, J., Mata, H.P., Vanderah, T.W., Lai, J., Porreca, F., Makriyannis, A. and Malan, T.P., Jr. (2003) Activation of CB2 cannabinoid receptors by AM1241 inhibits experimental neuropathic pain: pain inhibition by receptors not present in the CNS. Proc. Natl. Acad. Sci. USA 100, 10529–10533.

Idris, A.I., van 't Hof, R.J., Greig, I.R., Ridge, S.A., Baker, D., Ross, R.A. and Ralston, S.H. (2005) Regulation of bone mass, bone loss and osteoclast activity by cannabinoid receptors. Nat. Med. 11, 774–779.

Iskedjian, M., Bereza, B., Gordon, A., Piwko, C. and Einarson, T.R. (2007) Meta-analysis of cannabis based treatments for neuropathic and multiple sclerosis-related pain. Curr. Med. Res. Opin. 23, 17–24.

Jarai, Z., Wagner, J.A., Varga, K., Lake, K.D., Compton, D.R., Martin, B.R., Zimmer, A.M., Bonner, T.I., Buckley, N.E., Mezey, E., Razdan, R.K., Zimmer, A. and Kunos, G. (1999) Cannabinoid-induced mesenteric vasodilation through an endothelial site distinct from CB1 or CB2 receptors. Proc. Natl. Acad. Sci. USA 96, 14136–14141.

Karsak, M., Cohen-Solal, M., Freudenberg, J., Ostertag, A., Morieux, C., Kornak, U., Essig, J., Erxlebe, E., Bab, I., Kubisch, C., de Vernejoul, M.C. and Zimmer, A. (2005) Cannabinoid receptor type 2 gene is associated with human osteoporosis. Hum. Mol. Genet. 14, 3389–3396.

Karst, M., Salim, K., Burstein, S., Conrad, I., Hoy, L. and Schneider, U. (2003) Analgesic effect of the synthetic cannabinoid CT-3 on chronic neuropathic pain: a randomized controlled trial. JAMA 290, 1757–1762.

Kishimoto, S., Kobayashi, Y., Oka, S., Gokoh, M., Waku, K. and Sugiura, T. (2004) 2-Arachidonoylglycerol, an endogenous cannabinoid receptor ligand, induces accelerated production of chemokines in HL-60 cells. J. Biochem. (Tokyo) 135, 517–524.

Klein, T.W. (2005) Cannabinoid-based drugs as anti-inflammatory therapeutics. Nat. Rev. Immunol. 5, 400–411.

Klein, T.W. and Cabral, G. (2006) Cannabinoid-induced immune suppression and modulation of antigen-presenting cells. J. Neuroimmune Pharmacol. 1, 50–64.

Klein, T.W., Lane, B., Newton, C.A. and Friedman, H. (2000a) The cannabinoid system and cytokine network. Proc. Soc. Exp. Biol. Med. 225, 1–8.

Klein, T., Newton, C. and Friedman, H. (1998) Cannabinoid receptors and immunity. Immu nol. Today 19, 373–381.

Klein, T.W., Newton, C., Larsen, K., Lu, L., Perkins, I., Nong, L. and Friedman, H. (2003) The cannabinoid system and immune modulation. J. Leukoc. Biol. 74, 486–496.

Klein, T.W., Newton, C.A., Nakachi, N. and Friedman, H. (2000b) Δ^9-Tetrahydrocannabinol treatment suppresses immunity and early IFNγ, IL-12, and IL-12 receptor β2 responses to *Legionella pneumophila* infection. J. Immunol. 164, 6461–6466.

Klein, T.W., Newton, C., Widen, R. and Friedman, H. (1993) Δ^9- Tetrahydrocannabinol injection induces cytokine-mediated mortality of mice infected with *Legionella pneumophila*. J. Pharmacol. Exp. Ther. 267, 635–640.

Knoller, N., Levi, L., Shoshan, I., Reichenthal, E., Razon, N., Rappaport, Z.H. and Biegon, A. (2002) Dexanabinol (HU-211) in the treatment of severe closed head injury: a randomized, placebo-controlled, phase II clinical trial. Crit. Care Med. 30, 548–554.

Liu, J., Li, H., Burstein, S.H., Zurier, R.B. and Chen, J.D. (2003) Activation and binding of peroxisome proliferator-activated receptor gamma by synthetic cannabinoid ajulemic acid. Mol. Pharmacol. 63, 983–992.

Lu, T., Newton, C., Perkins, I., Friedman, H. and Klein, T.W. (2006) Role of cannabinoid receptors in Delta-9-tetrahydrocannabinol suppression of IL-12p40 in mouse bone marrow-derived dendritic cells infected with Legionella pneumophila. Eur. J. Pharmacol. 532, 170–177.

Lyman, W.D., Sonett, J.R., Brosnan, C.F., Elkin, R. and Bornstein, M.B. (1989) Δ^9-Tetrahydrocannabinol: a novel treatment for experimental autoimmune encephalomyelitis. J. Neuroimmunol. 23, 73–81.
Maas, A.I., Murray, G., Henney, H., 3rd, Kassem, N., Legrand, V., Mangelus, M., Muizelaar, J.P., Stocchetti, N. and Knoller, N. (2006) Efficacy and safety of dexanabinol in severe traumatic brain injury: results of a phase III randomised, placebo-controlled, clinical trial. Lancet Neurol. 5, 38–45.
Mackenzie, I.R., Hao, C. and Munoz, D.G. (1995) Role of microglia in senile plaque formation. Neurobiol. Aging 16, 797–804.
Malfait, A.M., Gallily, R., Sumariwalla, P.F., Malik, A.S., Andreakos, E., Mechoulam, R. and Feldmann, M. (2000) The nonpsychoactive cannabis constituent cannabidiol is an oral anti-arthritic therapeutic in murine collagen-induced arthritis [see comments]. Proc. Natl. Acad. Sci. USA 97, 9561–9566.
Marsicano, G., Goodenough, S., Monory, K., Hermann, H., Eder, M., Cannich, A., Azad, S.C., Cascio, M.G., Gutierrez, S.O., van der Stelt, M., Lopez-Rodriguez, M.L., Casanova, E., Schutz, G., Zieglgansberger, W., Di Marzo, V., Behl, C. and Lutz, B. (2003) CB1 cannabinoid receptors and on-demand defense against excitotoxicity. Science. 302, 84–88.
Massa, F., Storr, M. and Lutz, B. (2005) The endocannabinoid system in the physiology and pathophysiology of the gastrointestinal tract. J. Mol. Med. 26, 26.
Mathison, R., Ho, W., Pittman, Q.J., Davison, J.S. and Sharkey, K.A. (2004) Effects of cannabinoid receptor-2 activation on accelerated gastrointestinal transit in lipopolysaccharide-treated rats. Br. J. Pharmacol. 142, 1247–1254.
Matsuda, L.A., Lolait, S.J., Brownstein, M.J., Young, A.C. and Bonner, T.I. (1990) Structure of cannabinoid receptor and functional expression of the cloned cDNA. Nature 346, 561–564.
Mechoulam, R., Ben-Shabat, S., Hanus, L., Ligumsky, M., Kaminski, N.E., Schatz, A.R., Gopher, A., Almog, S., Martin, B.R., Compton, D.R., Pertwee, R.G., Griffin, G., Bayewitch, M., Barg, J. and Vogel, Z. (1995) Identification of an endogenous 2-monoglyceride, present in canine gut, that binds to cannabinoid receptors. Biochem. Pharm. 50, 83–90.
Mechoulam, R., Panikashvili, D. and Shohami, E. (2002) Cannabinoids and brain injury: therapeutic implications. Trends Mol. Med. 8, 58–61.
Milton, N.G. (2002) Anandamide and noladin ether prevent neurotoxicity of the human amyloid-beta peptide. Neurosci. Lett. 332, 127–130.
Mitchell, V.A., Aslan, S., Safaei, R. and Vaughan, C.W. (2005) Effect of the cannabinoid ajulemic acid on rat models of neuropathic and inflammatory pain. Neurosci. Lett. 382, 231–235.
Munro, S., Thomas, K.L. and Abu-Shaar, M. (1993) Molecular characterization of a peripheral receptor for cannabinoids. Nature 365, 61–65.
Nahas, G.G., Suciu-Foca, N., Armand, J.-P. and Morishima, A. (1974) Inhibition of cellular mediated immunity in marihuana smokers. Science 183, 419–420.
Newton, C.A., Klein, T.W. and Friedman, H. (1994) Secondary immunity to *Legionella pneumophila* and Th1 activity are suppressed by delta- 9-tetrahydrocannabinol injection. Infect. Immun. 62, 4015–4020.
Nogid, A. and Pham, D.Q. (2006) Role of abatacept in the management of rheumatoid arthritis. Clin Ther. 28, 1764–1778.
Ofek, O., Karsak, M., Leclerc, N., Fogel, M., Frenkel, B., Wright, K., Tam, J., Attar-Namdar, M., Kram, V., Shohami, E., Mechoulam, R., Zimmer, A. and Bab, I. (2006) Peripheral cannabinoid receptor, CB2, regulates bone mass. Proc. Natl. Acad. Sci. U S A. 103, 696–701.

Offertaler, L., Mo, F.M., Batkai, S., Liu, J., Begg, M., Razdan, R.K., Martin, B.R., Bukoski, R.D. and Kunos, G. (2003) Selective ligands and cellular effectors of a G protein-coupled endothelial cannabinoid receptor. Mol. Pharmacol. 63, 699–705.

Pacher, P., Batkai, S. and Kunos, G. (2006) The endocannabinoid system as an emerging target of pharmacotherapy. Pharmacol. Rev. 58, 389–462.

Pacifici, R., Zuccaro, P., Pichini, S., Roset, P.N., Poudevida, S., Farre, M., Segura, J. and De la Torre, R. (2003) Modulation of the immune system in cannabis users. JAMA 289, 1929–1931.

Panikashvili, D., Simeonidou, C., Ben-Shabat, S., Hanus, L., Breuer, A., Mechoulam, R. and Shohami, E. (2001) An endogenous cannabinoid (2-AG) is neuroprotective after brain injury. Nature 413, 527–531.

Pertwee, R.G. (2005) The therapeutic potential of drugs that target cannabinoid receptors or modulate the tissue levels or actions of endocannabinoids. AAPS J. 7, E625–E654.

Pinto, L., Izzo, A.A., Cascio, M.G., Bisogno, T., Hospodar-Scott, K., Brown, D.R., Mascolo, N., Di Marzo, V. and Capasso, F. (2002) Endocannabinoids as physiological regulators of colonic propulsion in mice. Gastroenterology 123, 227–234.

Porter, A.C., Sauer, J.M., Knierman, M.D., Becker, G.W., Berna, M.J., Bao, J., Nomikos, G.G., Carter, P., Bymaster, F.P., Leese, A.B. and Felder, C.C. (2002) Characterization of a novel endocannabinoid, virodhamine, with antagonist activity at the CB1 receptor. J. Pharmacol. Exp. Ther. 301, 1020–1024.

Pryce, G., Ahmed, Z., Hankey, D.J., Jackson, S.J., Croxford, J.L., Pocock, J.M., Ledent, C., Petzold, A., Thompson, A.J., Giovannoni, G., Cuzner, M.L. and Baker, D. (2003) Cannabinoids inhibit neurodegeneration in models of multiple sclerosis. Brain 126, 2191–2202.

Pryce, G. and Baker, D. (2007) Control of spasticity in a multiple sclerosis model is mediated by CB(1), not CB(2), cannabinoid receptors. Br. J. Pharmacol. 15, 15.

Ramirez, B.G., Blazquez, C., Gomez del Pulgar, T., Guzman, M. and de Ceballos, M.L. (2005) Prevention of Alzheimer's disease pathology by cannabinoids: neuroprotection mediated by blockade of microglial activation. J. Neurosci. 25, 1904–1913.

Rodan, G.A. and Martin, T.J. (2000) Therapeutic approaches to bone diseases. Science 289, 1508 1514.

Rog, D.J., Nurmikko, T.J., Friede, T. and Young, C.A. (2005) Randomized, controlled trial of cannabis-based medicine in central pain in multiple sclerosis. Neurology 65, 812–819.

Rousseaux, C., Thuru, X., Gelot, A., Barnich, N., Neut, C., Dubuquoy, L., Dubuquoy, C., Merour, E., Geboes, K., Chamaillard, M., Ouwehand, A., Leyer, G., Carcano, D., Colombel, J.F., Ardid, D. and Desreumaux, P. (2007) Lactobacillus acidophilus modulates intestinal pain and induces opioid and cannabinoid receptors. Nat. Med. 13, 35–37.

Salim, K., Schneider, U., Burstein, S., Hoy, L. and Karst, M. (2005) Pain measurements and side effect profile of the novel cannabinoid ajulemic acid. Neuropharmacology 48, 1164–1171.

Shohami, E., Gallily, R., Mechoulam, R., Bass, R. and Ben-Hur, T. (1997) Cytokine production in the brain following closed head injury: dexanabinol (HU-211) is a novel TNF-alpha inhibitor and an effective neuroprotectant. J. Neuroimmunol. 72, 169–177.

Shohami, E., Novikov, M. and Bass, R. (1995) Long-term effect of HU-211, a novel noncompetitive NMDA antagonist, on motor and memory functions after closed head injury in the rat. Brain Res. 674, 55–62.

Smith, S.R., Terminelli, C. and Denhardt, G. (2000) Effects of cannabinoid receptor agonist and antagonist ligands on production of inflammatory cytokines and anti-inflammatory interleukin-10 in endotoxemic mice. J. Pharmacol. Exp. Ther. 293, 136–150.

Steffens, S., Veillard, N.R., Arnaud, C., Pelli, G., Burger, F., Staub, C., Karsak, M., Zimmer, A., Frossard, J.L. and Mach, F. (2005) Low dose oral cannabinoid therapy reduces progression of atherosclerosis in mice. Nature. 434, 782–786.
Sumariwalla, P.F., Gallily, R., Tchilibon, S., Fride, E., Mechoulam, R. and Feldmann, M. (2004) A novel synthetic, nonpsychoactive cannabinoid acid (HU-320) with antiinflammatory properties in murine collagen-induced arthritis. Arthritis Rheum. 50, 985–998.
Sun, Y., Alexander, S.P., Kendall, D.A. and Bennett, A.J. (2006) Cannabinoids and PPARalpha signalling. Biochem. Soc. Trans. 34, 1095–1097.
Svendsen, K.B., Jensen, T.S. and Bach, F.W. (2004) Does the cannabinoid dronabinol reduce central pain in multiple sclerosis? Randomised double blind placebo controlled crossover trial. BMJ 329, 253.
Takeda, S., Elefteriou, F., Levasseur, R., Liu, X., Zhao, L., Parker, K.L., Armstrong, D., Ducy, P. and Karsenty, G. (2002) Leptin regulates bone formation via the sympathetic nervous system. Cell 111, 305–317.
Tomida, I., Pertwee, R.G. and Azuara-Blanco, A. (2004) Cannabinoids and glaucoma. Br. J. Ophthalmol. 88, 708–713.
Van Gaal, L.F., Rissanen, A.M., Scheen, A.J., Ziegler, O. and Rossner, S. (2005) Effects of the cannabinoid-1 receptor blocker rimonabant on weight reduction and cardiovascular risk factors in overweight patients: 1-year experience from the RIO-Europe study. Lancet 365, 1389–1397.
Vaney, C., Heinzel-Gutenbrunner, M., Jobin, P., Tschopp, F., Gattlen, B., Hagen, U., Schnelle, M. and Reif, M. (2004) Efficacy, safety and tolerability of an orally administered cannabis extract in the treatment of spasticity in patients with multiple sclerosis: a randomized, double-blind, placebo-controlled, crossover study. Mult. Scler. 10, 417–424.
Varga, K., Wagner, J.A., Bridgen, D.T. and Kunos, G. (1998) Platelet- and macrophage-derived endogenous cannabinoids are involved in endotoxin-induced hypotension. FASEB J. 12, 1035–1044.
Wade, D.T., Makela, P., Robson, P., House, H. and Bateman, C. (2004) Do cannabis-based medicinal extracts have general or specific effects on symptoms in multiple sclerosis? A double-blind, randomized, placebo-controlled study on 160 patients. Mult. Scler. 10, 434–441.
Wagner, J.A., Varga, K., Kunos, G. and Bridgen, D.T. (1998) Cardiovascular actions of cannabinoids and their generation during shock: platelet- and macrophage-derived endogenous cannabinoids are involved in endotoxin-induced hypotension. J. Mol. Med. 76, 824–836.
Wang, Y., Liu, Y., Ito, Y., Hashiguchi, T., Kitajima, I., Yamakuchi, M., Shimizu, H., Matsuo, S., Imaizumi, H. and Maruyama, I. (2001) Simultaneous measurement of anandamide and 2-arachidonoylglycerol by polymyxin B-selective adsorption and subsequent high-performance liquid chromatography analysis: increase in endogenous cannabinoids in the sera of patients with endotoxic shock. Anal. Biochem. 294, 73–82.
Wirguin, I., Mechoulam, R., Breuer, A., Schezen, E., Weidenfeld, J. and Brenner, T. (1994) Suppression of experimental autoimmune encephalomyelitis by cannabinoids. Immunopharmacolology 28, 209–214.
Wright, K., Rooney, N., Feeney, M., Tate, J., Robertson, D., Welham, M. and Ward, S. (2005) Differential expression of cannabinoid receptors in the human colon: cannabinoids promote epithelial wound healing. Gastroenterology 129, 437–453.
Yuan, M., Kiertscher, S.M., Cheng, Q., Zoumalan, R., Tashkin, D.P. and Roth, M.D. (2002) D9-tetrahydrocannabinol regulates Th1/Th2 cytokine balance in activated human T cells. J. Neuroimmunol. 133, 124–131.

Zajicek, J., Fox, P., Sanders, H., Wright, D., Vickery, J., Nunn, A. and Thompson, A. (2003) Cannabinoids for treatment of spasticity and other symptoms related to multiple sclerosis (CAMS study): multicentre randomised placebo-controlled trial. Lancet 362, 1517–1526.

Zajicek, J.P., Sanders, H.P., Wright, D.E., Vickery, P.J., Ingram, W.M., Reilly, S.M., Nunn, A.J., Teare, L.J., Fox, P.J. and Thompson, A.J. (2005) Cannabinoids in multiple sclerosis (CAMS) study: safety and efficacy data for 12 months follow up. J. Neurol. Neurosurg. Psychiatry 76, 1664–1669.

Zhu, W., Newton, C., Daaka, Y., Friedman, H. and Klein, T.W. (1994) Δ^9-Tetrahydrocannabinol enhances the secretion of interleukin 1 from endotoxin-stimulated macrophages. J. Pharmacol. Exp. Ther. 270, 1334–1339.

Zhu, L.X., Sharma, S., Stolina, M., Gardner, B., Roth, M.D., Tashkin, D.P. and Dubinett, S.M. (2000) Δ-9-Tetrahydrocannabinol inhibits antitumor immunity by a CB2 receptor-mediated, cytokine-dependent pathway. J. Immunol. 165, 373–380.

Zurier, R.B., Rossetti, R.G., Burstein, S.H. and Bidinger, B. (2003) Suppression of human monocyte interleukin-1beta production by ajulemic acid, a nonpsychoactive cannabinoid. Biochem. Pharmacol. 65, 649–655.

Zurier, R.B., Rossetti, R.G., Lane, J.H., Goldberg, J.M., Hunter, S.A. and Burstein, S.H. (1998) Dimethylheptyl-THC-11-oic acid. A nonpsychoactive antiinflammatory agent with a cannabinoid template structure. Arthritis Rheum. 41, 163–170.

Zygmunt, P.M., Petersson, J., Andersson, D.A., Chuang, H., Sorgard, M., Di Marzo, V., Julius, D. and Hogestatt, E.D. (1999) Vanilloid receptors on sensory nerves mediate the vasodilator action of anandamide. Nature 400, 452–457.

44

Micro- and Nanoparticle-Based Vaccines for Hepatitis B

Dhruba J. Bharali[1], Shaker A. Mousa[2], and Yasmin Thanavala[3]

[1] Roswell Park Cancer Institute, Department of Immunology, Buffalo, NY, USA
[2] Pharmaceutical Research Institute at Albany, Albany College of Pharmacy, Albany, NY, USA
[3] Roswell Park Cancer Institute, Department of Immunology, Buffalo, NY, USA, Yasmin.Thanavala@RoswellPark.org

Abstract. The incredible success of vaccinations in contributing to public health is undeniable. In fact, vaccines are the most cost-effective public health tool for disease prevention because their cost is less than the combined costs of treatment, hospitalization, and time loss from work. However, despite the availability of vaccines, cost per dose is a factor limiting the success of global vaccination campaigns, as are the limitations imposed by the need of delivering multiple vaccine doses. A number of approaches are being tested particularly for the delivery of subunit vaccines, and in recent years, a number of groups have devoted their efforts to develop nano/microparticles prepared from biodegradable and biocompatible polymers as vaccine delivery systems with the goal of inducing both humoral and cellular immune responses. Some important properties of biodegradable polymers are their documented safety history, biocompatibility, and an ability to provide controlled time/rate of antigen release and polymer degradation. The most extensively studied polymer used for encapsulating vaccine antigens is poly (lactide-*co*-glycolide acid) (PLGA). This chapter deals in brief with efforts targeting the use of PLGA micro-and nanoparticles for the delivery of hepatitis B surface antigen.

1. Introduction

Hepatitis B infection remains a serious global problem despite the availability of a safe and effective injectable vaccine. The situation is particularly serious in developing countries where morbidity and mortality are high due to the very high rate of infection. More than 350 million people around the world are infected by hepatitis B virus (HBV) and these individuals serve as the viral reservoir and can also transmit the virus and spread the disease (Davis 2005). The licensed hepatitis B vaccine consists of recombinant hepatitis B surface antigen (HBsAg) adsorbed to aluminum adjuvant (Gupta and Siber 1995; Clements and Griffiths 2002). The vaccine is administered in three doses over the course of 6 months and is both safe and effective with more than 90% vaccinees being protected upon completion of the full three-dose vaccination course (Thoelen et al. 2001). A major problem in many developing countries is that a significant number of vaccine recipients do not return for the booster doses, thus limiting the efficacy of the vaccination campaign. Single or multiple dosage of an oral or intranasal vaccine could significantly increase patient compliance, and this could in turn have a greater impact on global immunization programs.

In recent years, significant effort has been devoted to develop nano/microparticle-mediated vaccine delivery systems prepared from biodegradable and biocompatible polymers. A wide variety of polymeric materials has been utilized in efforts aimed at inducing both systemic and local immune response following vaccine administration by various routes. This includes the testing of poly(lactide-*co*-glycolide acid) (PLGA) (Spiers et al. 2000; Kavanagh et al. 2003; Jiang et al. 2005), starch (Wikingsson and Sjoholm 2002), chitosan (McNeela et al. 2000; Van der et al. 2003). Homo- and copolymers of lactic and polylactic glycolic acid have been extensively tested as components for antigen delivery systems (Kersten and Kaufmann 1996). Significant advantages of these biodegradable polymers are their long safety history, proven biocompatibility, and their property to control the time and rate of polymer degradation and antigen release (Kersten and Hirshberg 2004). PLGA copolymers can be manipulated to undergo noncatalytic hydrolysis to glycolic and lactic acids in the body (Shalaby 1995). The rate of hydrolysis and hence the rate of antigen release can be controlled by altering the ratio of lactic acid to glycolic acid in the copolymer (Park et al. 1993). The biodegradable poly-lactide-*co*-glycolide polymer has been used for encapsulation (Duncan et al. 1996; Eldridge et al. 1992) since the encapsulation process can be controlled to get particles smaller than 10 μm since particles of this size can be efficiently taken up by the M cells. Eldridge et al. (1989) studied how the particle size of polymer microspheres can influence the mucosal immune response. The immune response to Staphylococcus enterotoxin B was studied in BALB/c mice using PLGA microspheres. Microspheres <5 μm in diameter were effectively taken up by the Peyer's patches and disseminated to the mesenteric lymph nodes as well as the spleen. The corresponding humoral response was characterized by enterotoxin-specific IgM and IgG antibodies. Nonencapsulated controls, however, did not elicit any response. In contrast, microspheres with larger diameters ranging from 5 to 10 μm remained in the Peyer's patches for up to 35 days. Microspheres with diameters >10 μm were not taken up by the Peyer's patches. It was also noted in this study that uptake was higher for microspheres made with polymers of higher hydrophobicity. A brief overview of the different approaches using nano/microparticles for hepatitis B vaccine delivery following administration by various routes is described in this chapter.

2. Techniques for the Preparation of Nano/Microparticles

The most commonly used method for the preparation of antigen-encapsulated nano/microparticles is the solvent extraction or evaporation from a water-in-oil-in-water (W/O/W) emulsion (Gupta et al. 1997; O'Hagan et al. 1993; Cleland et al. 1997; McGee et al. 1997). A primary emulsion (W/O) is formed by homogenizing or sonicating an aqueous solution of the antigen with PLGA in an organic solvent. The organic solvents mainly used are methylene chloride or ethyl acetate. The ratio between the aqueous phases is variable and is an important factor controlling various parameters of the nanoparticles such as size and entrapment efficiency. The primary emulsion is subsequently dispersed in a larger volume of an aqueous emulsifier/stabilizer to form a secondary emulsion (W/O/W). The most commonly used stabilizer is polyvinyl alcohol. The organic solvent is then removed by evaporation using a rotary evaporator under low pressure allowing polymer hardening to form the nano/microparticles. Additionally, the use of co-solvents like alcohol or acetone can also facilitate the solvent extraction (Tambera et al. 2005). Various modifications and also alternative approaches like coacervation (Fong 1979) (or phase separation) and spray drying (Nuwayser and Nucefora 1986) have been reported, although the first step for the formation of the primary emulsion is similar in all the approaches.

3. Oral Vaccination

Mucosal immunization is an attractive alternative to parental immunization; as with the use of appropriate delivery system, it is possible to stimulate both humoral and cell-mediated responses and to induce mucosal and systemic immunity simultaneously (Vady and O'Hagan 1996). The most convenient way of administering a drug/vaccine is by the oral route. It affords high patient acceptability, compliance, and ease of administration when compared with traditional parental administration. Successful oral vaccination for hepatitis B could certainly have a greater impact on global immunization efforts. It is generally acknowledged that oral vaccination is largely ineffective due to substantial degradation of antigens by harsh acidic condition of the stomach and enzymatic degradation in gastrointestinal tract (Gabor et al. 2002; Singh et al. 2004; Lavelle et al. 2004). Additionally, poor absorption of the antigens by the gut-associated lymphoid tissue (GALT) contributes to the lower efficacy and thus requires larger doses of antigen to be administered to achieve desirable level of immunity comparable with systemic administration (Gupta et al. 2006). Several investigators have shown that micro/nanoparticle carrier systems are very effective in protecting the antigen in the gastrointestinal tract and are also capable of sustained antigen release and can achieve targeted antigen delivery with the help of selective ligands (Jain et al. 2005; Lavelle et al. 2001). Gupta et al (Gupta et al. 2006) have synthesized HBsAg-bearing PLGA nanoparticles by double emulsion a method using a protein stabilizer, trehalose. Furthermore, lectin from *Arachis hypogaea* was conjugated on the surface of these PLGA nanoparticles with the help of glutaraldehyde in order to increase the affinity toward antigen-presenting cells of the Peyer's patches. They found that plain nanoparticles had a moderate loading efficiency around 55% while lectin-conjugated nanoparticles had very low loading efficiency (~18%). However, the lectin-conjugated nanoparticles showed approximately four-fold increase in the degree of interaction in vitro with bovine submaxillary mucin (BSM) than the nonconjugated nanoparticles. Thus, these nanoparticles have the potential for targeted delivery to the antigen-presenting cells of Peyer's patches.

In efforts to develop a single-dose oral vaccine for hepatitis B, Rajkannan and colleagues (2006) encapsulated immunogenic peptide representing residues 127–145 of the immunodominant B cell epitope of HBsAg in PLGA microparticles and achieved a loading efficiency of ~65%. They reported that serum IgG and IgM anti-HBs antibodies were detected after a single oral dose in BALB/c mice.

4. Intranasal Delivery

Humoral and cellular responses to HBV antigens are believed to play a key role in the elimination of virus by the host. In a recent study, Jaganathan and Vyas (2006) showed that, following nasal administration, HBsAg encapsulated in surface-modified PLGA microspheres elicited titers of anti-HBs antibodies in serum and mucosal secretions. For their studies, they synthesized HBsAg-encapsulated PLGA microsphere of 1–10 μm size, with the surface of the nanoparticles being modified with chitosan (a biopolymer known for its mucoadhesive properties). They demonstrated that surface-modified PLGA (with chitosan) microsphere had significantly reduced the rate of clearance from nasal cavity of rabbits as compared with the unmodified PLGA microsphere. They reported that the serum anti-HBs antibody titer obtained after nasal administration of chitosan-modified PLGA microspheres was comparable with the titer achieved when alum-adsorbed HBsAg was administered subcutaneously. Modified PLGA microspheres (cationic microspheres) thus produced humoral (both systemic and mucosal) and cellular immune responses upon nasal administration. On the contrary, there was no significant response of the unmodified PLGA nanoparticles as it was rapidly cleared from the nasal cavity. These data support the high potential of modified PLGA microspheres for their use as a carrier adjuvant for subunit vaccines delivered by the intranasal route.

5. DNA-Based Vaccination

Various of genetic approaches are another novel strategy for the vaccination against hepatitis B infection particularly for chronic HBV. DNA-based vaccination can elicit significant and long-lasting humoral and cell-mediated immunity including cytotoxic T lymphocytes and cytokines in mice (Oka et al. 2001), ducks (Le et al. 2003), and chimpanzees (Davis et al. 1996). Zhou et al. (2003) demonstrated that intramuscular injection of pVAX-PS, a DNA vaccine constructed by inserting the gene encoding the middle (pre-S2 plus S) envelope protein of HBV into a plasmid vector pVAX1, induced strong humoral immune response in tree shrews. However, in human clinical trials even very high doses of HBV DNA vaccine provoked much lower levels of immune responses as compared with the small animals. Therefore, it is very essential to increase the efficacy of the DNA-based vaccines for successful human application. He et al. (2005) successfully encapsulated plasmid DNA encoding HBV HBsAg in PLGA nanoparticles using double emulsion solvent evaporation method. This DNA encoding HBsAg-encapsulated formulation was able to induce local and systemic HBsAg-specific immunity following a single oral dose in BALB/c mice. It was also found that PLGA–DNA micro-formulation also decreased the dose required to induce a comparable immune response to plasmid DNA in administered saline.

6. Perspectives

Despite encouraging results obtained from various approaches using PLGA-based nano/microparticle-mediated hepatitis B vaccination in animals, no successful human clinical trials have been reported to date. Other than PLGA, limited studies have been reported on any other nano/microparticulate formulation for hepatitis B vaccination. Systematic evaluation of the efficacy of the nanoparticle-mediated vaccine formulation step by step can help make better formulations. By using a combination of different polymers, tailor-made vaccines with variable release kinetics, can be developed. Entrapment efficiency, release kinetics, and other physical characteristics, such as morphology, porosity, and size distribution, which influence the efficacy of the formulations, can be controlled by using an appropriate combination of different polymers. Thus, an appropriate combination of various biodegradable and biocompatible polymers in nanoparticulate formulations may lead to a successful vaccine formulation for single-dose injectable or a multiple-dose oral/intranasal vaccines.

References

Cleland, J.L., Lim, A., Barron, L., Duenas, E.T. and Powell, M.F. (1997) Development of a single-shot subunit vaccine for HIV-1: Part 4. Optimizing microencapsulation and pulsatile release of MN rgp120 from biodegradable microspheres. J. Control Release 47, 135–150.

Clements, C.J. and Griffiths, E. (2002) The global impact of vaccines containing aluminium adjuvants. Vaccine 20, S24–S33.

Davis, J.P. (2005) Experience with hepatitis A and B vaccines. Am. J. Med. 118, 7S–15S.

Davis, H.L., McCluskie, M.J., Gerin, J.L. and Purcell, R.H. (1996) DNA vaccine for hepatitis B: evidence for immunogenicity in chimpanzees and comparison with other vaccines. Proc. Natl. Acad. Sci. USA 93, 7213–7218.

Duncan, J.D., Gilley, R.M., Schafer, D.P., Moldoveanu Z. and Mestecky J.F. (1996) Poly(lactide-co-glycolide) microencapsulation of vaccines for mucosal immunization. In: H. Kiyono, P.L. Ogra and J.R. McGhee (Eds.), *Mucosal Vaccines*, Academic Press, San Diego, pp. 159–174.

Eldridge, J., Gilley, R., Staas, R., Moldoveanu, Z., Muelbroek, J. and Tice T. (1989) Biodegradable microspheres: vaccine delivery system for oral immunization. Curr. Top. Microbiol. Immunol. 146, 59–66.

Eldridge, J.H., Staas, J.K., Tice, T.R. and Gilley, R.M. (1992) Biodegradable poly(lactide-co-glycolide) microspheres. Res. Immunol. 143, 557–563.

Fong, F.W. (1979) Microsphere production from particle dispersion in polymer solution by adding phase separation agent at low temperature. US Patent US 4, 166, 800.

Gabor, F., Scwarzbauer, A. and Wirth, M. (2002) Lectin mediated drug delivery: binding and uptake of BSA-WGA conjugates using the caco-2 model. Int. J. Pharm. 237, 227–239.

Gupta, R.K., Chang, A.C., Griffin, P., Rivera, R., Guo, Y.Y. and Siber G.R. (1997) Determination of protein loading in biodegradable polymer microspheres containing tetanus toxoid. Vaccine 15, 672–678.

Gupta, P.N., Mahor S., Rawat, A., Khatri, K., Goyal, A. and Vyas, S.P. (2006) Lectin anchored stabilized biodegradable nanoparticles for oral immunization 1. Development and in vitro evaluation. Int. J. Pharm. 318, 163–173.

Gupta, R.K. and Siber G.R. (1995) Adjuvants for human vaccines-current status, problems and future prospects. Vaccine 13, 1263–1276.

He, X., Wang, F., Jiang, L., Li, J., Liu, S., Xiao, Z., Jin, X., Zhang, Y., He, Y., Li, K., Guo, Y. and Sun, S. (2005) Induction of mucosal and systemic immune response by single-dose oral immunization with biodegradable microparticles containing DNA encoding HBsAg. J. Gen. Virol. 86, 601–661.

Jaganathan, K.S. and Vyas, S.P. (2006) Strong systemic and mucosal immune responses to surface-modifies PLGA microspheres containing recombinant Hepatitis B antigen administered intranasally. Vaccine 24, 4201–4211.

Jain, S., Singh, P., Mishra, V. and Vyas, S.P. (2005) Mannosylated niosomes as adjuvant–carrier system for oral genetic immunization against Hepatitis B. Immunol. Lett. 101, 41–49.

Jiang, W., Gupta, R.K., Deshpande, M.C. and Schwendeman, S.P. (2005) Biodegradable poly(lactic-co-glycolic acid) microparticles for injectable delivery of vaccine antigens. Adv. Drug Del. Rev. 57, 391–410.

Kavanagh, O.V., Early, B., Murray, M., Foster, C.J. and Adair, B.M. (2003) Antigenspecific IgA and IgG responses in calves inoculated intranasally with ovalbumin encapsulated in poly(dl-lactide-co-glycolide) microspheres. Vaccine 21, 4472–4480.

Kersten, G. and Hirshberg, H. (2004) Antigen delivery systems. Expert Rev. Vaccines 3, 453–462.

Kersten, G.F.A. and Kaufmann, G.B. (1996) Biodegradable microspheres as vehicles for antigens. In: W. deGruyter (Eds.) *Concepts in Vaccine Development.* New York, Springer Publ. Co., pp. 265–302.

Lavelle, E.C., Grant, G., Pfuller, U. and O'Hagan, D.T. (2004) Immunological implication of the use of plant lectins for drug and vaccine targeting to the gastrointestinal tract. J. Drug Target. 12, 89–95.

Lavelle, E.C., Grant, G., Pusztai, A., Fuller, U. and O'Hagan, D.T. (2001) Identification of plant lectin with mucosal adjuvant activity. Immunology 102, 77–86.

Le Guerhier, F., Thermet, A., Guerret, S., Chevallier, M., Jamard, C., Gibbs, C.S., Trepo, C., Cova, L. and Zoulim, F. (2003) Antiviral effect of adefovir in combination with a DNA vaccine in the duck hepatitis B virus infection model. J. Hepatol. 38, 328–334.

McNeela, E., O'Connor, D., Jabbal-Gill, I., Illum, L., Davis, S.S., Pizza, M., Peppoloni, S., Rappuoli, R. and Mills, K.H.G. (2000) A mucosally delivered vaccine against diphtheria: formulation of cross reacting material (CRM197) of diphtheria toxin with chitosan enhances local and systemic and Th2 responses following nasal delivery. Vaccine 19, 1188–1198.

McGee, J.P., Singh, M., Li, X.M., Qui, H. and O'Hagan, D.T. (1997) The encapsulation of a model protein in poly (lactide-co-glycolide) microparticles of various sizes; an evaluation of process reproducibility, J. Microencapsul. 14, 197–210.

Nuwayser, E.S. and Nucefora, W.A. (1986) Controlled release mi-croparticles comprising core of active ingredient and polymer and coating of the same polymer. US Patent US4, 623, 588.

O'Hagan, D.T., Jeffery H. and Davis S.S. (1993) The preparation and characterization of poly (lactide-co-glycolide) microparticles II. The entrapment of a model protein using a water-in-oil-inwater emulsion solvent evaporation technique. Pharm. Res. 10, 362–368.

Oka, Y., Akbar, S.M., Horiike, N., Joko, K. and Onji, M. (2001) Mechanism and therapeutic potential of DNA-based immunization against the envelope proteins of hepatitis B virus in normal and transgenic mice. Immunology 103, 90–97.

Park, K., Shalaby, W.S.W. and Park H. (1993) *Biodegradable Hydrogels for Drug Delivery*. Technomic Publishing Company, Inc., Pennsylvania.
Rajkannan, R., Dhanaraju, M.D., Gopinath, D., Selvaraj, D. and Jayakumar, R. (2006) Development of hepatitis B oral vaccine using B cell epitope loaded PLGA microparticles. Vaccine 24, 5149–5157.
Shalaby, W.S.W. (1995) Development of oral vaccines to stimulate mucosal andsystemic immunity: barriers and novel strategies. Clin. Immunol. Immunopathol. 74, 127–134.
Singh, P., Prabhakaran, D., Jain, S., Mishra, V., Jaganathan, K.S. and Vyas, S.P. (2004) Cholera toxin B conjugated bile salt stabilized vesicles (bilosomes) for oral immunization Int. J. Pharm. 278, 379–390.
Spiers, I.D., Eyles, J.E., Baillie, L.W.J., Williamson, E.D. and Oya Alpar, H. (2000) Biodegradable microparticles with different release profiles: effect on the immune response after a single administration via intranasal and intramuscular routes. J. Pharm. Pharmacol. 52, 1195–1201.
Tambera, H., Johansena, P., Merklea, H.P. and Gander, B. (2005) Formulation aspects of biodegradable polymeric microspheres for antigen delivery. Adv. Drug Del. Rev. 57, 357– 376.
Thoelen, S., Clercq, N.D. and Tornieporth, N.A. (2001) Prophylactic hepatitis B vaccine with a novel adjuvant system. Vaccine 19, 2400–2403.
Vady, M. and O'Hagan, D.T. (1996) Microparticles for intranasal immunization, Adv. Drug Del. Rev. 21, 33–47.
Van der, L.I.M., Kersten, G., Fretz, M.M., Beuvery, C., Verhoef, J.C. and Junginger, H.E. (2003) Chitosan microparticles for mucosal vaccination against diphtheria: oral and nasal efficacy studies in mice. Vaccine 21, 1400–1408.
Wikingsson, L. and Sjoholm, I. (2002) Polyacryl starch microparticles as adjuvant in oral immunization, inducing mucosal and systemic immune responses in mice. Vaccine 20, 3353–3363.
Zhou, F.J., Hu, Z.L., Dai, J.X., Chen, R.W., Shi, K., Lin, Y. and Sun, S.H. (2003) Protection of tree shrews by pVAX-PS DNA vaccine against HBV infection. DNA Cell Biol. 22, 475–478.

45

Mast Cells, T Cells, and Inhibition by Luteolin: Implications for the Pathogenesis and Treatment of Multiple Sclerosis

Theoharis C. Theoharides, Duraisamy Kempuraj, and Betina P. Iliopoulou

Departments of Pharmacology and Experimental Therapeutics; Internal Medicine and Biochemistry; Immunology Program, Tufts University School of Medicine and Tufts-New England Medical Center, Boston, MA, USA, theoharis.theoharides@tufts.edu

Abstract. Multiple sclerosis (MS) is a demyelinating disease of the central nervous system (CNS) mainly mediated by Th1, but recent evidence indicates that Th2 T cells, mostly associated with allergic reactions, are also involved. Mast cells are involved in allergic and inflammatory reactions because they are located perivascularly and secrete numerous pro-inflammatory cytokines. Brain mast cells are critically placed around the blood–brain barrier (BBB) and can disrupt it, a finding preceding any clinical or pathological signs of MS. Moreover, mast cells are often found close to MS plaques, and the main MS antigen, myelin basic protein (MBP), can activate human cultured mast cells to release IL-8, TNF-α, tryptase, and histamine. Mast cells could also contribute to T cell activation since addition of mast cells to anti-CD3/anti-CD28 activated T cells increases T cell activation over 30-fold. This effect requires cell-to-cell contact and TNF, but not histamine or tryptase. Pretreatment with the flavone luteolin totally blocks mast cell stimulation and T cell activation. Mast cells could constitute a new unique therapeutic target for MS.

1. Introduction

Multiple sclerosis (MS) is a demyelinating disease of the central nervous system (CNS) leading to severe disability in almost half a million young adults in the United states (Smith and McDonald 1999). MS is mediated by infiltration of $CD4^{+}$ Th1 cells and macrophages that results in chronic CNS inflammation and neurodegeneration. Recent evidence indicates that $CD4^{+}$ Th2 cells are also involved in MS, but it is not known how they enter the brain and get sensitized (Lassmann and Ransohoff 2004). $CD4^{+}$ Th1 cells produce mostly IL-2, IFN-γ, and TNF-β, while Th2 cells produce IL-4, IL-5, and IL-13. Th2 cytokines are associated with allergic reactions and maturation of mast cells and have recently been implicated in MS (Pedotti et al. 2003a).

Mast cells encircle microvessels (Figure 1) at the blood–brain barrier (BBB), especially at the third ventricle (Esposito 2002), and close to MS plaques (Ibrahim et al. 1996). Mast cells are also critical for the regulation of BBB permeability disruption of which precedes clinical or pathological signs of MS (Stone et al. 1995).

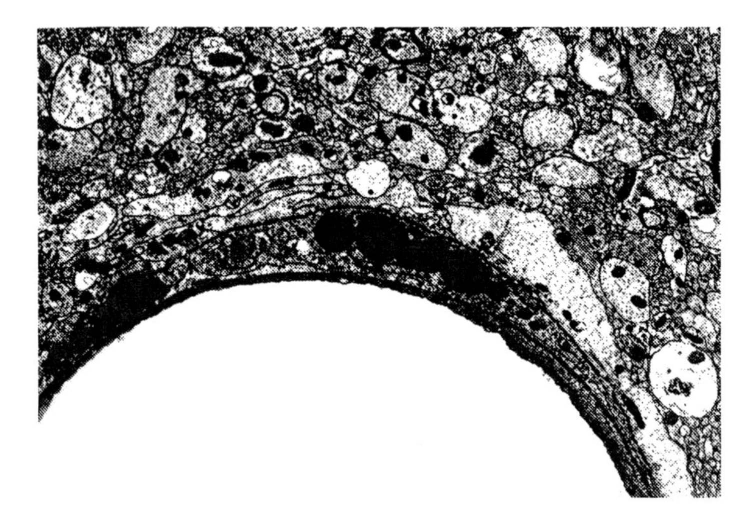

FIGURE 1. Electron photomicrograph of a mast cell encircling half the lumen of a rat diencephalic blood vessel in an area of demyelination.

Moreover, the genes found to be most upregulated in plaques from deceased patients with MS included those for the proteolytic enzyme tryptase, the IgE receptor (FcεRI) and the histamine-1 receptor, all associated with mast cells (Lock et al. 2002; Bomprezzi et al. 2003). Release of myelin fragments, such as myelin basic protein (MBP) or fragment P2, could induce mast cell activation (Johnson et al. 1988; Theoharides et al. 1993) and sensitize T cells (Figure 2). Mast cell tryptase is elevated in the cerebrospinal flind (CSF) of MS patients and can cause widespread inflammation by stimulating protease-activated receptors (PARs) (Molino et al. 1997). Mast cells are involved in allergy and inflammation, as well as in innate and acquired immunity

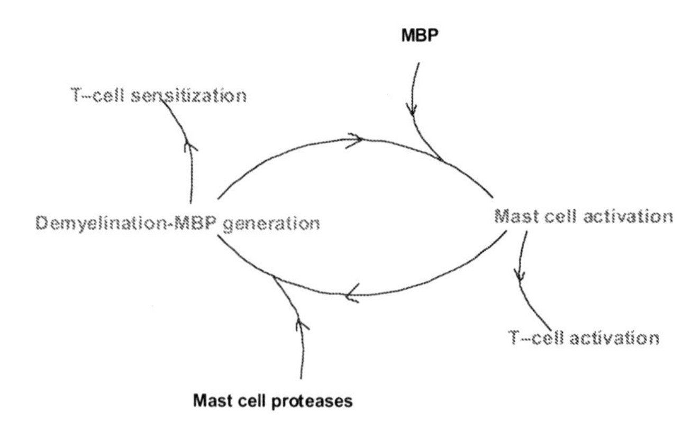

FIGURE 2. Schematic representation of mast cell involvement in multiple sclerosis (MS) pathogenesis.

(Galliet al. 2005), including T cell-mediated disorders. Mast cells and T cells could interact in a variety of acquired immune responses (Nakae et al. 2005; Pedotti et al. 2003a), including many inflammatory diseases.

2. Mast Cell Activation of T Cells

In order to investigate whether mast cells may affect T cell activation, human umbilical cord blood-derived cultured mast cells (hCBMCs) and T cells were co-cultured in 96-well plates and T cells were activated with anti-CD3/anti-CD28 (1 μg/ml each) when appropriate. Then, T cells were examined morphologically for signs of activation and by IL-2 release by ELISA (Figure 3). Out results indicate that hCBMCs can be stimulated by MBP and can activate T cells significantly. We also showed that T Cell–mast cell contact is necessary for the full T cell activation. Furthermore, our findings indicate that the flavonoid luteolin can inhibit mast cell and T cell activation.

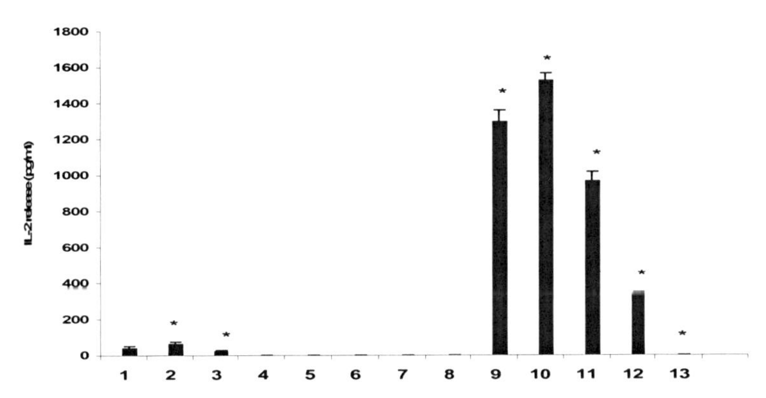

FIGURE. 3. Mast cells, myelin basic protein (MBP), and substance P enhance T cell activation induced by anti-CD3/anti-CD28. T cells were activated with anti-CD3/anti-CD28 and incubated for 48 h with MBP and human umblicel cultered mast cells (hCBMCs) ($n = 3$) in a 96-well plate. After 48 h IL-2 in the supernatant was measured by ELISA ($*p < 0.05$). Groups a as follows:

1. Jurkat + anti-CD3/anti-CD28;
2. Jurkat + anti-CD3/anti-CD28 + MBP 10 μM;
3. Jurkat + luteolin 10 μM + anti-CD3/anti-CD28 + MBP10 μM;
4. Jurkat;
5. Jurkat + MBP 10 μM;
6. human CBMC;
7. Jurkat + hCBMCs;
8. Jurkat + hCBMCs + MBP 10 μM;
9. Jukat + anti-CD3/anti-CD28 + hCBMCs;
10. Jurkat + anti-CD3/anti-CD28 + hCBMCs + MBP 10 μM;
11. Jurkat + luteolin 1 μM + anti-CD3/anti-CD28 + hCBMCs + MBP 10 μM;
12. Jurkat + luteolin 10 μM + anti-CD3/anti-CD28 + hCBMCs + MBP 10 μM;
13. Jurkat + luteolin 100 μM + anti-CD3/anti-CD28 + hCBMCs + MBP 10 μM.

It had previously been suggested that mast cells may be involved in MS (Theoharides and Cochrane 2004; Krüger 2001). Other evidence supports our findings. For instance, coculture of leukemic mast cells with activated T cells promoted adhesion of mast cells (Brill et al. 2004). Mast cells stimulated by FcεRI aggregation release TNF-α and activate T cells (Gregory et al. 2006), but direct contact is also required (Salamon et al. 2005). Soluble TNF can increase surface expression of costimulatory molecules on the surface of $CD3^+$ T cells (Nakae et al. 2005).

Myelin fragments can stimulate mast cells (Johnson et al. 1988; Theoharides et al. 1993). Mast cells were shown to affect T cell function (Redegeld and Nijkamp 2005; Nakae et al. 2005) and Th2 cytokines were considered important in MS pathogenesis (Pedotti et al. 2003a,b). EAE, an animal model of MS, was reduced and delayed in mast cell-deficient W/W^v mice (Brown et al. 2002), and mast cells were required for optimal T cell responses in this model. Development of EAE and MS is dependent on trafficking of activated, myelin-specific T cells into the CNS (Gregory et al. 2006). In fact, mast cells were also shown to be critical for the permeability of the BBB (Esposito et al. 2002) that precedes any clinical or pathological signs of MS (Minagar and Alexander 2003).

3. Mast Cell Mediators

Mast cells are mainly found at sites crucial in antigen entry and trafficked by $CD4^+$ T cells. Mast cells secrete a wide variety of potent chemical mediators (Galli et al. 2005) that can initiate and modulate several inflammatory pathways (Theoharides and Cochrane 2004). Mast cells represent a major potential source of TNF-α (Galli 1990; Nakae et al. 2005), which influences T cell recruitment and activation (Tartaglia et al. 1993). Mast cell chemokines MCP-1, MIP-1α, MIP-1β, RANTES, and IL-16 enhance T cell recruitment to the site of inflammation (Salamon et al. 2005), and leukotriene B4 regulates T cell migration (Ott et al. 2003). Histamine released from mast cells can promote Th1 and Th2 cell activation (Jutel et al. 2001). Mast cell-derived histamine and TNF-α increase microvascular permeability, leukocyte rolling, and adhesion, thus contributing to the infiltration of T cells and monocytes into the CNS. Mast cell-derived IL-4 worsens EAE severity by enhancing the encephalitogenic potential of Th1 cells (Gregory et al. 2006). Tryptase is elevated in CSF of MS patients (Rozniecki et al. 1995) and can damage myelin (Dietsch and Hinrichs 1991; Johnson et al. 1988). Tryptase can activate peripheral blood mononuclear cells (PBMCs) isolated from MS patients to release TNF-α, IL-1β, and IL-6, which are involved in the pathogenesis of MS (Martino et al. 2000). Tryptase also stimulates PAR-2 (Malamud et al. 2003) expressed on PBMCs, neutrophils, and Jurkat T cells (Dery et al. 1998; Macey et al. 1998), leading to intense inflammation. Mast cells could, therefore, participate at multiple steps in MS pathogenesis and could serve as the next therapeutic target for MS (Zappulla et al. 2002).

4. Effect of Flavonoids

Flavonoids are polyphenolic compounds found in fruits, vegetables, nuts, seeds, herbs, and spices with anti-oxidant, anti-allergic, and anti-inflammatory properties (Duthie et al. 1997; Herrmann 1976; Middleton et al. 2000). Certain flavones and flavonols also inhibit histamine release from mast cells (Foreman 1984; Middleton 1998) as well as IL-6 and TNF-α release from murine mast cells (Kimata et al. 2000). In particular, the flavone luteolin inhibits basophil release of IL-4 and IL-13 synthesis, as well as CD40 ligand expression (Hirano et al. 2006) and activation of AP-1. Luteolin also inhibits the IgE-mediated release of histamine, leukotrienes, prostaglandin D_2, and granulocyte–macrophage colony-stimulating factor from human cultured mast cells. Luteolin also inhibits IgE-mediated histamine IL-6 and TNF-α release from bone marrow-derived cultured murine mast cells (BMMCs) and rat peritoneal mast cells (Kimata et al. 2000).

Luteolin reduces CNS inflammation by preventing monocyte migration across the BBB (Hendriks et al. 2004) and, like quercetin (Aktas et al. 2004), can inhibit clinical symptoms of EAE (Hendriks et al. 2004); luteolin also inhibits macrophage myelin phagocytosis, as well as proliferation and activation of autoimmune T cells (Verbeek et al. 2004).

Flavonoids may be beneficial in neurodegenerative diseases (Halliwell 2001) because they can cross the BBB and are scavengers of reactive oxygen species (ROS), which participate in the pathogenesis of MS (Lu et al. 2000) and EAE (Ruuls et al. 1995). The unique subclasses of flavonoids, the flavones and the flavonols, inhibit human mast cell secretion of proinflammatory molecules in response to allergic triggers (Kempuraj et al. 2005) and IL-1 (Kandere-Grzybowska et al. 2006).

There are no curative therapies presently available for MS (Goldn 2007). The drugs used include corticosteroids, IFN-β, a copolymer, and a vascular adhesion molecule blocker, all of which are administered parenterally and have adverse effects (Goldn 2007). The ability of the naturally occurring flavonoid luteolin to inhibit mast cell and T cell activation presents a new approach to MS treatment.

Acknowledgments and Disclosure

This work was supported in part by Theta Biomedical Consulting and Development Co., Inc. (Brookline, MA). The use of flavonoids in MS is covered by US patents # 6, 689, 748; 6, 984, 667; EPO 1365777; and US patent pending application 10/811,829 awarded to TCT.

References

Aktas, O., Prozorovski, T., Smorodchenko, A., Savaskan, N.E., Lauster, R., Kloetzel, P.M., Infante-Duarte, C., Brocke, S. and Zipp, F. (2004) Green tea epigallocatechin-3-gallate mediates T cellular NF-kappa B inhibition and exerts neuroprotection in autoimmune encephalomyelitis. J. Immunol. 173, 5794–5800.

Bomprezzi, R., Ringner, M., Kim, S., Bittner, M.L., Khan, J., Chen, Y., Elkahloun, A., Yu, A., Bielekova, B., Meltzer, P.S., Martin, R., McFarland, H.F. and Trent, J.M. (2003) Gene expression profile in multiple sclerosis patients and healthy controls: identifying pathways relevant to disease. Hum. Mol. Genet. 12, 2191–2199.

Brill, A., Baram, D., Sela, U., Salamon, P., Mekori, Y.A. and Hershkoviz, R. (2004) Induction of mast cell interactions with blood vessel wall components by direct contact with intact T cells or T cell membranes in vitro. Clin. Exp. Allergy 34,1725–1731.

Brown, M.A., Tanzola, M. and Robbie-Ryan. M. (2002) Mechanisms underlying mast cell influence on EAE disease course. Mol. Immunol. 38:1373–1378.

Dery, O., Corvera, C.U., Steinhoff, M. and Bunnett, N.W. (1998) Proteinase-activated receptors: novel mechanisms of signaling by serine proteases. Am. J. Physiol. 274, C1429–C1452.

Dietsch, G.N. and Hinrichs, D.J, (1991) Mast cell proteases liberate stable encephalitogenic fragments from intact myelin. Cell. Immunol. 135, 541–548.

Duthie, S.J., Johnson, W. and Dobson, V.L. (1997) The effect of dietary flavonoids on DNA damage (strand breaks and oxidised pyrimidines) and growth in human cells. Mutat. Res. 390, 141–151.

Esposito, P., Chandler, N., Kandere-Grzybowska, K., Basu, S., Jacobson, S., Connolly, R., Tutor, D. and Theoharides, T.C. (2002) Corticotropin-releasing hormone (CRH) and brain mast cells regulate blood-brain-barrier permeability induced by acute stress. J. Pharmacol. Exp. Ther. 303, 1061–1066.

Foreman, J.C. (1984) Mast cells and the actions of flavonoids. J. Allergy Clin. Immunol. 73, 769–774.

Galli, S.J. (1990) New insights into "the riddle of the mast cells": microenvironmental regulation of mast cell development and phenotypic heterogeneity. Lab. Invest. 62, 5–33.

Galli, S.J., Nakae, S. and Tsai, M. (2005) Mast cells in the development of adaptive immune responses. Nat. Immunol. 6, 135–142.

Goldn, R. (2007) Towards individualized multiple sclerosis therapy. Lancet Neurol. 4, 693–694.

Gregory, G.D., Raju, S.S., Winandy, S. and Brown. M.A.(2006) Mast cell IL-4 expression is regulated by Ikaros and influences encephalitogenic Th1 responses in EAE. J. Clin. Invest. 116, 1327–1336.

Halliwell, B. (2001) Role of free radicals in the neurodegenerative diseases: therapeutic implications for antioxidant treatment. Drugs Aging 18, 685–716.

Hendriks, J.J., Alblas, J., van der Pol, S.M., van Tol, E.A., Dijkstra, C.D. and de Vries, H.E. (2004) Flavonoids influence monocytic GTPase activity and are protective in experimental allergic encephalitis. J. Exp. Med. 200, 1667–1672.

Herrmann, K. (1976) Flavonols and flavones in food plants. J. Food Technol. 11, 433–448.

Hirano, T., Higa, S., Arimitsu, J., Naka, T., Ogata, A., Shima, Y., Fujimoto, M., Yamadori , T., Ohkawara, T., Kuwabara, Y., Kawai, M., Matsuda, H., Yoshikawa M., Maezaki, N., Tanaka, T., Kawase, I. and Tanaka, T. (2006) Luteolin, a flavonoid, inhibits AP-1 activation by basophils. Biochem. Biophys. Res. Commun. 340, 1–7.

Ibrahim, M.Z.M., Reder, A.T., Lawand, R., Takash, W. and Sallouh-Khatib, S. (1996) The mast cells of the multiple sclerosis brain. J. Neuroimmunol. 70, 131–138.

Johnson, D., Seeldrayers, P.A. and Weiner, H.L. (1988) The role of mast cells in demyelination. 1. Myelin proteins are degraded by mast cell proteases and myelin basic protein and P_2 can stimulate mast cell degranulation. Brain Res. 444, 195–198.

Jutel, M., Watanabe, T., Klunker, S., Akdis, M., Thomet, O.A.R., Malolepszy, J., Zak-Nejmark, T., Koga, R., Kobayashi, T., Blaser, K. and Akdis, C.A. (2001) Histamine regulates T cell and antibody responses by differential expression of H1 and H2 receptors. Nature 413, 420–425.

Kandere-Grzybowska, K., Kempuraj, D., Cao, J., Cetrulo, C.L. and Theoharides, T.C. (2006) Regulation of IL-1-induced selective IL-6 release from human mast cells and inhibition by quercetin. Br. J. Pharmacol.148, 208–215.
Kempuraj, D., Madhappan, B., Christodoulou, S., Boucher, W., Cao, J., Papadopoulou, N., Cetrulo, C.L. and Theoharides, T.C. (2005) Flavonols inhibit proinflammatory mediator release, intracellular calcium ion levels and protein kinase C theta phosphorylation in human mast cells. Br. J. Pharmacol. 145, 934–944.
Kimata, M., Inagaki, N. and Nagai, H. (2000) Effects of luteolin and other flavonoids on IgE-mediated allergic reactions. Planta Med. 66, 25–29.
Krüger, P.G. (2001) Mast cells and multiple sclerosis: a quantitative analysis. Neuropathol Appl. Neurobiol. 27, 275–280.
Lassmann, H. and Ransohoff, R.M. (2004) The CD4-Th1 model for multiple sclerosis: a crucial re-appraisal. Trends Immunol. 25, 132–137.
Lock, C., Hermans, G., Pedotti, R., Brendolan, A., Schadt, E., Garren, H., Langer-Gould, A., Strober, S., Cannella, B., Allard, J., Klonowski, P., Austin, A., Lad, N., Kaminski, N., Galli, S.J., Oksenberg, J.R., Raine, C.S., Heller, R. and Steinman, L. (2002) Gene-microarray analysis of multiple sclerosis lesions yields new targets validated in autoimmune encephalomyelitis. Nat. Med. 8, 500–508.
Lu, F., Selak, M., O'Connor, J., Croul, S., Lorenzana, C., Butunoi, C. and Kalman, B. (2000) Oxidative damage to mitochondrial DNA and activity of mitochondrial enzymes in chronic active lesions of multiple sclerosis. J. Neurol. Sci. 177, 95–103.
Macey, M.G., Hou, L., Milne, T., Parameswaren, V., Howe, D., Cavenagh, J.D., Howells, G.L. and Newland, A.C. (1998) A CD4+ proliferation of large granular lymphocytes expresses the protease activated receptor-1. Br. J. Haematol. 101, 78–81.
Malamud, V., Vaaknin, A., Abramsky, O., Mor, M., Burgess, L.E., Ben-Yehudah, A. and Lorberboum-Galski, H. (2003) Tryptase activates peripheral blood mononuclear cells causing the synthesis and release of TNF-alpha, IL-6 and IL-1 beta: possible relevance to multiple sclerosis. J. Neuroimmunol. 138, 115–122.
Martino, G., Furlan R. and Poliani, P.L. (2000) [The pathogenic role of inflammation in multiple sclerosis]. Rev. Neurol. 30, 1213–1217.
Middleton, E., Jr. (1998) Effect of plant flavonoids on immune and inflammatory cell function. Adv. Exp. Med. Biol. 439, 175–182.
Middleton, E., Jr., Kandaswami, C. and Theoharides, T.C. (2000) The effects of plant flavonoids on mammalian cells: implications for inflammation, heart disease and cancer. Pharmacol. Rev. 52, 673–751.
Minagar, A. and Alexander, J.S. (2003) Blood-brain barrier disruption in multiple sclerosis. Mult. Scler. 9, 540–549.
Molino, M., Barnathan, E.S., Numerof, R., Clark, J., Dreyer, M., Cumashi, A., Hoxie, J.A., Schechter, N., Woolkalis, M. and Brass, L.F. (1997) Interactions of mast cell tryptase with thrombin receptors and PAR-2. J. Biol. Chem. 272, 4043–4049.
Nakae, S., Suto, H., Kakurai, M., Sedgwick, J.D., Tsai, M. and Galli, S.J. (2005) Mast cells enhance T cell activation: Importance of mast cell-derived TNF. Proc. Natl. Acad. Sci. USA 102, 6467–6472.
Ott, V.L., Cambier, J.C., Kappler, J., Marrack, P. and Swanson, B.J. (2003) Mast cell-dependent migration of effector CD8+ T cells through production of leukotriene B4. Nat. Immunol. 4, 974–981.
Pedotti, R., De Voss, J.J., Steinman, L. and Galli, S.J. (2003a) Involvement of both 'allergic' and 'autoimmune' mechanisms in EAE, MS and other autoimmune diseases. Trends Immunol. 24, 479–484.
Pedotti, R., DeVoss, J.J., Youssef, S., Mitchell, D., Wedemeyer, J., Madanat, R., Garren, H., Fontoura, P., Tsai, M., Galli, S.J., Sobel, R.A. and Steinman, L. (2003b) Multiple elements

of the allergic arm of the immune response modulate autoimmune demyelination. Proc. Natl. Acad. Sci. USA 100, 1867–1872.
Redegeld, F.A. and Nijkamp, F.P. (2005) Immunoglobulin free light chains and mast cells: pivotal role in T cell-mediated immune reactions? Trends Immunol. 24, 181–185.
Rozniecki, J.J., Hauser, S., Stein, M., Lincoln, R. and Theoharides, T.C. (1995) Elevated mast cell tryptase in cerebrospinal fluid of multiple sclerosis patients. Ann. Neurol. 37, 63–66.
Ruuls, S.R., Bauer, J., Sontrop, K., Huitinga, I., 't Hart, B.A. and Dijkstra, C.D. (1995) Reactive oxygen species are involved in the pathogenesis of experimental allergic encephalomyelitis in Lewis rats. J. Neuroimmunol. 56, 207–217.
Salamon, P., Shoham, N.G., Gavrieli, R., Wolach, B. and Mekori, Y.A. (2005) Human mast cells release interleukin-8 and induce neutrophil chemotaxis on contact with activated T cells. Allergy 60, 1316–1319.
Smith, K.J. and McDonald, W.I. (1999) The pathophysiology of multiple sclerosis: the mechanisms underlying the production of symptoms and the natural history of the disease. Philos. Trans. R Soc. Lond. B Biol. Sci. 1390, 1649–1673.
Stone, L.A., Smith, M.E., Albert, P.S., Bash, C.N., Maloni, H., Frank, J.A. and McFarland, H.F. (1995) Blood-brain barrier disruption on contrast-enhanced MRI in patients with mild relapsing-remitting multiple sclerosis: relationship to course, gender, and age. Neurology 45, 1122–1126.
Tartaglia, L.A., Goeddel, D.V., Reynolds, C., Figari, I.S., Weber, R.F., Fendly, B.M. and Palladino, M.A., Jr. (1993) Stimulation of human T cell proliferation by specific activation of the 75-kDa tumor necrosis factor receptor. J. Immunol. 151, 4637–4641.
Theoharides, T.C. and Cochrane, D.E. (2004) Critical role of mast cells in inflammatory diseases and the effect of acute stress. J. Neuroimmunol. 146, 1–12.
Theoharides, T.C., Dimitriadou, V., Letourneau, R.J., Rozniecki, J.J., Vliagoftis, H. and Boucher, W.S. (1993) Synergistic action of estradiol and myelin basic protein on mast cell secretion and brain demyelination: changes resembling early stages of demyelination. Neuroscience 57, 861–871.
Verbeek, R., Plomp, A.C., van Tol, E.A. and van Noort, J.M. (2004) The flavones luteolin and apigenin inhibit in vitro antigen-specific proliferation and interferon-gamma production by murine and human autoimmune T cells. Biochem. Pharmacol. 68, 621–629.
Zappulla, J.P., Arock, M., Mars, L.T. and Liblau, R.S. (2002) Mast cells: new targets for multiple sclerosis therapy? J. Neuroimmunol. 131, 5–20.

Index

Printed in the United States
84475LV00002B/64-96/A

9 780387 720043